# Chemical Biology of Carbohydrates

Kazunobu Toshima · Yukari Fujimoto
Editors

# Chemical Biology of Carbohydrates

Springer

*Editors*
Kazunobu Toshima
Department of Applied Chemistry
Keio University
Yokohama, Japan

Yukari Fujimoto
Department of Chemistry
Keio University
Yokohama, Japan

ISBN 978-981-95-5334-1 ISBN 978-981-95-5335-8 (eBook)
https://doi.org/10.1007/978-981-95-5335-8



This Springer imprint is published by the registered company Springer Nature Singapore Pte Ltd.
The registered company address is: 152 Beach Road, #21-01/04 Gateway East, Singapore 189721, Singapore

# Preface

Carbohydrates are not only the most abundant biomolecules on earth but also among the most structurally diverse and functionally sophisticated. Once regarded primarily as energy sources and structural materials, carbohydrates have now emerged as key players in a wide range of biological processes, including cell–cell recognition, signal transduction, protein folding, and immune responses. The intricate chemistry underlying these functions has established carbohydrates as indispensable subjects in the field of chemical biology.

The purpose of *Chemical Biology of Carbohydrates* is to provide a comprehensive overview of recent advances at the interface of carbohydrate chemistry and biology. This volume focuses on how chemical approaches—ranging from the synthesis, modification, and labeling of glycans to their analysis and biological evaluation—have contributed to elucidating complex glycan-mediated phenomena. Special attention is given to innovative glycosylation methodologies and the design of glycan-based probes and therapeutics.

This book has been written by leading Japanese experts who have long been at the forefront of glycoscience research. Their pioneering studies in carbohydrate chemistry and biology have significantly advanced our understanding of glycan function and have opened new avenues for applications in medicine and materials science. The chapters collected here reflect the depth and originality of Japanese contributions to the global field of chemical biology of carbohydrates.

We would like to express our sincere gratitude to all the contributing authors for their dedication, expertise, and enthusiasm in preparing their chapters. Their collective effort has made this volume both scientifically rich and accessible to readers from diverse backgrounds. We are also deeply grateful to the editorial staff and colleagues who provided invaluable assistance throughout the preparation of this book.

It is our hope that *Chemical Biology of Carbohydrates* will serve as a useful reference for researchers, educators, and students and that it will inspire further innovations and collaborations in the ever-expanding field of chemical biology of carbohydrates.

Yokohama, Japan

Kazunobu Toshima
Yukari Fujimoto

# Contents

**Precise Synthesis and Structure–Activity Relationship Studies of Low-Molecular-Weight Fucoidan Analogs** . . . . . . . . . . . . . . . . . . . . . . . 1
Daisuke Takahashi and Kazunobu Toshima

**Chemical Approach to Uncover the Biological Functions of Sialic Acid-Containing Glycans** . . . . . . . . . . . . . . . . . . . . . . . . . . . . . . . . . . . . 29
Hiromune Ando and Naoko Komura

**Synthetic Strategies for Di-, Oligo-, and Polysialic Acids with Applications as Siglec Ligands** . . . . . . . . . . . . . . . . . . . . . . . . . . . . . . . . 55
Hiroshi Tanaka and Chihiro Sato

**Chemical Biology of Core M3 *O*-Mannosyl Glycan** . . . . . . . . . . . . . . . . . . . 77
Jun-ichi Tamura, Kota Kotera, Haruto Taguchi, and Takahiro Tamura

**Chemical Approaches to Understand Physiological Functions of High-Mannose-Type Glycan-Related Proteins** . . . . . . . . . . . . . . . . . . . . . 103
Mitsuaki Hirose and Kiichiro Totani

**Pseudo-glycans as Molecular Tools for Chemical Glycobiology** . . . . . . . . . 133
Makoto Yoritate and Go Hirai

**Elucidating the Immune Functions of Glycans Through Synthesis-Based Approaches** . . . . . . . . . . . . . . . . . . . . . . . . . . . . . . . . . . . . . . . . 157
Koichi Fukase, Atsushi Shimoyama, and Yoshiyuki Manabe

**Chemical Approaches to Explore the Function of Bacterial Glycolipid MPIase** . . . . . . . . . . . . . . . . . . . . . . . . . . . . . . . . . . . . . . . . . . . . . . . . . . 177
Kohki Fujikawa, Tsukiho Osawa, Ken-ichi Nishiyama, and Keiko Shimamoto

**Glycolipid-Recognizing C-Type Lectin Receptors: Ligand Structures and Molecular Probes for Biofunctional Analysis** . . . . . . . . . . . 203
Takanori Matsumaru and Yukari Fujimoto

**Multivalent Probes for Exploring Carbohydrate Binding Proteins** ...... 227
Shione Kamoshita and Kaori Sakurai

**Chemical Probes for Analyzing Endoglycosidases** .................... 245
Nozomi Ishii and Ichiro Matsuo

**Modification Glycoside on IgG: Preparation of IgG with Homogeneous Glycan and Expansion to Antibody–Drug Conjugate** .................... 265
Shino Manabe

**Regulation of the Dynamics of Antibodies and Cell Membrane Molecules Through Synthetic Glycan Modification** .................. 283
Kazuya Kabayama and Yoshiyuki Manabe

**Peptide-Based Molecules Specific to Ganglioside Nanoclusters Toward Innovative Therapeutic Strategies** ......................... 299
Teruhiko Matsubara

**Glycomaterials for Biological Applications** ........................... 319
Yoshiko Miura, Masanori Nagao, and Tomonari Tanaka

**Atypical Protein Glycosylation: *C*-Mannosylation and Glucosyl-Galactosyl-Hydroxylation** ............................. 343
Kento Mori and Siro Simizu

# Precise Synthesis and Structure–Activity Relationship Studies of Low-Molecular-Weight Fucoidan Analogs

**Daisuke Takahashi and Kazunobu Toshima**

**Abstract** Fucoidan, a sulfated polysaccharide, is primarily obtained from marine organisms, particularly brown algae and certain invertebrate species. Structurally, it is characterized by repeating units of L-fucose linked via either α(1 → 3) or α(1 → 4) glycosidic bonds. Fucoidan has been reported to exhibit a broad spectrum of biological activities, including anticancer activities and antiviral activities against both the influenza virus and SARS-CoV-2. Moreover, its safety as an edible component has contributed to growing interest in its potential therapeutic applications. However, the structural heterogeneity of fucoidan—arising from variations in glycosidic linkages, sulfation patterns, and molecular weights depending on the biological source and extraction conditions—has hindered the identification of specific glycan motifs responsible for its biological activities. To address this issue, elucidation of the detailed structure–activity relationships (SAR) using structurally well-defined fucoidan analogs is highly warranted. In this context, we focused on three types of fucoidan categorized based on their linkage patterns: type I, composed of α(1 → 3)-linked fucose; type II, composed of alternating α(1 → 3) and α(1 → 4) linked fucose; and type III, composed of α(1 → 4)-linked fucose. A series of low-molecular-weight fucoidan analogs with well-defined glycan backbones and diverse sulfation patterns was synthesized accordingly. SAR studies were subsequently conducted to elucidate the key structural features governing their biological activities. In this chapter, we present our recent work on the precise chemical synthesis of structurally defined low-molecular-weight fucoidan analogs and SAR investigations focusing on their anti-proliferative effects against human cancer cells, inhibitory activity against influenza virus infection, and inhibitory activity against SARS-CoV-2 infection.

**Keywords** Fucoidan · Structure–activity relationship · Anti-proliferative activity · Anti-virus activity · Influenza virus · SARS-CoV-2 · Chemical library

D. Takahashi (✉) · K. Toshima (✉)
Department of Applied Chemistry, Faculty of Science and Technology, Keio University, Yokohama, Kanagawa, Japan
e-mail: dtak@applc.keio.ac.jp

K. Toshima
e-mail: toshima@applc.keio.ac.jp

 
K. Toshima and Y. Fujimoto (eds.), *Chemical Biology of Carbohydrates*,
https://doi.org/10.1007/978-981-95-5335-8_1

## 1 Introduction

Fucoidan is a type of complex, sulfated polysaccharide that is particularly rich in the sugar fucose. These polysaccharides are typically found in the fibrous cell walls and the internal matrices of brown seaweeds, also known as brown algae. In addition, fucoidan is also found in marine invertebrates, particularly in sea urchins and echinoderms [1–4]. Fucoidans are not uniform in their composition or structure, but they exhibit considerable diversity depending on a range of variables. This structural variability is influenced by the particular species of algae and marine invertebrates from which they are derived, as well as environmental and methodological factors such as the growing conditions, the specific marine habitat, the season or stage at which the algae are harvested, and the techniques employed to extract and purify the fucoidans. Broadly speaking, fucoidans are categorized into three major structural groups [5]. The first group, referred to as type I fucoidans, includes those obtained from algae species such as *Chorda filum* and *Cladosiphon okamuranus*. These polysaccharides are primarily composed of a linear backbone consisting of (1 → 3)-linked α-L-fucopyranosyl units. The second group, known as type II fucoidans, includes examples extracted from species like *Fucus vesiculosus* and *Ascophyllum nodosum*. These are characterized by alternating or repeating sequences of (1 → 3)- and (1 → 4)-linked α-L-fucopyranosyl units in their main chains, forming a more heterogeneous polymeric structure. The third group, referred to as type III fucoidans, includes examples isolated from *Arbacia lixula* and *Strongylocentrotus droebachiensis*. These sugar backbones are composed of a linear (1 → 4)-linked α-L-fucopyranosyl units (Fig. 1).

Fucoidans have garnered significant scientific interest owing to their diverse biological and pharmacological functions, including anti-inflammatory, anticoagulant, and antiviral activities, as well as anticancer activity that induces apoptotic cell death. In addition, fucoidans are edible components. Thus, fucoidans have attracted much attention in recent years [6, 7]. However, the intrinsic structural

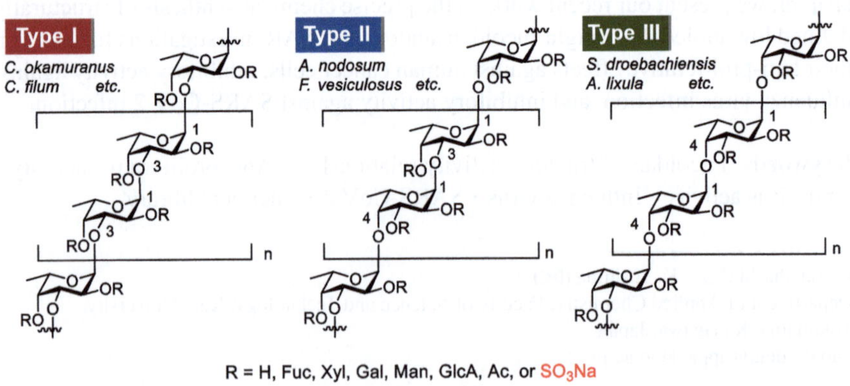


**Fig. 1** Three distinct homofucose backbone architectures (types I-III)

complexity and high degree of heterogeneity of fucoidans, particularly in terms of their branching patterns and the non-uniform distribution of sulfate groups, have significantly hindered efforts to elucidate the precise correlations between their molecular architecture and observed biological functions. One of the major challenges lies in the highly variable sulfation patterns, which can differ not only between species but also within samples of the same species, depending on environmental and extraction conditions. This variability makes it difficult to pinpoint which specific structural features are responsible for the biological activities of interest. Consequently, this lack of clarity has posed a substantial barrier to the rational design and development of fucoidan-based therapeutic agents or drugs for clinical use.

Numerous strategies have been employed to obtain low-molecular-weight fucoidans (LMWFs) from naturally occurring high-molecular-weight counterparts, given that LMWFs often show improved bioavailability and potentially enhanced biological activity. These strategies have included methods such as free radical-mediated depolymerization, chemical hydrolysis using acid or base treatments, and enzymatic hydrolysis with specific glycosidases [2, 3, 8–11]. Despite these efforts, it has proven extremely challenging to reproducibly generate LMWFs that are both structurally well-defined and chemically homogeneous. Batch-to-batch variations remain a problem, and the resulting products often contain mixtures of fragments with varying chain lengths, sulfation degrees, and monosaccharide linkages, complicating their characterization and the evaluation of their bioactivities. Against this background, although several excellent studies on the chemical synthesis of structurally well-defined and homogeneous oligofucosides bearing distinct sulfation patterns had already been described in the literature by the time we began our research, detailed structure–activity relationship (SAR) studies had not yet been sufficiently conducted [12–16]. In this context, we have undertaken the systematic chemical synthesis of low-molecular-weight fucoidan analogs featuring diverse glycan structures and sulfation patterns corresponding to types I–III. Comprehensive SAR studies were subsequently conducted to evaluate their inhibitory effects on cancer cell proliferation [17, 18], as well as their antiviral activities against the influenza virus [19] and SARS-CoV-2 infection [20, 21]. These findings are presented in the present chapter.

## 2 Systematic Synthesis of LMWF-Analogs and Their Anti-Proliferative Activities Against Human Cancer Cells [17, 18]

Recent studies have demonstrated that type I fucoidan, isolated from *C. okamuranus*, markedly suppresses the proliferation of peripheral blood mononuclear cells (PBMCs) derived from patients with adult T-cell leukemia, as well as T-cell lines infected with human T-cell leukemia virus type 1 (HTLV-1), while exhibiting minimal cytotoxicity toward normal PBMCs [22]. Furthermore, type II fucoidan obtained from *F. vesiculosus* has been reported to trigger apoptotic cell death in MCF-7 human breast cancer cells [23, 24] and HS-Sultan cells [25]. These findings highlight the therapeutic potential of structurally distinct fucoidan types. Considering this, we have undertaken the rational design and chemical synthesis of types I and II fucoidan analogs, followed by an evaluation of their anti-proliferative activities against MCF-7 and HeLa cell lines.

## *2.1 Systematic Synthesis of Fucoidan Analogs to Types I and II*

The structures of our synthesized fucoidan analogs **FA01-09** are illustrated in Fig. 2. These include type I analogs **FA01-04**, which differ in their sulfation patterns, namely, **FA01** (2,4-*O*-sulfated), **FA02** (4-*O*-sulfated), **FA03** (2-*O*-sulfated), and **FA04** (non-sulfated). In addition, type II analogs **FA05–09** were prepared, featuring various sulfation patterns such as **FA05** (2,3,4-tri-*O*-sulfated), **FA06** (2,3-*O*-sulfated), **FA07** (3,4-*O*-sulfated), **FA08** (4-*O*-sulfated), and **FA09** (non-sulfated). To enable an efficient and systematic synthesis of these analogs, we employed an orthogonal deprotection sequence based on common key intermediates **1** and **2** [17, 18]. These intermediates were strategically designed to bear three distinct types of protecting groups—benzoyl (Bz), benzyl (Bn), and *p*-methoxybenzyl (PMB)—positioned appropriately to allow selective manipulation. This approach minimized the total number of synthetic steps required and enabled the preparation of sufficient quantities of each analog for subsequent biological evaluation.

The synthetic routes for analogs **FA01-09** are outlined in Scheme 1. Following the chemical synthesis of the key intermediate **1**, a series of designed sulfated and non-sulfated analogs, **FA01-04**, were synthesized from this common precursor (Scheme 1a). Complete removal of the Bn and PMB groups in **1** under hydrogenolytic conditions, followed by methanolysis, yielded the non-sulfated analog **FA04**. Subsequent sulfation of **FA04** with the $SO_3 \cdot NEt_3$ complex in DMF afforded the 2,4-*O*-sulfated analog **FA01**. Alternatively, direct global deprotection of the Bn and PMB groups in **1**, followed by sulfation and saponification, led to the formation of the 2-*O*-sulfated analog **FA03**. In a separate route, selective removal of the PMB group in **1**, followed by methanolysis, sulfation, and final hydrogenolysis, yielded the 4-*O*-sulfated analog **FA02**.

In a similar manner, type II analogs **FA05-09** were synthesized from the common intermediate **2**, as illustrated in Scheme 1b. Briefly, complete removal of the Bn and PMB groups in compound **2** under hydrogenolytic conditions, followed by methanolysis, yielded the non-sulfated analog **FA09**. Subsequent sulfation of **FA09** using the $SO_3 \cdot NEt_3$ complex afforded the 2,3,4-*O*-sulfated analog **FA05**. In an alternative approach, removal of the Bn and PMB groups in **2**, followed by sulfation and methanolysis, resulted in the formation of the 2,3-*O*-sulfated analog **FA06**. Additionally, methanolysis of **2**, followed by sulfation and final hydrogenolysis, produced the 4-*O*-sulfated analog **FA08**. Finally, the methanolysis of **2**, followed by the selective

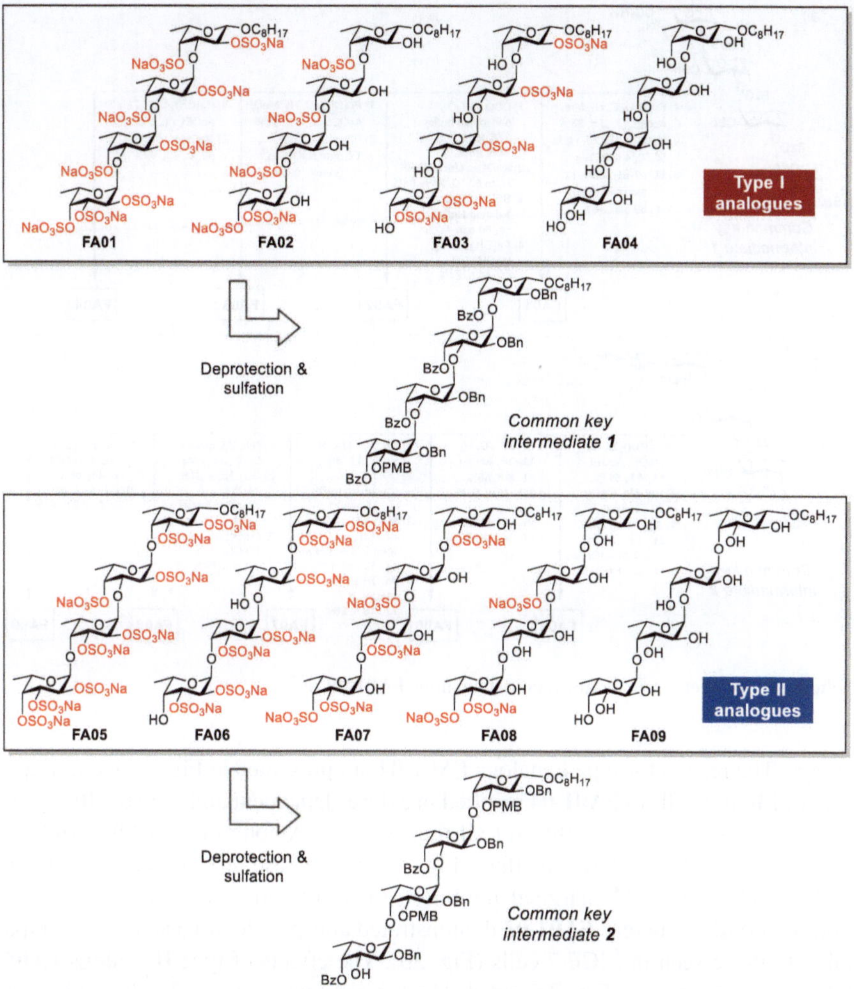


**Fig. 2** Chemical structures and retrosynthetic analysis of fucoidan analogs **FA01-09**

removal of the PMB groups using DDQ, sulfation, and subsequent hydrogenolysis, furnished the desired 3,4-*O*-sulfated analog, **FA07**.

## 2.2 *Anti-proliferation Activity of LMWF-Analogs Against Cancer Cells*

Having synthesized LMWF analogs **FA01-09**, their inhibitory effects on the proliferation of MCF-7 and HeLa cancer cells were evaluated using MTT-based cell viability

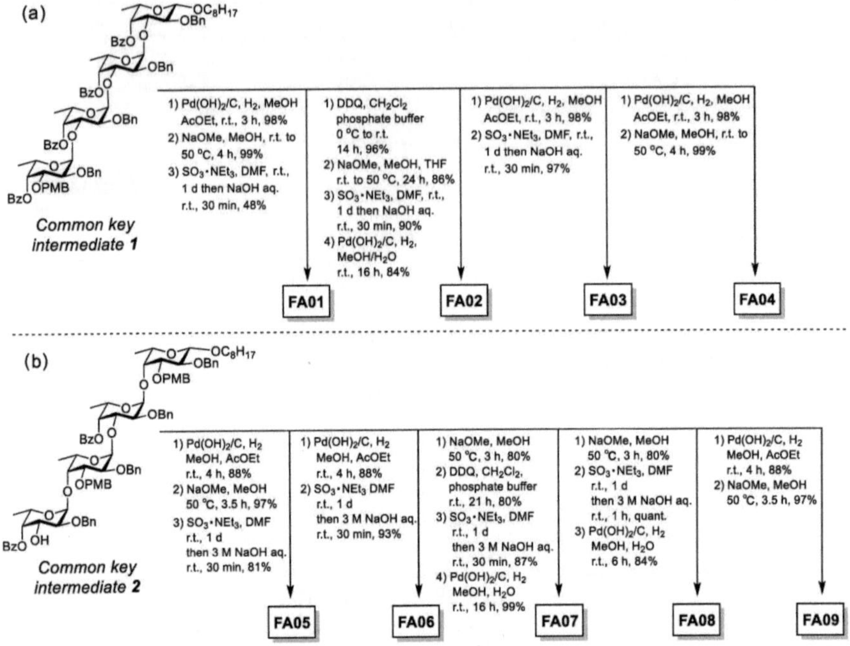


**Scheme 1** Synthetic scheme for fucoidan analogs **FA01-09**

assays. The results for type I analogs **FA01-04** are presented in Fig. 3a and b. Exposure of MCF-7 cells to **FA01-04** resulted in a dose-dependent and statistically significant reduction in cell viability for all four analogs. Among them, **FA01** exhibited the strongest anti-proliferative effect (Fig. 3a). In the case of HeLa cells, treatment with **FA04** led to only marginal inhibition of proliferation, even at a concentration of 800 μM, whereas **FA01-03** demonstrated anti-proliferative activities comparable to those seen in MCF-7 cells (Fig. 3b). The effects of type II analogs **FA05-09** are summarized in Fig. 3c and d. Upon treatment of MCF-7 cells with these compounds, **FA08** and **FA09** showed only weak anti-proliferative activity. In contrast, **FA05**, **FA06**, and **FA07** significantly suppressed cell proliferation in a concentration-dependent manner, with **FA07** exhibiting the greatest efficacy (Fig. 3c). An analogous trend was observed in HeLa cells treated with **FA05-09** (Fig. 3d). These findings suggest that subtle variations in both the glycan backbone and the sulfation pattern of analogs **FA01-09** critically influence their anti-proliferative properties.

As compounds **FA01-04** and **FA07** decreased the viability of MCF-7 and/or HeLa cells to less than 50%, their cytotoxicity was subsequently evaluated in normal human WI-38 lung fibroblast cells. The findings are shown in Fig. 4. Type I analogs **FA01-04** significantly decreased the viability of WI-38 cells in a concentration-dependent manner, indicating notable cytotoxicity toward normal cells. In contrast, **FA07**, which possesses a type II glycan backbone and a 3,4-*O*-sulfation pattern, exhibited neither

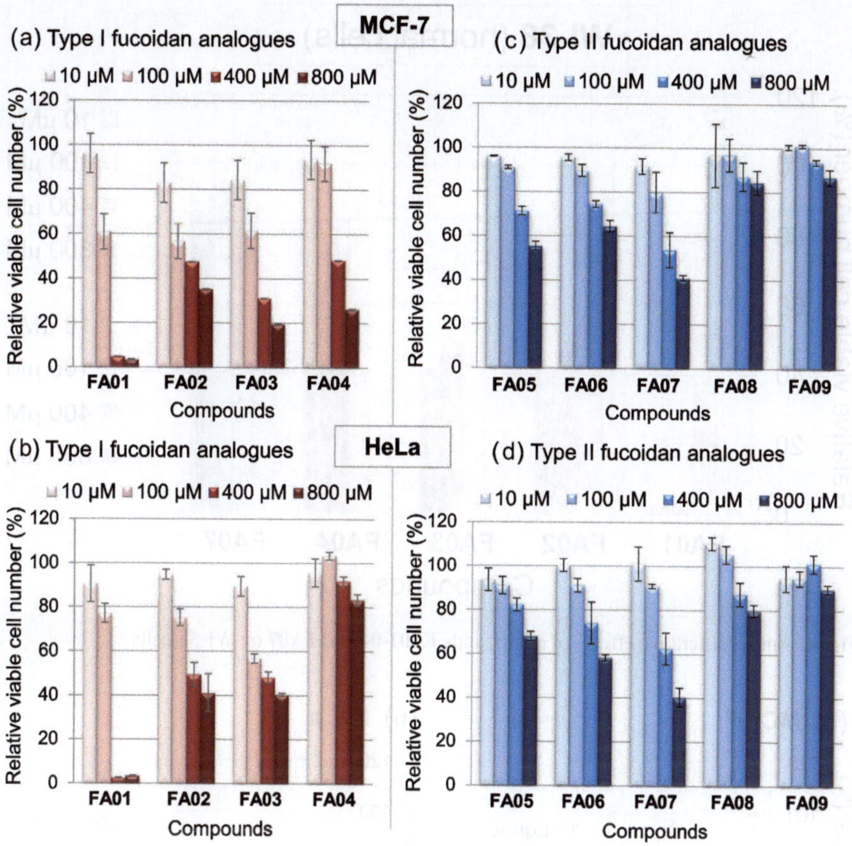


**Fig. 3** Anti-proliferative effects of compounds **FA01-09** on MCF-7 and HeLa cells

cytotoxic nor anti-proliferative effects against WI-38 cells across the tested concentrations. These findings clearly suggest that the 3,4-*O*-sulfated tetrafucoside **FA07**, derived from a type II fucoidan backbone, demonstrates selective anti-proliferative activity toward human cancer cells while sparing normal cells, indicating its promise as a candidate for anticancer drug discovery.

Given the strong and selective anti-proliferative activity of **FA07** against human cancer cells, we performed MTT assays to directly compare the activity of **FA07** with that of a naturally occurring high-molecular-weight fucoidan (HMWF, 600 kDa) **3**, obtained from *F. vesiculosus*, against MCF-7 and HeLa cells. Notably, fucoidan **3** demonstrated anti-proliferative effects comparable to those of **FA07** in both cancer cell lines as shown in Fig. 5. The results support the conclusion that 3,4-*O*-sulfation within a type II oligofucoside framework may be critical for the anticancer properties of natural fucoidans, highlighting the potential of **FA07** as a structurally simplified yet biologically active mimic of natural fucoidans.

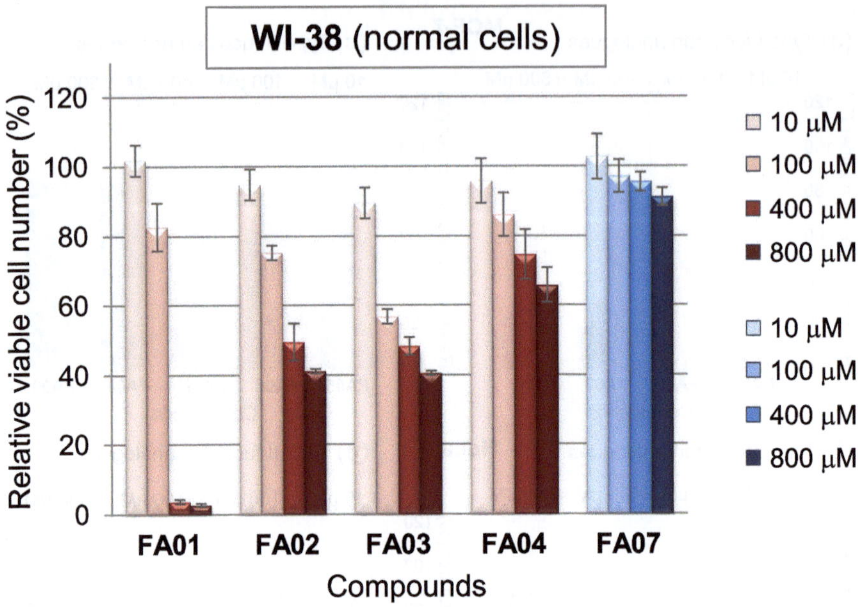


**Fig. 4** Anti-proliferative effects of compounds **FA01-04** and **FA07** on WI-38 cells

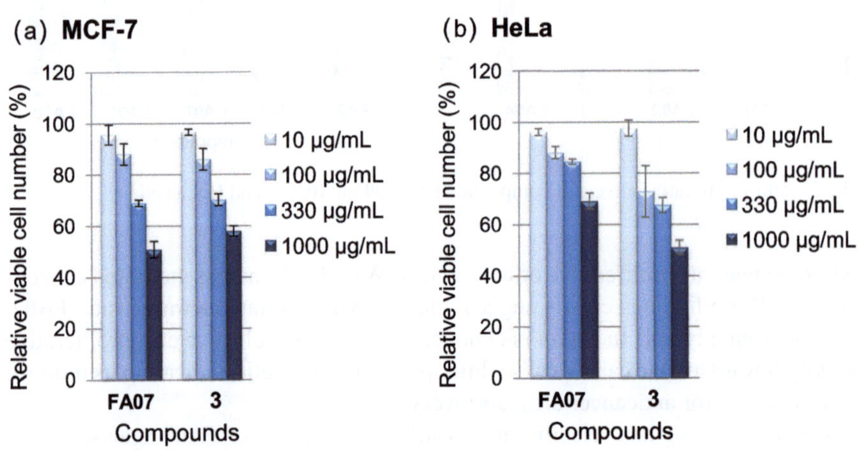


**Fig. 5** Anti-proliferative effects of compound **FA07** and fucoidan **3** on MCF-7 and HeLa cells

### *2.3 Mechanistic Study of the Anti-proliferative Effects of Fucoidan 3 and FA07 on Cancer Cells*

To investigate whether fucoidan **3** and compound **FA07** induce apoptotic cell death in cancer cells, nuclear morphology was assessed by Hoechst 33,342 [26] staining following treatment with compounds **3** or **FA07** using MCF-7 cells. As shown in Fig. 6, treatment with either compound resulted in chromatin condensation, evidenced by intense blue-white fluorescence within the nuclei compared to the untreated control (Fig. 6a–c). Notably, co-administration of Z-VAD-fmk (Z-VAD), a pan-caspase inhibitor, markedly reduced the number of apoptotic cells (Fig. 6d and e). These findings indicate that compounds **3** and **FA07** induce apoptosis in MCF-7 cells via a caspase-dependent mechanism. Similar results were observed in HeLa cells, confirming that the same apoptotic mechanism is also operative in this cell line.

To investigate the mechanistic contribution of caspases to apoptosis triggered by **3** or **FA07**, we assessed the activation status of caspase-8 and caspase-9, representative markers of the extrinsic and intrinsic pathways, respectively [27, 28], by western blotting analysis, and the results are summarized in Fig. 7. In MCF-7 cells, treatment with either compound induced proteolytic cleavage of caspase-8, resulting in the formation of its active fragments (43 and 41 kDa), whereas caspase-9 remained uncleaved (Fig. 7a and b). Caspase-8 activation is typically associated with apoptosis mediated by integrin engagement [29, 30] or death receptor signaling, including Fas, TNF receptor, and TRAIL receptor pathways [31]. These findings suggest that both fucoidan **3** and its low-molecular-weight analog **FA07** may interact with cell surface receptors on MCF-7 cells to initiate caspase-8 dependent apoptosis. In contrast, in HeLa cells, treatment with **3** or **FA07** did not induce caspase-8 cleavage, while caspase-9 was efficiently cleaved into its active fragments (37 and 35 kDa) (Fig. 7c and d). Since caspase-9 activation is typically triggered by mitochondrial release of cytochrome c in response to stress signaling pathways such as JNK/SAPK [32], these results suggest that compounds **3** and **FA07** activate the intrinsic apoptotic pathway in HeLa cells. Taken together, this study offers the first demonstration that a chemically defined sulfated tetrafucoside epitope is sufficient to induce apoptosis in MCF-7 and HeLa cells via activation of the extrinsic and intrinsic pathways, respectively.

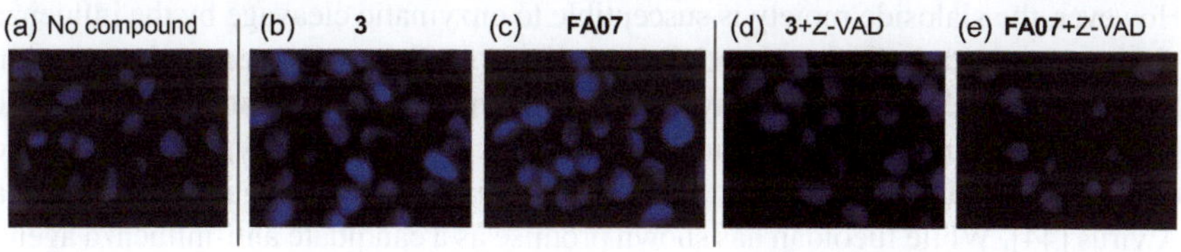


**Fig. 6** Fluorescence microscopy analysis of apoptotic cell death in MCF-7 cells based on nuclear morphology

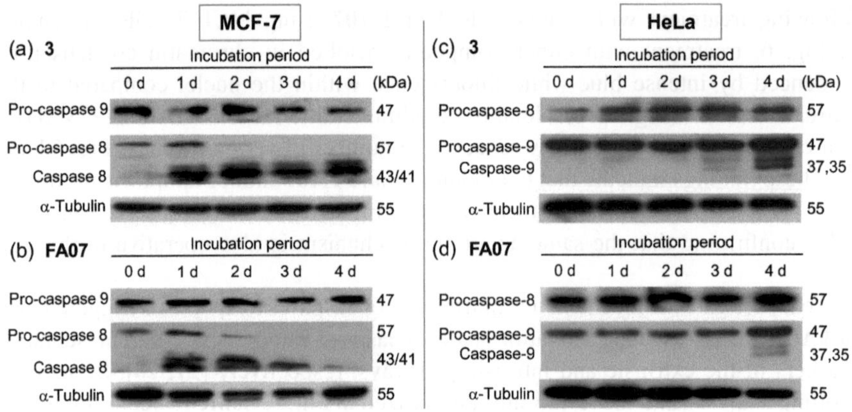


**Fig. 7** Western blot evaluation of apoptotic caspase activation in MCF-7 and HeLa cells treated with fucoidan **3** and **FA07**

# 3 Evaluation of Hemagglutinin Binding and Antiviral Activity Against Influenza Virus Infection [19]

Influenza A virus represents a critical zoonotic pathogen capable of infecting a broad range of avian and mammalian hosts, thereby posing substantial threats to both public health and global food security [33]. The viral envelope glycoprotein hemagglutinin (HA), a homotrimeric receptor, plays a pivotal role in viral entry [34] by recognizing and binding to multiple sialoside moieties on the host cell surface. Due to its essential function in the infection process, HA is considered a promising target for antiviral drug development. To date, various classes of HA-targeting compounds—referred to as HA inhibitors—have been identified [35–39]. Among them, sialic acid-containing oligomers and polymers have demonstrated potent inhibitory activity. However, the sialoside moiety is susceptible to enzymatic cleavage by the influenza viral neuraminidase (NA), potentially reducing its efficacy. Consequently, attention has shifted to polyanionic carbohydrates, such as heparin [40, 41] and dextran sulfate [42, 43], as alternative HA-binding inhibitors. Fucoidan, another type of polyanionic carbohydrate, has also been investigated for its antiviral potential against the influenza A virus [44]. While fucoidan has shown promise as a candidate anti-influenza agent, direct interaction between fucoidan and influenza virus HA has yet to be demonstrated [45–47]. Therefore, to identify novel and potent ligands for influenza virus HA, we evaluated the binding affinity of a series of synthesized fucoidan analogs **FA01-09** toward HA. Furthermore, to enhance inhibitory potency through multivalent interaction, a trimeric form of the selected ligand was designed and synthesized as a potential inhibitor of influenza virus infection.

### 3.1 Evaluation of Binding Affinity of Fucoidan Analogs Against Influenza Virus HA

To evaluate the hemagglutination inhibition (HAI) activity of compounds **FA01–09**, we assessed their ability to inhibit the agglutination of chicken red blood cells (RBCs) mediated by HAs derived from influenza A virus subtypes H1N1 (H1 HA) and H3N2 (H3 HA). The HAI titer was determined as the lowest concentration of compound capable of fully preventing hemagglutination of 0.5% chicken RBCs. Preliminary screening of **FA01-09** against H1 HA was carried out, and the results are shown in Table 1. Notably, among the tested analogs, only **FA05** exhibited HAI activity toward H1 HA, with a minimum concentration of 125 μM required for complete inhibition (Table 1, entries 1–9). Under the same conditions, type II fucoidan isolated from *F. vesiculosus* did not exhibit HAI activity (Table 1, entry 10). These findings suggest that subtle variations in both the sugar backbone structure and the sulfation pattern of the oligofucoside derivatives significantly influence HA binding. The natural fucoidan used in this study probably lacks the distinct sulfated oligofucoside structures essential for HA binding. As a positive control, fetuin [48]—a serum glycoprotein bearing 10–13 sialoside ligands—was evaluated and exhibited HAI activity at a sialoside concentration of 100–130 μM (corresponding to 10 μM as fetuin molecules; entry 11). Furthermore, **FA05** also demonstrated HAI activity against H3 HA, with a titer of 250 μM (entry 12). Taken together, these results indicate that the HAI activity of **FA05** is comparable to that of the natural sialoside ligand, highlighting **FA05** as a promising novel ligand for influenza virus HA.

Next, to evaluate both the chemical stability of compound **FA05** and its potential interaction with influenza virus neuraminidase (NA), we assessed its neuraminidase inhibition (NAI) activity using 2′-(4-methylumbelliferyl)-α-D-*N*-acetylneuraminic acid as a fluorogenic substrate, which is specifically cleaved by NA from the H1N1 strain A/California/04/2009 [49]. As a reference, the active form of oseltamivir (carboxylate) was employed as a positive control, exhibiting robust inhibition with 99% suppression of enzymatic activity at a concentration of 20 μM. In stark contrast, compound **FA05** failed to display measurable inhibitory activity, even at concentrations exceeding 500 μM. Furthermore, LC/MS analysis of the reaction mixture following enzymatic treatment revealed that 97% of compound **FA05** remained intact, indicating that it is resistant to hydrolysis by influenza NA and thus chemically stable under the assay conditions.

To enhance the binding affinity toward hemagglutinin (HA) trimers, we designed a $C_3$-symmetric trivalent ligand **4**, incorporating three sulfated oligofucoside moieties derived from **FA05**, spatially arranged on a benzene-1,3,5-tricarboxylic acid core [36, 50]. The synthetic scheme for compound **4** is outlined in Scheme 2. The key glycosylation step was performed by coupling disaccharide glycosyl donor **5** [17] with glycosyl acceptor **6** [51], followed by removal of the chloroacetyl (ClAc) group to afford **7**. Subsequent glycosylation of **7** and **5** in diethyl ether at –80 to –60 °C, followed by removal of the ClAc protecting group, furnished the target **8**. Removal of

**Table 1** HAI activities of **FA01-09** and fucoidan against H1 and H3 Has

| Entry | Compound | HA subtype[a] | HA titer[b] for H-A (μM) |
|---|---|---|---|
| 1 | **FA01** | H1 | ND[c] |
| 2 | **FA02** | H1 | ND[c] |
| 3 | **FA03** | H1 | ND[c] |
| 4 | **FA04** | H1 | ND[c] |
| 5 | **FA05** | H1 | 125 |
| 6 | **FA06** | H1 | ND[c] |
| 7 | **FA07** | H1 | ND[c] |
| 8 | **FA08** | H1 | ND[c] |
| 9 | **FA09** | H1 | ND[c] |
| 10 | Fucoidan (*F. vesicuiosus*) | H1 | ND[d] |
| 11 | Fetuin | H1 | 10 (100–130 μM as oligosacchalide residues) |
| 12 | **FA05** | H3 | 250 |

[a]H1 and H2 HAs were derived from Influenza viruses A/New Caledonia/20/99 (H1N1, 120 HA units) and A/New York/55/2004 (H3N2, 16 HA units), respectively
[b]The minimum concentration required for HAI activity
[c]Not detected (no activity at a concentration of 0.5 mM)
[d]Not detected (no activity at a concentration of 1 mg $mL^{-1}$)

the PMB and Bn groups in **8** under hydrogenolytic conditions, followed by hydrolysis, yielded desired **9**. Amidation of **9** and spacer **10**, followed by the removal of the Fmoc group in **11**, afforded intermediate **12**. Finally, amidation of **12** with benzene-1,3,5-tricarboxylic acid (**13**) using TBTU as a coupling reagent, followed by sulfation with a $SO_3 \cdot NEt_3$ complex and subsequent saponification, furnished the designed trivalent ligand **4**.

After synthesis of the trivalent ligand **4**, its HAI activity was evaluated to assess the impact of multivalency. Ligand **4** demonstrated significantly enhanced inhibitory activity against H1 HA, exhibiting an HAI titer of 2.94 μM (Scheme 2) in terms of oligosaccharide residues (corresponding to 0.98 μM as trimeric ligand), which is approximately 40-fold more potent than that of the monovalent counterpart **FA05** and the reference glycoprotein fetuin (Table 1, entries 5 and 11). Furthermore, trivalent ligand **4** also exhibited a strong binding affinity toward H3 HA, with inhibitory activity about tenfold greater than that of **FA05** on a per-oligosaccharide residue basis (Table 1, entry 12 and Scheme 2). These findings suggest that the sulfated oligosaccharide moieties of the ligands interact with the same binding pocket on HA as native sialosides.

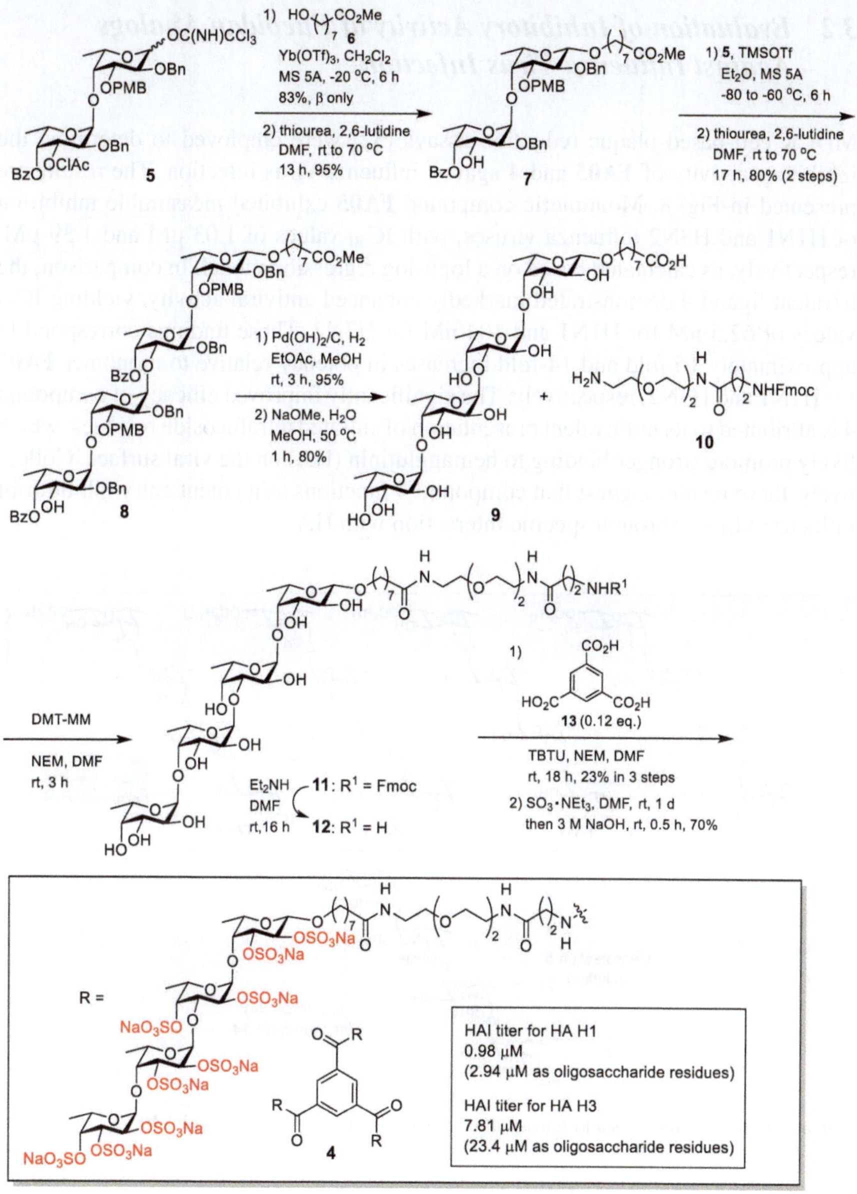


**Scheme 2** Synthetic scheme and HAI titer activities of trivalent ligand **4**

### 3.2 Evaluation of Inhibitory Activity of Fucoidan Analogs Against Influenza Virus Infection

MDCK cell-based plaque reduction assays [52] were employed to determine the inhibitory activity of **FA05** and **4** against influenza virus infection. The results are presented in Fig. 8. Monomeric compound **FA05** exhibited measurable inhibition of H1N1 and H3N2 influenza viruses, with $IC_{50}$ values of 1.03 μM and 1.39 μM, respectively, as calculated based on a logit-log regression model. In comparison, the trivalent ligand **4** demonstrated markedly enhanced antiviral activity, yielding $IC_{50}$ values of 62.5 nM for H1N1 and 104 nM for H3N2. These findings correspond to approximately 16-fold and 14-fold increases in potency relative to monomer **FA05** for H1N1 and H3N2, respectively. The significantly improved efficacy of compound **4** is attributed to its multivalent presentation of sulfated tetrafucoside residues, which likely promote stronger binding to hemagglutinin (HA) on the viral surface. Collectively, these results suggest that compound **4** functions as a potent entry inhibitor of influenza viruses through specific interaction with HA.

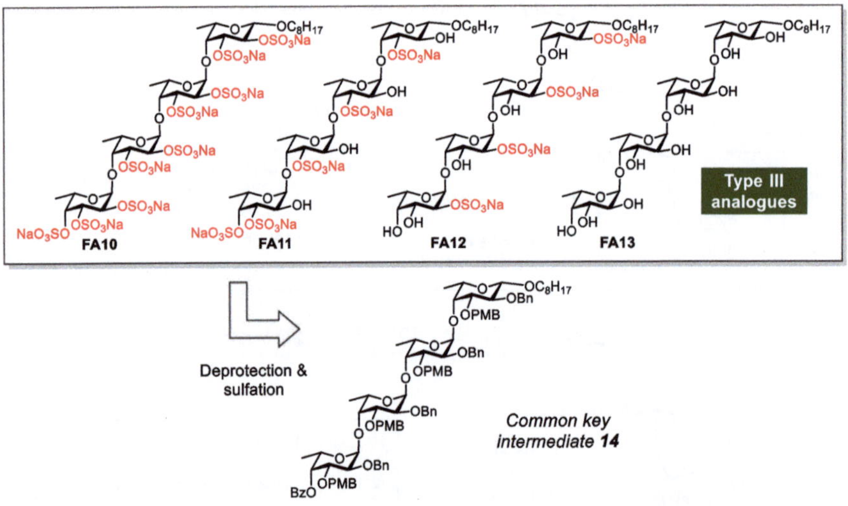


**Fig. 8** Chemical structures and retrosynthetic analysis of fucoidan analogs **FA10-13**

# 4 Synthesis of Fucoidan Analogs and Evaluation of Their Binding Affinity for SARS-CoV-2 Spike Protein and Inhibitory Activity Against SARS-CoV-2 Infection [20, 21]

The global outbreak of coronavirus disease 2019 (COVID-19), caused by the emergence of severe acute respiratory syndrome coronavirus 2 (SARS-CoV-2), has progressed with alarming speed, resulting in a significant mortality toll and contributing to widespread economic disruption. From the start of the pandemic until November 2024, the cumulative number of fatalities has surpassed 7 million worldwide [53]. Although the COVID-19 pandemic has officially been declared over, the potential threat posed by the emergence of novel variants, such as Omicron [54], remains a significant concern. Consequently, the development of innovative therapeutic strategies that can function as effective complements to existing vaccines remains a critical and pressing priority to ensure a prompt and effective response to future outbreaks. An increasing body of evidence indicates that the viral spike glycoprotein (SARS-CoV-2-S) plays a central role in mediating SARS-CoV-2 entry into host cells. Initial attachment occurs via interactions between the SARS-CoV-2-S and the cell surface heparan sulfate (HS). This interaction facilitates subsequent binding of the virus to its primary cellular receptor, angiotensin-converting enzyme 2 (ACE2), ultimately resulting in infection of host cells [55–60]. Certain fucoidans have been shown to associate with the SARS-CoV-2-S and block viral infection [61, 62]. Given this inhibitory potential, fucoidan has emerged as a promising antiviral candidate against SARS-CoV-2, prompting considerable interest in elucidating the detailed SAR governing its interaction with SARS-CoV-2-S [63]. In parallel, elevated activity of the endo-β-D-glucuronidase heparanase (HPSE), which enzymatically degrades HS [64], has been observed in patients with COVID-19 and is thought to exacerbate disease severity [65]. Moreover, HS degradation by HPSE may enhance viral dissemination by compromising the protective HS layer on the host cell surface. Consequently, synthetic HPSE inhibitors that can also bind to SARS-CoV-2-S and prevent viral entry have garnered increasing attention as potential therapeutic agents [66, 67]. To explore the development of a novel dual-targeting antiviral agent capable of inhibiting both HPSE and SARS-CoV-2-S, we designed and synthesized type III fucoidan analogs bearing distinct sulfation patterns. Thirteen analogs were evaluated in SAR studies for their ability to inhibit both SARS-CoV-2-S and HPSE. As a result, we successfully identified a fucoidan analog that exhibits potent inhibitory activity against both targets. Furthermore, by modifying the structure, we successfully identified a potential lead compound for the development of dual-action therapeutics targeting COVID-19.

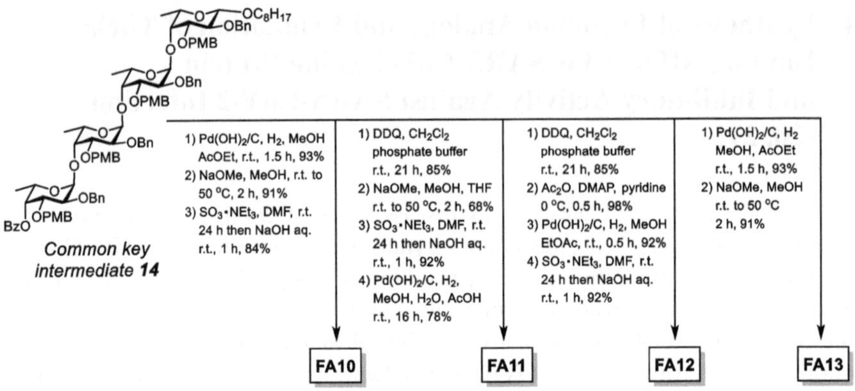


**Scheme 3** Synthetic scheme for fucoidan analogs **FA10-13**

## *4.1 Systematic Synthesis of Fucoidan Analogs to Types III*

The structures of our newly synthesized fucoidan analogs **FA10-13** are illustrated in Fig. 8. These include type III analogs **FA10-13**, which differ in their sulfation patterns, namely, **FA10** (2,3-*O*-sulfated), **FA11** (3-*O*-sulfated), **FA12** (2-*O*-sulfated), and **FA13** (non-sulfated). To enable an efficient and systematic synthesis of these analogs, we employed an orthogonal deprotection sequence using a common key intermediate **14** [20], as in the case of types I and II. The synthetic schemes for fucoidan analogs **FA10-13** are shown in Scheme 3. Complete removal of the Bn and PMB groups in compound **14** via hydrogenolysis, followed by methanolysis, yielded the non-sulfated analog **FA13**. Subsequent sulfation of **FA13** using the $SO_3{\cdot}NEt_3$ complex afforded the 2,3-*O*-sulfated analog **FA10**. PMB deprotection of **14** using DDQ, followed by methanolysis, sulfation with a $SO_3{\cdot}NEt_3$ complex in DMF, and hydrogenolytic cleavage, yielded the target 3-*O*-sulfated tetrafucoside **FA11**. On the other hand, PMB deprotection of **14**, followed by sequential acetylation, catalytic hydrogenation, sulfation, and final saponification to furnish the 2-*O*-sulfated tetrasaccharide **FA12**.

## *4.2 Evaluation of the Binding Affinity of Fucoidan Analogs FA01-13 for the SARS-CoV-2-S*

To assess the binding affinity of the thirteen synthesized fucoidan analogs for the SARS-CoV-2-S, we conducted bio-layer interferometry (BLI) analyses. In this assay, the ability of each compound (**FA01-13**) to inhibit the interaction between SARS-CoV-2-S and immobilized heparin on the BLI biosensor surface was quantified. As a reference compound, fondaparinux (FP) [68]—a clinically approved heparin-based anticoagulant known to bind SARS-CoV-2-S [60]—was employed. The results,

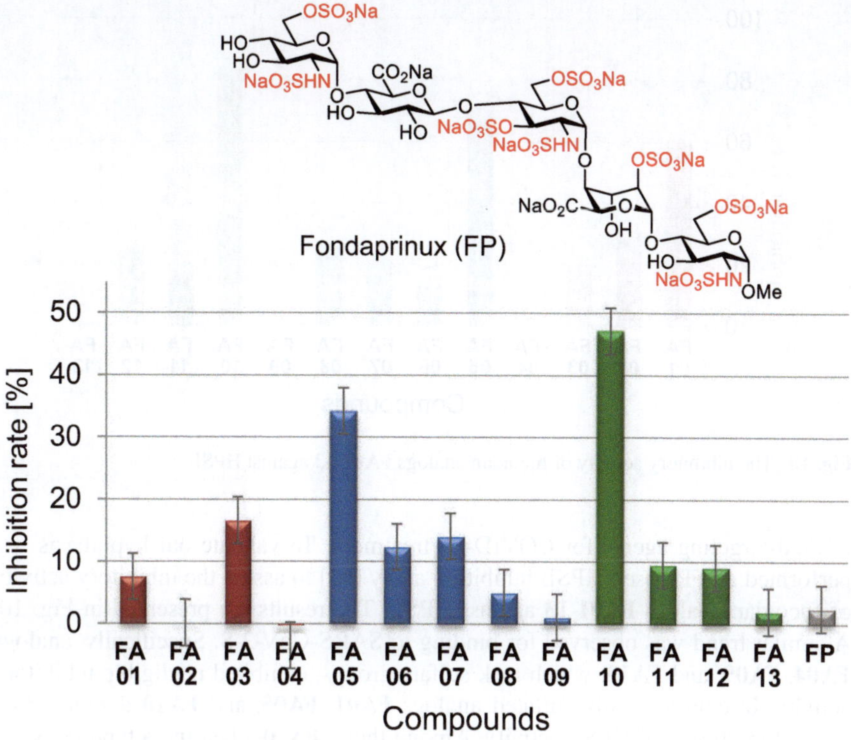


**Fig. 9** Chemical structure of FP and the binding affinity of fucoidan analogs **FA01-13** and FP for the SARS-CoV-2-S

summarized in Fig. 9, revealed that **FA10**, an all-sulfated derivative with a type III backbone, showed the strongest inhibitory effect. The second most potent compound was **FA05**, a fully sulfated analog possessing a type II backbone. In contrast, non-sulfated derivatives **FA04, FA09**, and **FA13** displayed negligible inhibitory activity. Notably, even **FA01**, an all-sulfated compound with a type I sugar backbone, showed only weak inhibition. These findings underscore the critical influence of both the sugar backbone architecture and the sulfation pattern on the inhibitory potency of these fucoidan analogs.

## *4.3 Evaluation of the Inhibitory Activity of Fucoidan Analogs FA01-13 Against HPSE*

Given that **FA10** demonstrated the strongest binding affinity toward SARS-CoV-2-S, we postulated that, should **FA10** also exhibit potent inhibitory activity against heparanase (HPSE), it could provide a valuable framework for the rational design

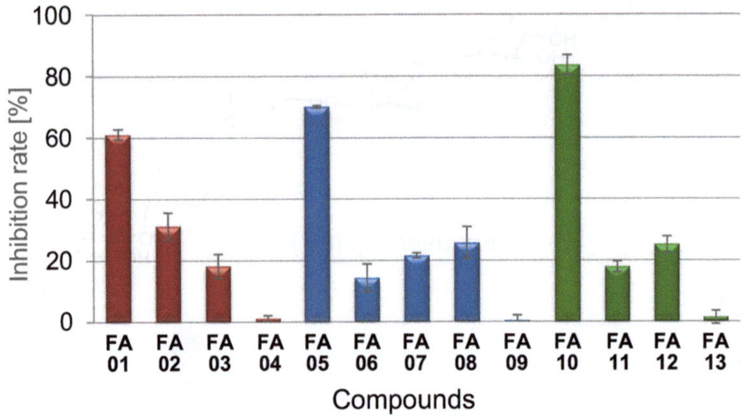


**Fig. 10** The inhibitory activity of fucoidan analogs **FA01-13** against HPSE

of dual-targeting agents for COVID-19 treatment. To validate our hypothesis, we performed an FP-based HPSE inhibition assay [69] to assess the inhibitory activity of fucoidan analogs **FA01-13** against HPSE. The results are presented in Fig. 10. A similar trend was observed for binding to SARS-CoV-2-S. Specifically, analogs **FA04**, **FA09**, and **FA13**, which lack sulfate groups, exhibited negligible inhibitory activity. In contrast, fully sulfated analogs **FA01**, **FA05**, and **FA10** demonstrated marked inhibition of HPSE activity. Among them, **FA10**—bearing a type III backbone—demonstrated the most potent activity, followed by **FA05** and **FA01**, which possess type II and type I backbones, respectively. These findings also suggest that both the degree and positional pattern of sulfation, as well as subtle variations in the sugar backbone structure, play critical roles in modulating HPSE inhibitory activity.

## *4.4 Design, Synthesis, and Biological Evaluation of Novel Fucoidan Analog FA14 Targeting SARS-CoV-2-S and HPSE*

With the optimal sugar backbone and sulfation profile established, attention was next directed toward modifying the aglycon moiety to further enhance both binding affinity for SARS-CoV-2-S and inhibitory activity against HPSE. To this end, we examined the literature on analogs of the established HPSE inhibitor PI-88 [70] analog **15**. Specifically, compound **15**, bearing a linear octyl chain as the aglycon, was compared with compound **16**, in which the aglycon is substituted with a bulkier hydrophobic cholestanyl group. Notably, compound **16** exhibited markedly improved potency, with a $K_i$ value of 5.9 nM—approximately tenfold stronger than that of **15** ($K_i = 58$ nM) [71]. In addition, PG545, a structurally related sulfated oligosaccharide also incorporating a cholestanyl moiety in its aglycon region, demonstrated

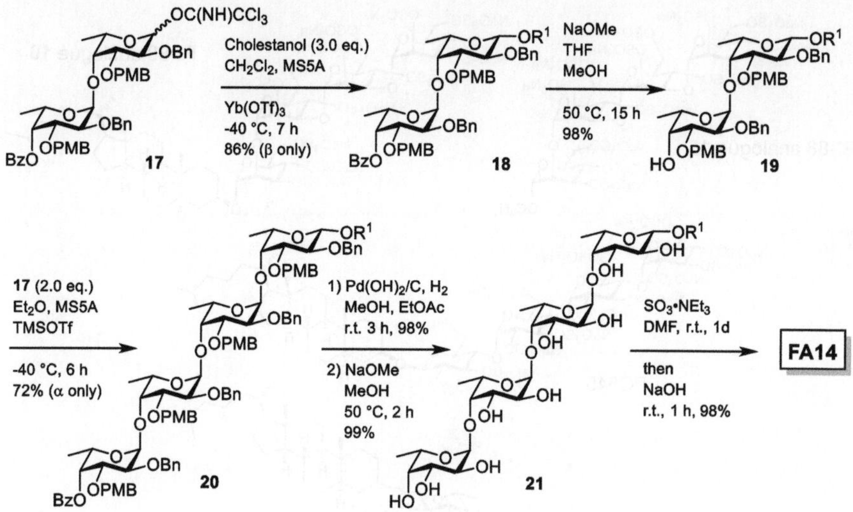


**Scheme 4** Synthetic scheme for fucoidan analog **FA14**

comparable inhibitory activity ($K_i = 6.1$ nM) [71]. These observations led us to hypothesize that introducing a cholestanyl group into the aglycon portion of **FA10** would enhance its inhibitory activity. Accordingly, we designed and synthesized novel fucoidan analog **FA14**, featuring a cholestanyl group in place of the octyl substituent present in **FA10**. The synthetic pathway employed to obtain **FA14** is outlined in Scheme 4.

Glycosylation of the trichloroacetimidate donor **17** [20] and cholestanol in the presence of a catalytic amount of $Yb(OTf)_3$ afforded compound **18** in high yield with exclusive 1,2-*trans*-β-stereoselectivity. Subsequent removal of the Bz protecting group from **18** yielded the corresponding acceptor **19**. A [2 + 2] glycosylation between acceptor **19** and donor **17**, promoted by TMSOTf, proceeded efficiently to furnish tetrasaccharide **20** with excellent 1,2-*cis*-α-selectivity. Subsequent hydrogenolysis of compound **20** to remove Bn and PMB groups, followed by methanolysis, yielded intermediate **21**. Sulfation of **21** using $SO_3{\cdot}NEt_3$, followed by saponification, successfully yielded the desired compound **FA14** with high overall efficiency (Fig. 11).

With fucoidan analog **FA14** successfully synthesized, its HPSE inhibitory activity was directly compared to that of **FA10**. Suramin (SU), a well-established HPSE inhibitor (Fig. 12a) [72], was included as a positive control. The $IC_{50}$ values for compounds SU, **FA10**, and **FA14** were determined to be 17.97 μM, 0.76 μM, and 0.25 μM, respectively, based on nonlinear regression analysis (Fig. 12). Furthermore, kinetic parameters were assessed by Michaelis–Menten analysis of FP hydrolysis by HPSE, and $K_i$ values for SU, **FA10**, and **FA14** were calculated to be 164.1 nM, 6.9 nM, and 2.3 nM, respectively. These data indicate that **FA14** exhibits approximately threefold greater inhibitory potency than **FA10**, and about 70-fold greater potency than SU, highlighting its potential as a highly effective HPSE inhibitor (Fig. 12a–c).

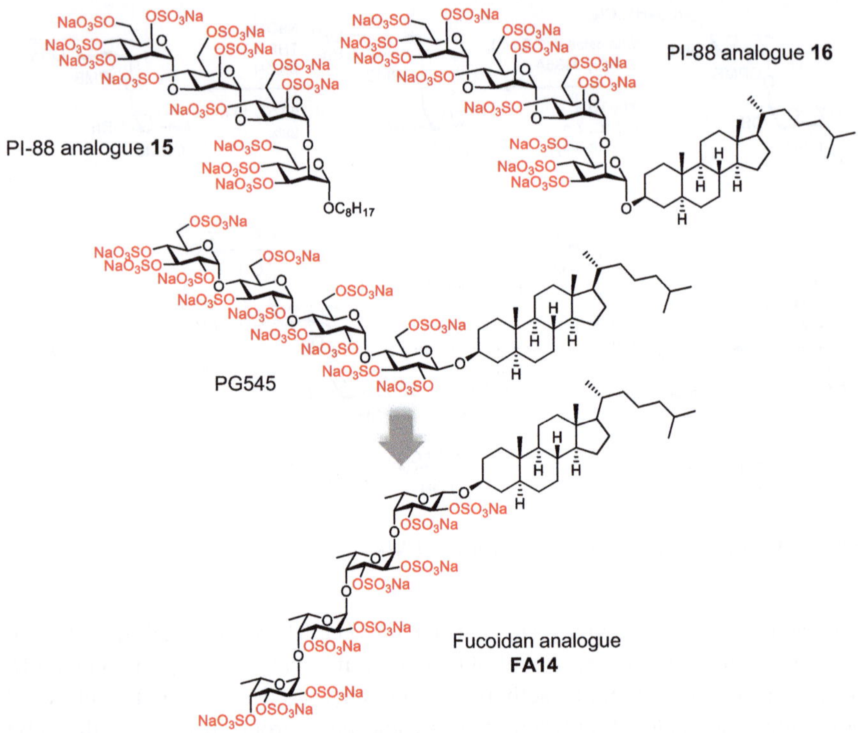


**Fig. 11** Chemical structures of PI-88 analogs and PG545, and the design of fucoidan analog **FA14**

Next, the binding affinities of **FA10** and **FA14** for the SARS-CoV-2-S were assessed using the heparin-competition BLI assay. PG545 was employed as a reference inhibitor. Nonlinear regression analysis yielded $IC_{50}$ values of 233 nM for PG545, 30.3 μM for **FA10**, and 75.8 nM for **FA14**. Using the known dissociation constant ($K_d = 36.0$ nM) of the heparin–SARS-CoV-2-S interaction, the calculated $K_i$ values were 8.08 nM for PG545, 1.05 μM for **FA10**, 2.63 nM for **FA14** (Fig. 13a–c). These results clearly demonstrate that compound **FA14** exhibits markedly enhanced binding affinity for SARS-CoV-2-S—approximately 400-fold greater than that of **FA10**. This dramatic improvement underscores the crucial role of the aglycon moiety in modulating interaction strength. Moreover, compound **FA14** also surpassed PG545 in potency by roughly threefold, highlighting the contribution of its unique glycan architecture to overall binding efficacy.

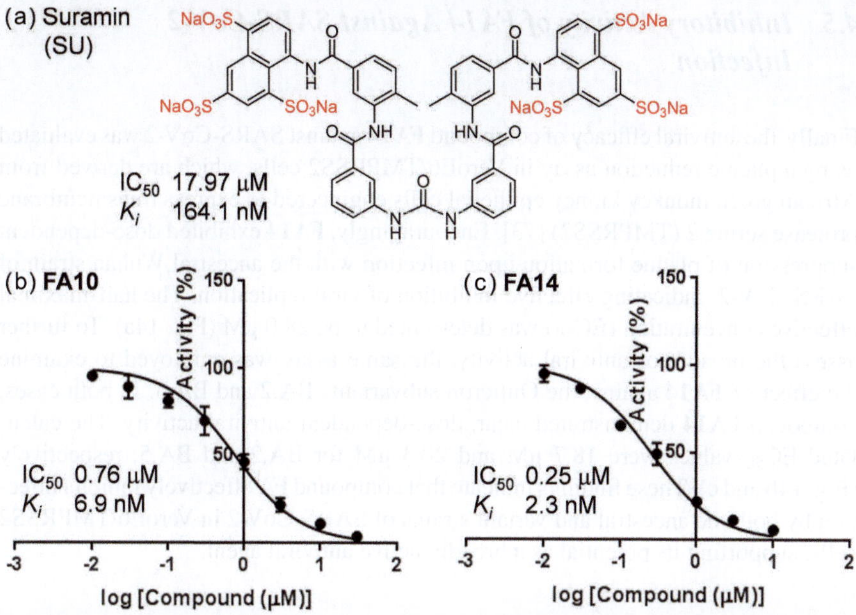


**Fig. 12** The inhibitory activity of SU (**a**), fucoidan analogs **FA10** (**b**), and **FA14** (**c**) against HPSE

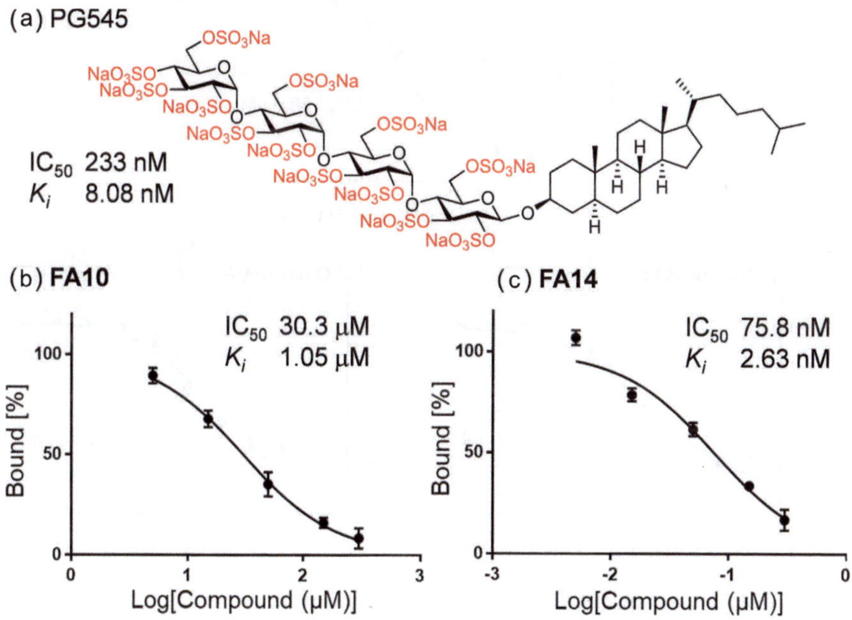


**Fig. 13** The binding affinity of PG545 (**a**), fucoidan analogs **FA10** (**b**), and **FA14** (**c**) for the SARS-CoV-2-S

## *4.5 Inhibitory Activity of FA14 Against SARS-CoV-2 Infection*

Finally, the antiviral efficacy of compound **FA14** against SARS-CoV-2 was evaluated using a plaque reduction assay in VeroE6/TMPRSS2 cells, which are derived from African green monkey kidney epithelial cells engineered to express transmembrane protease serine 2 (TMPRSS2) [73]. Encouragingly, **FA14** exhibited dose-dependent suppression of plaque formation upon infection with the ancestral Wuhan strain of SARS-CoV-2, indicating effective inhibition of viral replication. The half-maximal effective concentration ($EC_{50}$) was determined to be 28.0 μM (Fig. 14a). To further assess the breadth of antiviral activity, the same assay was employed to examine the effect of **FA14** against the Omicron subvariants BA.2 and BA.5. In both cases, compound **FA14** demonstrated clear, dose-dependent antiviral activity. The calculated $EC_{50}$ values were 18.7 μM and 20.3 μM for BA.2 and BA.5, respectively (Fig. 14b and c). These findings indicate that compound FA effectively inhibits infection by both the ancestral and variant strains of SARS-CoV-2 in VeroE6/TMPRSS2 cells, supporting its potential as a broadly active antiviral agent.

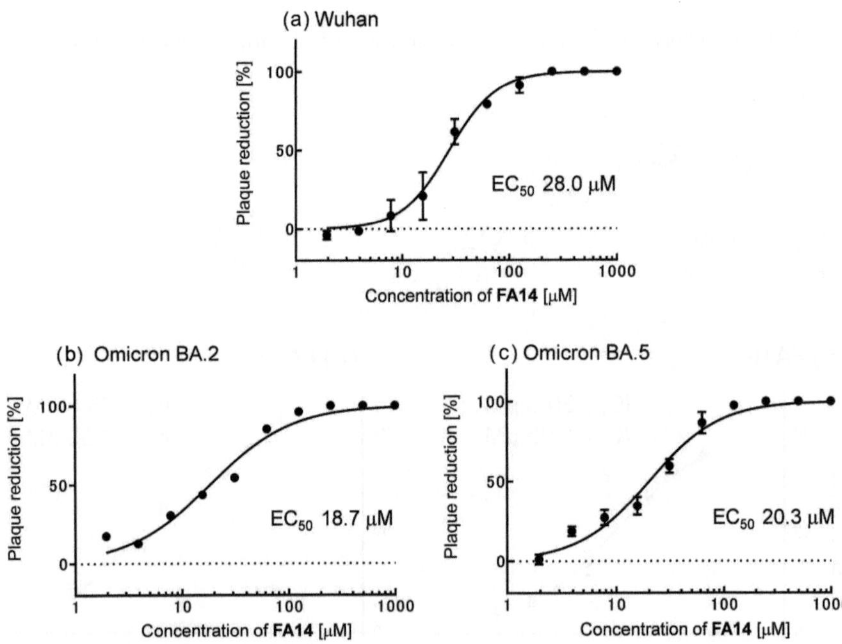


**Fig. 14** Inhibitory activity of **FA14** against SARS-CoV-2 infection

## 5 Conclusion

In summary, a set of type I and type II fucoidan analogs, **FA01-09,** with distinct sulfation profiles, were systematically synthesized to investigate their anti-proliferative properties. Among the type I analogs, **FA01-04** exhibited cytotoxic effects against both human cancer cell lines and normal WI-38 fibroblasts. In contrast, the type II analog **FA07** selectively suppressed the proliferation of MCF-7 and HeLa cancer cells in a dose-dependent manner while exhibiting minimal cytotoxicity toward WI-38 cells. Further mechanistic studies revealed that both natural fucoidan **3** and **FA07** promote apoptosis through activation of caspase-8 in MCF-7 cells and caspase-9 in HeLa cells. These findings suggest that **FA07** may serve as a promising lead structure for the development of sulfated oligosaccharide-based anticancer agents with reduced off-target toxicity compared to conventional chemotherapeutics. Next, among the fucoidan analogues **FA01-09**, only **FA05** demonstrated measurable binding affinity toward HAs from both H1 and H3 subtypes of influenza virus, as well as HAI activity against RBCs. Furthermore, comparison between the monomeric form of **FA05** and a synthetically constructed trivalent derivative **4** revealed that **4** exhibited substantially enhanced binding, with 65- and 70-fold increases in affinity toward H1 and H3 HAs, respectively. In a plaque reduction assay using MDCK cells, **4** significantly suppressed influenza virus infection, confirming its antiviral efficacy. These findings suggest that the multivalent presentation of fucoidan motifs significantly enhances HA binding and inhibitory activity, thereby providing a promising strategy for the development of potent influenza virus entry inhibitors targeting viral HAs. Next, a series of type III fucoidan analogs, **FA10-13,** with different sulfation patterns, were systematically synthesized. In addition, **FA10** was identified as the most potent HPSE inhibitor among derivatives **FA01-13** and also exhibited the highest binding affinity toward the SARS-CoV-2-S. Based on these findings, we hypothesized that a derivative structurally based on 10 could serve as a lead compound for dual-targeting COVID-19 therapy. To investigate this hypothesis, we designed and synthesized analog **FA14**, in which the octyl group of **FA10** was replaced with a cholestanyl moiety, and evaluated its biological activity. **FA14** demonstrated markedly enhanced HPSE inhibitory activity—approximately threefold increase over **FA10** and 70-fold enhancement relative to suramin (SU). Furthermore, **FA14** exhibited an approximately 400-fold increase in affinity compared to **FA10**, underscoring the critical role of the aglycon structure in mediating SARS-CoV-2-S recognition. Moreover, **FA14** suppressed viral infection in plaque assays using the original Wuhan strain of SARS-CoV-2 as well as the Omicron subvariants BA.2 and BA.5. Taken together, these findings establish **FA14** as a promising lead compound for dual-targeting therapeutics against COVID-19. Further in vivo studies and clinical evaluation will be essential to fully explore the therapeutic potential.

As demonstrated above, the integrated application of precise chemical synthesis and biological evaluation of fucoidan analogs—designed based on the structural motif of the natural sulfated polysaccharide fucoidan—holds great promise for the discovery of novel lead compounds for pharmaceutical development.

## References


1. Chevolot L, Foucault A, Chaubet F, Kervarec N, Sinquin C, Fisher A-M, Boisson-Vidal C (1999) Further data on the structure of brown seaweed fucans: relationships with anticoagulant activity. Carbohydr Res 319:154–165. https://doi.org/10.1016/S0008-6215(99)00127-5
2. Ale MT, Mikkelsen JD, Meyer AS (2011) Important determinants for fucoidan bioactivity: a critical review of structure-function relations and extraction methods for fucose-containing sulfated polysaccharides from brown seaweeds. Mar Drugs 9:2106–2130. https://doi.org/10.3390/md9102106
3. Ale MT, Meyer AS (2015) Fucoidans from brown seaweeds: an update on structures, extraction techniques and use of enzymes as tools for structural elucidation. RSC Adv 3:8131–8141. https://doi.org/10.1039/C3RA23373A
4. Pomn VH, Mourão PAS (2008) Structure, biology, evolution, and medical importance of sulfated fucans and galactans. Glycobiology 18:1016–1027. https://doi.org/10.1093/glycob/cwn085
5. Cumashi A, Ushakova NA, Preobrazhenskaya ME, D'Incecco A, Piccoli A, Totani L, Tinari N, Morozevich GE, Berman AE, Bilan MI, Usov AI, Ustyuzhanina NE, Grachev AA, Sanderson CJ, Kelly M, Rabinovich GA, Iacobelli S, Nifantiev NE (2007) A comparative study of the anti-inflammatory, anticoagulant, antiangiogenic, and antiadhesive activities of nine different fucoidans from brown seaweeds. Glycobiology 17:541–552. https://doi.org/10.1093/glycob/cwm014
6. Li B, Lu F, Wei X, Zhao R (2008) Fucoidan: structure and bioactivity. Molecules 13:1671–1695. https://doi.org/10.3390/molecules13081671
7. Wang Y, Xing M, Cao Q, Ji A, Liang H, Song S (2019) Biological activities of fucoidan and the factors mediating its therapeutic effects: a review of recent studies. Mar Drugs 17:183. https://doi.org/10.3390/md17030183
8. Nardella A, Chaubet F, Boisson-Vidal C, Blondin C, Durand P, Jozefonvicz (1996) Anticoagulant low molecular weight fucans produced by radical process and ion exchange chromatography of high molecular weight fucans extracted from the brown seaweed *Ascophyllum nodosum*. Carbohydr Res 289:201–208. https://doi.org/10.1016/0008-6215(96)00110-3
9. Rioux LE, Turgeon SL, Beaulieu M (2007) Characterization of polysaccharides extracted from brown seaweeds. Carbohydr Polym 69:530–537. https://doi.org/10.1016/j.carbpol.2007.01.009
10. Hahn T, Lang S, Ulber R, Muffler K (2012) Novel procedures for the extraction of fucoidan from brown algae. Process Biochem 47:1691–1698. https://doi.org/10.1016/j.procbio.2012.06.016
11. Jayawardena TU, Nagahawatta DP, Fernando IPS, Kim YT, Kim JS, Kim WS, Lee JS, Jeon YJ (2022) A review on fucoidan structure, extraction techniques, and its role as an immunomodulatory agent. Mar Drugs 20:755. https://doi.org/10.3390/md20120755
12. Hua Y, Gu G, Du Y (2004) Synthesis and biological activities of octyl 2,3,4-tri-*O*-sulfo-α-L-fucopyranosyl-(1→3)-2,4-di-*O*-sulfo-α-L-fucopyranosyl-(1→3)-2,4-di-*O*-sulfo-α-L-fucopyranosyl-(1→3)-2,4-di-*O*-sulfo-β-L-fucopyranoside. Carbohydr Res 339:867–872. https://doi.org/10.1016/j.carres.2003.12.014
13. Ustyuzhanina N, Krylov V, Grachev A, Gerbst A, Nifantiev N (2006) Synthesis, NMR and conformational studies of fucoidan fragments, 8: convergent synthesis of branched and linear oligosaccharides. Synthesis 4017–4031:950333. https://doi.org/10.1055/s-2006-950333
14. Zong C, Li Z, Sun T, Wang P, Ding N, Li Y (2010) Convenient synthesis of sulfated oligofucosides. Carbohydr Res 345:1522–1532. https://doi.org/10.1016/j.carres.2010.04.006
15. Krylov VB, Kaskova ZM, Vinnitskiy DZ, Ustyuzhanina NE, Grachev AA, Chizhov AO, Nifantiev NE (2011) Acid-promoted synthesis of per-O-sulfated fucooligosaccharides related to fucoidan fragments. Carbohydr Res 346:540–550. https://doi.org/10.1016/j.carres.2011.01.005

16. Ustyuzhanina NE, Ushakova NA, Preobrazhenskaya ME, Bilan MI, Tsvetkova EA, Krylov VB, Anisimova NA, Kiselevskiy MV, Krukovskaya NV, Li C, Yu G, Saran S, Saxena RK, Usov AI, Nifantiev NE (2014) Fucoidans as a platform for new anticoagulant drugs discovery. Pure Appl Chem 86:1365–1375. https://doi.org/10.1515/pac-2014-0404
17. Arafuka S, Koshiba N, Takahashi D, Toshima K (2014) Systematic synthesis of sulfated oligofucosides and their effect on breast cancer MCF-7 cells. Chem Commun 50:9831–9834. https://doi.org/10.1039/C4CC03544E
18. Kasai A, Arafuka S, Koshiba N, Takahashi D, Toshima K (2015) Systematic synthesis of low-molecular weight fucoidan derivatives and their effect on cancer cells. Org Biomol Chem 13:10556–10568. https://doi.org/10.1039/c5ob01634g
19. Kosono S, Kasai A, Komaba S, Matsubara T, Sato T, Takahashi D, Toshima K (2018) Novel hemagglutinin-binding sulfated oligofucosides and their effect on influenza virus infection. Chem Commun 54:7467–7470. https://doi.org/10.1039/c8cc03865a
20. Koike T, Sugimoto A, Kosono S, Komaba S, Kanno Y, Kitamura T, Anzai I, Watanabe T, Takahashi D, Toshima K (2018) Synthesis of low-molecular weight fucoidan derivatives and their binding abilities to SARS-CoV-2 spike proteins. RSC Med Chem 12:2016–2021. https://doi.org/10.1039/d1md00264c
21. Sugimoto A, Koike T, Kuboki Y, Komaba S, Kosono S, Aswathy M, Anzai I, Watanabe T, Toshima K, Takahashi D (2025) Synthesis of low-molecular-weight fucoidan analogue and its inhibitory activities against heparanase and SARS-CoV-2 infection. Angew Chem Int Ed 64:e202411760. https://doi.org/10.1002/anie.202411760
22. Haneji K, Matsuda T, Tomita M, Kawakami H, Ohshiro K, Uchihara JN, Masuda M, Takasu N, Tanaka Y, Ohta T, Mori N (2005) Fucoidan extracted from Cladosiphon okamuranus Tokida induces apoptosis of human T-cell leukemia virus type 1-infected T-cell lines and primary adult T-cell leukemia cells. Nutr Cancer 52:189–201. https://doi.org/10.1207/s15327914nc5202_9
23. Yamasaki-Miyamoto Y, Yamasaki M, Tachibana H, Yamada K (2009) Fucoidan induces apoptosis through activation of caspase-8 on human breast cancer MCF-7 cells. J Agric Food Chem 57:8677–8682. https://doi.org/10.1021/jf9010406
24. Yamasaki Y, Yamasaki M, Tachibana H, Yamada K (2012) Important role of $\beta_1$-integrin in fucoidan-induced apoptosis via caspase-8 activation. Biosci Biotechnol Biochem 76:1163–1168. https://doi.org/10.1271/bbb.111001
25. Aisa Y, Miyakawa Y, Nakazato T, Shibata H, Saito K, Ikeda Y, Kizaki M (2005) Fucoidan induces apoptosis of human HS-Sultan cells accompanied by activation of caspase-3 and down-regulation of ERK pathways. Am J Hematol 78:7–14. https://doi.org/10.1002/ajh.20182
26. Lizard G, Fournel S, Genestier L, Dhedin N, Chaput C, Flacher M, Mutin M, Panaye G, Revillard JP (1995) Kinetics of plasma membrane and mitochondrial alterations in cells undergoing apoptosis. Cytometry 21:275–283. https://doi.org/10.1002/cyto.990210308
27. Nagata S (1997) Apoptosis by death factor. Cell 88:355–365. https://doi.org/10.1016/S0092-8674(00)81874-7
28. Kuida K, Haydar TF, Kuan CY, Gu Y, Taya C, Karasuyama H, Su MSS, Rakic P, Flavell RA (1998) Reduced apoptosis and cytochrome c-mediated caspase activation in mice lacking caspase 9. Cell 94:325–337. https://doi.org/10.1016/S0092-8674(00)81476-2
29. Brassard DL, Maxwell E, Malkowski M, Nagabhushan TL, Kumar CC, L. Armstrong L (1999) Integrin $\alpha_v\beta_3$-mediated activation of apoptosis. Exp Cell Res 251:33–45. https://doi.org/10.1006/excr.1999.4559
30. Stupack DG, Puente XS, Boutsaboualoy S, Storgard CM, Cheresh DA (2001) Apoptosis of adherent cells by recruitment of caspase-8 to unligated integrins. J Cell Biol 155:459–470. https://doi.org/10.1083/jcb.200106070
31. Ashkenazi A, Dixit VM (1998) Death receptors: signaling and modulation. Science 281:1305–1308. https://doi.org/10.1126/science.281.5381.1305
32. Dhanasekaran DN, Reddy EP (2008) JNK signaling in apoptosis. Oncogene 27(48):6245–6251. https://doi.org/10.1038/onc.2008.301
33. Neumann G, Noda Y, Kawaoka Y (2009) Emergence and pandemic potential of swine-origin H1N1 influenza virus. Nature 459:931–939. https://doi.org/10.1038/nature08157

34. Steinhauer DA (1999) Role of hemagglutinin cleavage for the pathogenicity of influenza virus. Virology 258:1–20. https://doi.org/10.1006/viro.1999.9716
35. Yang J, Li M, Shen X, Liu S (2013) Influenza A virus entry inhibitors targeting the hemagglutinin. Viruses 5:352–373. https://doi.org/10.3390/v5010352
36. Waldmann M, Jirmann R, Hoelscher K, Wienke M, Niemeyer FC, Rehders D, Meyer B (2014) A nanomolar multivalent ligand as entry inhibitor of the hemagglutinin of avian influenza. J Am Chem Soc 136(2):783–788. https://doi.org/10.1021/ja410918a
37. Hatano K, Matsubara T, Muramatsu Y, Ezure M, Koyama T, Matsuoka K, Kuriyama R, Kori H, Sato T (2014) Synthesis and influenza virus inhibitory activities of carbosilane dendrimers peripherally functionalized with hemagglutinin-binding Peptide. J Med Chem 57:8332–8339. https://doi.org/10.1021/jm5007676
38. Matsubara T, Onishi A, Yamaguchi D, Sato T (2016) Heptapeptide ligands against receptor-binding sites of influenza hemagglutinin toward anti-influenza therapy. Bioorg Med Chem 24:1106–1114. https://doi.org/10.1016/j.bmc.2016.01.039
39. Nagao M, Fujiwara Y, Matsubara T, Hoshino Y, Sato T, Miura Y (2017) Design of glycopolymers carrying sialyl oligosaccharides for controlling the interaction with the influenza virus. Biomacromol 18:4385–4392. https://doi.org/10.1021/acs.biomac.7b01426
40. Suda Y, Zhang W, Takahashi Y, Yokoyama R, Nagatomo M, Aoyama K, Okuno T, Saito S, Morikawa S, Hiroi S, Kase T, Murakami N, Nishi J, Wakao M (2013) ACS Symp Ser 1135:331–350. https://doi.org/10.1021/bk-2013-1135.ch020
41. Skidmore MA, Kajaste-Rudnitski A, Wells NM, Gumond SE, Rudd TR, Yates EA, Vicenzi E (2015) Inhibition of influenza H5N1 invasion by modified heparin derivatives. Med Chem Commun 6:640–646. https://doi.org/10.1039/C4MD00516C
42. Romalho-Santos J, de Lima MC (2001) Fusion and infection of influenza and Sendai viruses as modulated by dextran sulfate: a comparative study. Biosci Rep 21:293–304. https://doi.org/10.1023/A:1013282015121
43. Lüscher-Mattli M, Glück R (1990) Dextran sulfate inhibits the fusion of influenza virus with model membranes, and suppresses influenza virus replication in vivo. Antiviral Res 14:39–50. https://doi.org/10.1016/0166-3542(90)90064-E
44. Cumashi A, Ushakova NA, Preobrazhenskaya ME, D'Incecco A, Piccoli A, Totani L, Tinari N, Morozevich GE, Berman AE, Bilan MI, Usov AI, Ustyuzhanina NE, Grachev AA, Sanderson CJ, Kelly M, Rabinovich GA, Iacobelli S, Nifantiev NE (2007) A comparative study of the anti-inflammatory, anticoagulant, antiangiogenic, and antiadhesive activities of nine different fucoidans from brown seaweeds. Glycobiology 17: 541–552. https://doi.org/10.1093/glycob/cwm014
45. Hayashi K, Lee JB, Nakano T, Hayashi T (2013) Anti-influenza A virus characteristics of a fucoidan from sporophyll of *Undaria pinnatifida* in mice with normal and compromised immunity. Microbes Infect 15:302–309. https://doi.org/10.1016/j.micinf.2012.12.004
46. Synytstya A, Bleha R, Synytstya A, Pohl R, Hayashi H, Yoshinaga K, Nakano T, Hayashi T (2014) Mekabu fucoidan: structural complexity and defensive effects against avian influenza A viruses. Carbohydr Polym 111:633–644. https://doi.org/10.1016/j.carbpol.2014.05.032
47. Wang W, Wu J, Zhang X, Hao C, Zhao X, Jiao G, Shan S, Tai W, Yu G (2017) Inhibition of influenza A virus infection by fucoidan targeting viral neuraminidase and cellular EGFR pathway. Sci Rep 7:40760. https://doi.org/10.1038/srep40760
48. Spiro RG (1960) Studies on fetuin, a glycoprotein of fetal serum. I. Isolation, chemical composition, and physicochemical properties. J Biol Chem 235:2860–2869. https://doi.org/10.1016/S0021-9258(18)64553-3
49. Potier M, Mameli L, Belisle M, Dallaire L, Melançon SB (1979) Fluorometric assay of neuraminidase with a sodium (4-methylumbelliferyl-α-D-*N*-acetylneuraminate) substrate. Anal Biochem 94:287–296. https://doi.org/10.1016/0003-2697(79)90362-2
50. Rao J, Lahiri J, Isaacs L, Weis RM, Whitesides GM (1998) A trivalent system from vancomycin·D-Ala-D-Ala with higher affinity than avidin·biotin. Science 280:708–711. https://doi.org/10.1126/science.280.5364.708

51. Corey EJ, Wright SW (1990) Total synthesis of colneleic acid. J Org Chem 55:1670–1673. https://doi.org/10.1021/jo00292a049
52. Nakazawa M, Kadowaki S, Watanabe I, Kadowaki Y, Takei M, Fukuda H (2008) PA subunit of RNA polymerase as a promising target for anti-influenza virus agents. Antiviral Res 78:194–201. https://doi.org/10.1016/j.antiviral.2007.12.010
53. C. disease (COVID-19) situation reports, WHO. https://www.who.int/emergencies/disease/novel-coronavirus-019/situation-reports
54. VanBlargan LA, Errico JM, Halfmann PJ, Zost SJ, Crowe JE, Purcell LA, Kawaoka Y, Corti D, Fremont DH, Diamond MS (2022) An infectious SARS-CoV-2 B.1.1.529 Omicron virus escapes neutralization by therapeutic monoclonal antibodies. Nat Med 28:490–495. https://doi.org/10.1038/s41591-021-01678-y
55. Dimitrov DS (2004) Virus entry: molecular mechanisms and biomedical applications. Nat Rev Microbiol 2:109–122. https://doi.org/10.1038/nrmicro817
56. Liu L, Chopra P, Li X, Bouwman KM, Tompkins SM, Wolfert MA, de Vries RP, Boons GJ (2021) Heparan sulfate proteoglycans as attachment factor for SARS-CoV-2. ACS Cent Sci 7:1009–1018. https://doi.org/10.1021/acscentsci.1c00010
57. Hu B, Guo H, Zhou P, Shi Z (2021) Characteristics of SARS-CoV-2 and COVID-19. Nat Rev Microbiol 19:141–154. https://doi.org/10.1038/s41579-020-00459-7
58. Kim SY, Jin W, Sood A, Montgomery DW, Grant OC, Fuster MM, Fu L, Dordick JS, Woods RJ, Zhang F, Linhardt RJ (2020) Characterization of heparin and severe acute respiratory syndrome-related coronavirus 2 (SARS-CoV-2) spike glycoprotein binding interactions. Antiviral Res 181:104873. https://doi.org/10.1016/j.antiviral.2020.104873
59. Clausen TM, Sandoval DR, Spliid CB, Pihl J, Perrett HR, Painter CD, Narayanan A, Majowicz SA, Kwong EM, McVicar RN, Thacker BE, Glass CA, Yang Z, Torres JL, Golden GJ, Bartels PL, Porell RN, Garretson AF, Laubach L, Feldman J, Yin X, Pu Y, Hauser BM, Caradonna TM, Kellman BP, Martino C, Gordts PLMS, Chanda SK, Schmidt AG, Godula K, Leibel SL, Jose J, Corbett KD, Ward AB, Carlin AF, Esko JD (2020) SARS-CoV-2 infection depends on cellular heparan sulfate and ACE2. Cell 183:1043–1057. https://doi.org/10.1016/j.cell.2020.09.033
60. Hao W, Ma B, Li Z, Wang X, Gao X, Li Y, Qin B, Shang S, Cui S, Tan Z (2021) Binding of the SARS-CoV-2 spike protein to glycans. Sci Bull 66:1205–1214. https://doi.org/10.1016/j.scib.2021.01.010
61. Kwon PS, Oh H, Kwon SJ, Jin W, Zhang F, Fraser K, Hong JJ, Linhardt RJ, Dordick JS (2020) Cell Discovery 6:50. https://doi.org/10.1038/s41421-020-00192-8
62. Tandon R, Sharp JS, Zhang F, Pomin VH, Ashpole NM, Mitra D, McCandless MG, Jin W, Liu H, Sharma P, Linhardt RJ (2021) Effective inhibition of SARS-CoV-2 entry by heparin and enoxaparin derivatives. J Virol 95:e01987-e2020. https://doi.org/10.1128/jvi.01987-20
63. Jin W, Zhang W, Mitra D, McCandless MG, Sharma P, Tandon R, Zhang F, Linhardt RJ (2020) The structure-activity relationship of the interactions of SARS-CoV-2 spike glycoproteins with glucuronomannan and sulfated galactofucan from *Saccharina japonica*. Int J Biol Macromol 163:1649–1658. https://doi.org/10.1016/j.ijbiomac.2020.09.184
64. Koganti R, Suryawanshi R, Shukla D (2020) Heparanase, cell signaling, and viral infections. Cell Mol Life Sci 77:5059–5077. https://doi.org/10.1007/s00018-020-03559-y
65. Buijsers B, Yanginlar C, de Nooijer A, Grondman I, Maciej-Hulme ML, Jonkman I, Janssen NAF, Rother N, de Graaf M, Pickkers P, Kox M, Joosten LAB, Nijenhuis T, Netea MG, Hilbrands L, van de Veerdonk FL, Duivenvoorden R, de Mast Q, van der Vlag J (2020) Increased plasma heparanase activity in COVID-19 patients. Front Immunol 11:1–9. https://doi.org/10.3389/fimmu.2020.575047
66. Xiang J, Lu M, Shi M, Cheng X, Kwakwa KA, Davis JL, Su X, S. Bakewell SJ, Zhang Y, Fontana F, Xu Y, Veis DJ, DiPersio JF, Ratner L, Sanderson RD, Noseda A, Mollah S, Li J, Weilbaecher KN (2022) Heparanase blockade as a novel dual-targeting therapy for COVID-19. J Virol 96:e00057-22. https://doi.org/10.1128/jvi.00057-22
67. Guimond SE, Mycroft-West CJ, Gandhi NS, Tree JA, Le TT, Spalluto CM, Humbert MV, Buttigieg KR, Coombes N, Elmore MJ, Wand M, Nyström K, Said J, Setoh YX, Amarilla AA,

Modhiran N, Sng JDJ, Chhabra M, Young PR, Rawle DJ, Lima MA, Yates EA, Karlsson R, Miller RL, Chen YH, Bagdonaite I, Yang Z, Stewart J, Nguyen D, Laidlaw S, Hammond E, Dredge K, Wilkinson TMA, Watterson D, Khromykh AA, Suhrbier A, Carroll MW, Trybala E, Bergström T, Ferro V, Skidmore MA, Turnbull JE (2022) Synthetic heparan sulfate mimetic pixatimod (PG545) potently inhibits SARS-CoV-2 by disrupting the spike–ACE2 interaction. ACS Cent Sci 8:527–545. https://doi.org/10.1021/acscentsci.1c01293

68. Bauer KA, Hawkins DW, Peters PC, Petitou M, Herbert JM, van Boeckel CAA, Meuleman DG (2002) Fondaparinux, a synthetic pentasaccharide: the first in a new class of antithrombotic agents—the selective factor Xa inhibitors. Cardiovasc Drug Rev 20:37–52. https://doi.org/10.1111/j.1527-3466.2002.tb00081.x
69. Hammond E, Li CP, Ferro V (2010) Development of a colorimetric assay for heparanase activity suitable for kinetic analysis and inhibitor screening. Anal Biochem 396:112–116. https://doi.org/10.1016/j.ab.2009.09.007
70. Parish CR, Freeman C, Brown KJ, Francis DJ, Cowden WB (1999) Identification of sulfated oligosaccharide-based inhibitors of tumor growth and metastasis using novel in vitro assays for angiogenesis and heparanase activity. Cancer Res 59:3433–3441
71. Ferro V, Liu L, Johnstone KD, Wimmer N, Karoli T, Handley P, Rowley J, Dredge K, Li CP, Hammond E, Davis K, Sarimaa L, Harenberg J, Bytheway I (2012) Discovery of PG545: a highly potent and simultaneous inhibitor of angiogenesis, tumor growth, and metastasis. J Med Chem 55:3804–3813. https://doi.org/10.1021/jm201708h
72. Nakajima M, Dechavigny A, Johnson CE, Hamada JI, Stein CA, Nicolson GL (1991) Suramin. a potent inhibitor of melanoma heparanase and invasion. J Biol Chem 266:9661–9666. https://doi.org/10.1016/S0021-9258(18)92871-1
73. Matsuyama S, Nao N, Shirato K, Kawase M, Saito S, Takayama I, Nagata N, Sekizuka T, Katoh H, Kato F, Sakata M, Tahara M, Kutsuna S, Ohmagari N, Kuroda M, Suzuki T, Kageyama T, Takeda M (2020) Enhanced isolation of SARS-CoV-2 by TMPRSS2-expressing cells. Proc Natl Acad Sci USA 117:7001–7003. https://doi.org/10.1073/pnas.2002589117

# Chemical Approach to Uncover the Biological Functions of Sialic Acid-Containing Glycans

**Hiromune Ando and Naoko Komura**

**Abstract** Sialic acid is critically involved in carbohydrate–protein interactions that regulate processes like cell growth, differentiation, adhesion, immune response, fertilization, cancer development, and viral infection. To detail the molecular basis on which sialic acid-containing glycans (glycoconjugates) exert their biological functions, chemical synthesis will play a vital role. Our group has been addressing the chemical synthesis of gangliosides and their functionalized analogs to clarify the unknowns about ganglioside's behavior in the live cell membrane. Robust chemical methods for α-sialylation and the conjugation of glycan with lipid allowed for the development of the fluorescently labeled ganglioside probes, which retain biophysical nature as lipid raft molecules. The single-molecule imaging of gangliosides revealed the specific interaction between gangliosides and lipid raft constituting a cluster of GPI-anchored protein CD59 and cholesterol. This observation showed the continual and dynamic exchange of gangliosides between lipid raft domain and the bulk domain, indicating that lipid raft domains are dynamic entities.

**Keywords** Sialic acid · Ganglioside · Lipid raft · Single-molecule imaging

## 1 Introduction

Sialic acids constitute a broad family of monosaccharides derived from 3-deoxy-non-2-ulosonic acid. The three primary sialic acids encountered in nature are the *N*-acetyl (Neu5Ac), *N*-glycolyl (Neu5Gc) derivatives of neuraminic acid (5-amino-3,5-dideoxy-D-*glycero*-D-*galacto*-non-2-ulosonic acid), and KDN (3-deoxy-D-*glycero*-D-*galacto*-non-2-ulosonic acid). Neu5Ac is synthesized through the condensation of Man-NAc-6-P with pyruvate, whereas KDN is produced via condensation of Man-6-P with pyruvate. Additionally, the *N*-acetyl group of CMP-Neu5Ac is oxidized to yield CMP-NeuGc, a reaction catalyzed by CMP-Neu5Ac hydrolase. The

H. Ando (✉) · N. Komura
Institute for Glyco-core Research, Gifu University, Gifu, Japan
e-mail: ando.hiromune.i0@f.gifu-u.ac.jp



K. Toshima and Y. Fujimoto (eds.), *Chemical Biology of Carbohydrates*,
https://doi.org/10.1007/978-981-95-5335-8_2

hydroxyl groups of sialic acid may undergo further modifications such as acetylation, lactylation, phosphorylation, sulfonation, or methylation [1].

In glycans attached to lipid or protein, sialic acids typically linked at the terminal positions through α-(2-6) and/or α-(2-3) linkages with Gal, GalNAc, Glc, or GlcNAc, or via α-(2-8), α-(2-9), or α-(2-4) linkages with other Neu residues, resulting in dimeric structures. The distinctive structural characteristics of these molecules confer diverse chemical and biological properties. The extensive variability arising from substitutions and modifications at several positions on the sialic acid backbone contributes significantly to the heterogeneity of cell surface glycoconjugates. This molecular diversity impacts cellular recognition processes and modulates susceptibility to enzymatic modification, as sialidases often exhibit specificity toward certain sialic acid derivatives. Consequently, the structural complexity of sialic acid-containing glycans underlies numerous biological functions and plays an important role in regulating cellular interactions and immune evasion mechanisms.

Sialic acid, located at the outermost position, is involved in carbohydrate–protein interactions that regulate processes like cell growth, differentiation, adhesion, immune response, fertilization, cancer development, and viral infection [2–5]. Its negative charge, from the carboxyl group, affects the cell surface's electrostatic environment and influences receptor-ligand binding. The presence and modifications of sialic acids on glycoproteins and glycolipids can dictate serum half-life of proteins, mediate cell–cell interactions, and serve as signals for developmental and immunological processes. In vertebrates, sialylation patterns vary by cell type and change during conditions like inflammation, cancer, and infection. Viruses such as influenza use sialylated glycans as entry receptors, showing their biological importance. Sialic acids appear mainly in deuterostomes and some bacteria, demonstrating their evolutionary specialization. Understanding how sialic acid structure relates to function is a key challenge in glycobiology, requiring new chemical methods for synthesizing and modifying specific sialo-glycans.

Gangliosides are a large family of glycolipids consisting of ceramide (Cer) and glycans that end with sialic acid. In the outer layer of the plasma membrane (PM), their lipid components anchor them through lipid-lipid interactions with other membrane lipids. Sialic acid displays several polar functional groups to the outside environment. As a result, gangliosides are involved in interactions with sugar-binding proteins. Gangliosides in mammals play crucial roles in various biological processes, such as cell adhesion and toxin binding, pathogen entry into cells [6–9]. Furthermore, gangliosides regulate the protein receptor's functions in the PM [10–12]. Echinoderm-derived gangliosides—such as those isolated from starfish and sea cucumbers—exhibited neuritogenic effects to model neuron cells (PC-12) in the presence of neural growth factors [13]. Notably, certain echinodermatous gangliosides (EGs) exhibit higher neuritogenic activity than ganglioside GM1, positioning EGs as promising candidates for the development of therapeutics targeting neural disorders. EGs have unique glycan and ceramide moieties. The glycan moieties consist of partially modified sialic acids and their oligomers. Moreover, sialic acids are often incorporated as internal residues in EGs. This arrangement was not observed in mammalian counterparts. To date, the structure of over 15 types of EGs have been defined, and their neuritogenic effect have been assessed. Nevertheless, the structure–activity relationships and molecular mechanisms responsible for EG-mediated potentiation of neurite outgrowth remain unclear. Our group has addressed the chemical synthesis of gangliosides to detail their biological functions.

## 2 Issues in the Chemical Synthesis of Gangliosides

One of the main challenges in synthesizing diverse gangliosides is achieving stereoselective incorporation of sialic acids into oligosaccharide moieties and attaching sialyl oligosaccharides to the Cer moiety (Fig. 1).

α-Glycosides of sialic acid (sialosides) can be produced by the reaction of an oxocarbenium ion formed from the sialic acid donor with the hydroxyl group of the glycosyl acceptor. However, the structure of sialic acid hampers its glycosylation. First, the electron-withdrawing carboxyl group at the anomeric center makes the oxocarbenium ion intermediate unstable, which is vulnerable to 1,2-elimination, giving 2,3-ene as a degradant. Second, without a hydroxyl group at C3, neighboring participation cannot be used to selectively produce α-glycoside, so the thermodynamically favored β-glycoside is mainly formed. In the past 50 years, enormous efforts have been paid to settling the issue of sialic acid glycosylation [14].

The challenges related to Cer introduction are primarily associated with the low reactivity of the 1-OH group. This reduced reactivity is linked to the presence of both the C3 amide moiety, which can form intra- or inter-molecular hydrogen bonds with the 1-OH group. Intra-molecular hydrogen bonding deteriorates the reactivity of the hydroxyl group, while inter-molecular hydrogen bonding facilitates the self-aggregation of Cer, together with hydrophobic interactions among the long alkyl chains. This aggregation hampers the glycosylation reaction of the C1 hydroxyl group of Cer, lowering the yield as the glycosyl donor length increases.

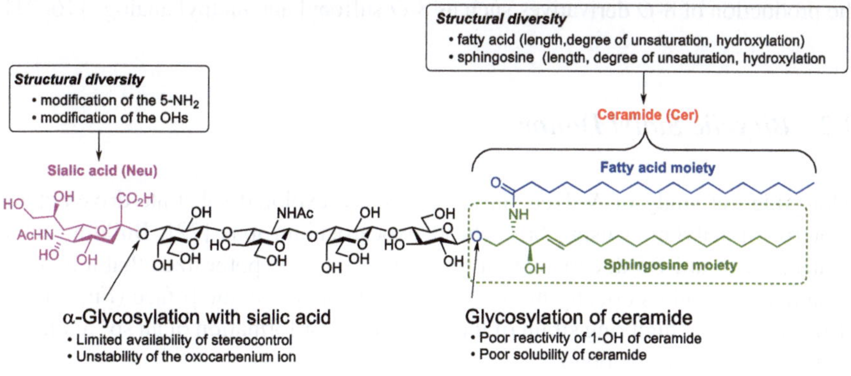


**Fig. 1** Issues to be overcome in the chemical synthesis of gangliosides

# 3 α-Sialylation

In sialic acid glycosidation, to primarily obtain α-glycoside products, it is necessary to block nucleophilic hydroxyl attack on the β-face of the oxocarbenium ion. Acetonitrile as solvent or nucleophilic groups at C1/C3 have been used to favor α-sialoside production. While the mechanism remains unclear, adding 4*O*,5*N*-oxazolizinone to the donor greatly enhances α-selectivity. However, substrate structure and reaction conditions continue to impact stereoselectivity across methods.

## *3.1* N-*Troc-Sialyl Donor*

The acetonitrile-assisted α-sialylation facilitates the synthesis of sialoglycans due to its straightforward procedure and scalability, producing homogenous sialoglycans and their analogs useful for biological study. The method's stereoselectivity is highly sensitive to reaction conditions, which restricts its application range. Additionally, the original method utilizing an *N*-acetyl sialic acid donor presents limitations; it poses challenges in converting the *N*-acetyl group into other amide groups, including *N*-glycolyl and *N*-glycosylglycolyl groups. To break the limitation, the C-5 amine was modified with a selectively removable 2,2,2-trichloroethoxycarbonyl (Troc) group, giving an *N*-Troc sialic acid donor [15]. This modification greatly shortened the reaction time: coupling of *N*-Troc donor **2** with a glycosyl acceptor in acetonitrile at − 50 °C finished in 20 min to give **4** in 54% (α/β = 4/1). By comparison, the conventional *N*-Ac derivative (**1**) took 13 h at − 25 °C (Fig. 2). The Troc group also improved solubility for glycosylation solvents and enhanced chromatographic separation of product anomers. *N*-Troc sialosides was used for the synthesis of various C5- or 8-*O*-modified analogs following selective Troc deprotection. Zinc/acetic acid treatment produces a 5-$NH_2$ derivative, which can then be converted into Neu5Ac, Neu5Gc, and 1,5-lactamized-Neu. Mild acid removal leads to *O*-8 acetyl migration, facilitating the production of 8-*O* derivatives such as 8-*O*-sulfonyl and methyl analogs [16–21].

## *3.2 Bicyclic Sialyl Donor*

Alongside our study of *N*-Troc sialyl donors, we explored substrate-driven stereocontrol in sialylation using bridgehead oxocarbenium ions [22]. Building upon Kraus et al.'s anti-Bredt carbocation research [23], we hypothesized that a bicyclic α-sialic acid would sterically hinder nucleophilic attack on the β-face (Fig. 3). We also expected the high ring strain of these ions to favor formation of an $sp^3$ anomeric center with a glycoside bond.

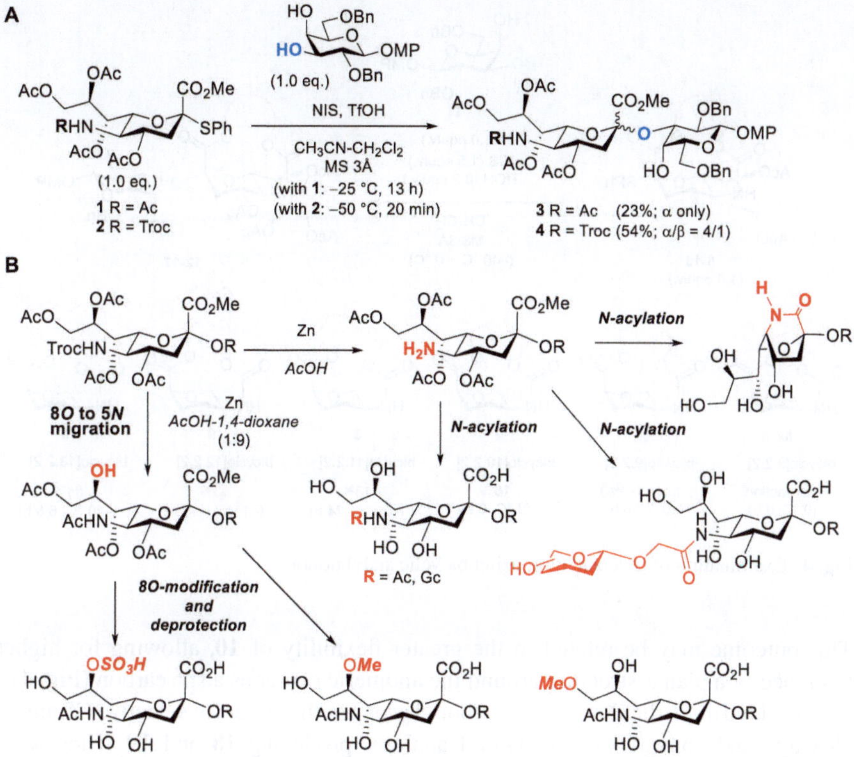


**Fig. 2** Glycosylation of *N*-Troc sialyl donor (**a**) and conversion into various sialoside analogs (**b**)

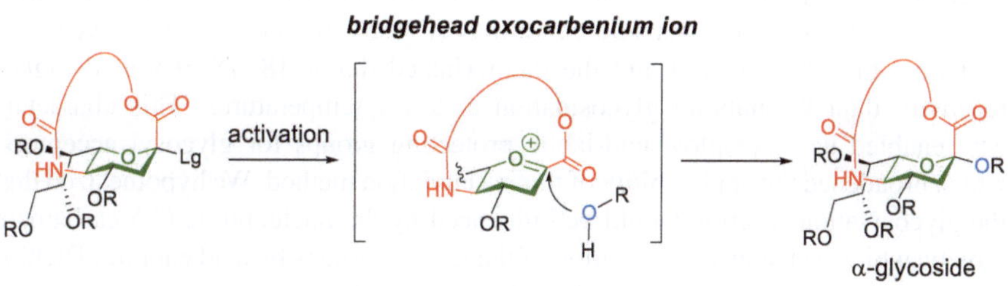


**Fig. 3** Concept of α-exclusive sialylation via a bridgehead oxocarbenium ion

The study of the glycosidation reaction of bicyclic sialyl donors with bicycle [8.2.2] to [13.2.2] systems indicated that bicyclic sialyl donors displayed complete α-selectivity. Among the bicyclic donors tested, donor **9** with a bicycle [12.2.2] system provided the highest yield (83%). Donor **10**, which contains a macroring one carbon longer than **9**, showed similar reactivity. However, **10** exhibited increased susceptibility to 1,2-elimination, resulting in a 2,3-ene derivative as the major byproduct.

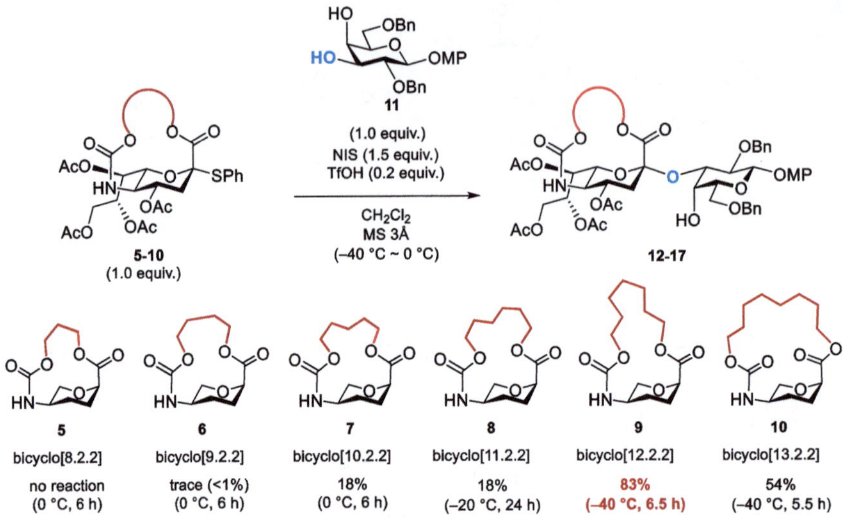


**Fig. 4** Examination of glycosidation using bicyclic sialyl donors

This outcome may be related to the greater flexibility of **10**, allowing for higher tolerance of a planar structure around the anomeric center as a $sp^2$ carbon (Fig. 4).

We added a 2,2-dichloroethoxycarbonyl group to the tether for selective C5 amine cleavage and conversion into natural analogs, producing **18** and **19**. They were glycosidated with various hydroxylated molecules—including alcohols (**26**), sugars (**27**, **28**), polycyclics (**29**), cholesterol (**30**), and oligosaccharides (**31**)—yielding single stereoisomers in high yields. Phosphate **19** also formed *C*-glycosides with carbon nucleophiles under acidic conditions (Fig. 5). In comparison between the prototype bicyclic donor **9** and the dichlorinated donor **18**, **18** exhibited higher reactivity than **9**, enabling glycosidation at lower temperatures. This characteristic enabled us to employ acid-labile protecting groups for glycosyl acceptors, which broadened the applicability of this α-sialylation method. We hypothesized that the glycosylation reaction would be influenced by the nucleophilic C-5 carbamate moiety, which can slow the activation of the leaving groups by acid capture. Dichlorinated moiety in the tether may reduce the nucleophilicity of the carbamate moiety, resulting in faster reactions for dichlorinated donors compared to the unsubstituted donor.

## 4 Glucosyl Ceramide Cassette Approach

As aforementioned, combining glycan and ceramide (Cer) moieties is challenging owing to the low reactivity of Cer's 1-OH group. Early syntheses, such as Shapiro et al.'s dihydrocerebrosides [24] and Ogawa et al.'s ganglioside GM3

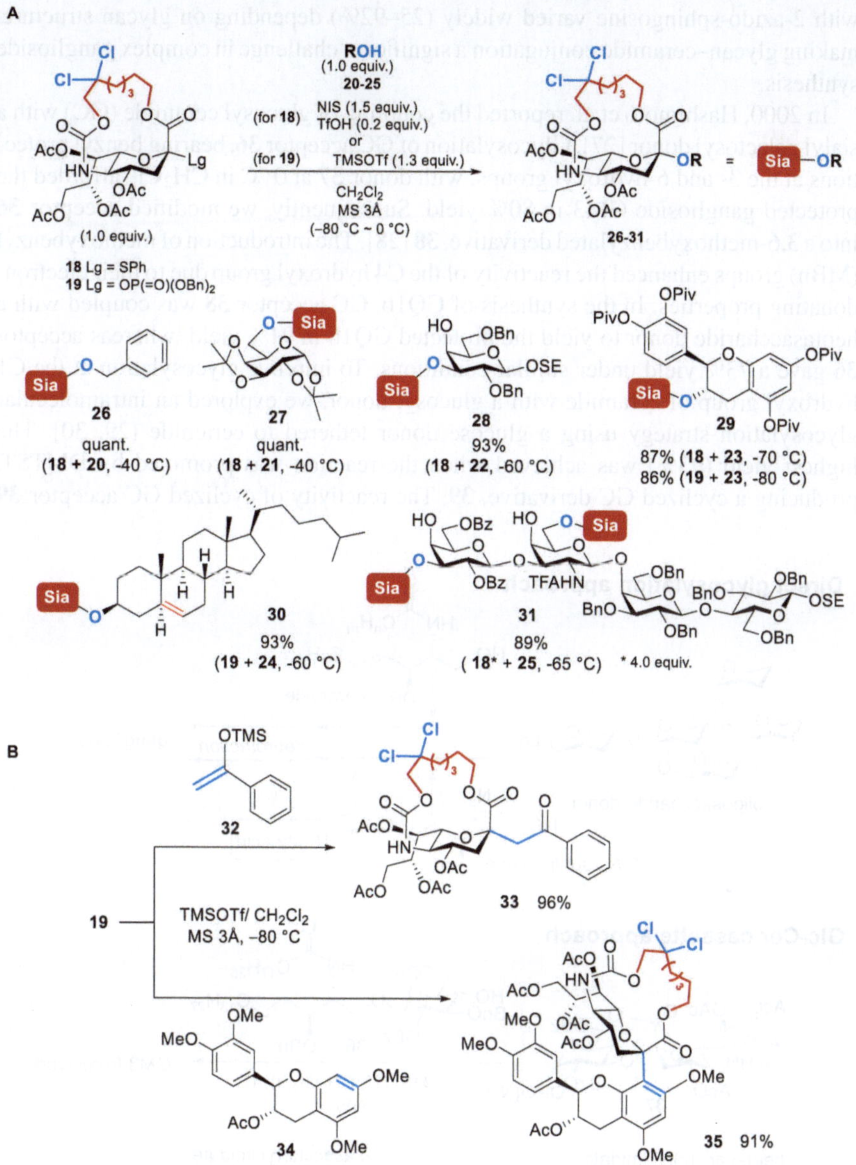


**Fig. 5** Substrate scope study

[25], relied on coupling glycan donors to Cer, but yields decreased with larger glycan donors—mainly because ceramide self-aggregation hinders reaction efficiency (Fig. 6). Schmidt addressed this by introducing 2-azido-sphingosine into glycans and then forming ceramide via azide reduction and fatty acid coupling, a method adopted for various gangliosides [26]. However, yields for coupling glycans

with 2-azido-sphingosine varied widely (25–92%) depending on glycan structure, making glycan–ceramide conjugation a significant challenge in complex ganglioside synthesis.

In 2000, Hashimoto et al. reported the coupling of glucosyl ceramide (GC) with a sialyl galactosyl donor [27]. Glycosylation of GC acceptor **36**, bearing benzyl protections at the 3- and 6-hydroxyl groups, with donor **37** at 0 °C in $CH_2Cl_2$ afforded the protected ganglioside GM3 in 80% yield. Subsequently, we modified acceptor **36** into a 3,6-methoxybenzylated derivative, **38** [28]. The introduction of methoxybenzyl (MBn) groups enhanced the reactivity of the C4 hydroxyl group due to their electron-donating properties. In the synthesis of GQ1b, GC acceptor **38** was coupled with a heptasaccharide donor to yield the protected GQ1b in 91% yield, whereas acceptor **36** gave a 73% yield under similar conditions. To improve glycosylation at the C1 hydroxyl group of ceramide with a glucosyl donor, we explored an intramolecular glycosylation strategy using a glucose donor tethered to ceramide [29, 30]. The highest yield (93%) was achieved when the reaction was promoted by DMTST, producing a cyclized GC derivative, **39**. The reactivity of cyclized GC acceptor **39**

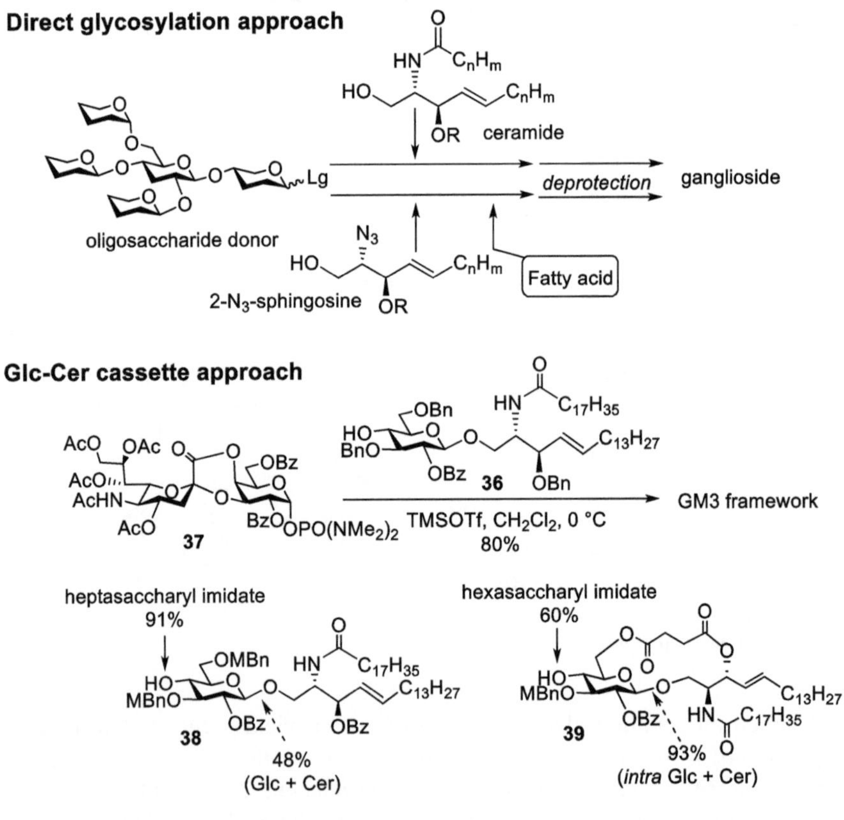


**Fig. 6** Approaches to the conjugation of glycan and Cer

was found to be comparable to that of GC **38**, as demonstrated by its successful coupling with various oligosaccharide donors. The utility of **39** was further exemplified in the synthesis of highly branched gangliosides [31]. These strategies employing GC acceptors are collectively referred to as the GC cassette approach. Given that mammalian gangliosides and other glycosphingolipids (GSLs) share a conserved substructure at their lipid tails—Galβ(1 → 4)Glcβ(1 → 1)Cer—the GC cassette approach represents a promising platform for the synthesis of mammalian GSLs.

## 5 Total Synthesis of Gangliosides Based on GC Cassette Approach

The aforementioned synthetic strategies for gangliosides—including α-selective sialylation and the GC cassette approach—enabled the construction of structurally diverse gangliosides. As an example, the total synthesis of ganglioside GP3 is illustrated in Fig. 7 [32]. GP3 features a highly branched glycan moiety embedded with sialic acid residues and a plant-type ceramide (phyto-Cer). An *N*-Troc sialyl donor (compound **40**) was employed to achieve stereoselective glycosylation at the anomeric position and to introduce a sugar moiety at the 4-hydroxyl group. Donor **40** reacted with galactose (Gal) acceptor **42** and galactofuranose (Galf) acceptor **41**, respectively, yielding α-sialosides in high efficiency. These derivatives were converted into glycosyl acceptors **43** and **44** via selective cleavage of the CAc protecting groups using 1-selenocarbamoylpiperidine. The resulting *N*-Troc sialyl Gal acceptors were subjected to [2 + 2] glycosylation reactions. As expected, both hydroxyl groups were sufficiently reactive to undergo coupling with disaccharide donors, affording tetrasaccharides in high yield. Subsequent manipulation of protecting groups furnished tetrasaccharides **45** and **46**, which were then coupled to form the octasaccharide intermediate **47**. Finally, an octasaccharyl donor derived from **47** was glycosylated with GC cassette **48**, yielding the protected form of GP3 in 76% yield. Complete deprotection afforded the target ganglioside GP3.

## 6 Late-Stage Sialylation of Glycolipid to Produce Ganglioside

Based on a fully stereoselective bicyclic sialyl donor which we developed, we have created a novel method for ganglioside synthesis. This method features late-stage sialylation to oligosaccharyl Cer, aiming to enhance accessibility to ganglioside analogs (Fig. 8). As a coupling partner of bicyclic sialyl donor, we have developed *tert*-butylbenzoylated (TBBz) oligosaccharyl Cer acceptors [22, 33]. The TBBz group increases the solubility in organic solvent by inhibiting π-π or CH-π interaction with the bulky *tert*-butyl group [34]. A disaccharyl Cer acceptor **49** was

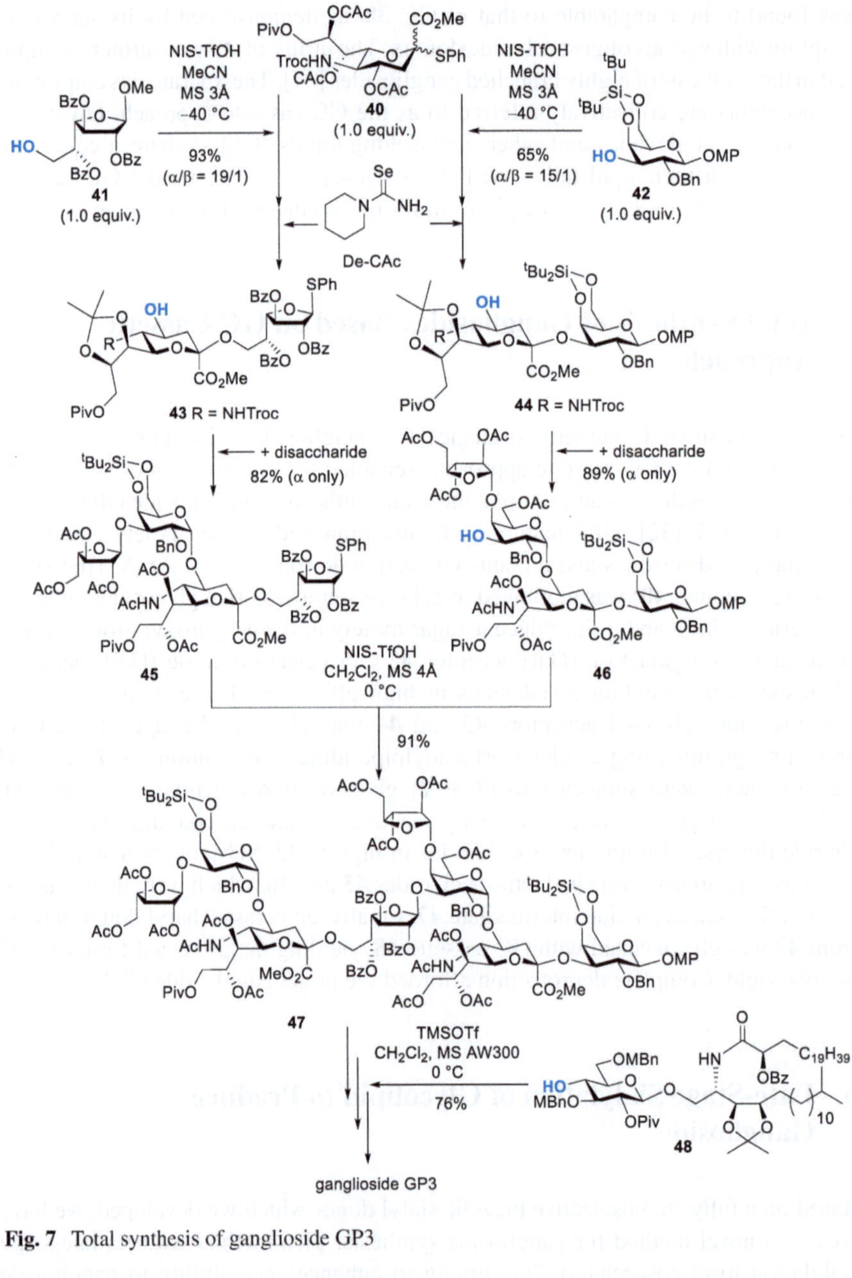


**Fig. 7** Total synthesis of ganglioside GP3

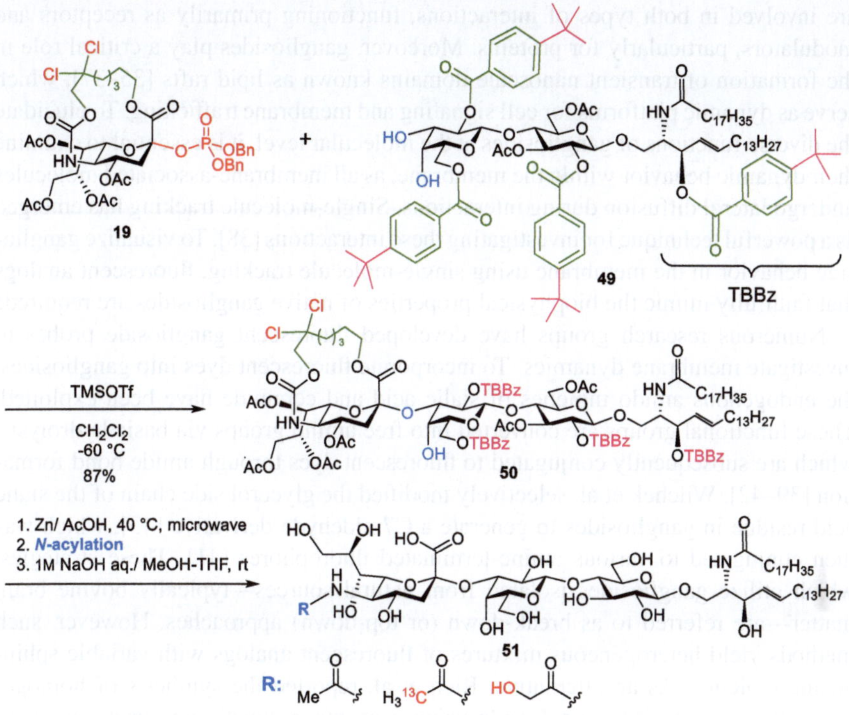


**Fig. 8** Synthesis of ganglioside GM3 analogs based on late-stage sialylation

found to be soluble in $CH_2Cl_2$ at reduced temperatures at − 60 °C. In contrast, the benzoylated analog exhibited minimal solubility under identical conditions. The coupling reactions of bicyclic sialyl dibenzylphosphate donor **19** and the glycolipid acceptor **49** was successfully proceeded at low temperatures (−60 °C), giving protected ganglioside GM3 **50** at high yield. Upon selective cleavage of the tether part, the protected GM3 was converted to C5-amino-modified analogs of GM3, such as *N*-acetyl, *N*-glycolyl a *N*-($^{13}$C)acetyl analogs.

## 7 Development of Fluorescent Probes of Gangliosides for Single-Molecule Imaging

Gangliosides embedded in the cell membrane participate in a variety of cell-surface interactions, acting as ligands for biomolecules such as proteins and glycoconjugates. These interactions are broadly categorized into trans and cis types. Trans interactions occur between membrane components and exogenous molecules or membrane constituents of neighboring cells, whereas cis interactions take place among molecules within the same membrane. In mammalian cells, gangliosides are involved in both types of interactions, functioning primarily as receptors and modulators, particularly for proteins. Moreover, gangliosides play a critical role in the formation of transient nanoscale domains known as lipid rafts [35–37], which serve as dynamic platforms for cell signaling and membrane trafficking. To elucidate the diverse functions of gangliosides at the molecular level, it is essential to examine their dynamic behavior within the membrane, as all membrane-associated molecules undergo lateral diffusion during interactions. Single-molecule tracking has emerged as a powerful technique for investigating these interactions [38]. To visualize ganglioside behavior in the membrane using single-molecule tracking, fluorescent analogs that faithfully mimic the biophysical properties of native gangliosides are required.

Numerous research groups have developed fluorescent ganglioside probes to investigate membrane dynamics. To incorporate fluorescent dyes into gangliosides, the endogenous amido moieties of sialic acid and ceramide have been exploited. These functional groups are converted into free amino groups via basic hydrolysis, which are subsequently conjugated to fluorescent dyes through amide bond formation [39–42]. Wilchek et al. selectively modified the glycerol side chain of the sialic acid residue in gangliosides to generate a C7-aldehyde derivative [43], which was then conjugated to various amine-terminated fluorophores [41]. These strategies, which utilize gangliosides isolated from natural sources—typically bovine brain matter—are referred to as break-down (or top-down) approaches. However, such methods yield heterogeneous mixtures of fluorescent analogs with variable sphingosine moieties. As an alternative, Rich et al. reported the synthesis of homogeneous fluorescent GM3 ganglioside using a chemoenzymatic approach involving glycoside hydrolase mutants known as glycosynthases [44]. This method enabled the construction of a C9-azido-lyso-GM3 derivative, which underwent chemoselective conjugation with fluorescent dyes to produce a GM3 molecule exhibiting dual fluorescence [45]. We have developed fluorescent ganglioside probes optimized for visualizing interactions between lipid rafts and gangliosides in plasma membranes (PMs) [46] (Fig. 9). In designing these probes, fluorescent dyes were incorporated into the glycan moieties of gangliosides to avoid disrupting the interaction between the ceramide moiety and membrane lipids—an interaction essential for lipid raft formation. For linear ganglioside GM3, dyes were attached at the central position (C6 of the inner galactose; G6) and the outermost position (C9 of sialic acid; S9). In the case of branched gangliosides such as GM1, GM2, and GD1b, the outermost positions (S9 and/or C6 of the terminal sugar residue) were selectively modified. The physical properties of these fluorescent gangliosides were evaluated based on their preferential incorporation into lipid rafts, a characteristic referred to as "raftphilicity."

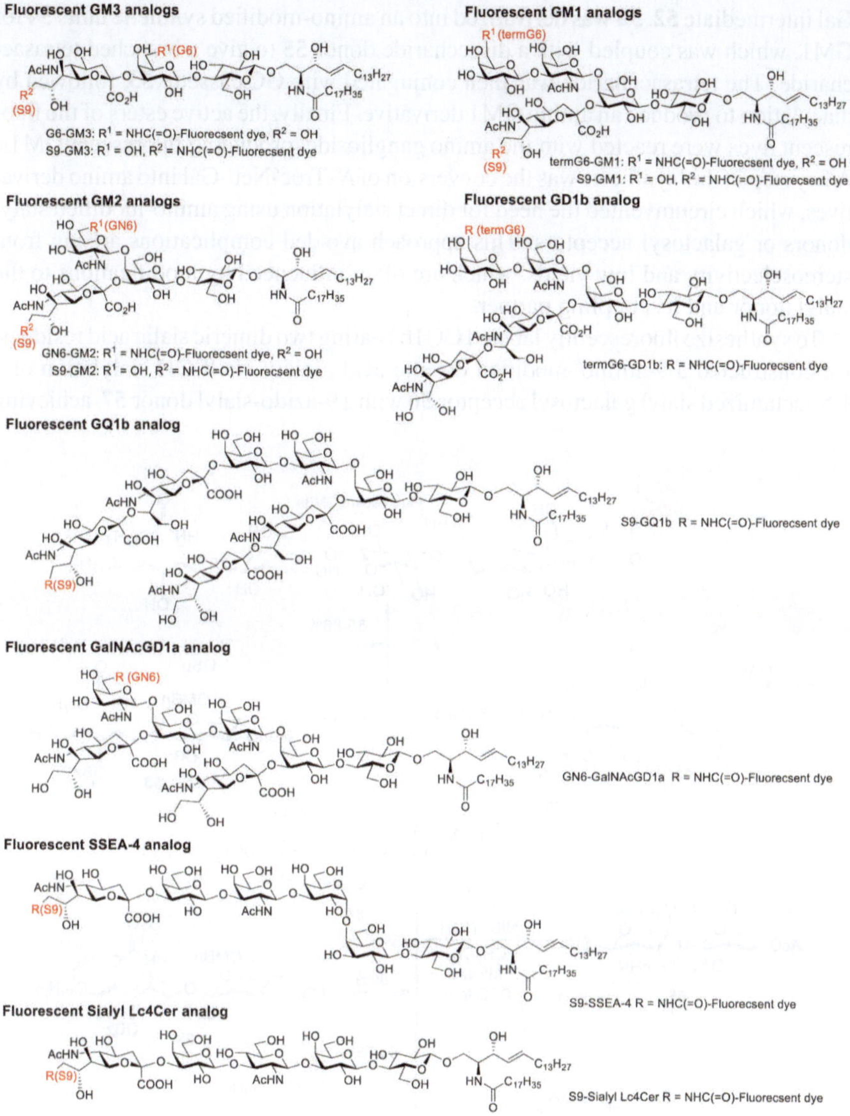


**Fig. 9** Representative fluorescent gangliosides for lipid raft study developed by the author's group

## *7.1 Chemical Synthesis of Ganglioside Analogs Modified with Fluorescent Dye at Their Glycan Moieties*

Our methods for synthesizing the entire structure of gangliosides were used to synthesize the fluorescently labeled gangliosides [46] (Fig. 10). The amino-modified analogs of gangliosides GM3 and GM1 were constructed with *N*-Troc-Neu α(2,3)

Gal intermediate **52**. **52** was derivatized into an amino-modified synthetic units **54** for GM1, which was coupled with a disaccharide donor **55** to give a branched tetrasaccharide. The tetrasaccharide was then conjugated with GC cassette **53**, followed by diacylation to produce an amino GM1 derivative. Finally, the active esters of the fluorescent dyes were reacted with the amino ganglioside, producing fluorescent GM1s. A key step in this synthesis was the conversion of *N*-Troc-Neu-Gal into amino derivatives, which circumvented the need for direct sialylation using amino-modified sialyl donors or galactosyl acceptors. This approach avoided complications arising from stereoselectivity and low yields, which are often influenced by modifications to the sialyl donor and its coupling partner.

To synthesize fluorescently labeled GQ1b bearing two dimeric sialic acid residues, we constructed a 9-amino–modified disialic acid moiety via 8-OH sialylation of a 1,5-lactamized sialyl galactosyl acceptor **56** with a 9-azido-sialyl donor **57**, achieving

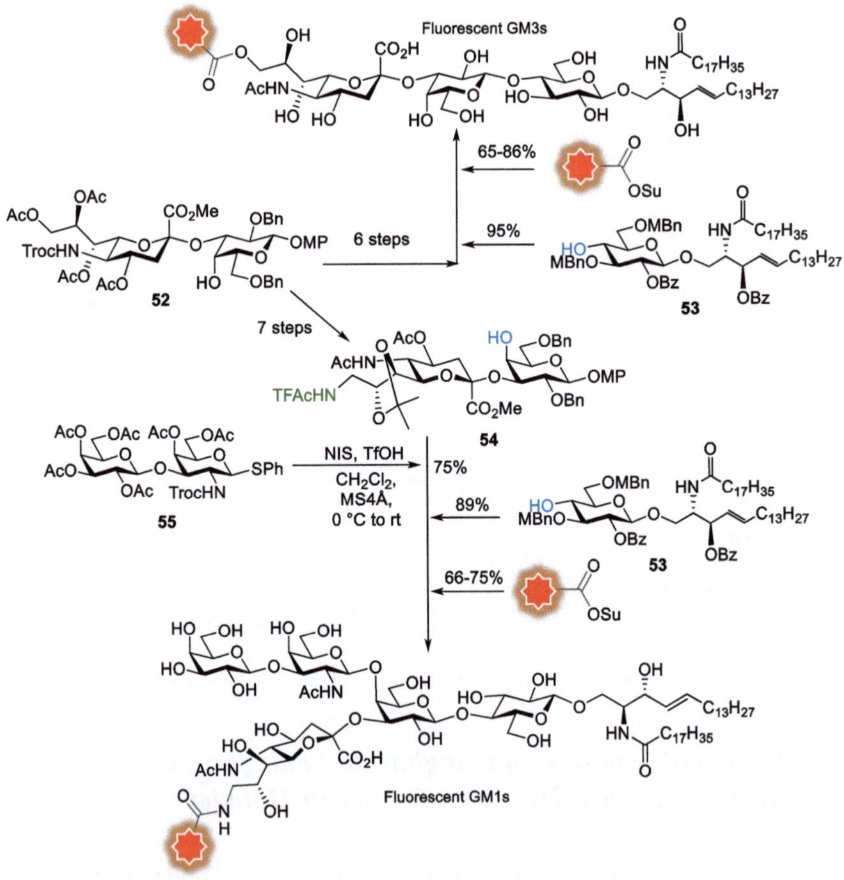


**Fig. 10** Outline of the synthesis of fluorescently labeled gangliosides GM3 and GM1

high yield. The resulting disialyl galactoside was converted into a glycosyl donor and subsequently coupled with a branched tetrasaccharide acceptor **58**, yielding the core glycan structure of GQ1b. Building on our GlcCer cassette strategy, the intermediate heptasaccharide **59** was extended to the complete GQ1b structure. Finally, the fully assembled ganglioside was conjugated with a fluorescent dye to produce the labeled GQ1b analog [47] (Fig. 11).

For the synthesis of the fluorescent lacto- and neolacto-series gangliosides, we employed the late-sialylation approach [48, 49]. To achieve the direct sialylation to the tetrasaccharyl Cer, highly TBBz-protected tetrasaccharyl Cer acceptor **60** was synthesized, which was confirmed to be solved in $CH_2Cl_2$ at − 70 °C (Fig. 12).

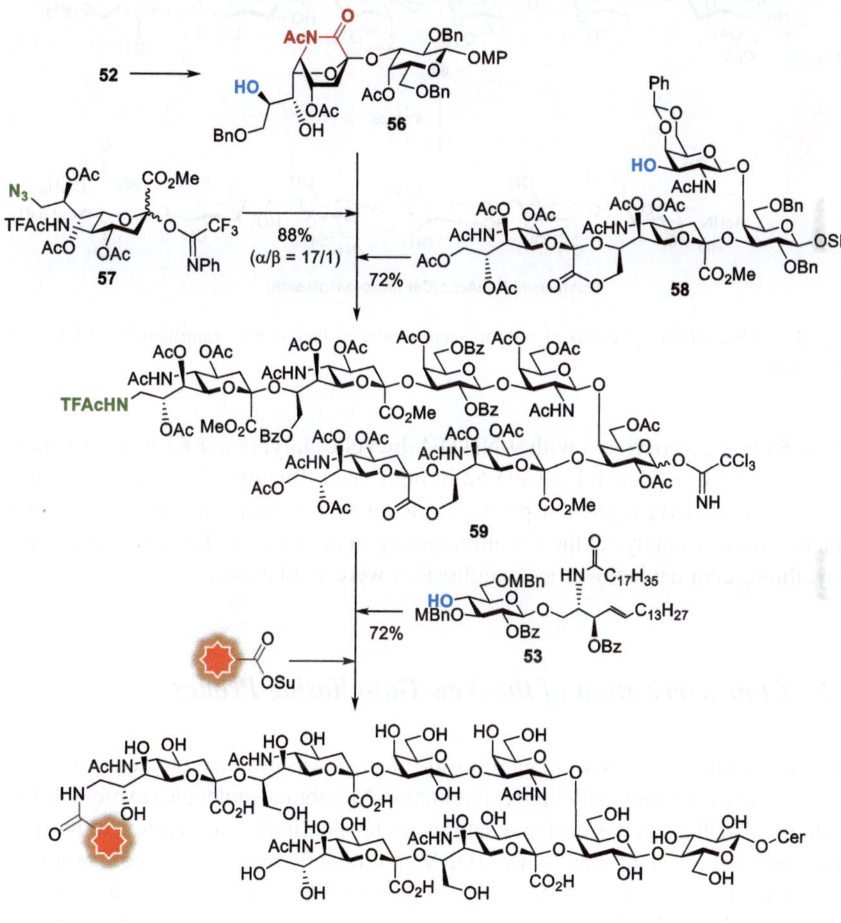


**Fig. 11** Outline of the synthesis of a fluorescently labeled ganglioside GQ1b

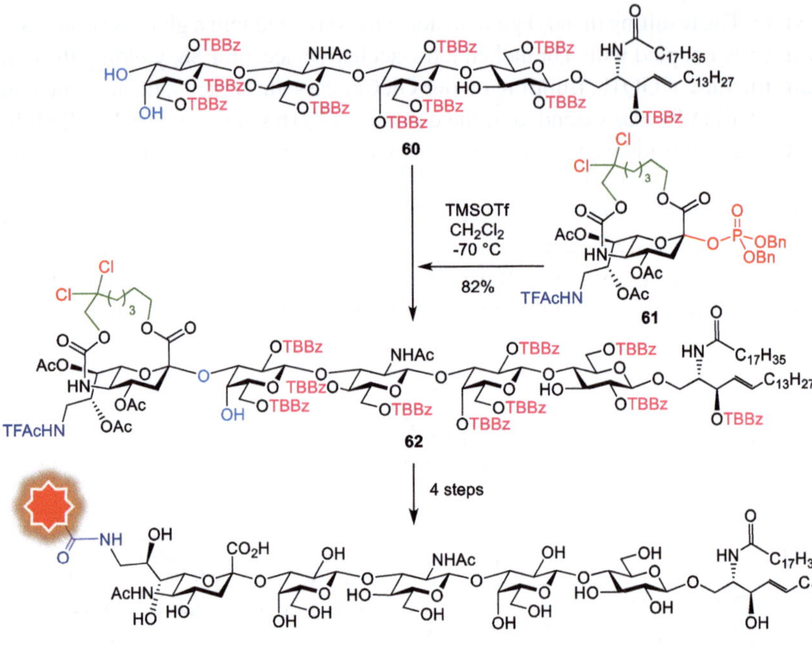


**Fig. 12** Outline of the synthesis of a fluorescently labeled lacto-series ganglioside via late-stage sialylation

Then, **60** was glycosylated with 9-NHTFA-bicyclic sialyl donor **61** to afford sialyl-lactotetraosyl Cer (NeuAcLc4Cer) **62** in high yield. Finally, **62** was derived into a fluorescent-labeled ganglioside probe via global deprotection and selective introduction of a fluorescent dye at the C9-amino group in the sialic acid residue. In a similar way, fluorescent neolacto-series gangliosides were synthesized.

## *7.2 Characterization of the New Ganglioside Probes*

The raftphilicity of fluorescent ganglioside analogs was evaluated to determine their preferential association with lipid raft domains. Membrane molecules exhibiting raftphilicity—such as cholesterol and saturated phospholipids—are known to partition into detergent-resistant membranes (DRMs), remaining insoluble in cold non-ionic detergents. In addition to DRM partitioning, raftphilic molecules are also enriched in liquid-ordered (Lo-like) phase domains within giant plasma membrane vesicles (GPMVs) [50, 51].

The cold-detergent solubility of fluorescent ganglioside analogs was assessed in human T24 epithelial cells. PMs, pre-incorporated with fluorescent ganglioside analogs and other fluorescent lipids, were treated with cold Triton X-100 and analyzed via epifluorescence microscopy [52]. These experiments revealed that analogs such as S9-GM3, S9-GM2, S9-GM1, termG6-GM1, GN6-GM2, and termG6-GD1b predominantly partitioned into detergent-resistant membranes (DRMs) (Fig. 13a and c). In contrast, tetramethylrhodamine (TMR)-labeled G6-GM3 was almost entirely solubilized by detergent, resembling the behavior of non-raft markers such as BodipyFL-C5-PC, ATTO594-DOPE, and ATTO647N-DOPE [46]. This suggests that the hydrophobic nature of TMR may interfere with DRM incorporation. Notably, fluorescent dyes with hydrophilic properties—such as ATTO594, ATTO488, and fluorescein (Fl)—when positioned at the terminal glycan moiety, preserved the analogs' preference for DRM partitioning.

The partitioning behavior of fluorescent ganglioside analogs into liquid-ordered (Lo-like) and liquid-disordered (Ld-like) domains in giant plasma membrane vesicles (GPMVs) was consistent with previous analyses of their distribution in detergent-resistant membranes (DRMs) (Fig. 13b and c) [46]. Taken together with the biophysical characterization of these analogs, we conclude that only ganglioside derivatives modified with hydrophilic fluorophores at the terminal position of the glycan moiety exhibit raft domain association comparable to that of native gangliosides.

Furthermore, we investigated the dissociation constants ($K_d$) between fluorescent ganglioside analogs and sugar-binding proteins. The binding affinities of cholera toxin subunit B (CTB) to fluorescent GM1 analogs, and those of wheat germ agglutinin (WGA) to fluorescent GM3 analogs, were determined via surface plasmon resonance (SPR) equilibrium analysis. ATTO594-conjugated GM1 and GM3 exhibited affinity constants comparable to their native counterparts: native GM1 ($K_d = 40 \pm 7$ nM), ATTO594-S9-GM1 ($K_d = 444 \pm 10$ nM); native GM3 ($K_d = 64 \pm 10$ μM), ATTO594-S9-GM3 ($K_d = 67 \pm 15$ μM) [46]. In contrast, a previously developed fluorescent GM1 analog bearing Alexa568 at the glycerol side chain (C7) of the sialic acid showed a ~ tenfold reduction in CTB binding affinity [41]. These findings underscore the significant impact of fluorophore positioning on ganglioside–protein interactions.

## 8 Single-Molecule Imaging of Ganglioside Probes in Live Cell Membrane

### *8.1 Single-Molecule Imaging of Ganglioside Probes, Both Inside and Outside Areas Stained with Raft-Associated GPI-Anchored Protein CD59 [46]*

Confocal fluorescence microscopy revealed that ganglioside probes co-localized with lipid rafts in the plasma membrane (PM). However, whether these probes dynamically entered or exited raft domains remained unresolved. Single-molecule fluorescence tracking demonstrated that ATTO594-S9-GM3 and GM1 diffused within both the bulk membrane and CD59-stained regions, the latter serving as markers for lipid

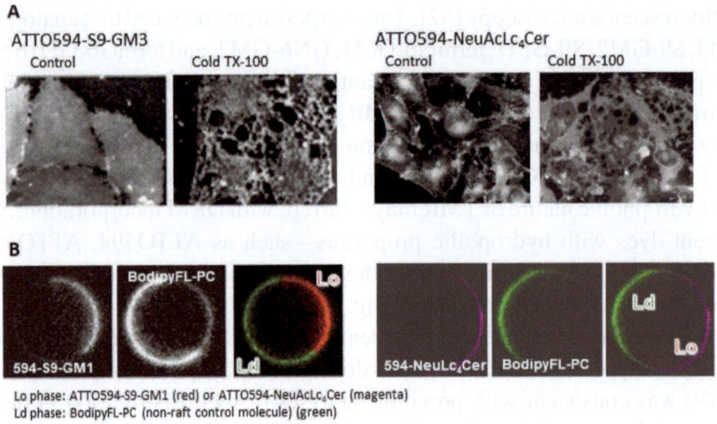


C

| Fluorescent ganglioside | DRM vs. Intact Ratio (mean ± SE) | Lo/Ld Ratio (mean ± SE) |
|---|---|---|
| Cy3-CTB | 1.2 ± 0.08 | 6.7 ± 0.49 |
| BodipyFL-GM1 | 0.042 ± 0.009 | 0.62 ± 0.05 |
| TMR-S9-GM1 | 1.0 ± 0.05 | 1.9 ± 0.25 |
| ATTO594-S9-GM1 | 1.1 ± 0.08 | 3.3 ± 0.11 |
| ATTO488-S9-GM1 | 1.1 ± 0.06 | 3.0 ± 0.23 |
| ATTO594-termG6-GM1 | 1.0 ± 0.05 | 4.7 ± 0.35 |
| ATTO594-S9-GM2 | 1.0 ± 0.05 | 4.1 ± 0.48 |
| ATTO594-GN6-GM2 | 1.1 ± 0.07 | 4.4 ± 0.43 |
| Fl-S9-GM3 | 1.1 ± 0.05 | 8.8 ± 1.5 |
| TMR-S9-GM3 | 1.0 ± 0.10 | 1.5 ± 0.08 |
| ATTO594-S9-GM3 | 1.0 ± 0.05 | 4.0 ± 0.14 |
| ATTO488-S9-GM3 | 1.1 ± 0.06 | 4.6 ± 0.58 |
| ATTO647N-S9-GM3 | 0.066 ± 0.008 | 0.26 ± 0.02 |
| TMR-G6-GM3 | 0.029 ± 0.006 | 1.7 ± 0.14 |
| ATTO594-G6-GM3 | 0.49 ± 0.04 | 1.7 ± 0.09 |
| ATTO594-termG6-GD1b | 1.0 ± 0.04 | 4.2 ± 0.40 |
| ATTO594-GD3 | 1.1 ± 0.08 | 4.8 ± 0.58 |
| ATTO594-GQ1b | 1.1 ± 0.07 | 4.3 ± 0.43 |
| ATTO594-NeuAcLc$_4$Cer | 1.03 ± 0.18 | 4.8 ± 0.07 |

**Fig. 13** Partitioning examination of fluorescent ganglioside analogs. **a** Partitioning into DRMs. **b** Partitioning into Lo- and Ld-like domains in the GPMVs. **c** Results of biophysical evaluation

rafts. Notably, each ganglioside probe remained confined within either the CD59 patch or the bulk region, without traversing between them.

In contrast, ATTO594-DOPE was predominantly excluded from CD59-rich domains and did not penetrate raft boundaries, distinguishing it from raft-associated probes. Quantitative analysis revealed that ganglioside probes were enriched 4- to tenfold in CD59-positive regions, whereas ATTO594-DOPE was largely absent. These findings corroborate confocal microscopy data, supporting the notion that CD59-rich domains transiently recruit raft-associated molecules such as gangliosides.

Within large CD59-stained regions, ATTO594-S9-GM3 and GM1 frequently exhibited temporal confinement lasting approximately 100 ms—a phenomenon termed *Temporal Arrest of LateraL diffusion* (TALL). Outside CD59 domains, however, the same probes remained mobile across imaging intervals of 33, 8, or 0.5 ms [53]. Upon actin depolymerization, the incidence of TALL within CD59 regions was significantly reduced, suggesting a mechanistic link between antibody-induced CD59 clustering and the actin-based membrane skeleton.

In our study [46], CD59 was crosslinked using primary and secondary antibodies, forming clusters smaller than 3 μm. Assuming average inter-CD59 distances of 9.6 nm or 6.0 nm and a molecular radius of 1.5 nm, these clusters occupied 7.7–20% of their respective areas. Higher CD59 densities corresponded with increased ganglioside probe accumulation and a marked reduction in diffusion coefficients, consistent with steric hindrance effects when obstacle spacing fell below 10 nm.

## 8.2 *Stabilization of GPI-Anchored Protein Cluster Rafts Through Ganglioside Recruitment*

Raft formation was investigated under physiological conditions. Upon ligation, CD59 forms stable oligomers that recruit and activate intracellular signaling molecules, including phospholipase Cγ (PLCγ), Src family kinases (SFKs), and trimeric G proteins [38, 54, 55], leading to $IP_3$-mediated $Ca^{2+}$ signaling. These oligomers, termed *CD59-cluster signaling rafts*, arise under specific physiological stimuli. Simultaneous two-color single-molecule imaging was performed using fluorescent ganglioside probes (GM1 and GM3) alongside ligation-induced CD59 oligomers. Both ATTO594-S9-GM1 and ATTO594-S9-GM3 were transiently recruited to CD59-cluster signaling rafts, exhibiting interaction durations of approximately 100 ms before dissociation (Fig. 14). In contrast, the non-raft control phospholipid ATTO594-DOPE displayed markedly shorter co-localization times with CD59 clusters. Cholesterol depletion or expression of the CD59-TM mutant reduced co-localization durations by roughly 50%, suggesting that ganglioside–CD59 interactions are mediated primarily through raft-lipid and alkyl chain compatibility rather than headgroup recognition. This interpretation is supported by recent findings involving sphingomyelin and saturated phosphatidylcholine probes [56].

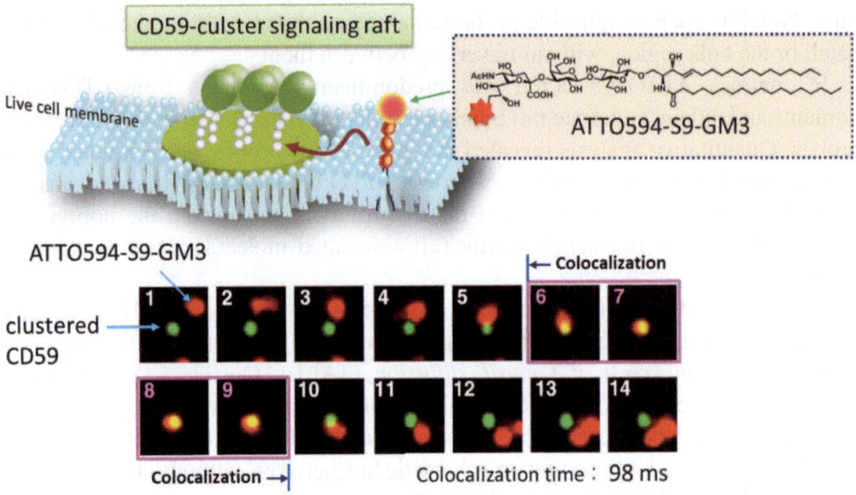


**Fig. 14** Single-molecule imaging of fluorescent GM3 probe co-localizing with CD59-cluster signaling raft in the live cell membrane

## *8.3 Stabilization of GPI-Anchored Protein Homodimers by Raft-Lipid Interactions*

In steady-state plasma membranes, CD59 transiently forms homodimers with an average lifetime of 160 ms [38, 57]. Cholesterol depletion or substitution of CD59's GPI anchor with a non-raft transmembrane domain reduced this lifetime by approximately half, indicating that raft-lipid interactions contribute to the stabilization of CD59 homodimers.

Single-molecule tracking revealed that ATTO594-S9-GM3 and GM1 were transiently recruited to CD59 homodimers for ~ 80 ms and to CD59 monomers for ~ 50 ms [46]. In contrast, the non-raft control phospholipid ATTO594-DOPE co-localized with CD59 monomers and homodimers for only ~ 40 ms, a shorter duration than observed for ganglioside probes. Likewise, ATTO594-S9-NeuAcLc4Cer and its asialo-analog ATTO594-Lc4Cer exhibited high affinity to CD59 homodimers rather than monomers [48]. To the homodimers, ATTO594-S9-NeuAcLc4Cer and ATTO594-Lc4Cer were recruited for 84 ms and 79 ms, respectively, while to the monomers, their recruitment continued for approximately 50 ms. These transient interactions likely arise from glycan–glycan or glycan–protein contacts, as well as lipid–lipid compatibility between gangliosides and CD59. Notably, ganglioside association with CD59-cluster signaling rafts persisted for ~ 100 ms—longer than with CD59 homodimers or monomers—suggesting that raft-lipid interactions are further stabilized by ganglioside recruitment. These findings align with previous reports that CD59 homodimers and cluster rafts are stabilized by lipid raft components, and that gangliosides may enhance this stabilization.

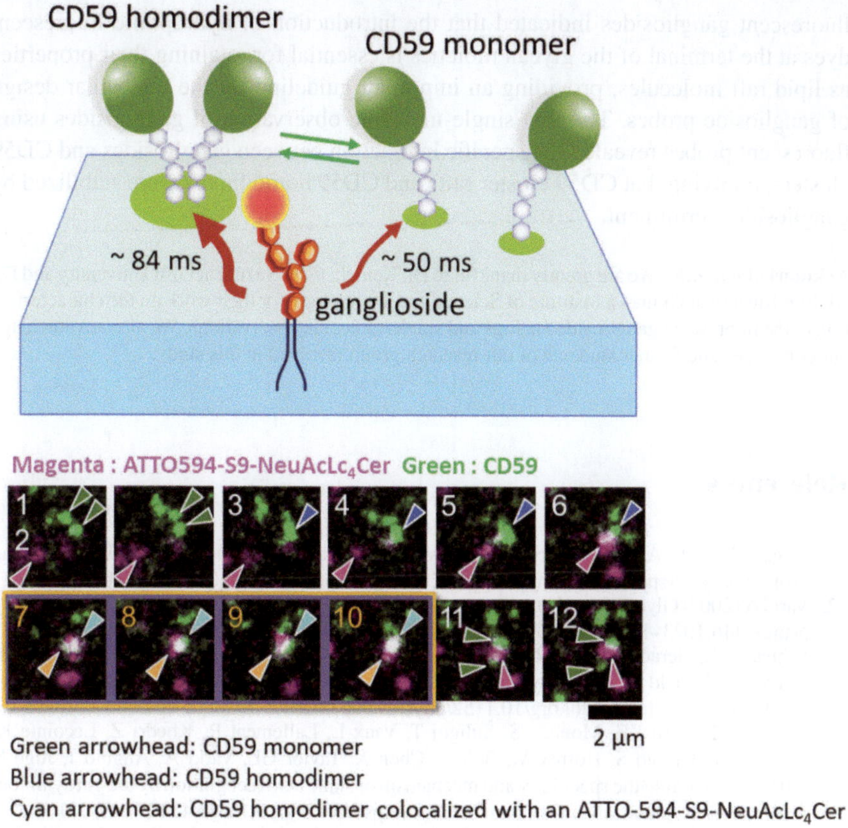


**Fig. 15** Transient co-localization of a ganglioside with CD59 monomers and homodimers

Historically, it was unclear whether clustered raft molecules could recruit additional raft components. However, simultaneous single-molecule imaging of CD59 and ganglioside probes demonstrated that non-crosslinked gangliosides are transiently recruited to a range of CD59 structures, including monomers, homodimers, signaling rafts, and raft-like regions. These results are consistent with prior studies showing that mGFP-GPI dimers formed via monoclonal antibody binding can transiently recruit other non-crosslinked GPI-anchored proteins [53, 58–61] (Fig. 15).

## 9 Conclusion

We have built robust methodology for the chemical synthesis of gangliosides, comprising efficient and stereoselective α-sialylation and GC cassette approach for the conjugation of glycan and Cer. The methodology enabled the chemical synthesis of a series of fluorescent ganglioside analogs. The biophysical examination of the fluorescent gangliosides indicated that the introduction of hydrophilic fluorescent dyes at the terminal of the glycan moieties is essential for retaining their properties as lipid raft molecules, providing an important guideline for the molecular design of ganglioside probes. The first single-molecule observation of gangliosides using fluorescent probes revealed the specific interaction between gangliosides and CD59 clusters, implying that CD59 cluster rafts and CD59 homodimers were stabilized by ganglioside recruitment.

**Acknowledgements** We are greatly thankful to Dr. Kenichi G.N. Suzuki at Gifu University and Dr. Akihiro Kusumi at Okinawa Institute of Science and Technology for their work on the characterization of the fluorescent ganglioside analogs and single-molecule observations. We also acknowledge our colleagues and former students of our research group involved in this study.

## References

1. Angata T, Varki A (2002) Chemical diversity in the sialic acids and related α-keto acids: an evolutionary perspective. Chem Rev 102:439–469. https://doi.org/10.1021/cr000407m
2. Varki A (2007) Glycan-based interactions involving vertebrate sialic-acid-recognizing proteins. Nature 446:1023–1029. https://doi.org/10.1038/nature05816
3. Schnaar RL, Gerady-Schahn R, Hildebrandt H (2014) Sialic acids in the brain: gangliosides and polysialic acid in nervous system development, stability, disease, and regeneration. Physiol Rev 94:461–518. https://doi.org/10.1152/physrev.00033.2013
4. Owen C, Tailford LE, Monaco S, Šuligoj T, Vaux L, Lallement R, Khedri Z, Lecointe K, Walshaw J, Tribolo S, Horrex M, Bell A, Chen X, Taylor GL, Varki A, Angulo J, Juge N (2017) Unravelling the specificity and mechanism of sialic acid recognition by the gut symbiont *Ruminococcus gnavus*. Nat Commun 8:2196. https://doi.org/10.1038/s41467-017-02109-8
5. Pagan JD, Kitaoka M, Anthony RM (2018) Engineered sialylation of pathogenic antibodies *in vivo* attenuates autoimmune disease. Cell 172:564-577.e13. https://doi.org/10.1016/j.cell.2017.11.041
6. Smith DC, Lord JM, Roberts LM, Johannes L (2004) Glycosphingolipids as toxin receptors. Semin Cell Dev Biol 15:397–408. https://doi.org/10.1016/j.semcdb.2004.03.005
7. Lopez PHH, Schnaar RL (2009) Gangliosides in cell recognition and membrane protein regulation. Curr Opin Struct Biol 19:549–557. https://doi.org/10.1016/j.sbi.2009.06.001
8. Fleming FE, Bohm R, Dang VT, Holloway G, Haselhost T, Madge PD, Deveryshetty J, Yu X, Blanchard H, von Itzstein M, Coulson BS (2014) Relative roles of GM1 ganglioside, *N*-acetylneuraminic acids, and α2β1 integrin in mediating rotavirus infection. J Virol 88:4558–4571. https://doi.org/10.1128/jvi.03431-13
9. Blaum BS, Frank M, Walker RC et al (2016) Complement factor H and simian virus 40 bind the GM1 ganglioside in distinct conformations. Glycobiology 26:532–539. https://doi.org/10.1093/glycob/cwv170
10. Kawashima N, Yoon SJ, Itoh K, Nakayama K-I (2009) Tyrosine kinase activity of epidermal growth factor receptor is regulated by GM3 binding through carbohydrate to carbohydrate interactions. J Biol Chem 284:6147–6155. https://doi.org/10.1074/jbc.M808171200
11. Fukuda Y, Fukui T, Hikichi C, Ishikawa T, Murate K, Adachi T, Imai H, Furuhata K, Ueda A, Kaplan AP, Mutoh T (2015) Neurotropin promotes NGF signaling through interaction of GM1 ganglioside with Trk neurotrophin receptor in PC12 cells. Brain Res 1596:13–21. https://doi.org/10.1016/j.brainres.2014.11.041
12. Jung WR, Kim HG, Shin MK, Park DI, Kim KL (2010) The effect of ganglioside GQ1b on the NMDA receptor signaling pathway in H19–7 cells and rat hippocampus. Neuroscience 165:159–167. https://doi.org/10.1016/j.neuroscience.2009.10.012

13. Kaneko M, Yamada K, Miyamoto T, Inagaki M, Higuchi R (2007) Neuritogenic activity of gangliosides from echinoderms and their structure–activity relationship. Chem Pharm Bull 55:462–463. https://doi.org/10.1248/cpb.55.462
14. Vibhute AM, Komura N, Tanaka H-N, Imamura A, Ando H (2021) Advanced chemical methods for stereoselective sialylation and their applications in sialoglycan syntheses. Chem Rec 21:3194–3223. https://doi.org/10.1002/tcr.202100080
15. Ando H, Koike Y, Ishida H, Kiso M (2003) Extending the possibility of an N-Troc-protected sialic acid donor toward variant sialo-glycoside synthesis. Tetrahedron Lett 44:6883–6886. https://doi.org/10.1016/S0040-4039(03)01707-6
16. Ando H, Koike Y, Koizumi S, Ishida H, Kiso M (2005) 1,5-Lactamized sialyl acceptors for various disialoside syntheses: Novel method for the synthesis of glycan portions of Hp-s6 and HLG-2 gangliosides. Ange Chem Int Ed 44:6759–6763. https://doi.org/10.1002/anie.200501608
17. Tamai H, Ando H, Tanaka H-N, Hosoda-Yabe R, Yabe T, Ishida H, Kiso M (2011) The total synthesis of the neurogenic ganglioside LLG-3 isolated from the starfish *Linckia laevigata*. Ange Chem Int Ed 50:2330–2333. https://doi.org/10.1002/anie.201006035
18. Tamai H, Ando H, Ishida H, Kiso M (2012) First synthesis of a pentasaccharide moiety of ganglioside GAA-7 containing unusually modified sialic acids through the use of N-Troc-sialic acid derivative as a key unit. Org Lett 14:6342–6345. https://doi.org/10.1021/ol303122w
19. Tamai H, Imamura A, Ogawa J, Ando H, Ishida H, Kiso M (2015) First total synthesis of ganglioside GAA-7 from starfish *Asterias amurensis versicolor*. Eur J Org Chem 23:5199–5211. https://doi.org/10.1002/ejoc.201500606
20. Goto K, Suzuki T, Tamai H, Ogawa J, Imamura A, Ando H, Ishida H, Kiso M (2015) Total synthesis and neuritogenic activity evaluation of ganglioside PNG-2A from the starfish *Protoreaster nodosus*. Asian J Org Chem 4:1160–1171. https://doi.org/10.1002/ajoc.201500282
21. Goto K, Tamai H, Takeda Y, Tanaka H-N, Mizuno T, Imamura A, Ishida H, Kiso M, Ando H (2019) Total synthesis of sialyl inositol phosphosphingolipids CJP-2, CJP-3, and CJP-4 isolated from feather star *Comanthus japonica*. Org Lett 21:4054–4057. https://doi.org/10.1021/acs.orglett.9b01229
22. Komura N, Kato K, Udagawa T, Asano S, Tanaka H-N, Imamura A, Ishida H, Kiso M, Ando H (2019) Constrained sialic acid donors enable selective synthesis of α-glycosides. Science 364:677–680. https://doi.org/10.1126/science.aaw4866
23. Kraus GA, Hon YS, Thomas PJ, Laramay S, Liras S, Hanson J (1989) Organic synthesis using bridgehead carbocations and bridgehead enones. Chem Rev 89:1591–1598. https://doi.org/10.1021/cr00097a013
24. Shapiro D, Flowers HM (1959) Synthetic studies on sphingolipids. V. the synthesis of dihydrocerebrosides. J Am Chem Soc 81:2023–2024. https://doi.org/10.1021/ja01517a060
25. Sugimoto M, Ogawa T (1985) Synthesis of a hematoside (GM3-ganglioside) and a stereoisomer. Glycoconjugate J 2:5–9. https://doi.org/10.1007/BF01225108
26. Zimmermann P, Sommer R, Bär T, Schmidt RR (1988) Azidosphingosine glycosylation in glycosphingolipid synthesis. J Carbohydr Chem 7:435–452. https://doi.org/10.1080/07328308808058935
27. Sakamoto H, Nakamura S, Tsuda T, Hashimoto S (2000) Chemoselective glycosidation strategy based on glycosyl donors and acceptors carrying phosphorus-containing leaving groups: a convergent synthesis of ganglioside GM3. Tetrahedron Lett 41:7691–7695. https://doi.org/10.1016/S0040-4039(00)01343-5
28. Imamura A, Ando H, Ishida H, Kiso M (2009) Ganglioside GQ1b: efficient total synthesis and the expansion to synthetic derivatives to elucidate its biological roles. J Org Chem 74:3009–3023. https://doi.org/10.1021/jo8027888
29. Fujikawa K, Imamura A, Ishida H, Kiso M (2008) Synthesis of a GM3 ganglioside analogue carrying a phytoceramide moiety by intramolecular glycosylation as a key step. Carbohydr Res 343:2729–2734. https://doi.org/10.1016/j.carres.2008.05.007

30. Fujikawa K, Nohara T, Imamura A, Ando H, Ishida H, Kiso M (2010) A cyclic glucosyl ceramide acceptor as a versatile building block for complex ganglioside synthesis. Tetrahedron Lett 51:1126–1130. https://doi.org/10.1016/j.tetlet.2009.12.121
31. Fujikawa K, Nakashima S, Konishi M, Fuse T, Komura N, Ando H, Ando T, Yuki N, Ishida H, Kiso M (2011) The first total synthesis of ganglioside GalNAc-GD1a, a target molecule for autoantibodies in Guillain-Barré syndrome. Chem Eur J 17:5641–5651. https://doi.org/10.1002/chem.201003357
32. Goto K, Sawa M, Tamai H, Imamura A, Ando H, Ishida H, Kiso M (2016) The total synthesis of starfish ganglioside GP3 bearing a unique sialyl glycan architecture. Chem Eur J 22:8323–8331. https://doi.org/10.1002/chem.201600970
33. Takahashi M, Shirasaki J, Komura N, Sasaki K, Tanaka H-N, Imamura A, Ishida H, Hanashima S, Murata M, Ando H (2020) Efficient diversification of GM3 gangliosides: via late-stage sialylation and dynamic glycan structural studies with $^{19}F$ solid-state NMR. Org Biomol Chem 18:2902–2913. https://doi.org/10.1039/D0OB00437E
34. Asano S, Tanaka H-N, Imamura A, Ishida H, Ando H (2019) *P-tert*-Butyl groups improve the utility of aromatic protecting groups in carbohydrate synthesis. Org Lett 21:4197–4200. https://doi.org/10.1021/acs.orglett.9b01372
35. Simons K, Ikonen E (1997) Functional rafts in cell membranes. Nature 387:569–572. https://doi.org/10.1038/42408
36. Simons K, Gerl MJ (2010) Revitalizing membrane rafts: new tools and insights. Nat Rev Mol Cell Biol 11:688–699. https://doi.org/10.1038/nrm2977
37. Sezgin E, Levental I, Mayor S, Eggeling C (2017) The mystery of membrane organization: composition, regulation and roles of lipid rafts. Nat Rev Mol Cell Biol 18:361–374. https://doi.org/10.1038/nrm.2017.16
38. Suzuki KGN, Kasai RS, Hirosawa KM, Nemoto YL, Ishibashi M, Miwa Y, Fujiwara TK, Kusumi A (2012) Transient GPI-anchored protein homodimers are units for raft organization and function. Nat Chem Biol 8:774–783. https://doi.org/10.1038/nchembio.1028
39. Marushchak D, Gretskaya N, Mikhalyov I, Johansson LBÅ (2007) Self-aggregation—an intrinsic property of $G_{M1}$ in lipid bilayers. Mol Membr Biol 24:102–112. https://doi.org/10.1080/09687860600995235
40. Eggeling C, Ringemann C, Medda R, Schwarzmann G, Sandhoff K, Polyakova S, Belov VN, Hein B, von Middendorff SA, Hell SW (2009) Direct observation of the nanoscale dynamics of membrane lipids in a living cell. Nature 457:1159–1162. https://doi.org/10.1038/nature07596
41. Chinnapen DJF, Hsieh W-T, te Welscher YM, Saslowsky DE, Kaoutzani L, Brandsma E, D'Auria L, Park H, Wagner JS, Drake KR, Kang M, Benjamin T, Ullman MD, Costello CE, Kenworthy AK, Baumgart T, Massol RH, Lencer WI (2012) Lipid sorting by ceramide structure from plasma membrane to ER for the cholera toxin receptor ganglioside GM1. Dev Cell 23:573–586. https://doi.org/10.1016/j.devcel.2012.08.002
42. Acquotti D, Sonnino S, Masserini M, Casella L, Fronza G, Tettamanti G (1986) A new chemical procedure for the preparation of gangliosides carrying fluorescent or paramagnetic probes on the lipid moiety. Chem Phys Lipids 40:71–86. https://doi.org/10.1016/0009-3084(86)90063-0
43. Wilchek M, Spiegel S, Spiegel Y (1980) Fluorescent reagents for the labeling of glycoconjugates in solution and on cell surfaces. Biochem Biophys Res Commun 92:1215–1222. https://doi.org/10.1016/0006-291X(80)90416-7
44. Rich JR, Cunningham AM, Gilbert M, Withers SG (2011) Glycosphingolipid synthesis employing a combination of recombinant glycosyltransferases and an endoglycoceramidase glycosynthase. Chem Commun 47:10806–10808. https://doi.org/10.1039/C1CC13885E
45. Yang GY, Li C, Fischer M, Cairo CW, Feng Y, Withers SG (2015) A FRET probe for cell-based imaging of ganglioside-processing enzyme activity and high-throughput screening. Angew Chem Int Ed 54:5389–5393. https://doi.org/10.1002/anie.201411747
46. Komura N, Suzuki KGN, Ando H, Konishi M, Koikeda M, Imamura A, Chadda R, Fujiwara TK, Tsuboi H, Sheng R, Cho W, Furukawa K, Furukawa K, Yamauchi Y, Ishida H, Kusumi A, Kiso M (2016) Raft-based interactions of gangliosides with a GPI-anchored receptor. Nat Chem Biol 12:402–410. https://doi.org/10.1038/nchembio.2059

47. Konishi M, Komura N, Hirose Y, Suganuma Y, Tanaka H-N, Imamura A, Ishida H, Suzuki KGN, Ando H (2020) Development of fluorescent ganglioside GD3 and GQ1b analogs for elucidation of raft-associated interactions. J Org Chem 85:15998–16013. https://doi.org/10.1021/acs.joc.0c02121
48. Takahashi M, Komura N, Yoshida Y, Yamaguchi E, Hasegawa A, Tanaka H-N, Imamura A, Ishida H, Suzuki KGN, Ando H (2022) Development of lacto-series ganglioside fluorescent probe using late-stage sialylation and behavior analysis with single-molecule imaging. RSC Chem Biol 3:868–885. https://doi.org/10.1039/D2CB00083K
49. Takahashi M, Komura N, Tanaka H-N, Imamura A, Ishida H, Ando H (2025) Ganglioside analog synthesis via late-stage sialylation and fatty acid introduction. Carbhydr Res 552:109460. https://doi.org/10.1016/j.carres.2025.109460
50. Baumgart T, Hammond AT, Sengupta P, Hess ST, Holowka DA, Baird BA, Webb WW (2007) Large-scale fluid/fluid phase separation of proteins and lipids in giant plasma membrane vesicles. Proc Natl Acad Sci USA 104:3165–3170. https://doi.org/10.1073/pnas.061135710
51. Levental I, Grzybek M, Simons K (2011) Raft domains of variable properties and compositions in plasma membrane vesicles. Proc Natl Acad Sci USA 108:11411–11416. https://doi.org/10.1073/pnas.1105996108
52. Kenworthy AK, Nichols BJ, Remmert CL, Hendrix GM, Kumar M, Zimmerberg J, Lippincott-Schwartz J (2004) Dynamics of putative raft-associated proteins at the cell surface. J Cell Biol 165:735–746. https://doi.org/10.1083/jcb.200312170
53. Huang H, Simsek MF, Jin W, Pralle A (2015) Effect of receptor dimerization on membrane lipid raft structure continuously quantified on single cells by camera based fluorescence correlation spectroscopy. PLoS One 10:e0121777. https://doi.org/10.1371/journal.pone.0121777
54. Suzuki KG, Fujiwara TK, Sanematsu F, Iino R, Edidin M, Kusumi A (2007) GPI-anchored receptor clusters transiently recruit Lyn and G alpha for temporary cluster immobilization and Lyn activation: single-molecule tracking study 1. J Cell Biol 177:717–730. https://doi.org/10.1083/jcb.200609174
55. Suzuki KG, Fujiwara TK, Edidin M, Kusumi A (2002) Dynamic recruitment of phospholipase C gamma at transiently immobilized GPI-anchored receptor clusters induces IP3-$Ca^{2+}$ signaling: single-molecule tracking study 2. J Cell Biol 177:731–742. https://doi.org/10.1083/jcb.200609175
56. Kinoshita M, Suzuki KG, Matsumori N, Takada M, Ano H, Morigaki K, Abe M, Makino A, Kobayashi T, Hirosawa KM, Fujiwara TK, Kusumi A, Murata M (2017) Raft-based sphingomyelin interactions revealed by new fluorescent sphingomyelin analogs. J Cell Biol 216:1183–1204. https://doi.org/10.1083/jcb.201607086
57. Suzuki KG, Kasai RS, Fujiwara TK, Kusumi A (2013) Single-molecule imaging of receptor-receptor interactions. Methods Cell Biol 117:373–390. https://doi.org/10.1016/B978-0-12-408143-7.00020-7
58. Kusumi A, Suzuki A (2005) Toward understanding the dynamics of membrane-raft-based molecular interactions. Biochim Biophys Acta 1746:234–251. https://doi.org/10.1016/j.bbamcr.2005.10.001
59. Tanaka KA, Suzuki KG, Shirai YM, Shibutani ST, Miyahara MS, Tsuboi H, Yahara M, Yoshimura A, Mayor S, Fujiwara TK, Kusumi A (2010) Membrane molecules mobile even after chemical fixation. Nat Methods 7:865–866. https://doi.org/10.1038/nmeth.f.314
60. Harder T, Scheiffele P, Verkade P, Simons K (1998) Lipid domain structure of the plasma membrane revealed by patching of membrane compartments. J Cell Biol 141:929–942. https://doi.org/10.1083/jcb.141.4.929
61. Mayor SR, Rothberg KG, Maxfield FR (1994) Sequestration of GPI-anchored proteins in caveolae traggered by cross-linking. Science 264:1948–1951. https://doi.org/10.1126/science.7516582

# Synthetic Strategies for Di-, Oligo-, and Polysialic Acids with Applications as Siglec Ligands

**Hiroshi Tanaka and Chihiro Sato**

**Abstract** In this study, we designed novel double carbonyl-protected sialyl donors that are able to control the conformation of glycerol side chains to enable highly α-selective α(2,8)-sialylation. Both*N*-acetyl and *N*-glycolyl derivatives were prepared and applied to the synthesis of structurally defined oligosialic acids. Using iterative cycles of 8-*O*-deprotection and glycosylation, we achieved efficient chain elongation in both disialic and octasialic acids in high yields, with complete α-selectivity at each step. Notably, the strategy was effective for both Neu5Ac- and Neu5Gc-containing sequences with loss of neither selectivity nor reactivity. Optimized deprotection conditions involving *S*-alkylthiocarbamate formation, selective oxidation, and mild basic hydrolysis allowed complete removal of protecting groups without *N*-acyl cleavage, which afforded α(2,8)-linked oligosialic acids in good yields. This method provides a robust platform for the synthesis of long-chain sialic acid oligomers, which includes the Neu5Gc-containing analogues that are valuable for studying biological functions and developing glycan-based probes for Siglec recognition.

**Keywords** Glycosylation · Polysaccharides · Carbonyl protecting group · Siglec-7 · Multivalent interaction

H. Tanaka (✉)
Faculty of Pharmacy, Juntendo University, Urayasu, Chiba, Japan
e-mail: h.tanaka.kn@juntendo.ac.jp

C. Sato
Integrated Glyco-Biomedical Research Center (iGMED), Institute for Glyco-core Research (iGCORE), Nagoya University, Nagoya, Japan

Graduate School of Bioagricultural Sciences, Nagoya University, Nagoya, Japan

C. Sato
e-mail: chi@agr.nagoya-u.ac.jp



K. Toshima and Y. Fujimoto (eds.), *Chemical Biology of Carbohydrates*,
https://doi.org/10.1007/978-981-95-5335-8_3

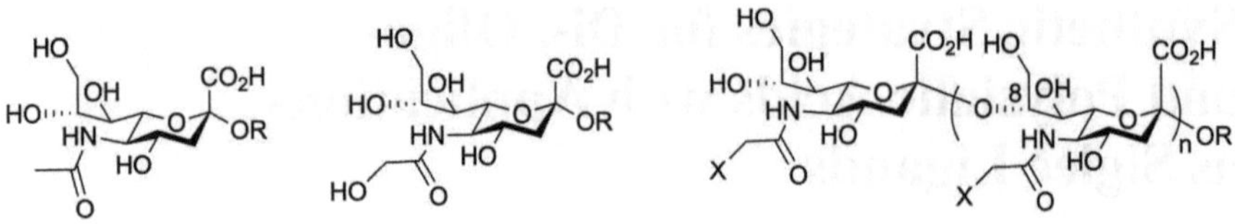


**Fig. 1** Chemical structures of sialic acids **1** and **2** containing *N*-acetyl and *N*-glycosyl groups and their oligomers **3-5**

# 1 Introduction

Sialic acids are a family of nine-carbon acidic monosaccharides that are predominantly located at the non-reducing termini of glycan chains in glycoproteins and glycolipids (Fig. 1). Among them, *N*-acetylneuraminic acid (Neu5Ac, **1**) and *N*-glycolylneuraminic acid (Neu5Gc, **2**) both are widely distributed in mammals and play essential roles in biological processes such as cell-to-cell interactions, immune regulation, viral infection, and neural development [1]. Furthermore, sialic acid dimers **3** and oligomers **4** and **5** linked via α(2,8) glycosidic bonds have been implicated in cancer progression and neuroplasticity. From the perspective of multimer classification, α(2,8)-linked sialic acids are referred to as disialic acids **3** when composed of two residues, oligosialic acids **4** when containing three to seven residues, and polysialic acids **5** when consisting of eight or more residues, as defined by their antibody-binding properties. Chains with eight or more residues are believed to adopt a characteristic conformation unique to polysialic acids. In light of these structural and functional features, this section is organized in two parts: the first outlines the development of synthetic methods for sialic acid oligomers, and the second presents functional studies of the Siglec-7 receptor using synthetic glycans as chemical probes.

# 2 Part I: Development of Synthetic Methods for α(2,8)-Linked Sialic Acid Oligomers

## *2.1 Synthetic Challenges in α(2,8) Sialylation: α-Selectivity and C8 Hydroxyl Reactivity*

The chemical synthesis of structurally well-defined α(2,8) sialic acid oligomers **8** is essential for elucidating their biological functions and exploring potential therapeutic applications. However, several challenges (Fig. 2) remain for the synthesis of α(2,8) sialic acids **8**:

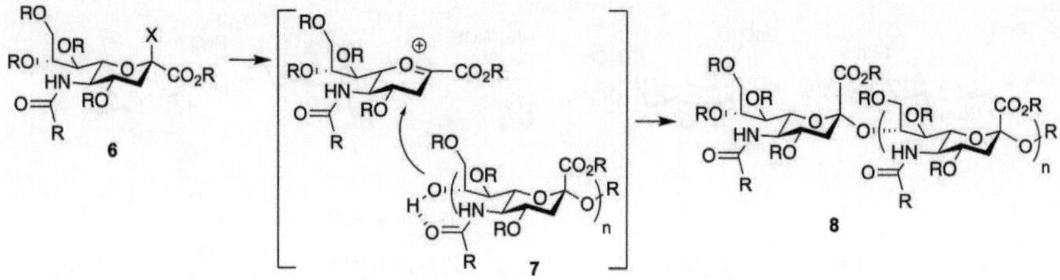

Fig. 2 Process of α(2,8) sialylation

- **Difficulty in controlling anomeric selectivity**, particularly in favor of the α-anomer, due to the lack of substitution at the C3 position of donor **6**.
- **Low reactivity of the C8 hydroxyl group** in acceptor **7**, which is required for α(2,8)-linkage formation.
- **Poor solubility of oligomeric intermediates 7** in organic solvents, which is caused by the presence of polar acylamino groups.

Many strategies have been explored to address these challenges; however, methods that could enable the controlled synthesis of longer oligomers and polymers, as well as glycans containing mixed NeuGc residues, have not been fully achieved [2].

## 2.2 *Development of 5N,4O-Carbonyl-Protected Sialyl Donors and Their Limitations*

In 2006, our group developed the 5*N*,4*O*-carbonyl-protected sialyl donor **9** (Fig. 3), in which the nitrogen atom at the C5 position and the oxygen atom at the C4 position are cyclized as a carbamate [3]. This design effectively enhanced the α-selectivity in glycosylation, even under conditions that did not require nitrile solvents. Furthermore, the conformational restriction imposed by the cyclic protecting group helped suppress the loss of reactivity at the C8 hydroxyl group of glycosyl acceptor **10**. As a result, this protecting system enabled the first chemical syntheses of α(2,8) tetrasialic acid (**13**) and the glycan portion of ganglioside GP1c [4], and thereby demonstrated its utility for the synthesis of α(2,9) oligosialosides [5, 6]. However, this protecting system exhibited two notable limitations. First, at the donor stage, it lacked *N*-acyl groups, necessitating post-glycosylation modification to introduce them. Such late-stage transformations increased the total number of synthetic steps and reduced overall efficiency. Second, in the synthesis of tetrasialoside **12**, the coupling yield decreased markedly, making further chain elongation difficult. These difficulties were attributed not only to the reduced reactivity caused by the increased size of the glycosyl acceptor, but also to its decreased solubility.

To address the first limitation, Crich and co-workers developed *N*-acetyl 5*N*,4*O*-carbonyl-protected sialyl donors **14** (Fig. 4) [7, 8]. This approach enabled direct

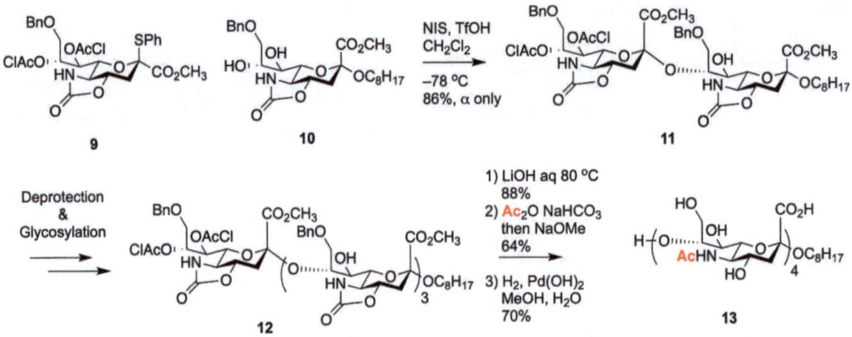


**Fig. 3** Synthesis of α(2,8) tetrasialic acid **13** by using the 5*N*,4*O*-carbonyl-protected sialyl donor **9**

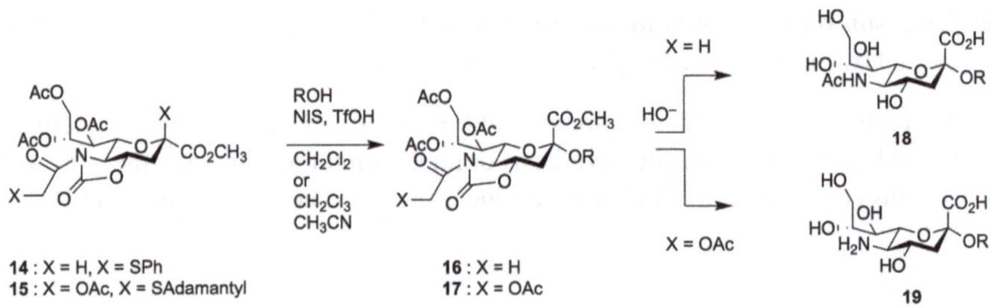


**Fig. 4** Glycosidation of the *N*-acyl 5*N*,4*O*-carbonyl-protected sialy donors **14** and **15**

access to *N*-acetyl sialic acids via mild basic hydrolysis. Moreover, the introduction of the *N*-acetyl group was expected to improve the solubility of the resultant intermediates by blocking the highly acidic amide hydrogen, which can act as a hydrogen-bond donor. Nevertheless, α-selectivity tended to decrease in glycosylations with sterically hindered alcohols. In particular, reactions with secondary alcohol acceptors required nitrile solvents to maintain adequate stereocontrol. Furthermore, the *N*-glycolyl sialoside **17** was found to undergo preferential de-glycolylation to **19** [9], which underscores the need for new strategies to overcome these challenges. Therefore, the development of efficient synthetic methods for extended oligo- and polysialic acids, as well as oligomers containing Neu5Gc residues, remains an important challenge.

## 2.3 *Protecting Side-Chain Hydroxy Groups Influences the Stereoselectivity of Sialylation*

To overcome the challenges associated with α(2,8) sialylation, we focused on the design of protecting groups on the glycerol side chain of sialyl donors. Traditionally,

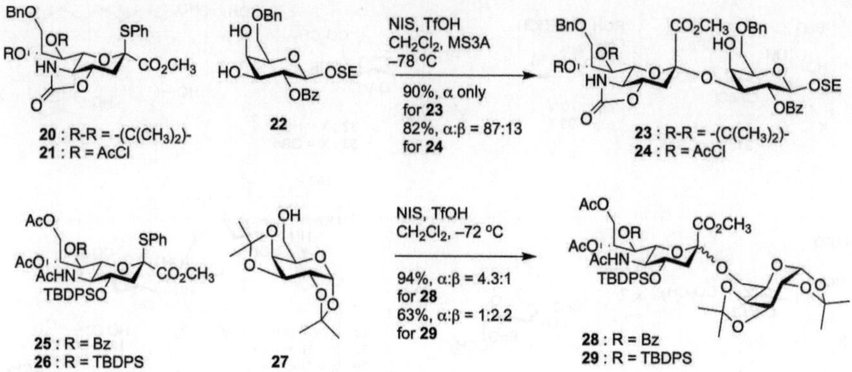


**Fig. 5** Effect that *O*-protecting groups of the side-chain exert on stereoselectivity

studies for improvement of α-selectivity had primarily concentrated on modifications near the anomeric center—such as at the C1 and C3 positions—or on protection of the C5 nitrogen substituent. In contrast, the influence of side-chain protection had received relatively little attention. In 2005, Nifantev and co-workers reported that a sialyl donor protected with 7,8,9-*O*-tris(trichloroacetyl) groups gave higher coupling yields and improved α-selectivity compared with a corresponding *O*-triacetyl donor (Fig. 5) [10]. In 2008, we demonstrated that donor **20**, bearing an isopropylidene group, exhibited higher α-selectivity in the glycosylation of galactoside **22** at the 3-hydroxyl position compared with donor **21**, which carried dichloroacetyl groups [8]. Later, in 2012, De Meo and co-workers conducted a more systematic study on hydroxyl protecting groups. They found that, in donors bearing a TBDPS group at the C4 position, replacing the C7 benzoyl group with a TBDPS group markedly decreased α-selectivity in dichloromethane [11]. Collectively, these reports demonstrated the potential influence of side-chain protecting groups and also revealed their limitations. However, a comprehensive strategy had not yet emerged.

In 2016, building on these observations, we undertook a systematic investigation of side-chain hydroxyl protection in *N*-acyl thiosialosides (Fig. 6) [12]. We found that donors **30** and **31**, in which the hydroxyl groups at C7 and C8 remained unprotected, underwent highly α-selective glycosylation. Notably, this stereoselectivity was achieved without requiring the solvent-directing effect of acetonitrile. In addition, we identified the *O*-chloroacetyl group as an effective protecting group for the hydroxyl at C4. Furthermore, by employing *N*-acetyl and *N*-glycolyl donors (**34** and **35**), we successfully synthesized four types of α(2,9)-disialosides containing both NeuAc and NeuGc residues. Collectively, these findings demonstrated that steric shielding of the α-face by side-chain protecting groups is a key factor leading to reduced α-selectivity. Building on these insights, we next designed a new protecting strategy for *N*-acyl 5*N*,4*O*-carbonyl-protected sialyl donors, which will be described in the following section.

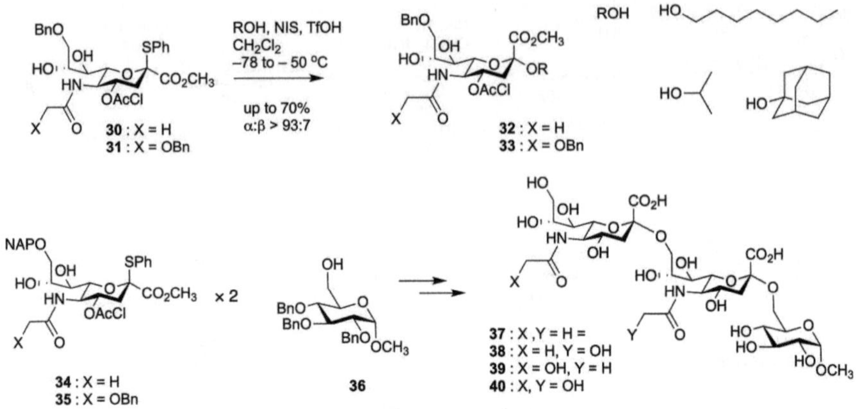


**Fig. 6** α-Selective glycosidation of partially unprotected sialyl donors

## 2.4 *Design of Double Carbonyl-Protected Sialyl Donor*

We designed the novel sialyl donor **41** incorporating an additional cyclic carbonyl protecting group at the C7 and C9 positions (Fig. 7) [13]. Introduction of the 7*O*,9*O*-carbonyl protection was expected to favor the anti-**41** conformation of the C7–C8 bond, in which the oxygen substituent at C8 and the pyranose ring adopt an anti-periplanar arrangement, rather than the gauche-**41** conformation, where both are oriented equatorially. Based on this conformational preference, we anticipated that the corresponding oxocarbenium ion **42** would expose the α-face by minimizing steric shielding from the side-chain substituents, and would thereby promote the desired α-selective glycosylation to give α-sialoside **43**. Moreover, in the anti-**41** conformation, the C8 hydroxyl group would not be sterically encumbered by the pyranose ring, and thus was expected to serve as an excellent glycosyl acceptor **44**.

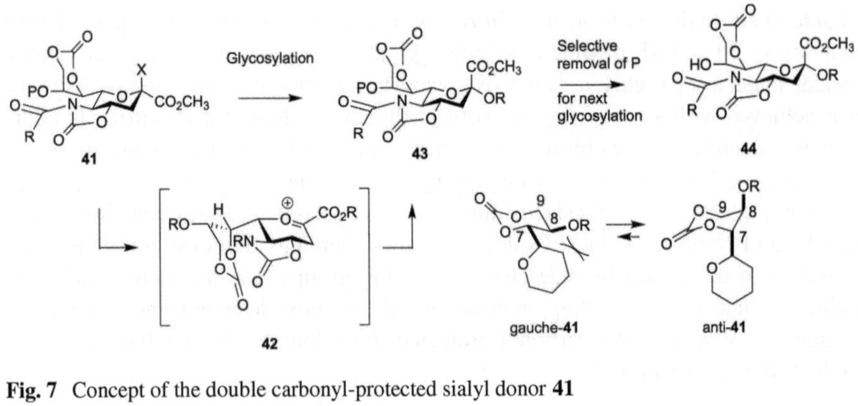


**Fig. 7** Concept of the double carbonyl-protected sialyl donor **41**

## 2.5 *Preparation of a Sialyl Donor Possessing* N-*Acetyl and* N-*Glycolyl Groups*

The synthesis commenced with phenyl thiosialoside **45** (Fig. 8) [14]. Treatment with methanesulfonic acid in methanol induced complete solvolysis of the acetyl groups, including the *N*-acetyl group. Subsequent reaction with *p*-nitrophenyl chloroformate in the presence of $NaHCO_3$ ($MeCN/H_2O$, 0 °C) afforded carbamate **46** in 60% overall yield in two steps. The C7 and C8 hydroxyl groups were then protected sequentially as *tert*-butyldimethylsilyl (TBS) ethers. Initial silylation with TBSCl and imidazole in $CH_2Cl_2$ at room temperature selectively afforded 9-*O*-silyl ether in 95% yield. Subsequent silylation at the C8 position with TBSOTf in pyridine gave disilyl ether **47** in 94% yield. In contrast, when the same step was performed in $CH_2Cl_2$, a mixture of the 7- and 8-silylated products was obtained. Selective deprotection of the C9 silyl group was accomplished using stoichiometric $BF_3{\cdot}OEt_2$. Notably, the use of TBAF—a commonly employed desilylating agent—led to concurrent cleavage of the C8 silyl group during later stages, which significantly lowered the yield of the desired diol. The resultant diol was then converted quantitatively into 7,9-cyclic carbonate **48** by treatment with disuccinimidyl carbonate. Finally, the corresponding acid chlorides were coupled in the presence of DMAP and a base to furnish the NeuAc- and NeuGc-type glycosyl donors **49** and **50**. Conformational analysis by $^1$H NMR revealed extremely small coupling constants between H7 and H8, which indicates that the C8 substituent and the pyranose ring adopt an anti-periplanar arrangement within the cyclic structure. Taken together, this synthetic route highlights the importance of precise stepwise protection and deprotection strategies to achieve the desired 7,9-cyclic carbonate motif. The successful preparation of both NeuAc- and NeuGc-type donors **49** and **50** established a reliable platform for probing the stereoelectronic effects that govern α(2,8) sialylation.

## 2.6 *Selective Hydrolysis of* N-*Acyl Carbamate to* N-*Acylamino Alcohol*

Deprotection of the double carbonyl-protected *N*-acyl sialosides **51** and **52** was achieved through selective hydrolysis of the cyclic carbamate (Fig. 9). The *N*-acyl sialosides **51** and **52** were prepared from the corresponding donors **49** and **50**. The *N*-acetyl derivative **51** smoothly underwent thiolysis to give the corresponding *S*-ethyl thiocarbonate in a good yield. In contrast, the *N*-glycolyl analogue **52** exhibited unexpected deglycolylation under the same conditions, which reflects the enhanced reactivity of both carbonyl groups induced by the electron-withdrawing oxygen atom of the glycoyl substituent. Careful optimization revealed that lowering the reaction temperature (– 20 °C) enabled selective formation of the desired thiocarbonate **54**. These observations highlight the phenomenon whereby the cyclic carbamate preferentially undergoes ring-opening, but the *N*-glycolyl group requires stricter control

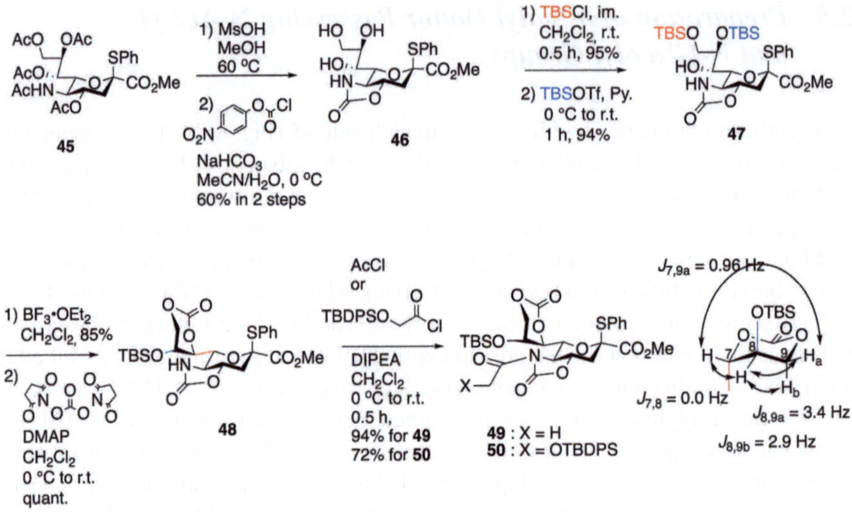

**Fig. 8** Preparation of sialyl donors **49** and **50** possessing *N*-acetyl and *N*-glycolyl groups

of the reaction conditions. Subsequent global deprotection delivered the fully unprotected *N*-acetyl and *N*-glycolyl α-sialosides **57** and **58** in high yields (83–90%), which establishes a robust method for preparing structurally defined N-acyl sialosides. Notably, when TBAF was used for desilylation, removal of the tetrabutylammonium salt proved difficult under solid-phase extraction (SPE) purification conditions. The development of this method thus establishes a robust platform for the preparation of various *N*-acyl sialosides with precise structural control.

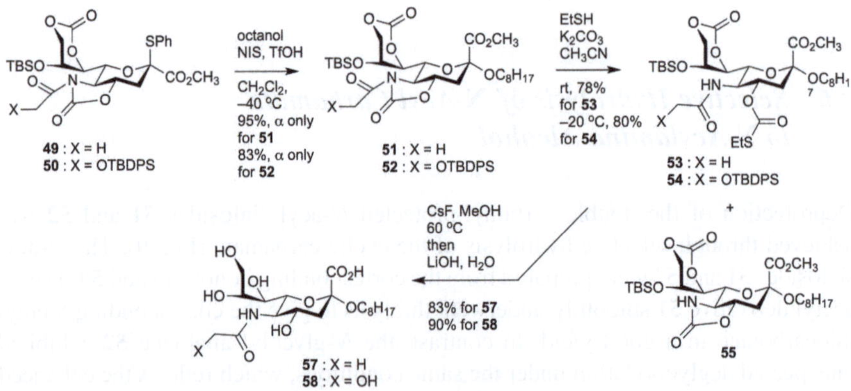

**Fig. 9** Deprotection of *N*-acyl double carbonyl-protected sialosides **49** and **50**

## 2.7 Synthesis of α(2,8) Disaccharides via the Use of Double Carbonyl Protected-Sialyl Donors

A comparative study was conducted alternatively using double carbonyl-protected sialyl donors and conventional glycosyl donors to synthesize α(2,8)-disialic acids (Fig. 10) [14]. The results clearly demonstrated that both *N*-acetyl and *N*-glycolyl donors **49** and **50** furnished the desired disaccharides in high yields with excellent α-selectivity. In contrast, the conventional tri-*O*-acetyl donor **60** afforded an anomeric mixture in which the β-isomer predominated, while the previously reported 5*N*,4*O*-carbonyl-protected donor **9** gave reduced yields. These findings indicate that the introduction of additional cyclic carbonyl protection not only enhances α-selectivity it but also improves the reactivity of the sialyl donor. Furthermore, the syntheses of α(2,8) disialosides **75-78** bearing azido groups were examined. The double carbonyl-protected donors **49** and **50** exhibited efficient reactivity and complete α-selectivity toward the *N*-acetyl and *N*-glycolyl acceptors **65** and **66**. No significant differences in reactivity were observed between the two series. In addition, selective ring opening of the cyclic carbamate using thiols was successfully achieved via precise temperature control. Subsequent global deprotection proceeded smoothly to afford structurally well-defined α(2,8)-disialic acids **75-78** in consistently high yields. Collectively, these results demonstrate that double carbonyl-protected donors represent a versatile and reliable tool for the stereoselective construction of sialic acid oligomers bearing α(2,8)-linkages, which provides a valuable foundation for further advances in sialoside chemistry and biology.

## 2.8 Synthesis of α(2,8) Octasialoside as a Minimum Structure of Polysialic Acid

Starting with disialic acid **61**, we explored the stepwise elongation of the sialic acid chain to an octamer, which represents the minimal structural unit of polysialic acid (Fig. 11) [13]. Repetitive cycles of the selective hydrolysis of C8-silyl ether were followed by glycosylation at the newly generated hydroxyl group to enable controlled extension of the oligosialic acid chain. Notably, the synthesis of tetramer **82** was markedly easier and more efficient than our first-generation method, and product **82** was afforded in good yield. We used the obtained oligomers as glycosyl acceptors to further elongate the chain. Acceptors up to the hexamer **87** showed sufficient solubility in dichloromethane and could be effectively employed in glycosylation reactions. Although these reactions required an excess amount of glycosyl donor, they proceeded smoothly to completion with exclusive α-selectivity. In contrast, the heptameric acceptor **89** exhibited reduced solubility, which necessitated the use of a dichloromethane–acetonitrile solvent system. Nevertheless, glycosylation proceeded efficiently to provide the corresponding product in a good yield with high selectivity. Finally, the resultant octamer **90** was subjected to global deprotection through a key

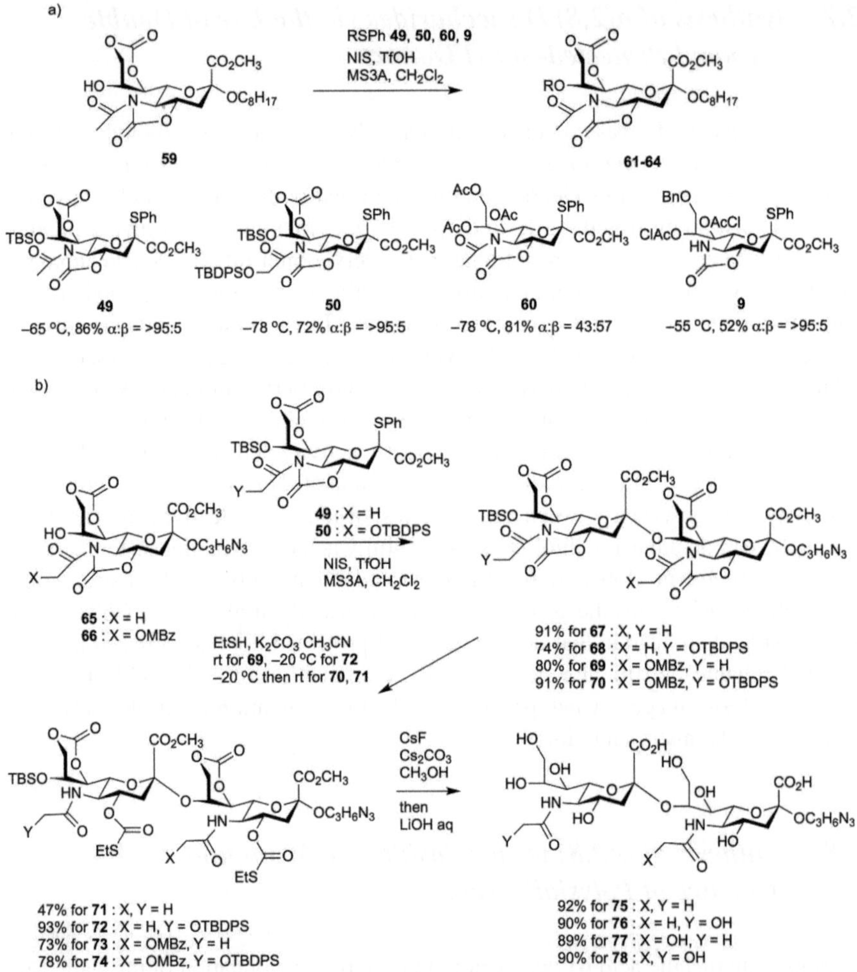


**Fig. 10** Synthesis of α(2,8) disaccharides via the use of double carbonyl-protected sialyl donors

sequence involving thiol-mediated ring-opening of the cyclic carbamate followed by oxidative hydrolysis of the thiocarbonate. This strategy culminated in the first chemical synthesis of polysialic acid **92**, which marks a milestone achievement in the field.

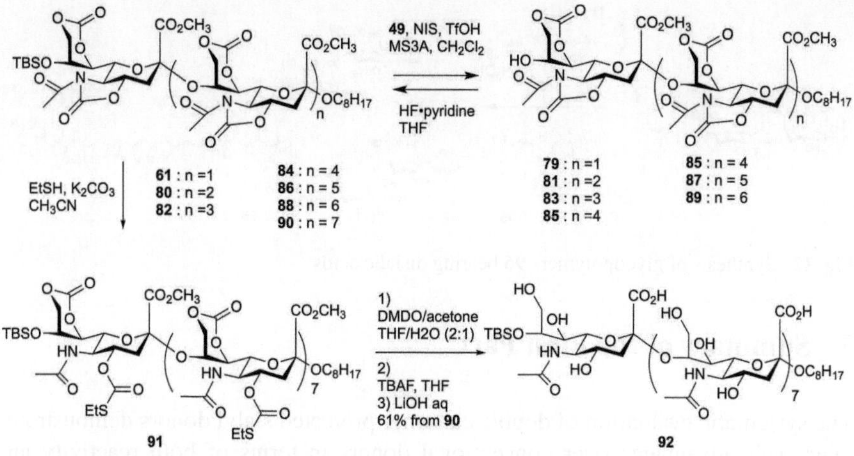


**Fig. 11** Synthesis of α(2,8) octasialoside **92**

## *2.9 Expansion to Glycopolymers: Synthesis of Glycopolymers Bearing α(2,8) Disialic Acids*

The next application of the synthetic α(2,8)sialosides involved the synthesis of chemical probes. Well-defined glycopolymers with precisely controlled degrees of polymerization serve as powerful chemical probes for elucidating glycan–biomolecule interactions and their biological functions [15], but achieving an polymerization of monomers bearing complex oligosaccharides has been a long-standing challenge. To address this, we established a synthetic strategy employing π-allyl nickel complex–mediated coordination polymerization of allene monomers to generate glycopolymers bearing α(2,8)-disialic acid residues (Fig. 12) [16, 17]. This method displays a high level of functional group tolerance and proceeds smoothly in aqueous media while maintaining in vivo characteristics, which offers unique advantages for glycopolymer synthesis [18]. In our design, allene monomers **93** incorporating a lactonized α(2,8)-disialic acid motif were polymerized using π-allyl nickel initiators **94** equipped with an azido group to afford glycopolymers with narrow molecular weight distributions (PDI ≈ 1.1). Ester-protection of the carboxyl groups proved essential for suppressing catalyst deactivation and ensuring efficient polymerization. Furthermore, the terminal azido functionality introduced by the initiator allowed post-polymerization modification, which was exemplified by the conjugation of fluorescent probes **95** to **96**. This approach enabled access to structurally defined glycopolymers that had been difficult to obtain through conventional stepwise glycosylation, and thereby opens new avenues for the creation of glycan-based macromolecular probes in chemical biology.

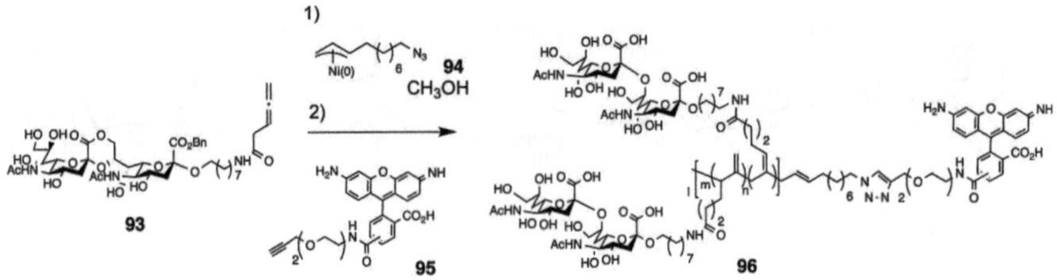


**Fig. 12** Synthesis of glycopolymers **96** bearing disialic acids

# 3 Summary of the First Part

The systematic evaluation of double carbonyl-protected sialyl donors demonstrated their clear advantages over conventional donors in terms of both reactivity and α-selectivity. N-acetyl and N-glycolyl derivatives afforded α(2,8)-linked disialosides and higher oligomers in consistently high yields with complete stereocontrol, whereas traditional donors such as triacetyl or 5*N*,4O-carbonyl variants suffered from diminished selectivity or reactivity. Importantly, the new protecting group strategy not only enhanced stereoselectivity but also facilitated efficient global deprotection, enabling access to structurally well-defined α(2,8)-sialic acid oligomers. Furthermore, the methodology was successfully extended to the first chemical synthesis of polysialic acid, underscoring its utility as a robust platform for oligosialic acid assembly.

# 4 Part II: Functional Studies Using Synthetic Sialic Acid Oligomers: Siglec-7 Recognition

## *4.1 Sialic Acid Recognizing Immunoglobulin-Like Lectins, Siglecs*

Sialic acid is closely associated with immune and nervous systems. A striking feature of sialic acid in the immune system is its role in identifying cells as "self." Interactions between sialic acid and sialic acid-binding immunoglobulin-like lectins (Siglecs) on immune cells are used to distinguish between self and non-self cells (Fig. 13) [19].

The discriminating system for non-self cells is also lectin, especially recognition at the first moment, and lectins include dectin, mincle, and so on [20]. They recognize Man-, Glc(NAc)-, and unusual glycans in mammals, and exclude non-self cells.

Siglecs have four domains: an N-terminal V-set domain, one to sixteen C2-set Ig domains, a transmembrane domain, and a cytosolic region (Fig. 14a).

In the cytosolic region, immunoreceptor tyrosine-based inhibition motif (ITIM) or ITIM-like motifs for inhibition are usually present [21]. The V-set domain is

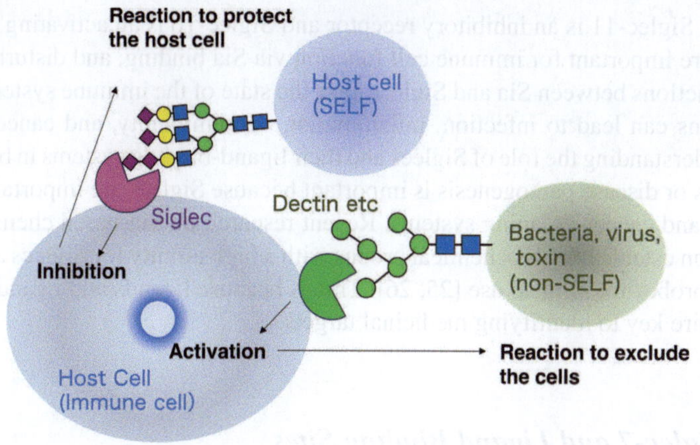


**Fig. 13** Recognition mechanism of lectins by host immune cell. Siglec (Purple) recognizes sialic acid on the host cell as a self tag and inhibits the activation of the immune cells. On the other hands, the non-self tag such as mannose-containing glycans are recognized by Dectin or other lectins (Green), leading to the excluding by activation of the immune cells

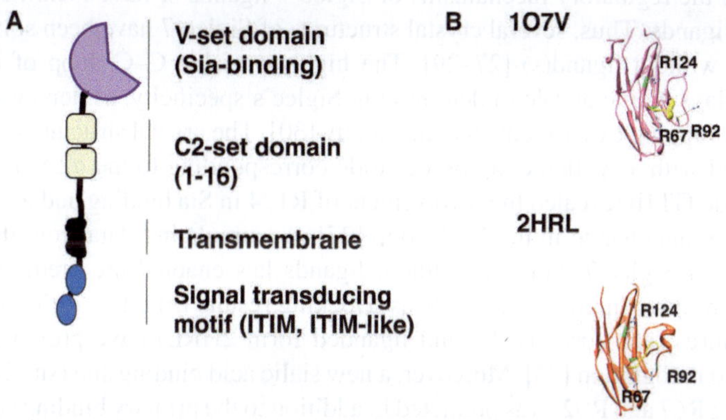


**Fig. 14** Structure of Siglec. **a** Schematic structure of Siglec. **b** Crystal structure of Siglec-7

important for binding to sialic acid (Sia). Furthermore, the arginine (R) residue on the F-strand of the V-set domain of Siglec is the critical sialic acid (Sia)-binding site [21, 22] (Fig. 14b; 1O7V and 2HRL), and the primary binding site is named site 1 [23]. The C2-set domain is characterized by the numbers 1–16. It is considered to function in oligomerization or display the V-set domain in the appropriate region. Almost all Siglecs have transmembrane and cytosolic regions with an ITIM and/or ITIM-like motifs that regulate immune signal inhibition. Several Siglecs do not contain cytosolic regions, such as Siglec-16, but they have a special R at the membrane region to pull the DAP protein that regulates signal activation [24]. Interestingly, paired Siglecs such as Siglec-11 and -16, have the same V-set domain sequences;

however, Siglec-11 is an inhibitory receptor and Siglec-16 is an activating receptor. Siglecs are important for immune cell function via Sia binding, and disturbances in the interactions between Sia and Siglecs alter the state of the immune system. These disruptions can lead to infection, inflammation, autoimmunity, and cancer escape [21]. Understanding the role of Siglecs and their ligand-binding systems in biological processes or disease pathogenesis is important because Siglecs are important for the immune and cancer-escaping systems. Recent research on Sia-based chemistry has focused on establishing biochemical probes with a high affinity for Siglecs as well as specific probes for clinical use [25, 26]. This is because ligands and ligand-binding systems are key to identifying medicinal targets.

## 4.2 Siglec-7 and Ligand Binding Sites

Siglec-7 (CDw328) belongs to the human CD33-related Siglec family [19]. Siglec-7 is known to localize to natural killer (NK) cells and monocytes, where it functions as an inhibitory receptor. Although Siglec-7 is a well-studied molecule, few reports have addressed the regulatory mechanisms of Siglec-7 ligands or have identified natural Siglec-7 ligands. Thus, several crystal structures of Siglec-7 have been solved, both with and without ligand(s) [27–29]. The highly variable C–C′ loop of the V-set domain plays a crucial role in determining Siglec's specificity, as demonstrated by domain-swapping experiments biochemically [30]. The crystal structure of Siglec-7 complexed with a synthetic oligosaccharide corresponding to the α2,8-disialylated ganglioside GT1b revealed the involvement of R124 in Sia binding and a significant conformational change in the C–C′ loop [27]. Accumulating data from the crystal structures of Siglec-7 with and without ligands has enabled the prediction of its ligands through computational analysis. Docking results from two different Siglec-7V structures (apo form 1O7V and liganded form 2HRL) have provided insight into ligand recognition [23]. Moreover, a new sialic acid binding site (site 2, a region containing R67 and R92) was predicted in addition to the primary binding site (site 1, a region containing R124) using in silico analysis. Interestingly, ligand-interaction experiments revealed the presence of a new binding site. It is important to prove the new binding site by biochemical analyses because in silico analyses are just prediction tools and in vitro analyses, such as equilibrium dialysis, mutation analyses, and inhibition analyses by double reciprocal plots using several chemically synthetic ligands for Siglec-7 tell the true story [18, 23, 31]. Interestingly, another arginine residue (R94, site 2-3, Fig. 14) is conserved among all Siglecs [32]. Mutual regulation of sialic acid-binding regions (sites 1 and 2) may be the key mechanism underlying the complexity of these ligands.

### 4.3 *Synthetic Glycan Ligands for Siglec-7*

The ligand presence of Siglec-7 was shown by the interaction between Siglec-7-expressing COS-1 cells and erythrocytes [33] and it was shown to have a Siglec-7 ligand on erythrocytes. However, the cell surface is typically covered with a complex glycocalyx, which is difficult to analyze and poorly understood due to its complex nature. Therefore, chemically defined glycans are typically used [24]. Biotinylated polyacrylamide (PAA) glycoconjugates containing Neu5Acα2,3Galβ1,4Glc, Neu5Acα2,6Galβ1,4Glc, and Galβ1,4Glc were firstly used as synthetic ligands. The binding specificity of Siglec-7 has been analyzed using an array system containing various peripheral structures. In 2002, ganglioside glycans were used for binding analysis [34]. The oligosaccharide was attached to bovine serum albumin to mimic the multivalent ganglioside glycans. Interestingly, Siglec-9, similar molecule to Siglec-7, did not bind to these multivalent Sia-probes, suggesting that multivalency is a feature of Siglec-7 binding. In an inhibition assay, free GD3 oligosaccharides inhibited the binding of Siglec-7 to gangliosides with $IC_{50}$ values of 5–10 mM and monovalent binding affinity was not good. The binding of GD3, GT1b, and GQ1b was demonstrated using Siglec-7-expressing CHO cells and Siglec-7 itself. Newly synthesized oligosialic acid (oligoSia)-containing glycolipids (oligoSia-PEs) were also analyzed for their ability to interact with Siglec-7. Di-, tri-, and tetra-sialic acid (Sia)-containing phosphatidylethanolamine (PE) were found to interact with Siglec-7 even before sialidase treatment [23].

### 4.4 *Multivalency of Siglec-7 Ligand Using Synthetic Chemical Compounds*

Multivalency as a glycan ligand is important when considering ligand-receptor interactions. In the case of Siglec-7, Sia clustering effects using α2,3-linked Neu5Ac-PAA and α2,6-linked Neu5Ac-PAA have been shown [24]. The effectiveness of the clustering effect on the delivery system was analyzed using ligands coated with liposomal nanoparticles in Siglec-7-expressing cells. DiSia-coated nanoparticles could bind to mouse Siglec-E, an orthologues of Siglec-7 and inhibit inflammation in human monocytes and macrophages [35]. The IC50 value of the GD3 epitope for the GD3-Siglec-7 interaction was 5–10 mM as described above. We first measured the affinity of the synthetic diSia structure for Siglec-7 [18] who showed that the IC50 values of fluorescently labeled poly-diSia (degree of polymerization [DP] = 20), fluorescently labeled poly-diSia (DP = 50), and diSia were 10 mM, 3 mM, and greater than 1 mM, respectively (Fig. 15). This was the first quantitative analysis to demonstrate that diSia multivalency effectively changed affinity. After the development of dextran-based new glycopolymers (Dex), the interaction ($K_D$) between diSia-Dex and Siglec-7 was measured by biolayer interference methods and found to be $6.0 \times 10^{-10}$ M [12] (Fig. 3). When calculated as the diSia unit, the mono-diSia

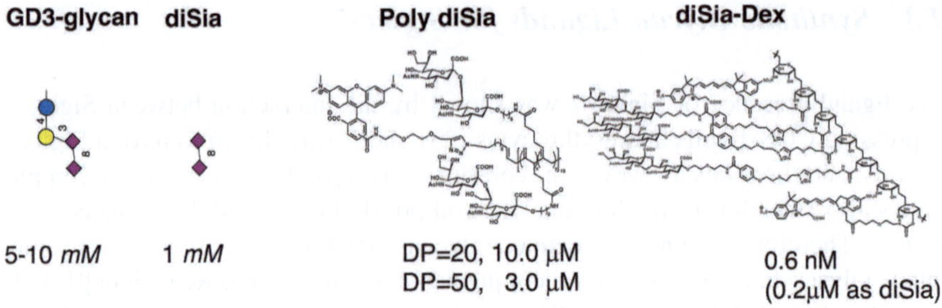


**Fig. 15** Binding affinity between Siglec-7 and ligands. The binding affinity between Siglec-7 and monovalent ligands or multivalent ligands were shown

contribution was almost $2 \times 10^{-7}$ M. The affinity between diSia-Gal and Siglec-7 was $2 \times 10^{-6}$ M when measured using the equilibrium dialysis method [23]. Based on these comprehensive kinetic values, the multivalent effect on the affinity was approximately tenfold.

This observation suggests that the multivalency effects are synergistic. We also analyzed the inhibitory effect of polySia on the diSia–Siglec-7 interaction and found that the $IC_{50}$ of diSia-Dex was 1 nM [31] and it is the same as $K_D$. Next, we considered the multivalency of the tandem diSia. PolySia consists of Sia (colominic acid) with a DP of ~ 40 and can be considered a polymer of diSia (DP = 20). In addition, the non-reducing terminal ends of polySia and diSia may be identical. However, colominic acid did not inhibit the Siglec-7–GD3 interaction, indicating that the diSia epitope on polySia, even at the non-reducing terminal end of colominic acid (polySia), was different from the glycotope of the diSia structure. In addition, the inhibitory effects of acidic glycopolymers, such as chondroitin, hyaluronan, heparan sulfate, chondroitin sulfate-6S, chondroitin sulfate-4S, keratan sulfate, and dermatan sulfate, were not observed, indicating that any negative charge, except for diSia, is not the cause of the interaction in the case of Siglec-7 [24]. In addition, other neutral glycopolymers such as mannose polymer (mannan), glucose polymer (glucan), and GlcN polymer (chitosan) do not bind to or inhibit Siglec-7 [24].

## 4.5 *Natural Ligands for Siglec-7*

Disialylgalactosylgloboside (DSGG), IV3Neu5AcIII6Neu5AcLc4 (DSLc4), and IV4GalNAcIV3Neu5AcIII6Neu5AcLc4 (GalNAcDSLc4), which are associated with lung metastasis, were initially identified in renal cell carcinoma [36]. From the perspective of the genes responsible for synthesizing the diSia epitope, ST8SIA1 has been identified [33] and GD3 has been shown to be a ligand for Siglec-7. In 2004, gene analysis identified diSiaLe$^a$ as a natural Siglec-7 ligand [37], although the authors did not purify the glycoconjugates; therefore, whether the epitope is linked to glycolipids or glycoproteins remains unknown.

The natural ligands between NK and K562 cells were partially analyzed. NK cells are typically masked by surface-expressed Sia-containing glycoconjugates; however, the underlying mechanisms remain unclear. After unmasking, Siglec-7 can bind to tumor cells (K562 cells) as an inhibitory receptor. In 2021, proximity-labeled molecules in K562 cells were affinity-purified using a Siglec-7 column. The specifically bound molecules were then eluted using a high-affinity multivalent ligand, diSia-Dex. Following mass spectrometry analysis, the natural ligand of K562 cells was identified as sialophorin (SPN) [38]. At the same timing, genetically engineered cell-based experiments revealed that SPN is a Siglec-7 ligand [39]. Ganglioside ligand expression in K562 cells was simultaneously analyzed using Siglec-7 blotting and thin-layer chromatography [32]. No binding was observed despite the detection of the diSia epitope-containing ganglioside GD3, although GD3 was identified as a Siglec-7 ligand using glycan arrays and enzyme-linked immunosorbent assays. These data suggest that the ganglioside on K562 cells may not be a Siglec-7 ligand and that SPN is a trans-natural ligand. During self-cell surveillance, Siglec-7 on NK cells masked with GD3 on NK cells as a cis-ligand binds to the cells to be surveyed (Fig. 16). At this point, a clustering effect was required. SPN, also known as CD43, is a type I mucin-type glycoprotein composed of a signal peptide, an extracellular 255-amino-acid region with 80 *O*-glycans, and an *N*-glycan near the transmembrane region (N239). The *O*-glycans on SPN showed a clustering effect as trans-ligands (Fig. 16).

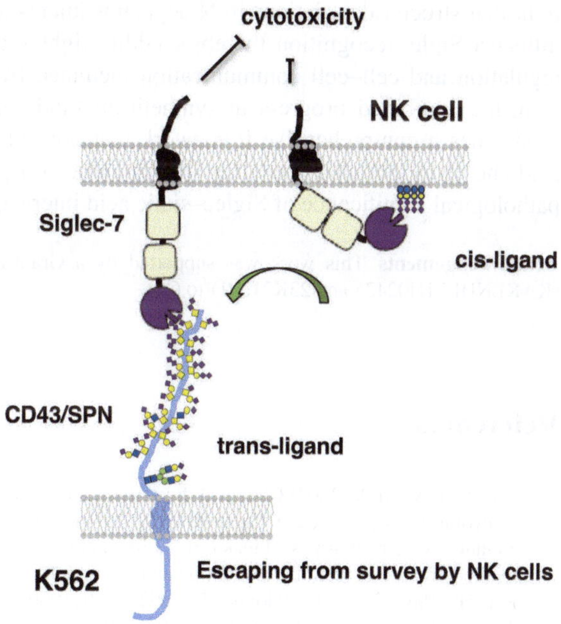


**Fig. 16 Cis- and trans-ligand for Siglec-7.** Siglec-7 on NK cells is normally masked via cis-ligands of GD3 synthesized by ST8Sia1. GD3 is clustered in the raft fraction. Sialophorin or CD43 on K562 inhibited the cytotoxicity of NK cells by binding to Siglec-7. The diSia clustering effect on CD 43 is greater than that of GD3 in raft, indicating that the Siglec-7 binds to K562. Although K562 is cancer cells, it escapes from the NK cell's surveillance. Site 1 and site 2 in Siglec-7 are important for the cis- and trans- recognition system

## 5 Summary of the Second Portion

Multivalency is a key feature for understanding both natural ligands and drug targeting. Under natural conditions, the ligands on glycoproteins and glycolipids differ. For glycoproteins, the natural ligand is localized on the *O*-linked glycan of mucin-type glycoproteins, such as CD43/SPN. For glycolipids, especially GD3, the ceramide moiety is important for localization and raft formation should be considered because of clustering. Binding parameters and clustering ligands were evaluated using valuable synthetic glycopolymers. Another point to consider is the presence of at least two Sia-binding sites in Siglec-7. Attempts to overcome this affinity have been successful using chemical synthesis, and materials have been used in some cases for clinical therapy [40]. Understand the underlying mechanisms of ligand binding, which may lead to new applications in disease treatment is important.

## 6 Future Perspective

The synthetic advances achieved with double carbonyl-protected donors provided the foundation for biological studies on sialic acid–Siglec interactions. By supplying structurally defined sialyl oligomers and polysialic acids, this methodology enabled precise evaluation of their binding properties and functional roles. The results highlight that structural variations in N-acyl substituents and oligomer length critically influence Siglec recognition, thereby shedding light on the molecular basis of immune regulation and cell–cell communication mediated by polysialic acids. In conclusion, the combined progress in synthetic methodology and biological evaluation establishes a comprehensive framework that not only advances the field of sialic acid chemistry but also opens new opportunities for probing the physiological and pathological significance of Siglec–sialic acid interactions.

**Acknowledgements** This work was supported by a Grant-in-Aid for Scientific Research(B) (KAKENHI 21H02425 and 23K21291)(to CS).

## References

1. Angata T, Varki A (2002) Chemical diversity in the sialic acids and related alpha-keto acids: an evolutionary perspective. Chem Rev 102:439–469. https://doi.org/10.1021/cr000407m
2. Vibhute AM, Komura N, Tanaka H-N, Imamura A, Ando H (2021) Advanced chemical methods for stereoselective sialylation and their applications in sialoglycan syntheses. Chem Rec 21:3194–3223. https://doi.org/10.1002/tcr.202100080
3. Tanaka H, Nishiura Y, Takahashi T (2006) Stereoselective synthesis of oligo-α-(2,8)-sialic acids. J Am Chem Soc 128:7124–7125. https://doi.org/10.1021/ja0613613
4. Tanaka H, Nishiura Y, Takahashi T (2008) An efficient convergent synthesis of GP1c ganglioside epitope. J Am Chem Soc 130:17244–17245. https://doi.org/10.1021/ja0807821t

5. Tanaka H, Tateno Y, Nishiura Y, Takahashi T (2009) Efficient synthesis of an α(2,9) trisialic acid by one-pot glycosylation and polymer-assisted deprotection. Org Lett 10:5597–5600. https://doi.org/10.1021/ol802207e
6. Tanaka H, Nishiura Y, Takahashi T (2009) Stereoselective synthesis of α(2,9) di- to tetrasialic acids, using a 5,4-*N,O*-carbonyl protected thiosialoside. J Org Chem 74:4383–4386. https://doi.org/10.1021/jo900176e
7. Crich D, Li W (2007) *O*-sialylation with *N*-acetyl-5-*N*,4-*O*-carbonyl-protected thiosialoside donors in dichloromethane: facile and selective cleavage of the oxazolidinone ring. J Org Chem 72:2387–2391. https://doi.org/10.1021/jo062431r
8. Crich D, Li W (2007) α-Selective sialylations at -78 °C in nitrile solvents with a 1-adamantanyl thiosialoside. J Org Chem 72:7794–7797. https://doi.org/10.1021/jo7012912
9. Crich D, Wu B (2008) One-pot synthesis of *N*-glycolylneuraminic acid-containing oligosaccharides. Org Lett 10:4033–4035. https://doi.org/10.1021/jo801548k
10. Tsvetkov YE (2005) Nifantev NE (2005) Enhanced sialylating activity of *O*-chloroacetylated 2-thioethyl sialosides. Synlett 9:1375–1380. https://doi.org/10.1021/s-2005-868514
11. Premathilake HD, Gobble CP, Pornsuriyasak P, Hardimon T, Demchenko AV, De Meo C (2012) How *O*-substitution of sialyl donors affects their stereoselectivity. Org Lett 14:1126–1129. https://doi.org/10.1002/ol3000475
12. Aoyagi T, Ohira S, Fuse S, Uzawa J, Yamaguchi Y, Tanaka H (2016) The α-glycosidation of partially unprotected *N*-acetyl and *N*-glycolyl sialyl donors in the absence of a nitrile solvent effect. Chem Eur J 22:6968–6973. https://doi.org/10.1002/chem.201601031
13. Koinuma R, Tohda K, Aoyagi T, Tanaka H (2020) Chemical synthesis of α(2,8) octasialosides, the minimum structure of polysialic acids. Chem Commun 56:12981–12984. https://doi.org/10.1039/d0cc05901c
14. Takeuchi Y, Tohda K, Tanaka H (2024) Syntheses of α(2,8) Sialosides containing NeuAc and NeuGc by using double carbonyl-protected *N*-acyl sialyl donors. Chem Eur J 30:e202400883. https://doi.org/10.1039/chem.202400883
15. Ladmiral V, Melia E, Haddleton DM (2004) Synthetic glycopolymers: an overview. Eur Polym J 40:431–449. https://doi.org/10.1016/j.eurpolymj.2003.10.019
16. Tomita, Y. Kondo Y, Takagi K, Endo T (1994) A novel living coordination polymerization of methoxyallene by π-allylnickel catalyst. Macromolecules 27:4413–4414. https://doi.org/10.1021/ma00093a052
17. Wang J, Tomita I, Endo T (2001) Synthesis of well-defined glycopolymers by π-allylnickel-catalyzed living coordination polymerization. Macromolecules 34:4294–4295. https://doi.org/10.1021/ma001344c
18. Ohira S, Yasuda Y, Tomita I, Kitajima K, Takahashi T, Sato C, Tanaka H (2017) Synthesis of end-functionalized glycopolymers containing α(2,8) disialic acids via π-allyl nickel catalyzed coordinating polymerization and their interaction with Siglec-7. Chem Commun 53:553–556. https://doi.org/10.1039/c6cc07115e
19. Angata T, Varki A (2023) Discovery, classification, evolution and diversity of Siglecs. Mol Aspects Med 90:101117. https://doi.org/10.1016/j.mam.2022.101117
20. Watanabe M, Motooka D, Yamasaki S (2023) The kinetics of signaling through the common FcRγ chain determine cytokine profiles in dendritic cells. Sci Signal 16:eabn9909. https://doi.org/10.1126/scisignal.abn9909
21. Crocker PR, Paulson JC, Varki A (2007) Siglecs and their roles in the immune system. Nat Rev Immunol 7:255–266. https://doi.org/10.1038/nri2056
22. Alphey MS, Attrill H, Crocker PR, van Aalten DM (2003) High resolution crystal structures of Siglec-7. J Biol Chem 278:3372–3377. https://doi.org/10.1074/jbc.M210602200
23. Yamakawa N, Yasuda Y, Yoshimura A, Goshima A, Crocker PR, Vergoten G, Nishiura Y, Takahashi T, Hanashima S, Matsumoto K, Yamaguchi Y, Tanaka H, Kitajima K, Sato C (2020) Discovery of a new sialic acid binding region that regulates Siglec-7. Sci Rep 10:8647. https://doi.org/10.1038/s41598-020-64887-4
24. Morishita S, Hane M, Kitajima K, Sato C (2023) Substantial basis for glyco-assembly: Siglec7 and synthetic sialylpolymers. In: Furukawa K, Fukuda M (eds) Glycosignals in cancer. Springer, Singapore. https://doi.org/10.1007/978-981-19-7732-9_3.

25. Büll C, Collado-Camps E, Kers-Rebel ED, Heise T, Søndergaard JN, den Brok MH, Schulte BM, Boltje TJ, Adema GJ (2017) Metabolic sialic acid blockade lowers the activation threshold of moDCs for TLR stimulation. Immunol Cell Biol 95:408–415. https://doi.org/10.1038/icb.2016.105
26. Hudak JE, Canham SM, Bertozzi CR (2014) Glycocalyx engineering reveals a Siglec-based mechanism for NK cell immunoevasion. Nat Chem Biol 10:69–75. https://doi.org/10.1038/nchembio.1388
27. Alphey M, Attrill H, Crocker P, van Aalten D (2003) High resolution crystal structures of Siglec-7. Insights into ligand specificity in the Siglec family. J Biol Chem 278:3372–3377. https://doi.org/10.1074/jbc.M210602200
28. Attrill H, Takazawa H, Witt S, Kelm S, Isecke R, Brossmer R, Ando T, Ishida H, Kiso M, Crocker PR, van Aalten DM (2006) The structure of siglec-7 in complex with sialosides: leads for rational structure-based inhibitor design. Biochem J 397:271–278. https://doi.org/10.1042/BJ20060103
29. Dimasi N, Moretta L, Biassoni R, Mariuzza RA (2003) Expression, crystallization and preliminary crystallographic analysis of the extracellular IgV-like domain of the human natural killer cell inhibitory receptor p75/AIRM1. Acta Crystallogr D Biol Crystallogr 59:1856–1858. https://doi.org/10.1107/s0907444903018146
30. Yamaji T, Mitsuki M, Teranishi T, Hashimoto Y (2005) Characterization of inhibitory signaling motifs of the natural killer cell receptor Siglec-7: attenuated recruitment of phosphatases by the receptor is attributed to two amino acids in the motifs. Glycobiology 15:667–676. https://doi.org/10.1093/glycob/cwi048
31. Yamaguchi S, Yoshimura A, Yasuda Y, Mori A, Tanaka H, Takahashi T, Kitajima K, Sato C (2017) Chemical synthesis and evaluation of a disialic acid-containing dextran polymer as an inhibitor for the interaction between Siglec 7 and its ligand. ChemBioChem 18:1194–1203. https://doi.org/10.1002/cbic.201600694
32. Yoshimura A, Hatanaka R, Tanaka H, Kitajima K, Sato C (2021) The conserved arginine residue in all Siglecs is essential for Siglec-7 binding to sialic acid. Biochem Biophys Res Commun 534:1069–1075. https://doi.org/10.1016/j.bbrc.2020.10.023
33. Nicoll G, Avril T, Lock K, Furukawa K, Bovin N, Crocker P (2003) Ganglioside GD3 expression on target cells can modulate NK cell cytotoxicity via siglec-7-dependent and -independent mechanisms. Eur J Immunol 33:1642–1648. https://doi.org/10.1002/eji.200323693
34. Yamaji T, Teranishi T, Alphey MS, Crocker PR, Hashimoto Y (2002) A small region of the natural killer cell receptor, Siglec-7, is responsible for its preferred binding to alpha 2,8-disialyl and branched alpha 2,6-sialyl residues. A comparison with Siglec-9. J Biol Chem 277:6324–6332. https://doi.org/10.1074/jbc.M110146200
35. Spence S, Greene MK, Fay F, Hams E, Saunders SP, Hamid U, Fitzgerald M, Beck J, Bains BK, Smyth P, Themistou E, Small DM, Schmid D, O'Kane CM, Fitzgerald DC, Abdelghany SM, Johnston JA, Fallon PG, Burrows JF, McAuley DF, Kissenpfennig A, Scott CJ (2015) Targeting Siglecs with a sialic acid-decorated nanoparticle abrogates inflammation. Sci Transl Med 7:303ra140. https://doi.org/10.1126/scitranslmed.aab3459.
36. Ito A, Handa K, Withers DA, Satoh M, Hakomori S (2001) Binding specificity of siglec7 to disialogangliosides of renal cell carcinoma: possible role of disialogangliosides in tumor progression. FEBS Lett 504:82–86. https://doi.org/10.1016/s0014-5793(01)02476-0
37. Miyazaki K, Ohmori K, Izawa M, Koike T, Kumamoto K, Furukawa K, Ando T, Kiso M, Yamaji T, Hashimoto Y, Suzuki A, Yoshida A, Takeuchi M, Kannagi R (2004) Loss of disialyl Lewis(a), the ligand for lymphocyte inhibitory receptor sialic acid-binding immunoglobulin-like lectin-7 (Siglec-7) associated with increased sialyl Lewis(a) expression on human colon cancers. Cancer Res 64:4498–4505. https://doi.org/10.1158/0008-5472.CAN-03-3614
38. Yoshimura A, Asahina Y, Chang LY, Angata T, Tanaka H, Kitajima K, Sato C (2021) Identification and functional characterization of a Siglec-7 counter-receptor on K562 cells. J Biol Chem 296:100477. https://doi.org/10.1016/j.jbc.2021.100477
39. Wisnovsky S, Möckl L, Malaker SA, Pedram K, Hess GT, Riley NM, Gray MA, Smith BAH, Bassik MC, Moerner WE, Bertozzi CR (2021) Genome-wide CRISPR screens reveal a specific

ligand for the glycan-binding immune checkpoint receptor Siglec-7. Proc Natl Acad Sci USA 118:e2015024118. https://doi.org/10.1073/pnas.2015024118
40. Xiao H, Woods EC, Vukojicic P, Bertozzi CR (2016) Precision glycocalyx editing as a strategy for cancer immunotherapy. Proc Natl Acad Sci USA 113:10304–10309. https://doi.org/10.1073/pnas.1608069113

# Chemical Biology of Core M3 *O*-Mannosyl Glycan

**Jun-ichi Tamura, Kota Kotera, Haruto Taguchi, and Takahiro Tamura**

**Abstract** Facile structure of the core M3 *O*-mannosyl glycan has recently been elucidated. The glycan stabilizes muscle tissue by binding to dystroglycan and laminin at the reducing and non-reducing end, respectively. We succeeded to chemically synthesize pivotal parts of the core M3 *O*-mannosyl glycan including tandem ribitol phosphate and matriglycan. The former acted as primer of the enzymatic glycan elongation. The latter exhibited superior affinity for laminin. Our synthesized glycans may serve as medicine to stabilize the muscle tissue.

**Keywords** Core M3 *O*-mannosyl glycan · Matriglycan · Muscular dystrophy · Tandem ribitol phosphate · Glycosylation

## 1 Structure of Core M3 *O*-Mannosyl Glycan and the Biological Synthesis

The physiological and biochemical functions of glycans are actively involved in numerous cellular and systemic pathways, and the glycans also play a role in stabilizing muscle fibers by mediating the binding of proteins. Dystroglycan (DG) is a glycoprotein identified in skeletal muscle as a component of the dystrophin-glycoprotein complex [1]. Myofibrils constitute the basic building blocks of the detailed organization in muscle fibers, and the cell membranes are covered by a basement membrane composed of extracellular matrix (ECM) components such as type IV collagen. Figure 1 shows an enlarged view of the muscle cell membrane and the basement membrane. DG is a transmembrane glycoprotein that spans the muscle cell membrane and consists of two subunits: α and β [2]. α-DG is glycosylated, while β-DG is a transmembrane subunit that binds to dystrophin in the cell [3, 4]. DG binds

J. Tamura (✉) · K. Kotera · T. Tamura
United Graduate School of Agricultural Science, Tottori University, Tottori, Japan
e-mail: jtamura@tottori-u.ac.jp

J. Tamura · H. Taguchi
Graduate School of Sustainable Sciences, Tottori University, Tottori, Japan

 
K. Toshima and Y. Fujimoto (eds.), *Chemical Biology of Carbohydrates*,
https://doi.org/10.1007/978-981-95-5335-8_4

to laminin, an ECM component, with the glycans, playing a role in anchoring the basement membrane to the cytoskeleton and stabilizing muscle fibers.

Endo and his co-workers discovered *O*-mannosyl glycan (*O*-MG) having Siaα2-3Galβ1-4GlcNAcβ1-2Manα sequence, the first *O*-MG identified in mammals, on α-DG [5]. In 2016, Kanagawa et al. identified a tandem ribitol phosphate consisting of two consecutive D-ribitol-5-phosphate (Rbo5P) units, revealing the complete structure and biosynthetic pathway of core M3 *O*-MG. The tandem ribitol phosphate was discovered in mammalian at the first time [6]. To date, 18 types of glycosyltransferases have been identified as being involved in the biosynthesis of core M3 *O*-MG (Fig. 2).

First Rbo5P is added to the trisaccharide [6] at the reducing terminal by fukutin (FKTN). Subsequent transfer of the next RboP occurs by fukutin-related protein (FKRP) [6]. β-D-Xylose (Xyl) residue is first attached to *O*-4 of Rbo at the non-reducing side [7]. Then, β-D-glucuronic acid residue (GlcA) is added to *O*-4 of Xyl [8, 9]. The "matriglycan." which acts as the binding site for laminin, is formed by

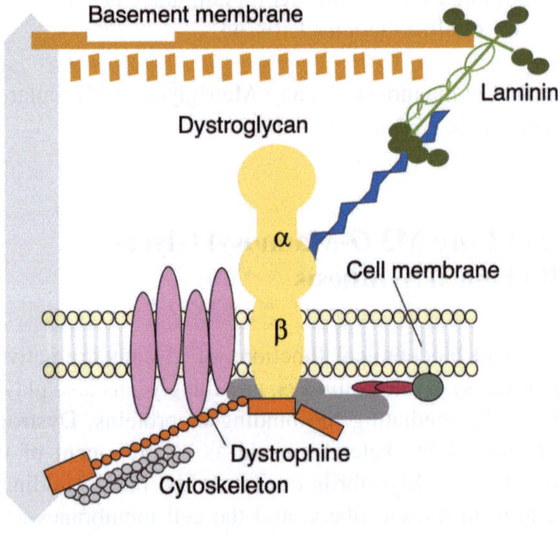


**Fig. 1** Enlarged view of muscle cell membrane and the basement membrane

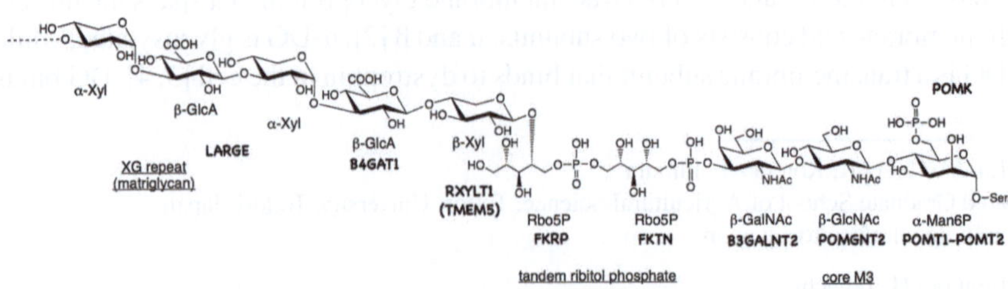


**Fig. 2** Glycan structure of core M3 *O*-MG and the corresponding transferases

the alternating addition of α-Xyl and β-GlcA by like-acetyl-glucosaminyltransferase (LARGE) [10]. The core M3 *O*-MG is responsible for the linkage between the ECM and cytoskeleton that stabilizes muscle tissue. The glycosyltransferases which form the core M3 *O*-MG exhibit extremely high substrate specificity, and their genetic disorder leads to impaired glycan elongation. As a result, muscle fibers become destabilized, leading to a collapse of skeletal muscle tissue and the onset of glycosylation-defective muscular dystrophies (α-dystroglycanopathies). For example, Fukuyama congenital muscular dystrophy (FCMD) occurs when FKTN, which transfers the first CDP-ribitol phosphate, is deficient. Core M3 *O*-MG containing matriglycan is considered a highly attractive target for drug development, and utilizing chemically synthesized matriglycan or partial structures of core M3 *O*-MG holds potential for improving symptoms of muscular dystrophies. While matriglycan, which interacts with laminin, is also enzymatically elongated by LARGE, this chapter summarizes recent studies on the total chemical synthesis of the glycan—which allows for precise structural control—as well as its function as a primer for glycan elongation and its affinity for laminin.

## 2 Synthesis of Matriglycan Repeating Disaccharide Unit (Xylα1-3GlcA)

Xyl is a minor sugar residue in mammals and has been identified at the non-reducing terminus of the matriglycan in core M3 *O*-MG, as well as in the xyloglucosyl trisaccharide (Xylα1-3Xylα1-3Glcβ1-*O*-Ser) found on serine residues of EGF-like domains in blood coagulation proteins [11, 12], and in the linkage region tetrasaccharide (GlcAβ1-3Galβ1-3Galβ1-4Xyl-*O*-Ser) attached to the core proteins of proteoglycans [13, 14]. In contrast, plants contain large amounts of Xyl, primarily as components of xylan, xyloglucan, and xylogalacturonan.

Since xylopyranose lacks a *C*-6 carbon and readily undergoes ring inversion, it is difficult to achieve 1,2-*cis* glycoside formation based on the anomeric effect during glycan synthesis. In the glycosylation reactions using thioglycosides or halo sugars protected with Bn groups at the 2-, 3-, and 4-hydroxyl positions as donors, there are few examples demonstrating good stereoselectivity [15–18]. There are also examples in which glycoside donors, protected with Bn groups at the 2- and 4-hydroxyl positions and an Ac group at the 3-position to promote remote participation by the 3-OAc group, were used [19, 20].

When xylosyl imidates are synthesized, the Xyl residues are sometimes isolated as conformational isomers; C1 and 1C forms. However, the product after the glycosylation showed only the C1 conformation [21]. In the synthesis of the tandem ribitol phosphate moiety described later, the attached Xyl residue also frequently undergoes ring inversion [22]. Thus, the xylopyranosyl ring readily flips the ring due to the influence of protecting groups on the sugar residue.

To chemically synthesize the matriglycan with a defined chain length, a repeating disaccharide unit is required. In the synthesis, the order of Xyl (X) and GlcA (G) in the disaccharide unit is important. We adopted the glycan-sequence as "X–G" for the following two reasons: (1) difficult 1,2-*cis* glycosidic bond formation can be addressed in the initial stage of the disaccharide synthesis; (2) the formation of β-GlcA can be fully stereochemically controlled by the neighboring group effect during oligomerization. Regarding the formation of 1,2-*cis*-α-glycosides, Demchenko and Kan have each carried out the syntheses using a xylosyl donor bearing a 4-*O*-picolyl group [23, 24]. On the other hand, the intramolecular aglycon delivery (IAD) reaction developed by Ito is superior in terms of complete stereocontrol [25]. Our group fixed the pyranose ring conformation by acetal ring formation at 3,4-hydroxyl groups of the sugar donor (**3α, 3β**) using 2,3-butanedione. We then coupled the donor with the sugar acceptor **5** using DDQ to form the mixed acetal **6**. Intramolecular glycosylation was carried out using NIS and AgOTf to obtain **7**, which was subsequently hydrolyzed with TFA to give the disaccharide triol **8α** regio- and stereo-selectively. As a result, we obtained **9** as a pentaacetate of **8α** from **5** in 4 steps with an overall yield of 20% (Scheme 1). In order to improve the yield, we did not insist on the selective synthesis of the 1,2-*cis*-α-glycoside, but instead aimed to obtain it as a diastereomeric mixture favoring the α-anomer; the desired 1,2-*cis* glycoside can be obtained by subsequent separation of the isomers.

As a first step, we examined a coupling of fully benzylated thioglycosyl donor **14** and the acceptor **5**. The β/α ratio of **14** was 64:36, and we expected that the nucleophilic attack would occur from the side opposite to the leaving group when the reaction proceeded via an SN2 mechanism to give the preferred α-glycoside **16α**. However, the stereochemistry at the C1 position of the Xyl residue in the obtained product did not reflect that of the donor. The ratio of **16β/16α** were 83:17 and 82:18 at 0 and − 20 °C, respectively. This result suggests the formation of the intermediate **15** (carbocation) of which cation at the anomeric position may adopt more stable $^4H_3$ conformation (Scheme 2). Bn groups of **15** at positions 2 and 4 take diaxial, which hinders the nucleophilic attack from α-face and facilitates the attack from β-face, resulting in preferential formation of the β-glycoside. We concluded that fixing the ring in the C1 conformation is important to form the α-glycoside [26].

Glycosylation using donor **3αβ** in dichloromethane showed an α/β ratio of the products as 2:1 (Entries 1 and 2, Table 1). This indicates that the reaction also proceeded via an SN1 mechanism. On the other hand, 1,4-dioxane increased the α-ratio. Even when the donors with different α/β ratios were used, the product still showed an α/β ratio of 5:1, suggesting that the SN1 mechanism was operative in these cases as well (Entries 3 and 4, Table 1). This can be explained by the coordination of 1,4-dioxane to the β-face of the carbocation intermediate due to the reverse anomeric effect. This coordination facilitates nucleophilic attack from the α-face, resulting in a higher proportion of α-glycoside compared to when dichloromethane was used as the solvent.

However, the isomers of **17αβ** could not be separated to each other. We quantitatively removed the protecting group on xylose residue to give **8αβ** which was also inseparable by silica gel chromatography. Interestingly, these anomeric isomers

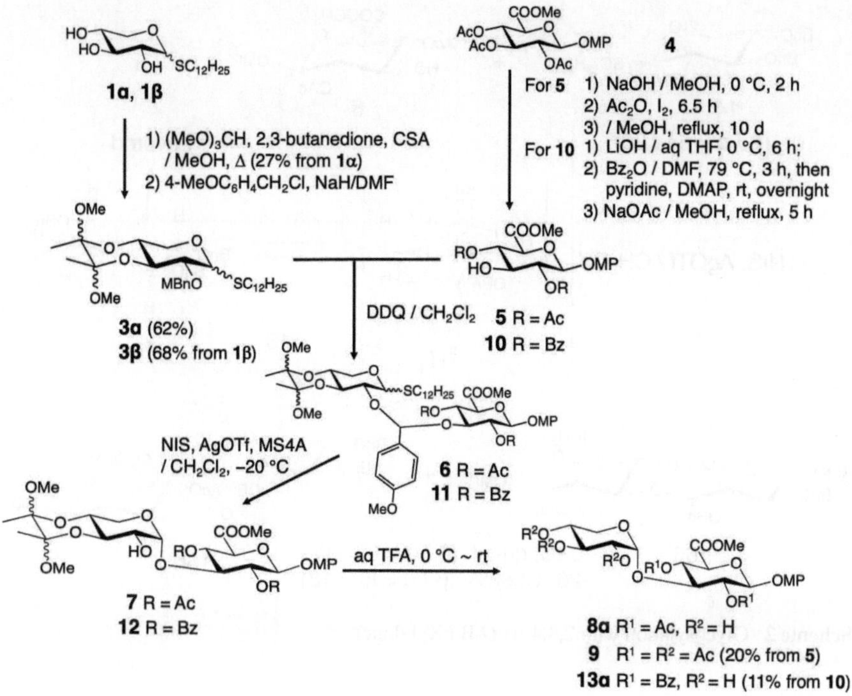


**Scheme 1** Synthesis of Xylα-GlcAβ unit via IAD method. Abbreviations: CSA, camphorsulfonic acid; DDQ, 2,3-dichloro-5,6-dicyano-1,4-benzoquinone; NIS, *N*-iodosuccinimide; MBn, 4-methoxybenzyl; MP, 4-methoxyphenyl; MS, molecular sieves; AgOTf, silver trifluoromethanesulfonate; TFA, trifluoroacetic acid

exhibited solvent-specific solubility (Scheme 3). Specifically, the α-anomer **8α** was completely insoluble in chloroform, whereas the β-anomer **8β** dissolved readily in chloroform. Utilizing this phenomenon, the desired α-anomer **8α** was almost quantitatively separated. Although the glycosylation did not exhibit complete stereoselectivity, this novel separation method enabled both a reduction in the number of steps and an improvement the yield [26]. Subsequently, triol **8α** was converted to **18** in 4 steps overall yield of 22%. Unfortunately, corresponding disaccharide donor and acceptor derived from **18** failed to undergo condensation with each other [27].

Meanwhile, the glycosylation was similarly carried out using acceptor **10**, in which the 2,4-hydroxy groups of GlcA were protected with benzoyl (Bz) groups (Scheme 4). The coupling reaction in dichloromethane gave a disaccharide with an α/β ratio of 1/2 and 45% yield, while the same coupling in toluene/1,4-dioxane (1:1) afforded the disaccharide **19αβ** with higher α-selectivity (**19α/β** = 2/1) and a better yield of 64%. Nevertheless, neither **19** nor the corresponding triol **13** allowed separations of their isomeric forms. Although both isomers dissolved in chloroform, we discovered that **13β** had significantly higher solubility in dichloromethane compared to **13α** [27].

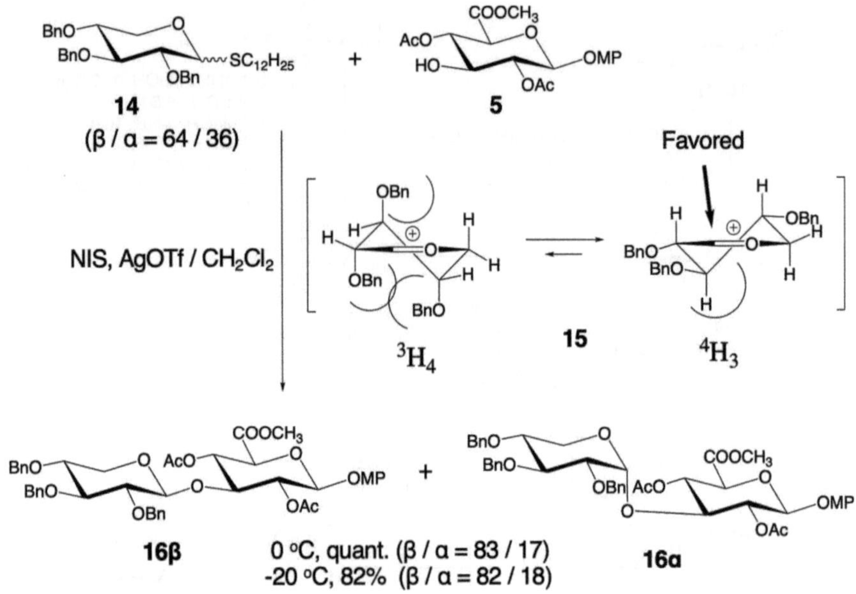


**Scheme 2** Glycosylation with 2,3,4-tri-*O*-Bn Xyl donor

**Table 1** Direct glycosylation with sterically fixed donor

| Entry | Donor (D/A) | Solvent | Time (h) | Yield of 17 (%) | α/β |
|---|---|---|---|---|---|
| 1 | **3β** (2/1) | $CH_2Cl_2$ | 2 | 55 | 2/1 |
| 2 | **3αβ** (α/β = 1/2) (2/1) | $CH_2Cl_2$ | 5 | 64 | 2/1 |
| 3 | **3β** (1.5/1) | Toluene: 1,4-dioxane = 1: 1 | 2 | 72 | 5/1 |
| 4 | **3α** (1.5/1) | Toluene: 1,4-dioxane = 1: 1 | 2 | 70 | 5/1 |

By taking advantage of this phenomenon, we successfully re-isolated the desired α-anomer selectively. Although a small amount of the α-anomer was slightly soluble in dichloromethane, this unique separation method enabled the isolation of the target α-anomer in 39% yield over two steps (compared to 11% yield over three steps by the IAD method), resulting in a reduction in the number of steps and an improvement in overall yield.

As shown in Scheme 5, the triol **13α** was subjected to the stannylene acetal method to afford the 2,4-bis-*O*-MBz derivative **20** in 64% yield over two steps. As by-products, 2,3,4-tris-*O*-MBz derivative **21** was obtained in 10% yield. The free hydroxyl group of **20** was quantitatively protected with a levulinoyl (Lev) group, affording the elongatable disaccharide donor **23** in 82% yield over two steps. The disaccharide donor **23** was coupled with the ethylene glycol linker in 93% yield, and

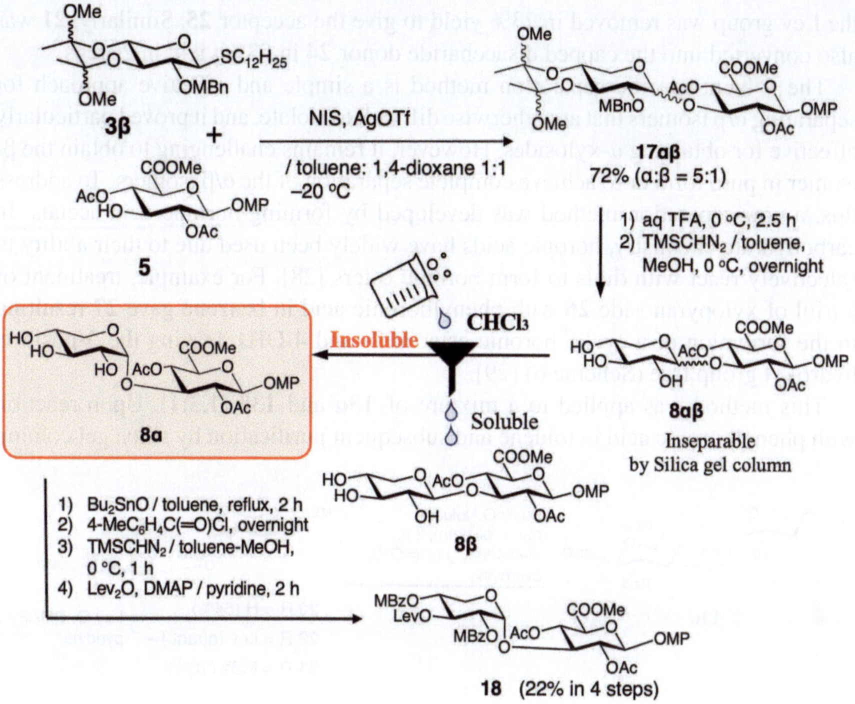


**Scheme 3** Synthesis of protected Xylα1-3GlcA (Ac type). Abbreviations: Lev, $CH_3COC_2H_4CO$-; MBz, 4-methylbenzoyl

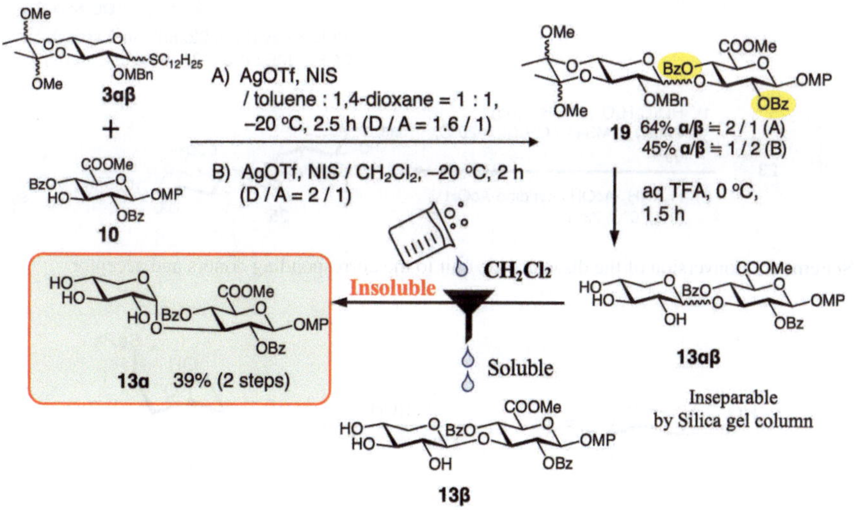


**Scheme 4** Synthesis of protected Xylα1-3GlcA (Bz type)

the Lev group was removed in 73% yield to give the acceptor **25**. Similarly, **21** was also converted into the capped disaccharide donor **24** in 58% yield in 2 steps.

The solvent-specific separation method is a simple and effective approach for separating α/β isomers that are otherwise difficult to isolate, and it proved particularly effective for obtaining α-xylosides. However, it remains challenging to obtain the β-isomer in pure form or to achieve complete separation of the α/β isomers. To address this, a new separation method was developed by forming boronic acid acetals. In carbohydrate chemistry, boronic acids have widely been used due to their ability to selectively react with diols to form boronic esters [28]. For example, treatment of a triol of xylopyranoside **26** with phenylboronic acid in benzene gave **27** resulting in the formation of a cyclic boronic ester at 2- and 4-OHs, leaving the 3-position hydroxyl group free (Scheme 6) [29].

This method was applied to a mixture of **13α** and **13β** (1.3:1). Upon reaction with phenylboronic acid in toluene and subsequent purification by silica gel column

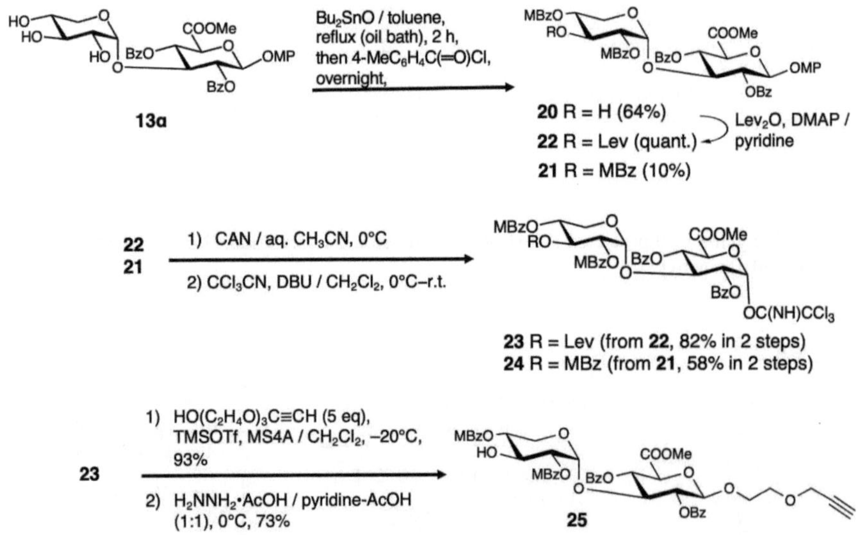


**Scheme 5** Conversion of the disaccharide unit to the corresponding donors and acceptor

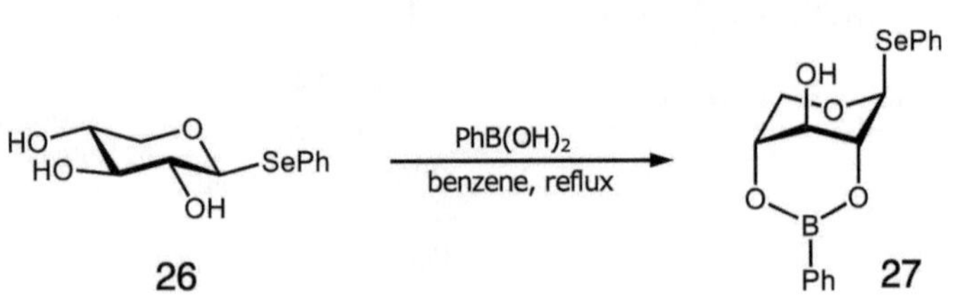


**Scheme 6** Cyclic boronic ester formation [29]

chromatography, **13α** was intriguingly recovered. In contrast, under the same conditions, the β-anomer formed a stable cyclic boronic ester and was isolated as **28β**. This method thus enabled complete separation of the diastereomeric mixture (**13α** and **13β**) (Scheme 7).

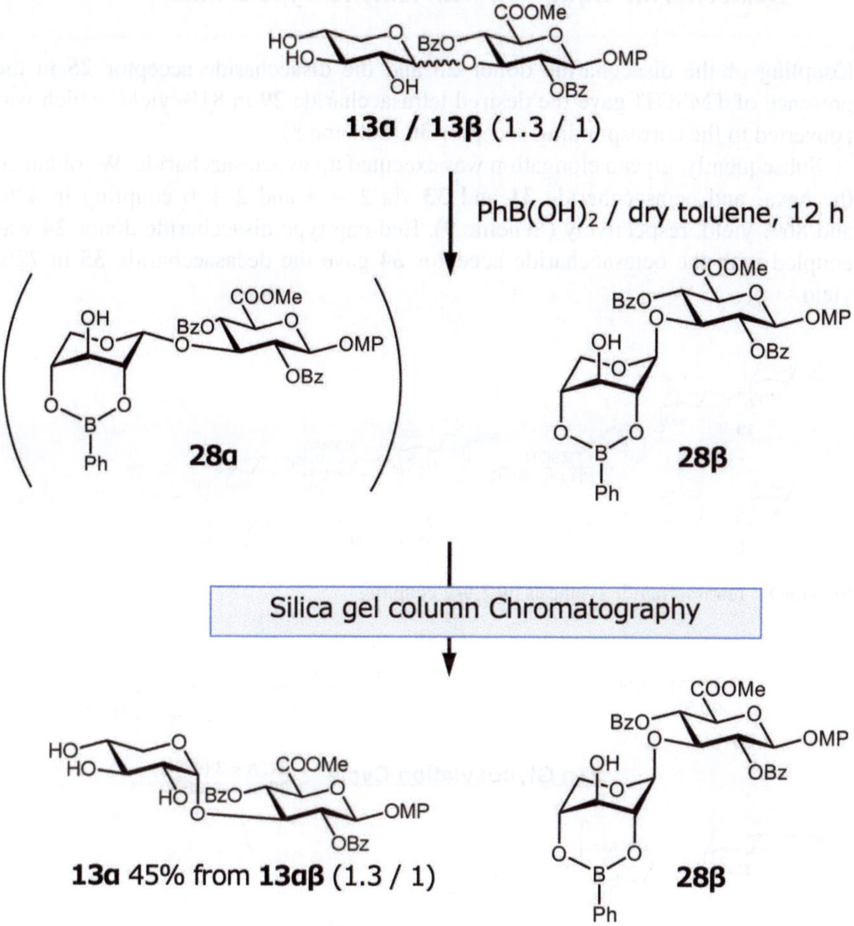


**Scheme 7** Protection at *O*-2,4 of Xyl with phenyboronic acid

## 3 Oligomerization of the Matriglycan Repeating Disaccharide

### 3.1 Synthesis of 9-mer of the Matriglycan Repeating Disaccharide Equipped with Alkyne-Type Linker

Coupling of the disaccharide donor **23** and the disaccharide acceptor **25** in the presence of TMSOTf gave the desired tetrasaccharide **29** in 81% yield, which was converted to the corresponding acceptor **30** (Scheme 8).

Subsequently, glycan elongation was executed up to octasaccharide. We obtained the hexa- and octasaccharide **31** and **33** via 2 + 4 and 2 + 6 coupling in 42% and 86% yield, respectively (Scheme 9). End-cap-type disaccharide donor **24** was coupled with the octasaccharide acceptor **34** gave the decasaccharide **35** in 72% yield.

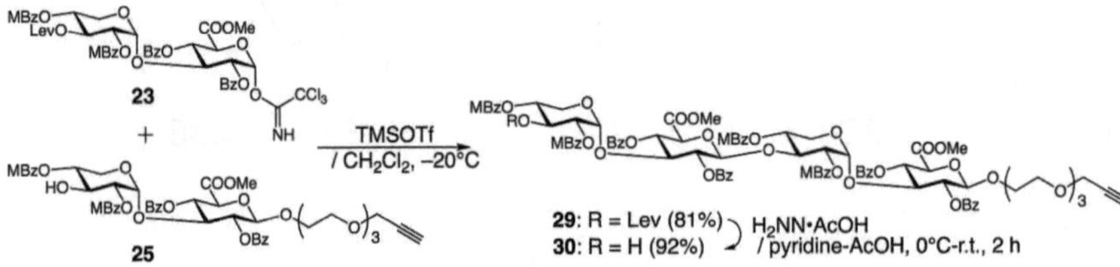


**Scheme 8** Tetrasaccharide synthesis by 2 + 2 coupling

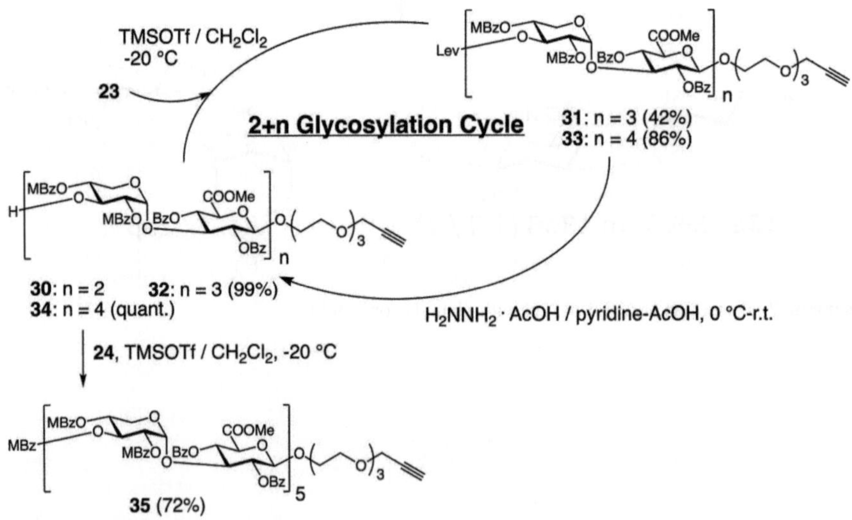


**Scheme 9** Repetitive glycan elongation employing disaccharide donor **23**

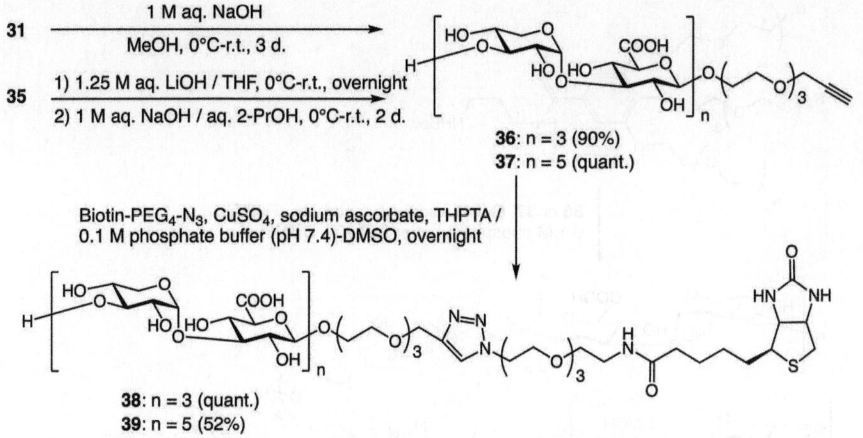


**Scheme 10** Conjugation of the matriglycan oligosaccharides with biotin via hydrophilic linker. Abbreviation: THPTA, Tris(3-hydroxypropyltriazolylmethyl)amine

Deprotection of the hexa- and decasaccharide **31** and **35** was performed by saponification to give the hexa- and decasaccharide **36** and **37** both in quantitative yield, respectively, without any elimination reactions. Subsequent click reaction of **36** and **37** with biotin linker afforded corresponding biotinylated glycan **38** and **39** in quantitative and 52% yield, respectively (Scheme 10) [30].

## *3.2 Synthesis of the Glycodendrimer Composed of Matriglycan Oligosaccharide*

Affinity of the matriglycan for laminin increases along with the elongation of the chain length [31]. Although the chemical elongation of the matriglycan oligosaccharide is a challenging strategy, precise chain length of the glycan is valuable to evaluate the binding affinity for laminin. We designed and synthesized three branched glycodendrimer equipped with the matriglycan oligosaccharide.

Matriglycan hexa- and decasaccharide **36** and **37** were coupled with the dendrimer-core **40** by click reaction to give the three branched glycodendrimers **41** and **42** in 46 and 57% yields, respectively (Scheme 11). $PEG_4$-biotin was finally coupled with glycol-dendrimers to give the desired **43** and **44** in 66 and 94% (in 2 steps), respectively [30].

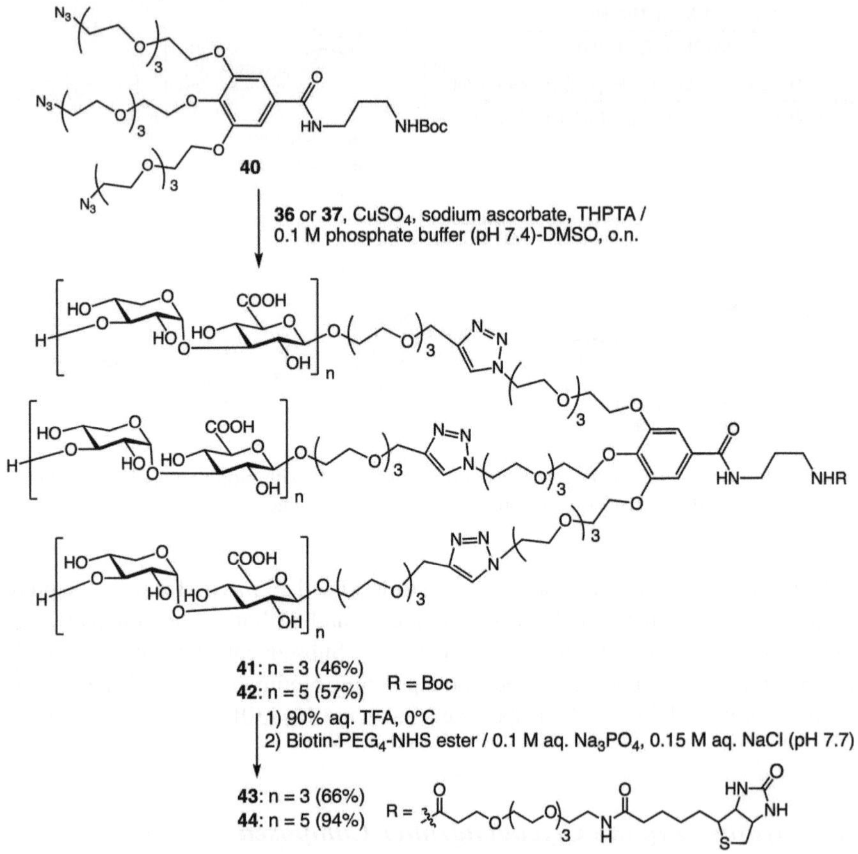


**Scheme 11** Dendrimer formation of the matriglycan oligosaccharides

# 4 Synthesis of the Tandem Ribitol Phosphate and the Non-reducing-side Glycan

## *4.1 Synthesis of Xylosyl Ribitol and the Extended Glycans to the Non-reducing End*

In the tandem ribitol phosphate region, two ribitol phosphate (RboP) moieties are conjugated. The reducing-side RboP is transferred to the OH-3 of GalNAc by FKTN, utilizing CDP-ribitol as a donor. The non-reducing-side RboP is subsequently transferred by FKRP, also employing CDP-ribitol [6]. Furthermore, RXYLT1 (TMEM5), which catalyzes the transfer of Xyl onto RboP, acts exclusively on the non-reducing-side RboP [7]. Since RXYLT1 is the only known xylosyltransferase that acts on Rbo, the core M3 *O*-MG substructures **45~48**, which contain the enzymatically challenging Xylβ1-4Rbo linkage as their backbone, were systematically synthesized.

### 4.1.1 Synthesis of Xylβ1-4Rbo, Xylβ1-4Rbo5P, and Xylβ1-4Rbo5P-1Rbo

As shown in Scheme 12, the xylose donor **49** and the ribitol acceptor **50** were subjected to glycosylation, affording the β-selective product in 87% yield. Interestingly, the resulting glycoside was a 2:1 mixture of two conformational isomers, with the Xyl moiety adopting both C1 **51** and 1C **51′** conformations, respectively. The same conformational mixture was observed after removal of the allyl group (**52, 52′**) and the TBDPS group (**53, 53′**). Upon removal of the Lev group, however, the Xyl moiety of **54** exclusively adopted C1 conformation. Finally, deprotection of the Bn groups from **54** afforded the target Xylβ1-4Rbo **45** in 93% yield in 3 steps. The Xyl residue in the final product **45** exhibited $J_{1,2} = 7.80$ Hz, consistent with a β-anomeric linkage in the C1 conformation.

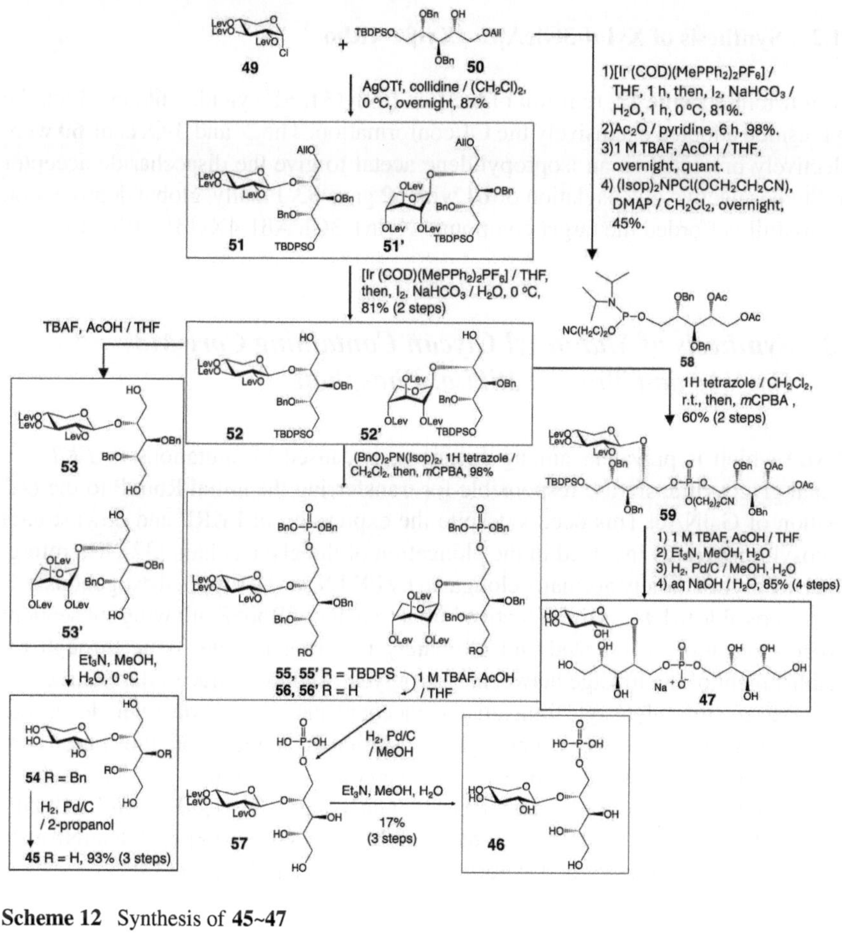


**Scheme 12** Synthesis of **45~47**

Meanwhile, condensation of (**52, 52′**) with $(BnO)_2PN(iPr)_2$ followed by oxidation afforded the phosphate esters (**55, 55′**) in 98% yield as a mixture of the conformational isomers of xylopyranose ring. Subsequent removal of the TBDPS, Bn, and Lev groups gave Xylβ1-4RboP **46** in 17% yield in 3 steps. The Xyl moiety of **57** adopted exclusively the C1 conformation. While removal of the Lev group from (**53, 53′**) led to the Xyl residue only in the C1 conformation, the same conformational preference was observed in **57** after the removal of Bn group from (**56, 56′**). We propose that this conformational shift is driven by steric repulsion between substituents on the xylopyranose ring and/or between the ring substituents and the aglycone moiety.

Coupling of (**52, 52′**) and **58**, obtained from **50** over 4 steps, followed by oxidation afforded **59** in 60% yield (2 steps) of which Xyl moiety exclusively adopted C1 conformation. Subsequent removal of the TBDPS, Lev, and Bn protecting groups, followed by saponification, yielded the target **47** in 85% yield in 4 steps.

#### 4.1.2 Synthesis of Xylαl-3GlcAβ1-4Xylβ1-4Rbo

As shown in Scheme 13, removal of the Lev from (**51, 51′**) yielded **60**, in which the Xyl residue adopted exclusively the C1 conformation. The 2- and 3-OHs of **60** were selectively protected as an isopropylidene acetal to give the disaccharide acceptor **61**. Stereoselective glycosylation of **61** with **62** gave **63**. Finally, global deprotection successfully afforded the target compound Xylα1-3GlcAβ1-4Xylβ1-4Rbo **48**.

### *4.2 Synthesis of Mannosyl Glycan Containing Core M3 GalNAc and Tandem Ribitol Phosphate*

FCMD which is prevalent among Japanese, is caused by mutations in *FKTN*, a pivotal glycosyltransferase responsible for transferring the initial Rbo5P to the *O*-3 position of GalNAc. This occurs despite the expression of FKRP and downstream glycosyltransferases involved in the elongation of the glycan chain [32]. Therefore, when the glycan moiety normally elongated by FKTN can be supplied exogenously, it may be possible to bypass the defective initial transfer of Rbo5P, allowing subsequent glycan elongation to proceed and ultimately restoring muscle tissue through re-establishment of the linkage between dystroglycan and the extracellular matrix.

However, the glycosyl transferring mechanisms in a molecular level are still unclear because we cannot get a sufficient amount of the pure core M3 *O*-MG from natural sources. To overcome this issue, we successfully synthesized the oligosaccharides: Rbo5P-3GalNAcβ1-$OC_2H_4NHAc$ **65**, Rbo5P-1Rbo5P-3GalNAcβ1-$OC_2H_4NHAc$ **69**, and Xylβ1-4Rbo5P-1Rbo5P-3GalNAcβ1-$OC_2H_4NHAc$ **70**, in regio- and stereo-controlled manners (Fig. 3) [33].

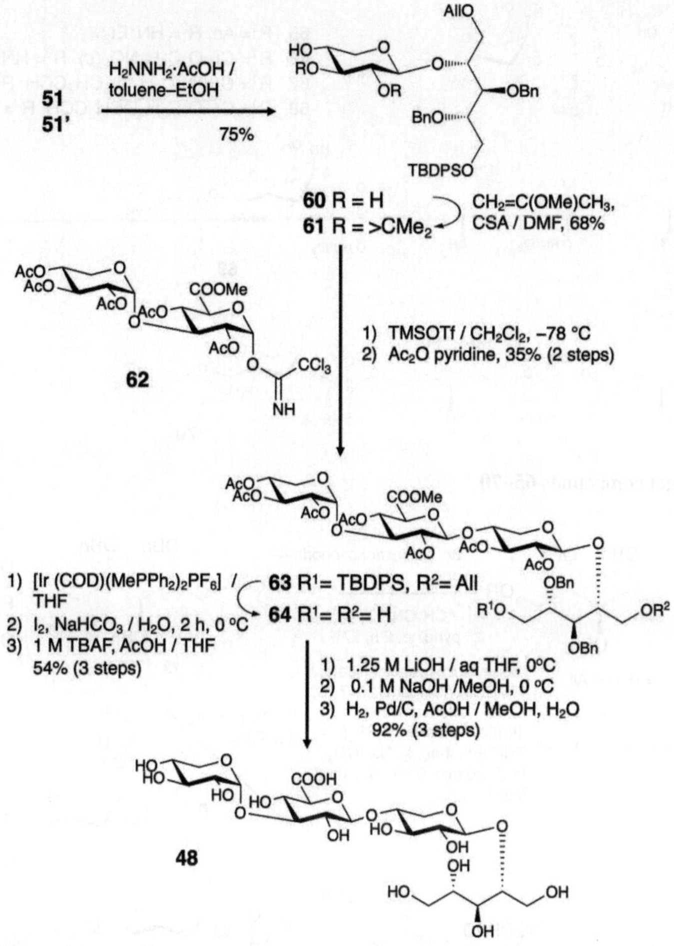


**Scheme 13** Synthesis of **48**

## 4.3 *Synthesis of Rbo5P-3GalNAc*

Ribitol derivative **73** obtained from **50** was activated with PivCl in pyridine and reacted with $GalN_3$ acceptors **74** to give a phosphonate diester **75** in 62% yield, which were then oxidized with $I_2$ to yield **76** in 98% yield. Subsequent removal of the TBDPS group quantitatively afforded **77** (Scheme 14).

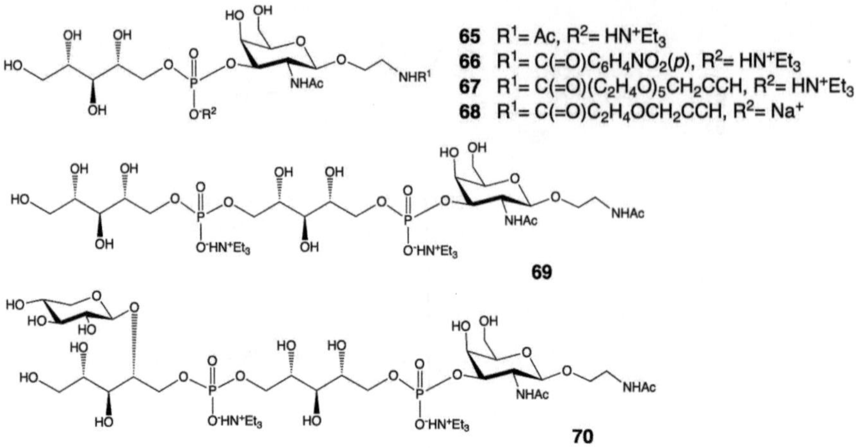


**Fig. 3** Target compounds **65~70**

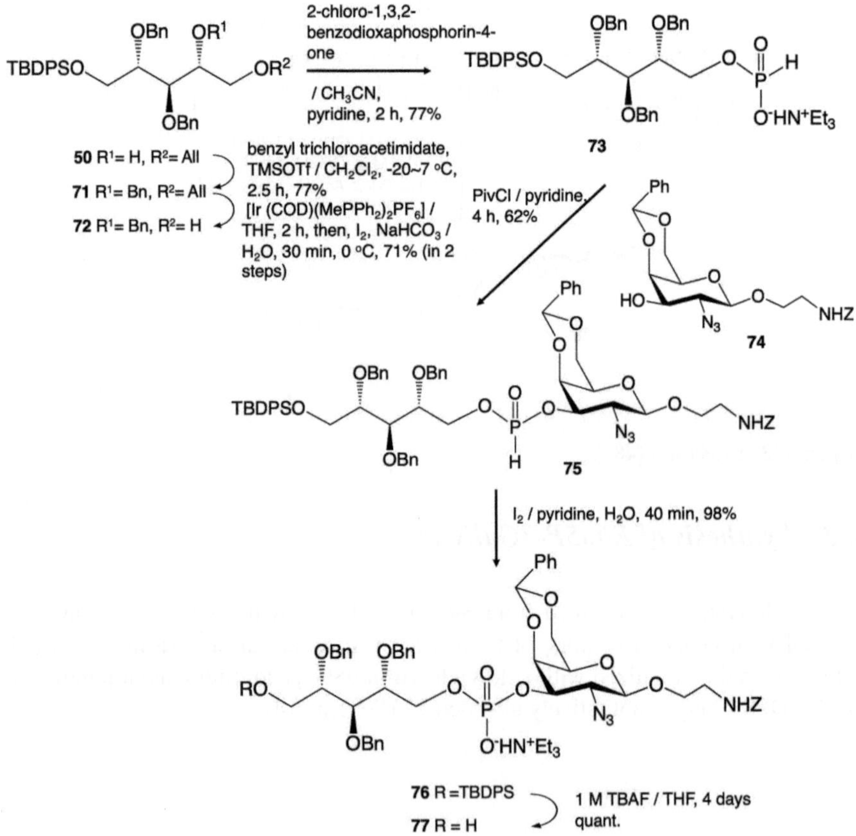


**Scheme 14** Synthesis of the Rbo5P-3GalN$_3$-type disaccharide acceptor **77**

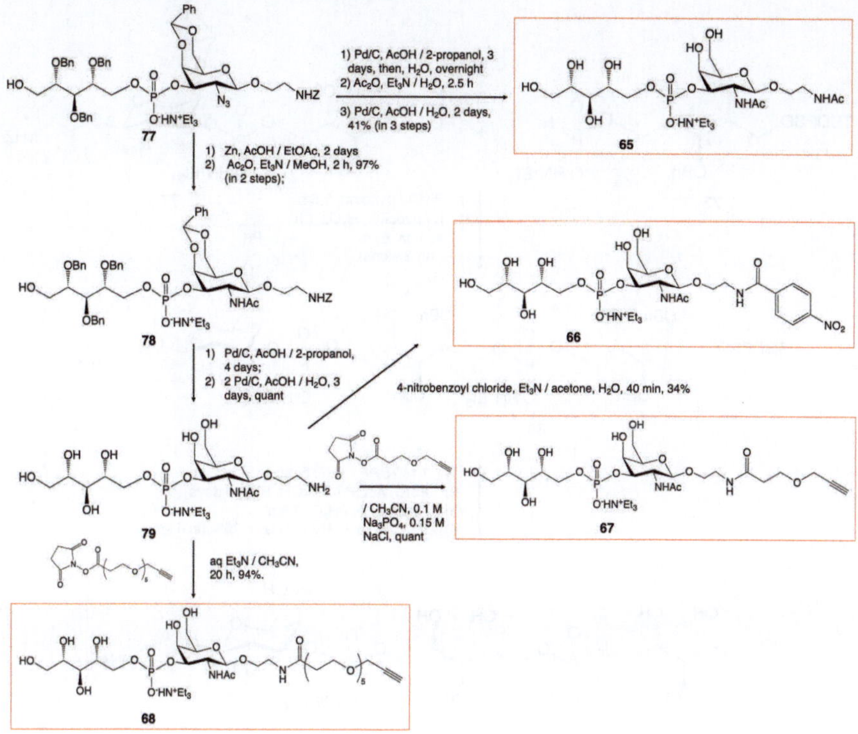


**Scheme 15** Synthesis of the target compounds **65~68**

## 4.4 *Synthesis of Rbo5P-3GalNAc with Different Aglycon*

Scheme 15 illustrates the synthesis of Rbo5P-3GalNAc derivatives bearing different aglycones using building block **77**. Compound **77** was subjected to hydrogenolysis followed by acetamide formation to yield **65**. On the other hand, selective reduction of the azide group in **77**, followed by *N*-acetylation, gave **78**, which furnished **79** upon hydrogenolysis. Using the common intermediate **79**, compounds **66~68** were individually synthesized [33].

## 4.5 *Synthesis of Rbo5P-1Rbo5P-3GalNAc and Xylβ1-4Rbo5P-1Rbo5P-3GalNAc*

Coupling of the disaccharide acceptor **73** with **77** followed by oxidation with $I_2$ afforded **80** (Scheme 16). Subsequent removal of the TBDPS group, hydrogenolysis, and *N*-acetylation yielded Rbo5P-1Rbo5P-3GalNAc **69**.

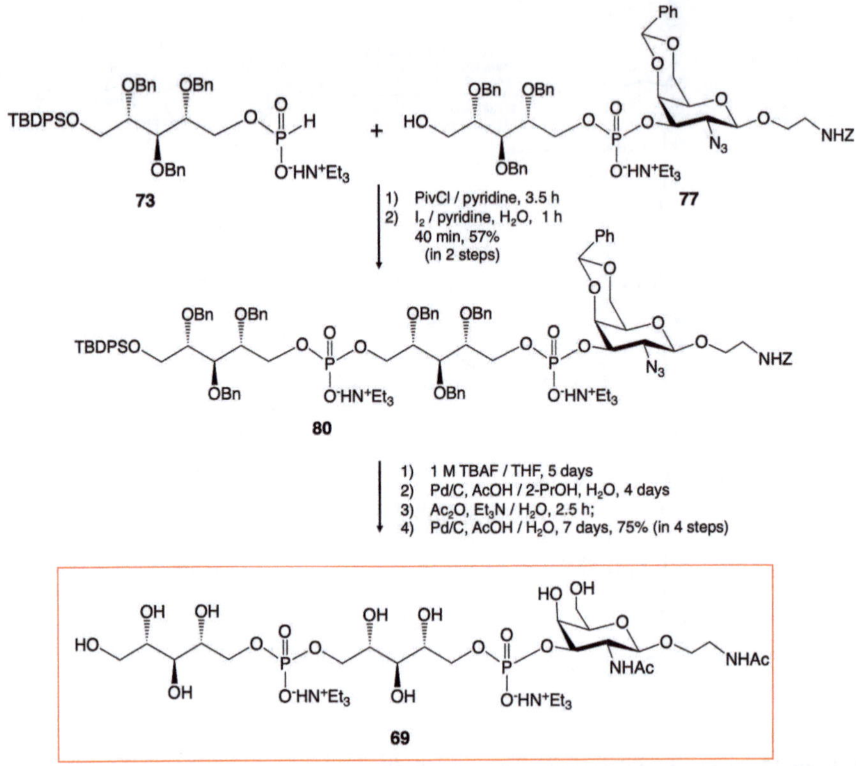


**Scheme 16** Synthesis of the target compound **69**

Xylβ1-4Rbo derivative (**52, 52′**) was converted to **81** which was activated with PivCl and subsequently coupled with **77**. Oxidation with $I_2$ afforded **82** (Scheme 17). Finally, global deprotection yielded Xylβ1-4Rbo5P-1Rbo5P-3GalNAc **70**. The chemically obtained RboP-GalNAc derivatives **65~70** can serve as primers for glycan elongation. These compounds hold therapeutic potential by supplementing glycan structures that cannot be biosynthesized due to the gene defects, thereby contributing to the restoration of muscle tissue.

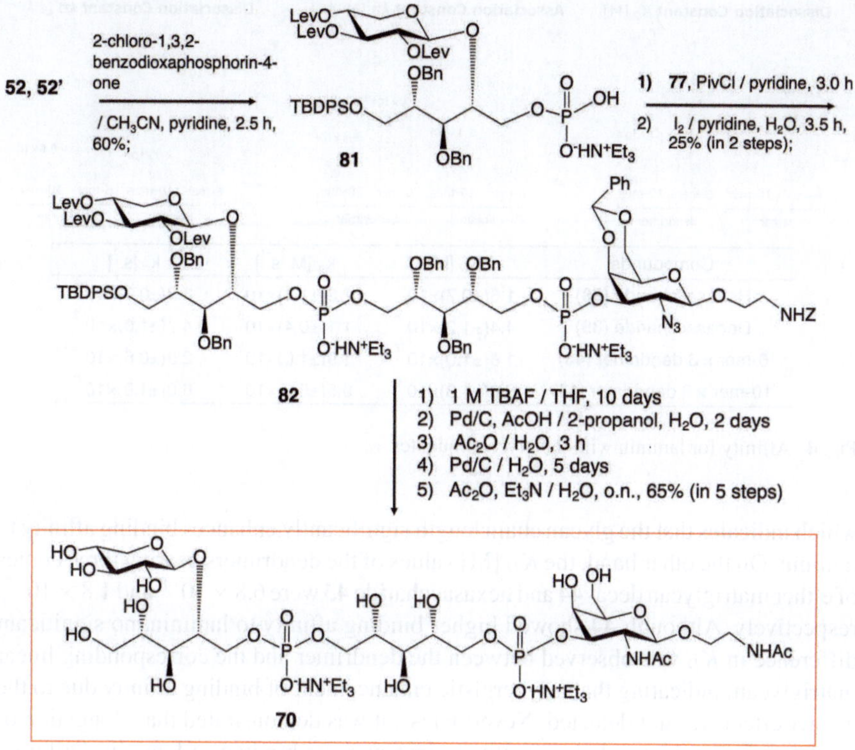


**Scheme 17** Synthesis of the target compound **70**

# 5 Evaluation of the Biological Activity of Synthetic Matriglycan Derivatives

## *5.1 Binding Affinity of Matriglycan Oligosaccharides to Laminin*

Matriglycan plays a pivotal role to stabilize the muscle fibers through the interaction with the LG4-5 domain of laminin. To enhance the binding affinity of the matriglycan derivatives with laminin we have synthesized several linear glycoconjugates **38** and **39** as well as the glycodendrimers **43** and **44** as mentioned last part.

Yamaguchi's group analyzed the glycoconjugates by biolayer interferometry (BLI) method to examine the effects for the binding affinity to laminin-α2 LG4-5 concerning the chain length and dendrimer formation. Four compounds were immobilized on a sensor chip coated with streptavidin (StAv). Laminin α2 LG4-5 (monomer) was used as the interacting analyte.

As shown in Fig. 4, the dissociation constants ($K_D$ [M]) of the linear matriglycans were $4.4 \times 10^{-8}$ and $1.6 \times 10^{-7}$ for the deca- **39** and hexasaccharide **38**, respectively,

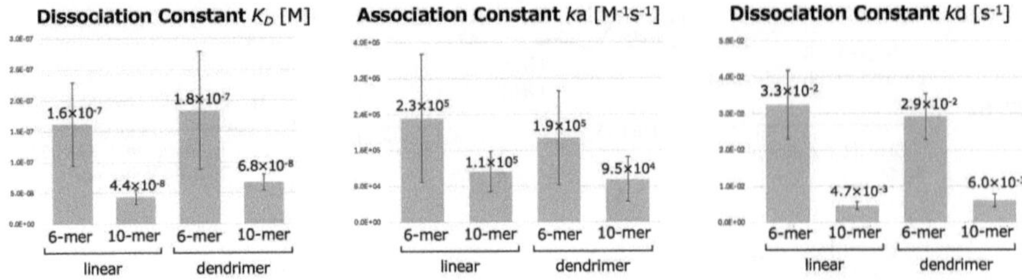


| Compounds | $K_D$ [M] | $k_a$ [$M^{-1}s^{-1}$] | $k_d$ [$s^{-1}$] |
|---|---|---|---|
| Hexasaccharide (**38**) | $1.6(\pm0.7)\times10^{-7}$ | $2.3(\pm1.4)\times10^{5}$ | $3.3(\pm0.9)\times10^{-2}$ |
| Decasaccharide (**39**) | $4.4(\pm1.2)\times10^{-8}$ | $1.1(\pm0.4)\times10^{5}$ | $4.7(\pm1.0)\times10^{-3}$ |
| 6-mer x 3 dendrimer (**43**) | $1.8(\pm1.0)\times10^{-7}$ | $1.9(\pm1.0)\times10^{5}$ | $2.9(\pm0.6)\times10^{-2}$ |
| 10-mer x 3 dendrimer (**44**) | $6.8(\pm1.3)\times10^{-8}$ | $9.5(\pm5.0)\times10^{4}$ | $6.0(\pm1.8)\times10^{-3}$ |

**Fig. 4** Affinity for laminin with the glycoconjugates

which indicates that the glycan chain length significantly enhances binding affinity to laminin. On the other hand, the $K_D$ [M] values of the dendrimers bearing three copies of either matriglycan deca- **44** and hexasaccharide **43** were $6.8 \times 10^{-8}$ and $1.8 \times 10^{-7}$, respectively. Although **44** showed higher binding affinity to laminin, no significant difference in $K_D$ was observed between the dendrimer and the corresponding linear matriglycan, indicating that a synergistic enhancement of binding affinity due to the cluster effect was not detected. Nevertheless, it was demonstrated that elongation of the linear matriglycan chain effectively improves its binding to laminin. Different type dendrimer and the longer glycan are anticipated to improve the affinity for laminin.

### 5.1.1 Enzymatic Addition of Rbo5P at 1Rbo5P-3GalNAcβ

Compound **66** was subjected to enzymatic elongation by FKRP using cytidine diphosphate ribitol (CDP-Rbo) as a donor substrate (Scheme 18). We analyzed the FKRP products by HPLC and high-resolution electrospray ionization mass spectrometry (HR ESI-MS), which shows that FKRP transfers Rbo5P from CDP-Rbo to **66**. Considering to the substrate specificity of FKRP [6], Rbo5P was transferred to the C1 position of the first Rbo5P residue. NMR analysis of the FKRP product confirmed that the FKRP product to be **83**.

Kanagawa et al. previously demonstrated that FKRP recognized Rbo5P-3GalNAcβ1-3GlcNAcβ-$PhNO_2$ (*p*) as the acceptor [6]. Kuwabara et al. reported the substrate recognition mechanism of FKRP using Rbo5P-3GalNAcβ1-3GlcNAcβ1-4Man6Pα-peptide as the acceptor [34]. The dimeric structure of FKRP is important for enzymatic activity from the structure-based functional studies, and the two subunits in the protomeric dimer recognize the acceptor glycan. The phosphate group on Man has an interaction with His252 and Lys256 in the stem domain of one subunit, and the phosphate group on the other side of the acceptor glycan, which

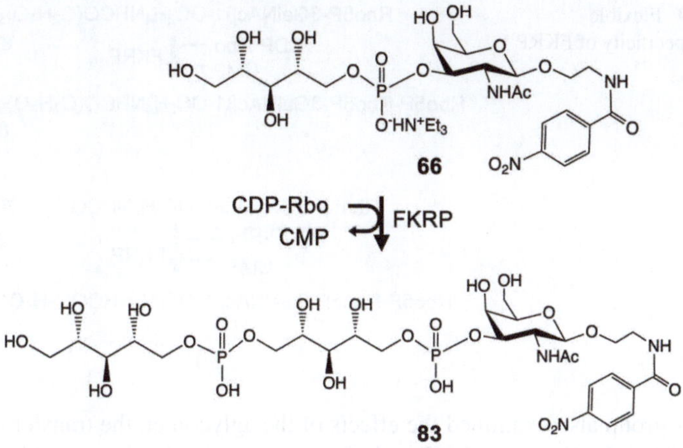


**Scheme 18** Addition of Rbo5P to 3-OH of GalNAc by FKRP

is the phosphate group of Rbo5P, interacts with Arg295 in the catalytic domain of another subunit (Fig. 5) [34]. Endo's group further examined the roles of H252, K256, and R295 in Rbo5P transfer activity when **66** was used as the acceptor [33]. Mutated FKRPs (H252A, K256A, and R295A) were used for the enzymatic transfer of Rbo5P. H252A and K256A mutants gave the same results as that of the wild type. However, the R295A mutant did not show any transferase activity. These results indicated that FKRP recognized the phosphate on Rbo5P by R295, which was significant for the catalytic reaction of Rbo5P transfer. In contrast, H252 and K256 in FKRP, which recognized the phosphate on Man, were not important for catalytic activity. Thus, Rbo5P-3GalNAcβ can be the minimal glycan structure required for the Rbo5P transfer by FKRP as illustrated in Fig. 5. Compound **66** can be useful for the elongation of the core M3 *O*-MG without the formation of the Rbo5P-3GalNAcβ1-3GlcNAcβ1-4Man6Pα structure on α-dystroglycan.

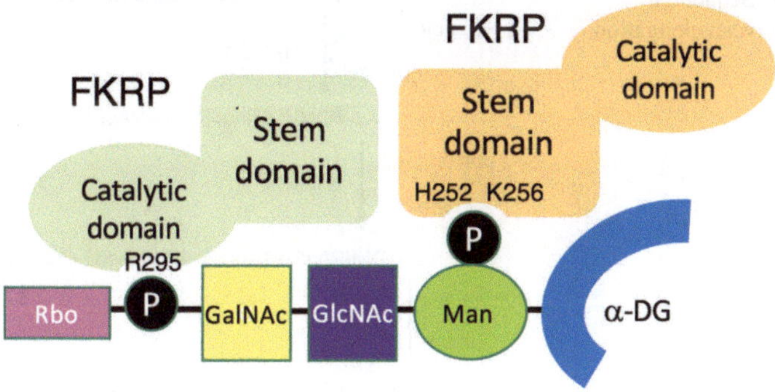


**Fig. 5** Recognition site in the transfer of Rbo5P by FKRP

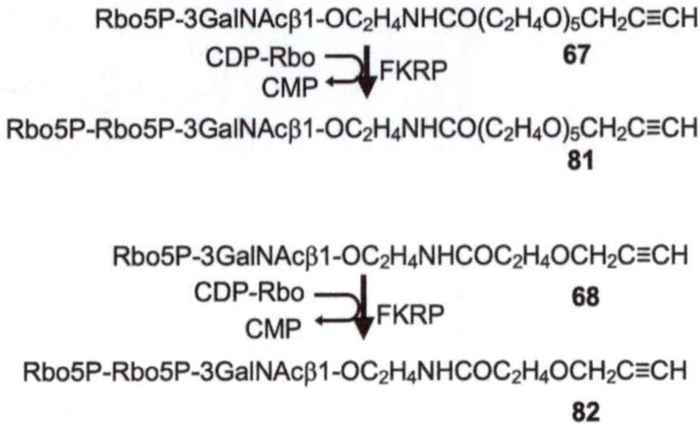


**Scheme 19** Flexible substrate specificity of FKRP

Endo's group also examined the effects of the aglycon on the transfer efficiency (Scheme 19) [33]. Compounds **67** and **68** were both identified as suitable acceptors of FKRP similar to the *p*-nitrobenzoyl derivative **66**. This result suggests that the substrate specificity of FKRP is flexible on the aglycon moiety of Rbo5P-3GalNAcβ glycosides.

### 5.1.2 Sequential Enzymatic Glycan Elongation to Rbo5P-1Rbo5P-3GalNAcβ

Endo and his co-workers tried further glycan elongation to **83** obtained as the FKRP product of **66** by the subsequent glycosyltransferases [33]. They incubated **83** with uridine diphosphate xylose (UDP-Xyl), UDP-GlcA, and UDP-Xyl as glycosyl donors by RXYLT1, B4GAT1, and LARGE, respectively (Scheme 20). HR ESI-MS of the respective products confirmed successful glycosyl transfer as expected.

The results indicated that the Rbo5P-3GalNAcβ derivative acted as a good primer for further elongation by the corresponding glycosyltransferases; FKRP, RXYLT1,

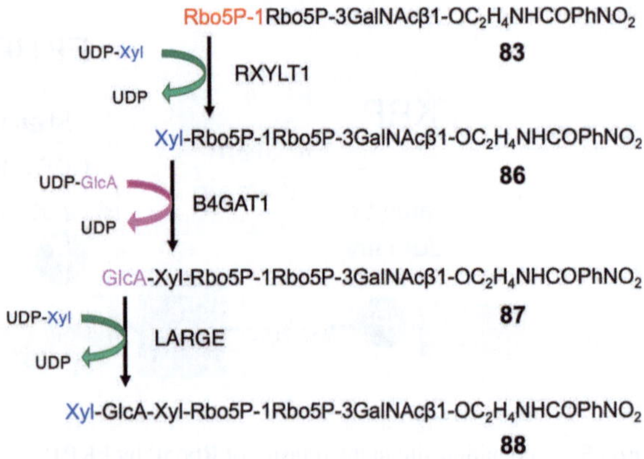


**Scheme 20** Sequential enzymatic glycan elongation

B4GAT1, and LARGE in vitro. It is noteworthy that the enzymatically elongated Rbo5P-3GalNAcβ derivative may also be effective for patients with POMT1, POMT2, POMGNT2, B3GALNT2, and POMK mutations who cannot synthesize the precursor structure (GalNAcβ1-3GlcNAcβ1-4Man6Pα1-Thr) for FKTN. In addition, Rbo5P-1Rbo5P-3GalNAcβ1-$OC_2H_4NHAc$ **69** and Xylβ1-4Rbo5P-1Rbo5P-3GalNAcβ1-$OC_2H_4NHAc$ **70** can be useful for the treatment of patients who are defecting the corresponding glycosyltransferases, FKRP and RXYLT1, respectively.

## 6 Summary

We have chemically synthesized several oligosaccharides related to the core M3 *O*-mannosyl glycan including tandem ribitol phosphate and matriglycan. The patients of Fukuyama type congenital muscular dystrophy cannot synthesize the pivotal transferase, FKTN, and thus the subsequent glycan is not synthesized. We synthesized some Rbo5P-3GalNAcβ1-OR type precursors which acted as substrates for further glycan elongation by skipping the FKTN-mediated step. We also succeeded to get matriglycan oligosaccharide efficiently by full-chemical synthesis, which showed superior interaction with laminin. These results may open the gate to improve the symptom of the muscular dystrophy.

## References

1. Ervasti JM, Ohlendieck K, Kahl SD, Gaver MG, Campbell KP (1990) Deficiency of a glycoprotein component of the dystrophin complex in dystrophic muscle. Nature 345:315–319. https://doi.org/10.1038/345315a0
2. Ibraghimov-Beskrovnaya O, Ervasti JM, Leveille CJ, Slaughter CA, Sernett SW, Campbell KP (1992) Primary structure of dystrophin-associated glycoproteins linking dystrophin to the extracellular matrix. Nature 355:696–702. https://doi.org/10.1038/355696a0
3. Barresi R, Campbell KP (2006) Dystroglycan: from biosynthesis to pathogenesis of human disease. J Cell Sci 119:199–207. https://doi.org/10.1242/jcs.02814
4. Moore CJ, Winder SJ (2012) The inside and out of dystroglycan post-translational modification. Neuromuscul Disord 22:959–965. https://doi.org/10.1016/j.nmd.2012.05.016
5. Chiba A, Matsumura K, Yamada H, Inazu T, Shimizu T, Kusunoki S, Kanazawa I, Kobata A, Endo T (1997) Structures of sialylated O-linked oligosaccharides of bovine peripheral nerve alpha-dystroglycan. The role of a novel O-mannosyl-type oligosaccharide in the binding of alpha-dystroglycan with laminin. J Biol Chem 272:2156–2162. https://doi.org/10.1074/jbc.272.4.2156
6. Kanagawa M, Kobayashi K, Tajiri M, Manya H, Kuga A, Yamaguchi Y, Akasaka-Manya K, Furukawa J, Mizuno M, Kawakami H, Shinohara Y, Wada Y, Endo T, Toda T (2016) Identification of a post-translational modification with ribitol-phosphate and its defect in muscular dystrophy. Cell Rep 14:2209–2223. https://doi.org/10.1016/j.celrep.2016.02.017
7. Manya H, Yamaguchi Y, Kanagawa M, Kobayashi K, Tajiri M, Akasaka-Manya K, Kawakami H, Mizuno M, Wada Y, Toda T, Endo T (2016) The muscular dystrophy gene TMEM5 encodes a ribitol β1,4-Xylosyltransferase required for the functional glycosylation of dystroglycan. J Biol Chem 291:24618–24627. https://doi.org/10.1074/jbc.M116.751917

8. Willer T, Inamori K, Venzke D, Harvey C, Morgensen G, Hara Y, de Bernabé DBV, Yu L, Wright KM, Campbell KP (2014) The glucuronyltransferase B4GAT1 is required for initiation of LARGE-mediated α-dystroglycan functional glycosylation. elife 3:e03941. https://doi.org/10.7554/eLife.03941
9. Praissman JL, Live DH, Wang S, Ramiah A, Chinoy ZS, Boons G-J, Moremen KW, Wells L (2014) B4GAT1 is the priming enzyme for the LARGE-dependent functional glycosylation of α-dystroglycan. elife 3:e03943. https://doi.org/10.7554/eLife.03943
10. Inamori K, Yoshida-Moriguchi T, Hara Y, Anderson ME, Yu L, Campbell KP (2012) Dystroglycan function requires xylosyl- and glucuronyltransferase activities of LARGE. Science 335:93–96. https://doi.org/10.1126/science.1214115
11. Omichi K, Aoki K, Minamida S, Hase S (1997) Presence of UDP-D-Xylose:β-D-Glucoside α-1,3-D-Xylosyltransferase involved in the biosynthesis of the Xylαl-3Glcβ-Ser structure of glycoproteins in the human hepatoma cell line HepG2. Eur J Biochem 245:143–146. https://doi.org/10.1111/j.1432-1033.1997.00143.x
12. Takeuchi H, Kantharia J, Sethi MK, Bakker H, Haltiwanger RS (2012) Site-specific *O*-glucosylation of the epidermal growth factor-like (EGF) repeats of notch: efficiency of glycosylation is affected by proper folding and amino acid sequence of individual EGF repeats. J Biol Chem 287:33934–33944. https://doi.org/10.1074/jbc.M112.401315
13. Lindahl U, Roden L, Gottschalk A (ed) (1972) Glycoproteins, pp 491–517. Elsevier, New York
14. Thorsheim K, Siegbahn A, Johnsson RE, Stålbrand H, Manner S, Widmalm G, Ellervik U (2015) Chemistry of xylopyranosides. Carbohydr Res 418:65–88. https://doi.org/10.1016/j.carres.2015.10.004
15. Sakai K, Nakahara Y, Ogawa T (1990) Total synthesis of nonasaccharide repeating unit of plant cell wall xyloglucan: an endogenous hormone which regulates cell growth. Tetrahedron Lett 31:3035–3038. https://doi.org/10.1016/S0040-4039(00)89019-X
16. Hayakawa S, Koide R, Hinou H, Nishimura S (2016) Synthetic human NOTCH1 EGF modules unraveled molecular mechanisms for the structural and functional roles of calcium ions and O-glycans in the ligand-binding region. Biochem 55:776–787. https://doi.org/10.1021/acs.biochem.5b01284
17. Kunz H, Sager W (1985) Stereoselective glycosylation of alcohols and silyl ethers using glycosyl fluorides and boron trifluoride etherate. Helv Chim Acta 68:283–287. https://doi.org/10.1002/hlca.19850680134
18. Nishizawa M, Kan Y, Yamada H (1989) Thermal glycosidation with benzylated glycosyl chlorides: a very simple procedure for O-glycosidation. Chem Pharm Bull 37:565–567. https://doi.org/10.1248/cpb.37.565
19. Ustyuzhanina N, Komarova B, Zlotina N, Krylov V, Gerbst A, Tsvetkov Y, Nifantiev N (2006) Stereoselective α-glycosylation with 3-O-acetylated *D-Gluco* donors. Synlett 6:921–923. https://doi.org/10.1055/s-2006-939037
20. Okamoto N, Kanie O, Huang YY, Fujii R, Watanabe H, Shimamura M (2005) Synthetic α-mannosyl ceramide as a potent stimulant for an NKT cell repertoire bearing the invariant Vα19-Jα26 TCR α chain. Cell Chem Biol 12:677–683. https://doi.org/10.1016/j.chembiol.2005.04.014
21. Rio S, Beau J-M, Jacquinet J-C (1993) Total synthesis of the carbohydrate-protein linkage region common to several mammalian proteoglycans. Carbohydr Res 244:295–313. https://doi.org/10.1016/0008-6215(83)85009-5
22. Tamura T, Tamura J (2019) Regio- and stereo-controlled synthesis of β-Xyl(1–4)Rbo-5P1-Rbo, the partial structure of O-mannosyl glycan. Tetrahedron Lett 60:465–468. https://doi.org/10.1016/j.tetlet.2018.12.065
23. Mannino MP, Yasomanee JP, Demchenko AV (2018) Investigation of the H-bond-mediated aglycone delivery reaction in application to the synthesis of β-glucosides. Carbohydr Res 470:1–7. https://doi.org/10.1016/j.carres.2018.09.003
24. Cai D, Bian Y, Wu S, Ding K (2021) Conformation-controlled hydrogen-bond-mediated aglycone delivery method for α-xylosylation. J Org Chem 86:9945–9960. https://doi.org/10.1021/acs.joc.1c00187

25. Ito Y, Ogawa T (1994) A novel approach to the stereoselective synthesis of β-mannosides. Angew Chem Int Ed Engl 33:1765–1767. https://doi.org/10.1002/anie.199417651
26. Tamura T, Omura Y, Tamura J (2020) Stereo- and regioselective synthesis of O-mannosyl glycan containing matriglycan and a part of the tandem ribitol phosphate. J Org Chem 85:12935–12946. https://doi.org/10.1021/acs.joc.0c01569
27. Tamura T, Omura Y, Kotera K, Ito R, Ohno S, Manabe N, Yamaguchi Y, Tamura J (2022) Synthesis of the matriglycan hexasaccharide, -3Xylα1-3GlcAβ1- trimer and its interaction with laminin. Org Biomol Chem 20:8489–8500. https://doi.org/10.1039/D2OB01388F
28. Mancini RS, Lee JB, Taylor MS (2017) Boronic esters as protective groups in carbohydrate chemistry: processes for acylation, silylation and alkylation of glycoside-derived boronates. Org Biomol Chem 15:132–143. https://doi.org/10.1039/C6OB02278B
29. Abe H, Shuto S, Matsuda A (2001) Highly α- and β-selective radical C-glycosylation reactions using a controlling anomeric effect based on the conformational restriction strategy. A study on the conformation—anomeric effect—stereoselectivity relationship in anomeric radical reactions. J Am Chem Soc 123:11870–11882. https://doi.org/10.1021/ja011321t
30. Kotera K, Miyamoto R, Mochizuki G, Tamura T, Manabe N, Yamaguchi Y, Tamura J (2025) Chemical extension and glycodendrimer formation of the matriglycan decasaccharide, -$(3Xyl\alpha 1\text{-}3GlcA\beta 1)_5$- and its affinity for laminin. Carbohydr Res 547:109328. https://doi.org/10.1016/j.carres.2024.109328
31. Sheikh MO, Capicciotti CJ, Liu L, Praissman J, Ding D, Mead DG, Brindley MA, Willer T, Campbell KP, Moremen KW, Wells L, Boons G-J (1998) Cell surface glycan engineering reveals that matriglycan alone can recapitulate dystroglycan binding and function. Nat Commun 13:3617. https://doi.org/10.1038/s41467-022-31205-7
32. Kobayashi K, Nakahori Y, Miyake M, Matsumura K, Kondo-Iida E, Nomura Y, Segawa M, Yoshioka M, Saito K, Osawa M, Hamano K, Sakakihara Y, Nonaka I, Nakagome Y, Kanazawa I, Nakamura Y, Tokunaga K, Toda T (1998) An ancient retrotransposal insertion causes Fukuyama-type congenital muscular dystrophy. Nature 394:388–392. https://doi.org/10.1038/28653
33. Tamura J, Tamura T, Hoshino S, Imae R, Kato R, Nagase M, Ohno S, Manabe N, Yamaguchi Y, Manya H, Endo T (2021) Chemical and chemo-enzymatic syntheses of glycans containing ribitol phosphate scaffolding of matriglycan. ACS Chem Biol 17:1513–1523. https://doi.org/10.1021/acschembio.2c00181
34. Kuwabara N, Imae R, Manya H, Tanaka H, Mizuno M, Tsumoto H, Kanagawa M, Kobayashi K, Toda T, Senda T, Endo T, Kato R (2020) Crystal structures of fukutin-related protein (FKRP), a ribitol-phosphate transferase related to muscular dystrophy. Nat Commun 11:303. https://doi.org/10.1038/s41467-019-14220-z

# Chemical Approaches to Understand Physiological Functions of High-Mannose-Type Glycan-Related Proteins

**Mitsuaki Hirose and Kiichiro Totani**

**Abstract** High-mannose-type glycans on glycoproteins play pivotal roles in protein folding, secretion, and degradation in the endoplasmic reticulum (ER). Chemical approaches using structurally defined synthetic glycan probes and inhibitors have enabled the precise elucidation of the functions of glycan-related proteins involved in ER glycoprotein quality control (ERQC), including UGGT1, glucosidase II, calreticulin (CRT), and EDEM family proteins. This chapter highlights how synthetic high-mannose-type glycans have revealed the molecular-level functions of these proteins, including substrate recognition dependent on the branching patterns and aglycone hydrophobicity. Our study demonstrated that UGGT1 preferentially glucosylates glycan substrates with hydrophobic aglycones and recognizes specific glycan branching patterns, whereas glucosidase II exhibits distinct reactivities toward glycan isomers. Moreover, synthetic trisaccharides clarified EDEM's preference for branched chains, and selective inhibitors elucidated discrete mannose-trimming pathways leading to the production of secretion or degradation signals. The reconstructed glycan profile method further revealed disease-related alterations in ER glycan processing. These findings underscore the power of chemical glycan probes in dissecting the mechanisms of ERQC at the molecular level and provide insights into the relationship between glycan structures and protein folding states, advancing our understanding of ER-associated diseases.

**Keywords** High-mannose-type glycan · Chemical probe · UGGT1 · Glucosidase II · Calreticulin · EDEM

---

M. Hirose · K. Totani (✉)
Department of Science and Technology, Seikei University, 3-3-1 Kichijoji-kitamachi, Musashino-shi, Tokyo 180-8633, Japan
e-mail: ktotani@st.seikei.ac.jp

 
K. Toshima and Y. Fujimoto (eds.), *Chemical Biology of Carbohydrates*,
https://doi.org/10.1007/978-981-95-5335-8_5

# 1 Introduction

## 1.1 Biosynthetic and Trimming Pathways of High-Mannose-Type Glycan on Glycoproteins

*N*-linked glycans play key roles in biological events, such as major signal regulators for cell–cell interactions, viral infection, and the reflection of the glycoprotein folding state [1]. The *N*-linked glycans are mainly of three types: high-mannose-, complex-, and hybrid-type glycans. In particular, high-mannose-type glycans play essential roles in glycoprotein folding, secretion, and degradation, thereby maintaining cellular homeostasis [2–4]. High-mannose-type glycans are assembled starting from the outer leaflet of the endoplasmic reticulum (ER) membrane, followed by the inner leaflet of the ER membrane. The glycans are assembled through sequential addition to the backbone of a dolichol pyrophosphate (PP-Dol) on the ER membrane by a series of glycosyltransferases encoded by asparagine-linked glycosylation genes (Fig. 1, cytosol) [5]. Through the biosynthetic pathway, mammalian high-mannose-type glycans finally reach $Glc_3Man_9GlcNAc_2$ (G3M9) as a common precursor oligosaccharide before being transferred to proteins. During G3M9 assembly, the $Man_{5iso}GlcNAc_2$ (M5iso) precursor with PP-Dol flips across the ER membrane from the cytosolic lumen to the ER lumen [6–8].

In the ER lumen, G3M9 is transferred to asparagine (Asn) residue in the consensus sequence (Asn-Xaa-Ser/Thr; Xaa = any amino acid residues except proline) of polypeptide by oligosaccharyltransferase (OST), producing G3M9 type glycopeptides (Fig. 1, ER lumen) [9]. Subsequently, glucosidase I and glucosidase II trim two glucoses from G3M9 to yield $Glc_1Man_9GlcNAc_2$ (G1M9) type glycopeptides. The polypeptide moiety of G1M9-polypeptides is assisted in folding by lectin chaperones, calnexin/calreticulin (CNX/CRT) [10, 11], with its complex partners, the protein disulfide isomerase (PDI) family, such as ER-resident protein 57 (ERp57) [12], ER-resident protein 29 [13], and cyclophilin B [14]. After these chaperones capture the G1M9-polypeptide, the polypeptide moiety is folded into a compact structure. Thus, folding is accelerated in the presence of chaperone complexes. Folded G1M9-type glycoproteins are hydrolyzed by glucosidase II to generate $Man_9GlcNAc_2$ (M9) glycoproteins [15, 16]. Based on the folding states of M9 glycoproteins, the fate of glycoproteins branches into three pathways: secretion, degradation, and refolding. Properly folded M9 glycoproteins are trimmed by ER-degradation enhancing α-1,2 mannosidase-like protein 2 (EDEM2) complexed with thioredoxin domain containing 11 (TXNDC11) *via* disulfide bond, producing $Man_{8B}GlcNAc_2$ (M8B) glycoproteins lacking terminal mannose on the B-branch of M9 as a glycosignal leading to the secretion pathway [17]. After M8B glycoproteins are captured by VIP36-like (VIPL) and ER–Golgi intermediate compartment protein of 53 kDa (ERGIC53), they are transported to the Golgi apparatus [18]. In contrast, partially folded M9 glycoproteins are glucosylated at the terminal end of the A-branch by uridine 5'-diphosphate (UDP)-glucose:glycoprotein glucosyltransferase 1 (UGGT1) to regenerate G1M9 glycoproteins as refolding intermediates [19]. UGGT1

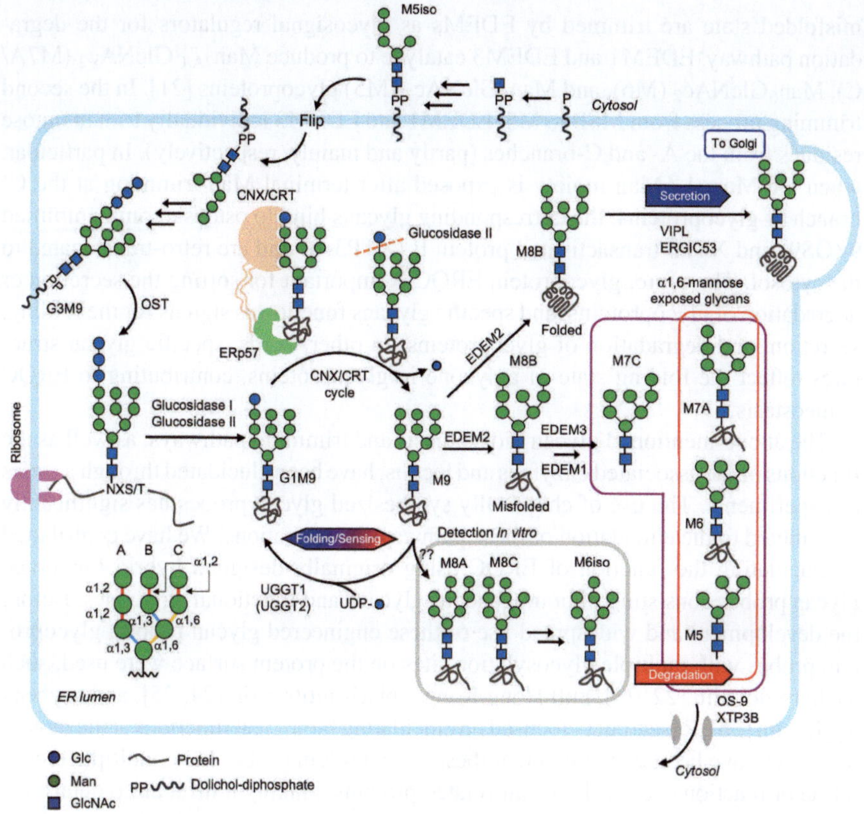


**Fig. 1** Schematic diagram of the glycoprotein quality control system. In the endoplasmic reticulum (ER) lumen and cytosol, glycans on glycoproteins are trimmed and recognized by specific enzymes and lectins. In particular, the CNX/CRT cycle facilitates the refolding of G1M9- and M9-glycoproteins, whereas mannose trimming by EDEMs directs glycoproteins toward secretion or degradation. Misfolded glycoproteins that persist despite the CNX/CRT cycle are further trimmed by EDEMs to produce glycans, such as M6 and M5, with exposed α1,6-mannose residues, after which they are transported to the cytosol. The red box highlights glycan structures exposing α1,6-mannose on the C-branch, which signal transport to the cytosol. The purple box shows the degradation glycan signals. The gray box indicates a possible trimming pathway identified through *in vitro* assays of mannosidases in ER-enriched fractions from the mouse liver. A-, B-, and C-branches are shown as Man-α1,2-Man-α1,2-Man, Man-α1,2-Man-α1,3-Man, Man-α1,2-Man-α1,6-Man, respectively

has a sensing ability to monitor the folding state of aglycone [20]. The resulting G1M9 glycoproteins are given a refolding opportunity by CNX/CRT(-PDIs). The de-glucosylation/re-glucosylation process between G1M9 and M9 glycoproteins, called the CNX/CRT cycle, contributes to maximizing the folding, which refers to the glucose residue-dependent folding acceleration, check, and refolding fundamental system in endoplasmic reticulum glycoprotein quality control (ERQC). Despite M9 glycoproteins being refolded through this cycle, glycoproteins still showing misfolded state are trimmed by EDEMs as glycosignal regulators for the degradation pathway. EDEM1 and EDEM3 catalyze to produce $Man_{7A/C}GlcNAc_2$ (M7A/C), $Man_6GlcNAc_2$ (M6), and $Man_5GlcNAc_2$ (M5) glycoproteins [21]. In the second trimming process from M8B to M5, EDEM1 and EDEM3 individually trim mannose residues from the A- and C-branches (partly and mainly, respectively). In particular, when the Manα1-6Man moiety is exposed after terminal Man-trimming at the C-branch of glycoproteins, the corresponding glycans bind to osteosarcoma-amplified 9 (OS9) and XTP3 transactivated protein B (XTP3-B) and are retro-translocated to the cytosol. Therefore, glycoprotein ERQC is important for sorting the secretion or degradation of glycoproteins, and specific glycans function as signals for the folding, secretion, and degradation of glycoproteins. In other words, specific glycan structures reflect the folding state of aglycone in glycoproteins, contributing to ERQC homeostasis.

The above-mentioned glycan biosynthesis and trimming pathways, as well as the functions of the associated enzymes and lectins, have been elucidated through a series of experiments. The use of chemically synthesized glycan probes has significantly contributed to the elucidation of these pathways and functions. We have contributed to elucidating the function of ERQC using originally designed hybrid functional glycan probes consisting of homogeneous glycans and functional aglycones. Before the development and widespread use of these engineered glycan probes, glycoprotein probes with multiple glycosylation sites on the protein surface were used, such as thyroglobulin [22, 23], null Hong Kong alpha1-antitrypsin [24, 25], and soybean agglutinin [26]. Researchers focused on elucidating biological functions at the molecular level have been concerned that these glycoprotein probes have multiple recognition or reaction sites with glycan-related proteins, making it difficult to determine which glycosylation sites a target glycan processing enzyme or lectin recognizes. Therefore, understanding the glycan function of glycoproteins with multiple glycosylation sites is limited. Moreover, even single-glycosylated glycoprotein probes prepared or purified from biological sources exhibit glycan heterogeneity. These undesirable properties of glycoprotein probes often affect analytical results. To overcome these issues, chemical synthesis procedures for a series of high-mannose-type glycans were first established by exploring efficient synthetic strategies and pathways in the early 2000s [27, 28]. Structurally defined, chemically synthesized high-mannose-type glycan probes are useful for determining the precise glycan specificity of glycan-related proteins at the molecular level [29, 30]. Additionally, inhibition assays and the application of high-mannose-type glycan probes have contributed to revealing the recognition and functions of various enzymes in ERQC.

This chapter summarizes our contributions to the elucidation of glycan as well as aglycone specificity and recognition mechanisms of glycan-recognizing enzymes and chaperones in the ER, such as UGGT1, glucosidase II, CRT, and EDEMs, using chemical approaches with chemically synthetic glycan probes and inhibitors.

## *1.2 Overlooked Aspects Regarding High-Mannose-Type Glycan Probes for Biologists*

Before the development of the chemical biology area, biological events and phenomena were analyzed using biological approaches. At that time, most biological approaches did not consider the molecular-level understanding of glycan structures, including differences in structural isomers, because biological interest was focused on understanding these observed biological events as a series of life phenomena. Thus, these life science studies have not focused on understanding the molecular structural level. In contrast, chemical biology approaches focus on understanding the substrate specificity of glycan processing enzymes and lectin recognition at the molecular structural level, including structural isomers, during biological events. Hence, chemical biology approaches have greatly contributed to paradigm shifts in the life sciences, including glycobiology.

One example of research that has been overlooked by biologists is the correction of the substrate specificity of glucosidase II. In 1980, differences in the glycan specificity of glucosidase II were interpreted as depending only on the number of mannose residues in the glycan [31]. Grinna and Robbins first proposed that the reactivity of glucosidase II gradually decreased with decrease of the number of mannoses in monoglucosylated high-mannose-type glycans; i.e., G1M9 > G1M8 > G1M7 (Sect. 5.3.2). However, this proposal lacked an understanding of the reactivity of glucosidase II, considering structural isomers; i.e., glycans with the same number of sugar residues but different branching patterns. In fact, analysis of glucosidase II with chemically synthesized monoglucosylated glycoprobes revealed that the glycan specificity follows the order of G1M9 = G1M8B >> G1M8C = G1M7 [32]. Thus, the reactivity of glucosidase II depends on the differences in branching patterns due to structural isomers rather than on the number of mannoses. This is a classic example of the effectiveness of chemical approaches.

Another often overlooked aspect concerns glycan labeling methods, such as pyridylamination (PA) and 2-aminobenzamidation (AB), which are commonly used in glycobiology research [33, 34]. These glycan labeling methods, such as PA and AB, have a serious disadvantage in that the chemical structure of the reducing end GlcNAc is modified. Namely, the reductive ring-opening of the reducing end GlcNAc occurs upon labeling with PA or AB (Fig. 2). The labeling of the GlcNAc structure at the reducing end is completely different from that of the natural GlcNAc structure. Of course, it would be extremely beneficial to use these labeling methods for the analysis of glycans derived from biological resources. However, these modified GlcNAc-containing glycans with unnatural GlcNAc are often interpreted by biologists unfamiliar with molecular structures as equivalent to native glycan structures and have frequently been used as substrates or ligands for elucidating the function of glycan processing enzymes and lectins. Such misidentification of molecular structures can be seriously misleading in the analysis of glycan processing enzymes and lectins that recognize the reducing end GlcNAc structure. In fact, UGGT1 [23, 35] and ubiquitin-ligase Fbs1 [36, 37] have been reported to recognize the reducing

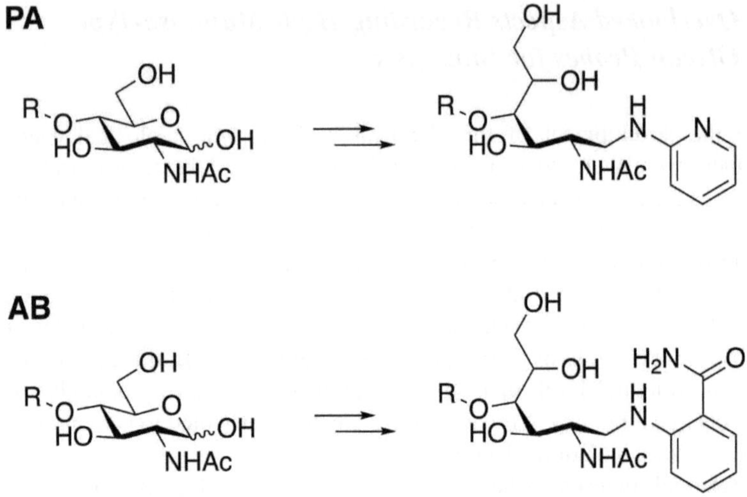


**Fig. 2** Modification of the chemical structure of the reducing end GlcNAc residue following glycan labeling with PA or AB (R = glycan)

end GlcNAc residue of high-mannose-type glycoproteins. The authors believe that a chemical approach that focuses on molecular structure is essential to truly understand biological phenomena at the molecular structure level. In the next section, we introduce our research to elucidate the function of glycoprotein quality control using chemical approaches.

# 2 Functional Analysis of the Folding Sensor Enzyme UGGT1

## *2.1 Glycan Specificity of UGGT1*

UGGT1 plays a key role in the CNX/CRT cycle as a folding sensor that recognizes both the glycan structure and aglycone moiety of glycoproteins. UGGT1 transfers one glucose to incompletely folded M9 glycoproteins, regenerating G1M9 glycoproteins for CNX/CRT refolding. This system maximizes the opportunity to obtain properly folded glycoproteins. As a key enzyme for elucidating the overall ERQC mechanism, many researchers have attempted to elucidate UGGT1 specificity. However, detailed analysis of UGGT1 has been problematic due to its unique substrate specificity. UGGT1 accepts only incompletely folded glycoproteins with a compact molten globule-like conformation. In other words, UGGT1 rejects properly folded and viciously misfolded glycoproteins [23, 38, 39]. Therefore, until the early 2000s, only qualitative analyses of UGGT1 had been reported, mainly using heat-denatured glycoproteins, trypsin-digested (glyco)peptides, or neoglycoproteins possessing a single-glycosylated domain as substrates for UGGT1 [38–40]. As the branched point of the UGGT1 analysis, we developed an M9-methotrexate conjugate (**M9-G-MTX**) as a synthetic non-proteinic substrate (Fig. 3), opening up a new avenue for the quantitative analysis of UGGT1 [41]. Initially, the utility of **M9-G-MTX** was not well understood. Here is one anecdote: A collaborating biologist used **M9-G-MTX** as a substrate and after the first assays told me that it was much less reactive than denatured thyroglobulin, which is known as a universal substrate. However, due to the lack of understanding of the molecular structure, the setting up of enzyme assay conditions with substrate concentrations based on the number of glycans on the protein (20 glycans) was overlooked. Subsequent optimization of the enzyme assay conditions based on chemist suggestions demonstrated that the **M9-G-MTX** substrate is more useful than the previously widely used glycoprotein substrates in terms of activity against a single glycan. Thus, the MTX structure would function as a suitable hydrophobic patch mimicking the molten globule-like state. Moreover, an assay using Glycan-G-MTXs with various glycan structures revealed that the reactivity of UGGT1 was in the order of M9 > M8B = M8C > M7 [41], indicating that UGGT1 has a substrate specificity for different glycans.

## 2.2 *Recognition Motifs of UGGT1 Focused on Aglycone Structures*

The development of Glycan-G-MTX allowed us to elucidate the substrate specificities of UGGT1, focusing on glycan structures among M9, M8B, M8C, and M7. However, the precise recognition motifs of UGGT1 derived from both glycan and aglycone structures have not yet been revealed. In this section, we discuss the recognition motifs of UGGT1, focusing on the aglycone structure. To this end, a series of synthetic substrates possessing various aglycone moieties were used in UGGT1 assay *in vitro*. MTX, 9-fluorenylmethyloxycarbonyl (Fmoc), boron-dipyrromethene (BODIPY), and carboxytetramethylrhodamine (TAMRA) were used as typical aglycone hydrophobic motifs for UGGT1 substrates (Fig. 4) [42]. Comparing MTX with Fmoc, the UGGT1 reactivity of **M9-G-MTX** was almost the same as that of **M9-G-Fmoc**. In contrast, **M9-G-E-Fmoc**, in which the M9-Gly moiety and Fmoc group were linked by a glutamic acid (Glu) residue, showed approximately four-fold higher reactivity than **M9-G-MTX**. These results show that reactivity does not only depend on the chemical structure of MTX. Because our standard UGGT1 substrate **M9-G-MTX** contains the Gly-Glu (G-E) dipeptide as a linker, UGGT1 reactivity was compared in terms of various aglycones with the G-E linker. **M9-G-E-5TAMRA**, **M9-G-E-BODIPY**, and **M9-G-E-BODIPY(493)** showed relatively high reactivity against UGGT1, while **M9-G-E-6TAMRA** exhibited only one-third reactivity of these substrates. Since the structures of the highly reactive **M9-G-E-5TAMRA** and less reactive **M9-G-E-6TAMRA** differ only in the linked angle of the TAMRA

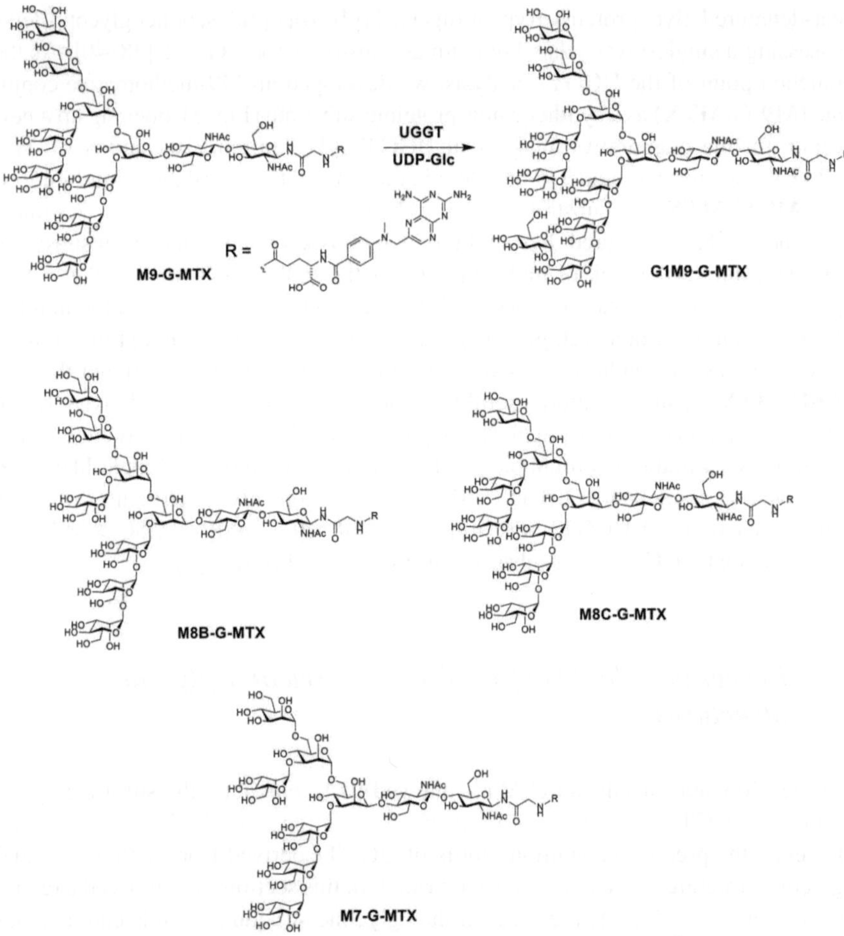


**Fig. 3** Chemical structures of UGGT1 substrates (**M9-G-MTX, M8B-G-MTX, M8C-G-MTX,** and **M7-G-MTX**) and the product (**G1M9-G-MTX**) possessing MTX

moiety, UGGT1 affinity for the substrate was found to depend on the orientation of the aglycone hydrophobic patches.

Synthetic **M9-G-BODIPY**, **M9-G-EG4-BODIPY**, **M9-G-EG8-BODIPY**, and **M9-G-EG12-BODIPY** bearing ethylene glycol (EG)-type linkers of various lengths were used as UGGT1 substrates to understand the effects of the distance between the glycan and aglycone BODIPY as a recognition motif (Fig. 4). The reactivity gradually decreased with an increase in linker length, suggesting that UGGT1 favors the presence of a hydrophobic substituent adjacent to the reducing end of the glycans. In addition, the order of glucose transfer using purified UGGT1 was the same as that using an enriched ER fraction containing UGGT1 [43].

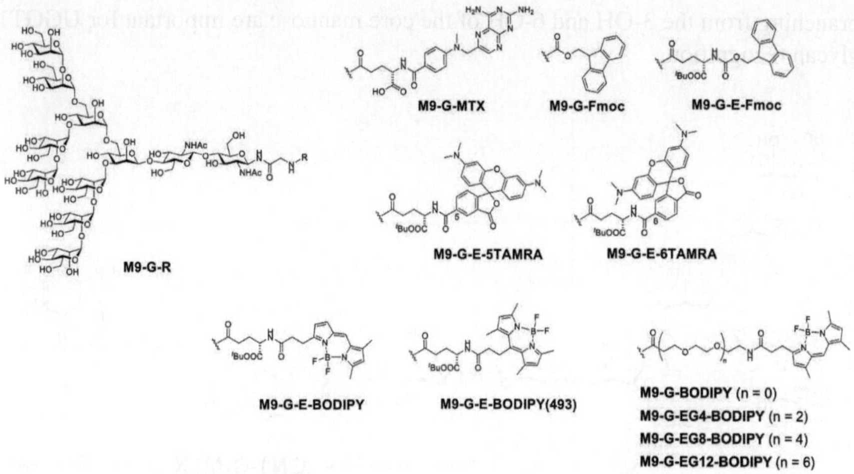


**Fig. 4** Chemical structures of UGGT1 substrates possessing various aglycones

## *2.3 Recognition Motifs of UGGT1 Focused on Glycan Structures*

In recent years, chemical biology has been developed to elucidate biological events and their mechanisms at the molecular level using synthetic compounds that target biomolecules. Inhibitor assays are one of the major chemical biology methods that can reveal the recognition motifs of target enzymes based on the decrease in reaction rate in the presence of inhibitors and substrates. We applied an inhibitor assay to elucidate the recognition motifs of UGGT1, focusing on glycan structures. We selected **GN1-G-MTX, GN2-G-MTX, M1-G-MTX, M3-G-MTX, M5A-G-MTX,** and **M7A-G-MTX** as inhibitors to identify the glycan motifs recognized by UGGT1 (Fig. 5). In these inhibitors, the mannose(α1-2)mannose residue in the A-branch terminal is designed to prevent the transfer of glucose to inhibitors during the UGGT1 reaction. In the presence of a tenfold concentration of **GN1-G-MTX** to the substrate, UGGT1 glucosylation was inhibited by ca. 10%. The disaccharide **GN2-G-MTX** and trisaccharide **M1-G-MTX** showed similar inhibition as **GN1-G-MTX,** indicating that the contribution of the second GlcNAc and core mannose is marginal. Based on the fact that **M8BGN1-G-MTX,** which lacks the innermost GlcNAc, showed no glucosylation activity as an acceptor substrate [42] and no inhibitory activity toward UGGT1, the distance between the innermost GlcNAc and the reaction site is an important factor for UGGT1 recognition. In contrast, **M3-G-MTX,** corresponding to the core pentasaccharide of high-mannose-type oligosaccharides, showed much higher inhibition (~50%) than **GN1-**, **GN2-**, and **M1-G-MTX**. Surprisingly, **M5A-G-MTX** and **M7A-G-MTX** exhibited lower inhibition than **M3-G-MTX**, implying that a bulky B- or C- arm negatively affect the binding of UGGT1. Taken together, these results indicate that the reducing end GlcNAc residue and the mannose residues branching from the 3-OH and 6-OH of the core mannose are important for UGGT1 glycan recognition.

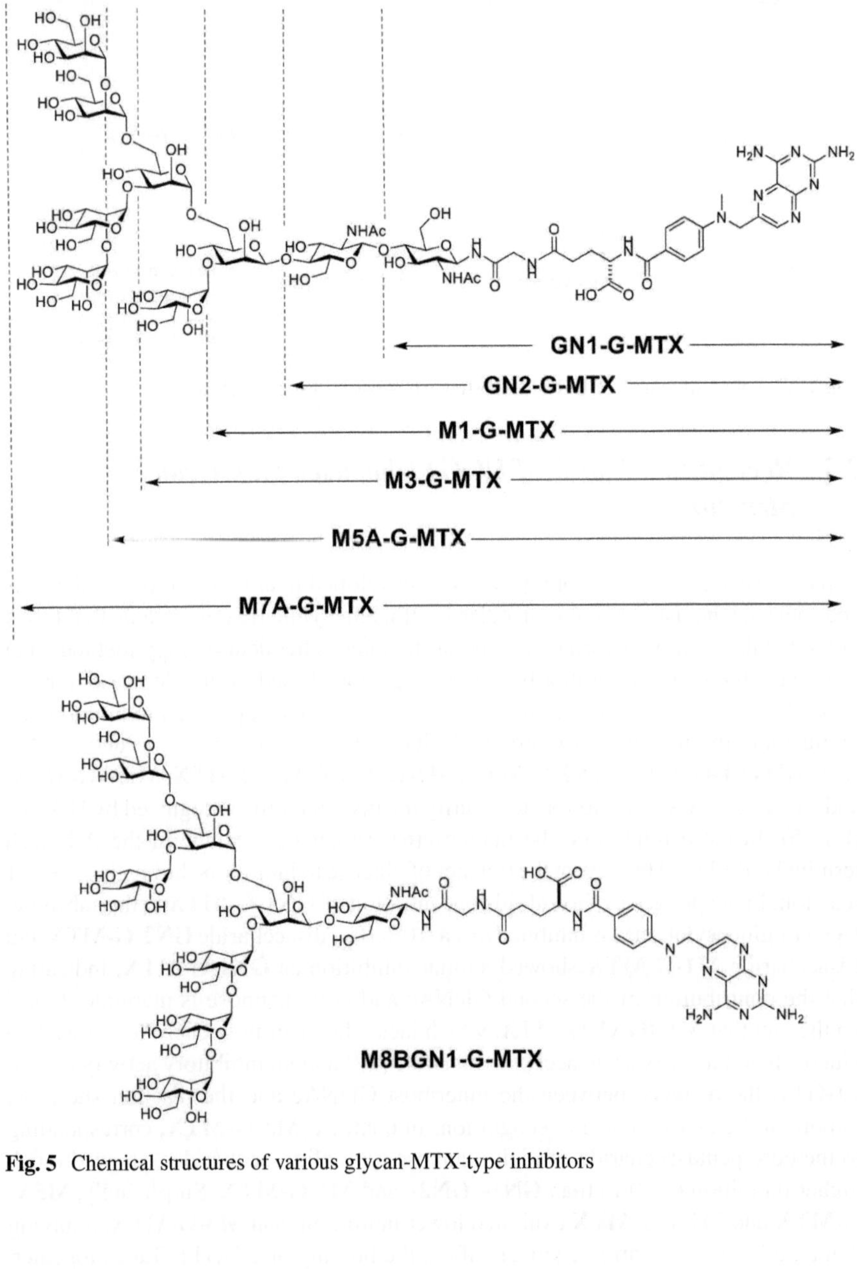


**Fig. 5** Chemical structures of various glycan-MTX-type inhibitors

### 2.4 *Influence of Consensus Sequence for* N*-Glycosylation in UGGT1 Recognition*

UGGT1 recognizes not only aglycone hydrophobic patches such as MTX, TAMRA, Fmoc, and BODIPY, but is also sensitive to the distance between the glycan moiety and aglycone. Specifically, the glucosylation activity of **M9-G-E-Fmoc** was approximately fourfold higher reactivity than that of **M9-G-Fmoc** [42]. Hence, UGGT1 has the potential to exhibit peptide specificity in natural aglycones. The peptide specificity of UGGT1 was examined using an inhibitor assay with chitobiose pentapeptides (Fig. 6) [44]. A series of chitobiose-pentapeptides was systematically designed by bonding hydrophobic leucine (Lue) and/or hydrophilic serine (Ser) at the *C-* and *N-termini* of the central Asn residue. The inhibition activity of these inhibitors for UGGT1 glucosylations were in the order of Ser-Ser-Asn(GlcNAc$_2$)-Ser-Ser > Leu-Leu-Asn(GlcNAc$_2$)-Ser-Ser >> Ser-Ser-Asn(GlcNAc$_2$)-Leu-Leu > Lue-Lue-Asn(GlcNAc$_2$)-Lue-Lue > Lue-Ser-Asn(GlcNAc$_2$)-Lue-Ser > Ser-Ser-Asn(GlcNAc$_2$)-Leu-Lue. Although UGGT1 is believed to recognize a hydrophobic patch near the glycosylation site, the observed inhibition patterns surprisingly indicate that hydrophilic Ser residues located near the glycosylation site are recognized more than hydrophobic Leu residues. Applying this UGGT1 peptide specificity to the consensus sequence for the *N*-glycosylation site (Asn-Xaa-Ser/Thr), glycoproteins with the Asn-Ser-Ser type consensus sequence could potentially be better recognized by UGGT1 than glycoproteins with other consensus sequences. This suggests that UGGT1 recognizes the amino acid sequence around the glycosylation site.

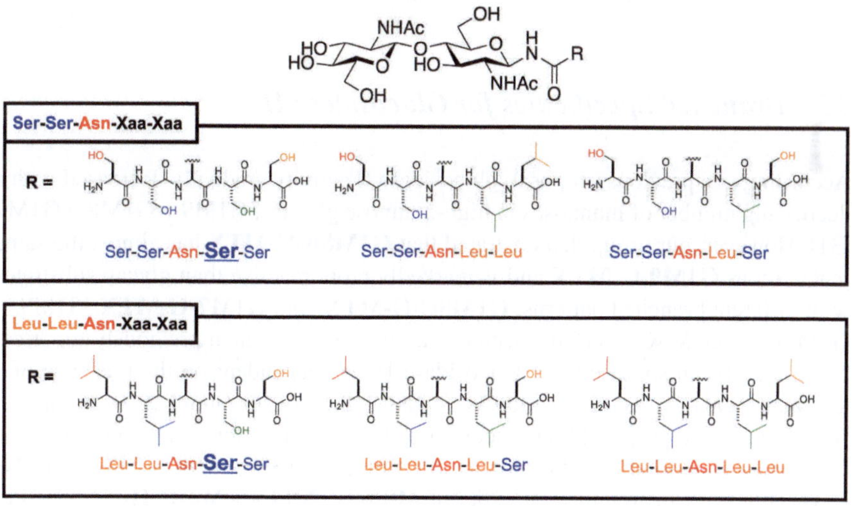


**Fig. 6** Chemical structures of chitobiose-pentapeptide-type UGGT1 inhibitors

# 3 Functional Analysis of Glucosidase II

## *3.1 Overlooked Aspects of First Proposal for Glucosidase II Specificity*

Glucosidase I/II catalyzes the trimming of G3M9 glycoproteins during initial biosynthesis in the ER lumen. Specifically, glucosidase II can trim two glycosidic bonds of Glc(α1-2)Glc in G2M9 and Glc(α1-3)Man in G1M9 glycoproteins [45]. This enzyme plays a key role in regulating the entry and exit of glycoproteins in the central machinery of ERQC, called the CNX/CRT cycle.

In 1980, Grinna and Robbins first reported the glycan specificity of glucosidase II, focusing on the second cleavage depending on the branched oligomannose patterns of the B- and C-arms, using various high-mannose-type glycan derivatives digested by endo-β-*N*-acetylglucosaminidase H [31]. However, there are two serious issues with this study in terms of the glycan substrates used. First, glycans lacking the innermost GlcNAc residue were used as substrates, whereas glucosidase II might recognize the natural reducing end GlcNAc. Second, the activities of glucosidase II could not be compared between the structural isomers G1M8B and G1M8C because they focused only on the number of mannose residues in the glycan substrate, i.e., a mixture of G1M8 structures was utilized as the substrate. To overcome these issues, we applied a series of chemically synthesized glycans with MTX (Fig. 7) were used as substrates to elucidate the substrate specificities of glucosidase II by quantitative analysis (see next section) [32]. Chemical approaches can correct previous proposals, revealing the true enzyme function in ERQC.

## *3.2 Branched Specificities for Glucosidase II*

According to a previous proposal, glucosidase II activity gradually decreased with a decreasing number of mannoses in high-mannose glycans: G1M9 > G1M8 > G1M7 [31]. However, our group demonstrated that **G1M8B-G-MTX** has almost the same reactivity as **G1M9-G-MTX** and is markedly more reactive than glycan substrates with different branched patterns, **G1M8C-G-MTX,** and **G1M7-G-MTX**: G1M9 = G1M8B >> G1M8C = G1M7, which is a different order than previously reported. Thus, the glycan specificity of glucosidase II was dependent on the presence of a terminal mannose residue at the C-arm rather than on the number of mannoses. Additionally, the proposal that glucosidase II recognizes a terminal mannose at the C-arm was supported by inhibition assays with **M9-**, **M8B-**, **M8C-**, **M7-**, and **M4-$C_3H_7$** (Fig. 8) with an inhibition order of **M8B- >> M8C- > M4-$C_3H_7$** [32]. Moreover, the inhibition order of **M8B-** and **M7- > M9-$C_3H_7$** suggests that a terminal mannose at the B-arm negatively affects glucosidase II. Chemically synthesized probes allowed us to reveal the true relative activity of glucosidase II toward G1M9, G1M8B, G1M8C, and G1M7 in terms of branch specificity.

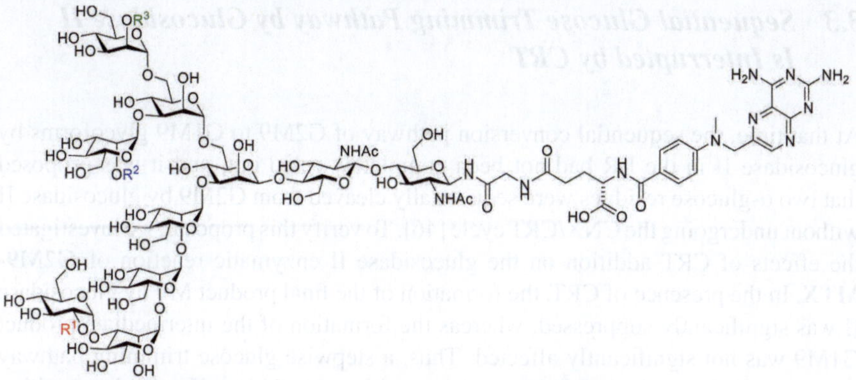


| Glycan structure | R1 | R2 | R3 |
|---|---|---|---|
| G2M9-G-MTX | 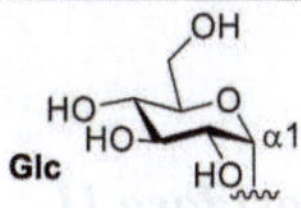 Glc | 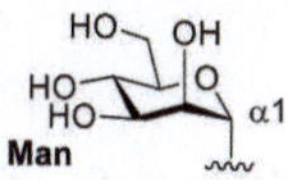 Man | 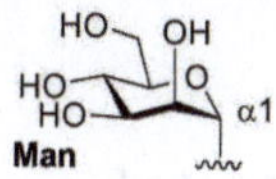 Man |
| G2M7-G-MTX | Glc | H | H |
| G1M9-G-MTX | H | Man | Man |
| G1M8B-G-MTX | H | H | Man |
| G1M8C-G-MTX | H | Man | H |
| G1M7-G-MTX | H | H | H |

**Fig. 7** Chemical structures of the glycan substrates for glucosidase II

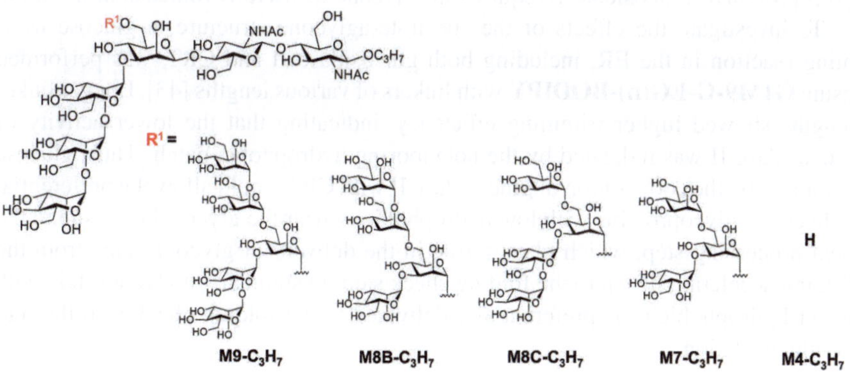


**Fig. 8** Chemical structures of glycan inhibitors of glucosidase II

### *3.3 Sequential Glucose Trimming Pathway by Glucosidase II Is Interrupted by CRT*

At that time, the sequential conversion pathway of G2M9 to G1M9 glycoforms by glucosidase II in the ER had not been completely ruled out, and it was proposed that two α-glucose residues were sequentially cleaved from G2M9 by glucosidase II without undergoing the CNX/CRT cycle [46]. To verify this proposal, we investigated the effects of CRT addition on the glucosidase II enzymatic reaction of **G2M9-MTX**. In the presence of CRT, the formation of the final product M9 by glucosidase II was significantly suppressed, whereas the formation of the intermediate product G1M9 was not significantly affected. Thus, a stepwise glucose trimming pathway was supported, in which G2M9 was trimmed by glucosidase II, and the resulting G1M9 was trapped by CRT.

### *3.4 Aglycone Effects for Glucosidase II*

The effect of aglycone on glucosidase II was examined using various glycan-MTXs grafted onto dihydrofolate reductase (DHFR), designated as glycan-DHFR [32]. Interestingly, **G1M8B-DHFR** was more reactive than **G1M9-DHFR**, which differed from the results described above, where the reactivity of **G1M9-MTX** was equivalent to that of **G1M8B-MTX**. Since MTX is strongly captured by correctly folded DHFR, glycan-DHFR is considered a model of a mature glycoprotein with a fully folded aglycone. When G1M8B is attached to a fully folded protein, it may be more susceptible to glucosidase II than G1M9 and is more likely to actively generate the secretion signal, M8B. Thus, we propose that the trimming of glucose from G1M9 glycoprotein by glucosidase II requires an aglycone with a fully folded conformation.

To investigate the effects of the substrate-aglycone structure, a glucose trimming reaction in the ER, including both glucosidase II and CRT, was performed using **G1M9-G-EG(n)-BODIPY** with linkers of various lengths [43]. Longer linker lengths showed higher trimming efficiency, indicating that the lower activity of glucosidase II was indicated by the neighboring hydrophobic patch. Thus, glucose trimming by the cooperation of glucosidase II with CRT would allow the preferential delivery of glycoproteins with low hydrophobicity near the glycosylation site to the next processing step, which plays a role in the delivery of glycoproteins from the folding acceleration step to the folding check step. In summary, a glycoprotein with lower hydrophobicity is preferentially delivered to the folding check step through de-glucosylation.

# 4 Functional Analysis of CRT

## 4.1 Overview of CRT Functions

CRT is a soluble lectin-like molecular chaperone mainly localized in the ER [47–49], which promotes glycoprotein folding and maintains $Ca^{2+}$ homeostasis [50]. The functions of CRT depend on three domains: a globular N-domain, which is a lectin domain that contributes to the capture of monoglucosylated high-mannose-type glycoproteins. A flexible P-domain is a chaperone domain that promotes client protein folding. A C-domain functions as a $Ca^{2+}$-storage domain. Moreover, in the CNX/CRT cycle, CRT recognizes G1M9-type glycoproteins and promotes the folding of nascent glycoproteins by corroborating with the folding sensor enzyme UGGT1 [42]. Therefore, CRT plays an essential role in the central machinery of the ERQC system. To precisely understand the recognition mechanism of CRT, the development of CRT-targeting ligands and inhibitors is essential. As the first synthetic ligand for CRT, Matsuo et al. succeeded in preparing **G1M9-$C_3H_7$** *via* convergent and stereoselective synthetic routes [51]. The interaction between **G1M9-$C_3H_7$** and CRT was observed using $^1$H NMR spectroscopy. Moreover, the binding constant was $K_b = \sim 5.2 \times 10^6$ $M^{-1}$, as measured using isothermal titration calorimetry [52, 53]. Interestingly, CRT did not interact with the non-natural β-linked **G1M9-$C_3H_7$** isomer bearing a β1-3 bond between glucose and mannose on the A-branch. Thus, CRT can distinguish between α- and β-glycosidic linkages at the terminal glucose residue. In addition, di-, tri-, and tetrasaccharides, which are truncated versions of the glucosylated A-branch of G1M9, bind to CRT [54, 55]. The increased number of saccharide residues gradually increased the binding constant, demonstrating that CRT recognizes the Glc(α1-3)Man(α1-2)Man(α1-2)Man structure of ligands. These ligands were used to identify the key binding sites, Asp317 and Trp319 of CRT. Taken together, structurally defined CRT ligands enable the elucidation of the CRT recognition mechanism.

## 4.2 CRT Recognition Depends on Hydrophobicity of Aglycone and Its Spatial Arrangement

Since CRT contains a chaperone domain, this lectin-like molecular chaperone may recognize not only the glycan moiety but also the hydrophobic aglycone of misfolded G1M9 glycoproteins. In other words, whether CRT recognizes both the glycan and aglycone of the G1M9-glycoprotein or only recognizes the glycan remains controversial. To clarify this issue, we selected a G1M9-possesing glycoprotein, immunoglobulin Y (**G1M9-IgY**), as a CRT ligand [56]. The protein moiety of **G1M9-IgY** was denatured to examine the changes in CRT-binding intensity. The strength of CRT binding increased with the degree of IgY denaturation. However, because denatured IgY randomly exposes hydrophobic patches at different distances from the

*N*-glycosylation site upon thermal denaturation, our experiments could not reveal the precise aglycone specificity of CRT. To determine whether CRT shows specificity for aglycone hydrophobicity, we used a series of synthetic G1M9 ligands possessing different aglycones: **G1M9-OH** [27], **G1M9-Gly-NH$_2$** [57], and **G1M9-Gly-Glu-$^t$Bu** (Fig. 9). The hydrophobicity of these ligands was in the order **G1M9-Gly-Glu-$^t$Bu > G1M9-Gly-NH$_2$ > G1M9-OH**, which corresponds to the order of CRT-binding ability for these three ligands. Moreover, the glucose transfer activity of UGGT1 showed the same aglycone specificity, suggesting that the similar discrimination properties of CRT and UGGT1 mediate the recognition of the aglycone moiety in the CNX/CRT cycle. Thus, the surface hydrophilicity of the glycoprotein aglycone would be the driving force for the release of the folded glycoprotein from the CNX/CRT cycle.

We also investigated the appropriate distance between the glycan and aglycone hydrophobic patch using G1M9 derivatives bearing different linker lengths connected to BODIPY as the hydrophobic patch, **G1M9-G-BODIPY**, **G1M9-G-EG4-BODIPY**, **G1M9-G-EG8-BODIPY**, and **G1M9-G-EG12-BODIPY** (Fig. 9) [43]. **G1M9-G-BODIPY**, which has the shortest linker, showed the highest affinity for CRT among the other ligands. The binding strength of CRT is inversely proportional to the distance between the reducing end of G1M9 and the hydrophobic substituents.

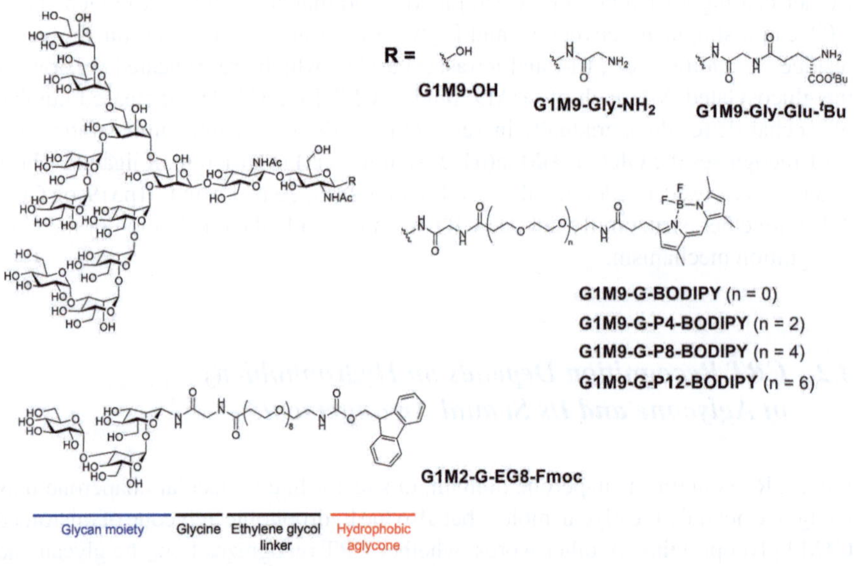


**Fig. 9** Chemical structures of G1M9 derivatives with different aglycones as CRT ligands and our rationally designed ligand, **G1M2-G-EG8-Fmoc**, based on the hybrid-binding concept

### 4.3 CRT-Targeting Ligand Based on the Hybrid-Binding Concept

The development of CRT ligands has been useful for understanding not only glycan specificity but also aglycone specificity. However, the synthesis of the native G1M9 glycan of CRT has been labor-intensive because it requires complicated multi-step reactions. Thus, there is still room for improvement in CRT ligand design. In this section, a CRT ligand based on the hybrid-binding concept is introduced [58]. The hybrid-binding concept serves as a guideline for designing ligands that can simultaneously bind to both the lectin and chaperone domains of CRT. As a molecular design based on the hybrid-binding concept, the CRT ligand **G1M2-G-EG8-Fmoc** was developed (Fig. 9). The trisaccharide **G1M2** moiety is the minimal glycan structure for CRT binding. In addition, the Fmoc group mimics the hydrophobic surface of misfolded glycoproteins. The glycan moiety was connected to the Fmoc group *via* an octapolyethylene glycol linker (EG8), which was estimated to be approximately the same as the overall length of **G1M9-G-Fmoc** as an effective CRT ligand. The binding affinities of **G1M2-G-EG8-Fmoc** and its substructures for CRT were compared using isothermal titration calorimetry (Table 1). Comparing **G1M2-G-EG8-Fmoc** with **G1M2**, the affinity of **G1M2-G-EG8-Fmoc** ($K_b = 5.0 \times 10^6\ M^{-1}$) was higher than that of **G1M2** ($K_b = 1.4 \times 10^5\ M^{-1}$) which only had the glycan moiety, suggesting that Fmoc group could enhance the affinity to CRT. Additionally, the binding of the Fmoc group to the CRT chaperone domain was supported by experiments in which **G1M2-G-EG8-Fmoc** inhibited CRT chaperone activity, whereas **G1M2** and **G1M3** [59] did not inhibit the chaperone activity. The Fmoc group may interact with the aromatic amino acid residues of CRT *via* $\pi$–$\pi$ interactions. **HOOC-EG8-Fmoc**, which lacks a glycan moiety, and **Maltotriose**, which has a different saccharide sequence, showed no affinity for CRT, indicating that G1M2 binding to the lectin domain of CRT is important for the initial stage of ligand capture by CRT. The affinity of **G1M2-G-EG8-Fmoc,** enhanced by the hybrid-binding concept, was comparable to that of the native-type glycan ligand **G1M9-$C_3H_7$** ($K_b = \sim 5.2 \times 10^6\ M^{-1}$) [52]. Therefore, our proposed hybrid-binding concept provides a useful strategy for designing and developing ligands and inhibitors of CRT.

**Table 1** Binding affinity of synthetic ligands to CRT

| Ligand | $K_b$ ($M^{-1}$) |
|---|---|
| G1M2 | $1.4 \times 10^5$ |
| G1M2-G-EG8-Fmoc | $5.0 \times 10^6$ |
| HOOC-EG8-Fmoc | n.d.[a] |
| Maltotriose | n.d.[a] |
| G1M9-$C_3H_7$ | $\sim 5.2 \times 10^6$ |

[a] Not determined

# 5 Functional Analysis of EDEMs

## 5.1 *Historical Changes in the Interpretation of EDEMs Functions*

Mannose trimming of glycans on proteins is a critical step in distinguishing protein folding states and determining the pathway sorting of properly folded and misfolded high-mannose-type glycoproteins in ERQC [60]. ER-degradation enhancing mannosidase-like proteins 1, 2, and 3 (EDEM1, EDEM2, and EDEM3) are currently interpreted to catalyze the trimming of α-1,2-mannosidic linkages on the A-, B-, and C-branched chains of high-mannose-type glycoprotein glycans by reflecting the folding state of glycoproteins [3]. EDEM (later renamed EDEM1) was first discovered in 2001 and was suggested to be a lectin that recognizes high-mannose-type glycoproteins because it does not exhibit mannosidase activity in vitro [61]. Later, some researchers found that overexpressed EDEM1 promoted mannose trimming from M9 to M8A/B and M8B to M7A/C [62, 63]. This result supports the notion that EDEM1 has mannosidase activity. Moreover, EDEM2 and EDEM3 were discovered by comparing their primary amino acid sequence homology with that of EDEM1 [64]. Initially, EDEM2 showed no mannosidase activity, whereas EDEM3 showed mannosidase activity [65, 66]. Thus, at this point, it was unclear which of these three EDEMs were mannosidases and which were lectins. In 2014, Ninagawa et al. reported that EDEM2-KO cells accumulated more M9-proteins than EDEM1-KO or EDEM3-KO cells [67], and proposed that EDEM2 contributes to the conversion of M9- to M8B-proteins in the mannose-trimming pathway. Although EDEM2 monomer showed no mannosidase activity in vitro [68], EDEM2-TXNDC11 complex with disulfide bonding catalyzed the mannose trimming of M9 to M8B in vitro [17]. EDEM1 and EDEM3 monomers are proposed to contribute to mannose trimming from M8B to M5 *via* M7A/C and M6, respectively [21]. ER mannosidase I (ERManI, later renamed MAN1B1) was first proposed to cooperate with EDEMs as an ER-resident mannosidase and was later demonstrated to localize in the Golgi apparatus [69–71]. In the above-mentioned EDEMs controversy, our group demonstrated three issues of EDEMs: (1) effects of aglycone in the mannose trimmings, (2) inhibitor-based validation of mannose-trimming pathways, and (3) recognition mechanism for each branch of high-mannose-type glycans.

## 5.2 *Effects of Aglycone in the Mannose Trimmings*

UGGT1 shows higher activity for an M9-derivative with a hydrophobic patch positioned closer to the reducing end, demonstrating that *N*-glycan processing reactions can be regulated by the aglycone structure in the ER [42]. However, it remains unclear whether EDEMs are affected by the aglycone structure of *N*-glycans during ER glycan processing. To examine the effect of aglycone, we used M9 derivatives possessing hydrophobic BODIPY and different lengths of linkers (oligoethylene glycol) as substrates, which are mimics of molten globule-like glycoproteins (Fig. 4) [43]. Under conditions of enriched ER fractions containing EDEMs, the oligoethylene glycol additive had no effect on the calculated total mannose-trimming activities. Thus, oligoethylene glycol moieties do not affect the substrate specificity of EDEMs. In contrast, the total mannose-trimming activity of **M9-G-EG(0–12)-BODIPY** gradually decreased with increasing linker length. Thus, the presence of a hydrophobic substituent close to the reducing end of the glycan accelerates mannose trimming.

### *5.3 Inhibitor-Based Validation of Mannose-Trimming Pathways*

As a proposal of the biosynthetic pathway of glycoprotein glycans in the ER, M8B, which is generated by trimming a mannose from the B-branch of M9, is believed to act as a secretion signal, whereas M8A/C and their isomers, M7, M6, and M5, which are in turn generated from trimming the A- or C-branches of M9, are proposed to be degradation signals [72]. The contribution of EDEMs to mannose trimming has been investigated using various approaches, and it has been proposed that multiple α-mannosidases (EDEMs) works in parallel in the ER [62, 63, 67]. However, at that time, no in vitro mannosidase activity had been reported for purified EDEM1, EDEM2, or EDEM3. Thus, little is known about how these mannose-trimmed products are formed and whether distinct trimming pathways lead to secretion or degradation signals. To overcome this issue, Kuribara/Totani and co-workers analyzed mannose trimming of glycans in the enriched ER lysate containing EDEMs using reciprocally selective inhibitors [73].

Most previous inhibitory studies on *N*-glycan processing have been conducted using high concentrations of inhibitors. Representative *N*-glycan processing inhibitors are shown in Fig. 10: Mannoimidazole (**ManIm**) and derivatives (**2-phenyl ManIm**, **2-phenethyl ManIm**, and **2-octyl ManIm**) as inhibitors for GH47 family mannosidases; Mannostatin A (**ManA**) and swainsonine (**SW**) as Golgi mannosidase II inhibitors; kifunensine (**Kif**) and 1-deoxymannojirimycin (**dMJ**) as ERManI inhibitors; and dMJ derivatives (***N*-butyl dMJ** and ***N*-decyl dMJ**) and their analogs 1-deoxynojirimycin (**dNJ**), 1-deoxygalactonojirimycin (**dGJ**), 1-deoxyfuconojirimycin (**dFJ**). We used **M9-G-BODIPY** as a substrate to investigate the selectivity of the inhibitors in glycan processing within the ER fraction and examined the glycan processing status under low inhibitor concentrations. The $IC_{50}$ values of the representative inhibitors for M8B (as a secretion signal) and M8A/C (as degradation signals) production are shown in Table 2. **ManIm** showed no significant selective inhibition. At the strictly selected inhibitor concentrations, **Kif** and **dNJ** could selectively inhibit M8B production: i.e., **M9-G-BODIPY** was favored to proceed *via* the degradation pathway in the presence of **Kif** or **dNJ**. In contrast,

**ManA**, **SW**, and **dMJ** could selectively inhibit M8A/C production: i.e., secretion pathway was preferentially selected in the presence of **ManA**, **SW**, or **dMJ**. ER glycan processing assays using reciprocally selective inhibitors demonstrated that discrete pathways for glycosignal production must exist, mediated by sets of regioselective α-1,2-mannosidases, such as EDEMs.

Reciprocally selective iminosugar-type inhibitors **dMJ** and **dNJ** were selected for a detailed analysis of the mannose-trimming pathways in the ER owing to their structural relationship, selectivity, and commercial availability. The Dixon plot of **dNJ** and **dMJ** for M8 isomer production showed $K_i$ values of 3 mM and 0.16 μM, respectively, as competitive inhibitors. Furthermore, these selective inhibitors were used to determine the individual glycan production pathways of **M9-G-BODIPY** in the ER fraction (Fig. 11). In the absence of inhibitors, M9 was trimmed to produce M8A/B/C → M7A/B/C → M6A/iso → M5. In the presence of 5 μM **dMJ**, the first mannose trimming from M9 selectively generated M8B. Further mannose trimmings proceeded *via* dMJ-insensitive mannosidase(s) to provide M7A → M6 →

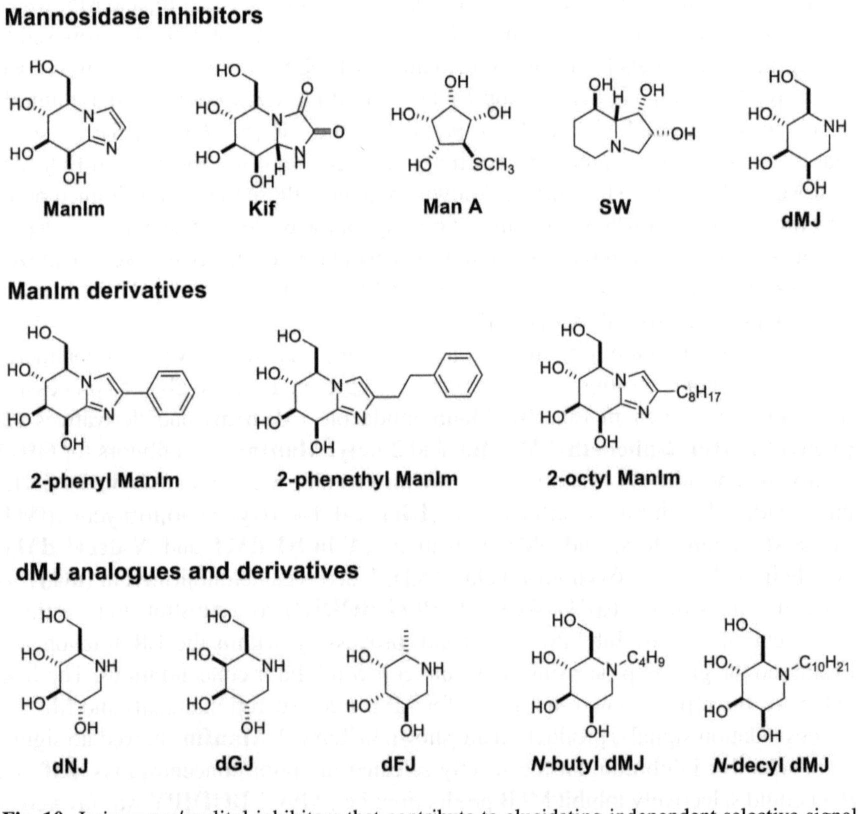


**Fig. 10** Iminosugar/cyclitol inhibitors that contribute to elucidating independent selective signal production pathways for M8B- or M8A/C-type glycans using glycan-BODIPY-type substrates

**Table 2** $IC_{50}$ values of representative inhibitors for M8B and M8A/C production

| Inhibitors | $IC_{50}^{M8B}$ (μM) | $IC_{50}^{M8A/C}$ (μM) |
|---|---|---|
| ManIm | $37 \pm 16$ | $60 \pm 13$ |
| Kif | $0.090 \pm 0.013$ | $0.40 \pm 0.080$ |
| ManA | $1.9 \times 10^3 \pm 0.66 \times 10^3$ | $0.27 \times 10^3 \pm 0.090 \times 10^3$ |
| SW | $> 1.0 \times 10^4$ | $3.3 \times 10^3 \pm 0.66 \times 10^3$ |
| dMJ | $3.0 \pm 0.28$ | $0.43 \pm 0.039$ |
| dNJ | $5.3 \times 10^3 \pm 0.86 \times 10^3$ | $> 5.0 \times 10^4$ |

The Table is adapted from [73]

M5 as products. In contrast, in the presence of 50 mM **dNJ**, M8A was preferentially generated as the first product from M9, followed by M7B → M6iso → M5 as dNJ-insensitive mannosidase(s) products. These findings provide molecular evidence for the regioselective production of ER signal glycoforms, supporting the notion that specific glycoforms play important roles in the traffic control of glycoprotein secretion and degradation [74–76].

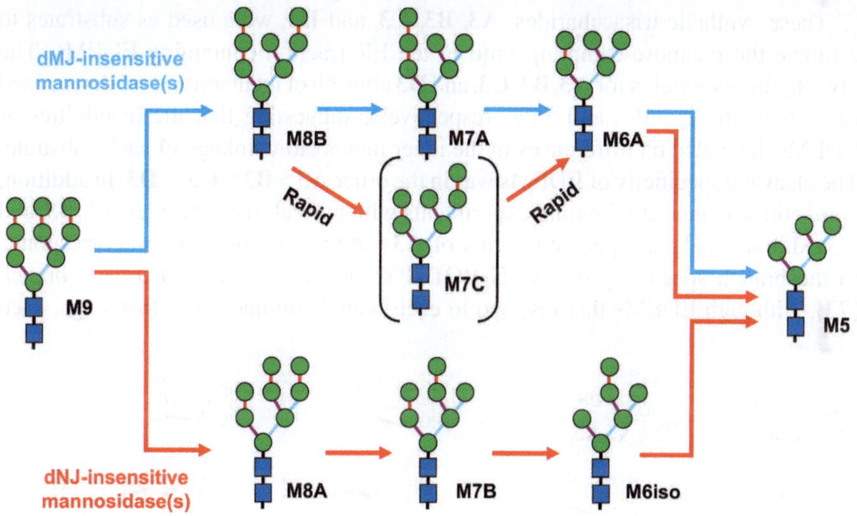


**Fig. 11** Observed mannose-trimming pathways in the cell-free mouse ER system

### *5.4 Recognition of Each Branch of High-Mannose-Type Glycan by EDEMs*

EDEM family proteins (EDEM1, EDEM2, and EDEM3) are important for glycoprotein quality control in the ER. EDEMs trim α1,2-mannose residue from M9-glycoprotein and produce secretion and degradation glycan signals. In the biosynthetic pathway, EDEM2 is proposed to produce M8B-glycan, with further A- and/or C-branch mannose trimming(s) thought to be mediated by EDEM3 and EDEM1 [3]. Interestingly, the terminal mannosidic linkage of the three branches (A-, B-, and C-branches) of the M9 structure are all α1,2-form, implying that, in principle, all EDEMs are able to produce all M8-type glycans; however, in practice, EDEM2 specifically trims the B-branch of M9. In contrast, the branch specificity of EDEM1 and EDEM3 for M9 remains unclear. To address this issue, we focused on the differences in the internal glycosidic bond patterns of the A-, B-, and C-branches as a factor for the branch discrimination of EDEMs. Specifically, we synthesized trisaccharides corresponding to the A-, B-, and C-branches of the M9 glycan, named **A3**, **B3**, and **C3** with innermost Man(α1,2)Man, Man(α1,3)Man, and Man(α1,6)Man linkages, respectively, and an unnatural **D3** with a Man(α1,4) linkage as chemical tools to elucidate the branch discrimination of EDEMs (Fig. 12) [77].

These synthetic trisaccharides, **A3**, **B3**, **C3**, and **D3,** were used as substrates to estimate the mannose-trimming ratio in the ER fraction containing EDEMs. The glycohydrolysis yields for **A3**, **B3**, **C3**, and **D3** after 8 h of incubation were determined to be 66%, 69%, 35%, and 35%, respectively, suggesting that the reactivities of EDEMs depended on differences in the inner mannosidic linkage of each substrate. The cleavage specificity of EDEMs was in the order **A3** > **B3** > **C3** = **D3**. In addition, competitive mannose-trimming experiments with natural trisaccharides **A3**, **B3**, and **C3** exhibited a glycan specificity order of **A3** > **B3** > **C3**, which almost corresponds to the branch specificity of **M9-G-BODIPY**: A-branch = B-branch > C-branch [73]. Although EDEMs that respond to each branch are unclear at this stage, each

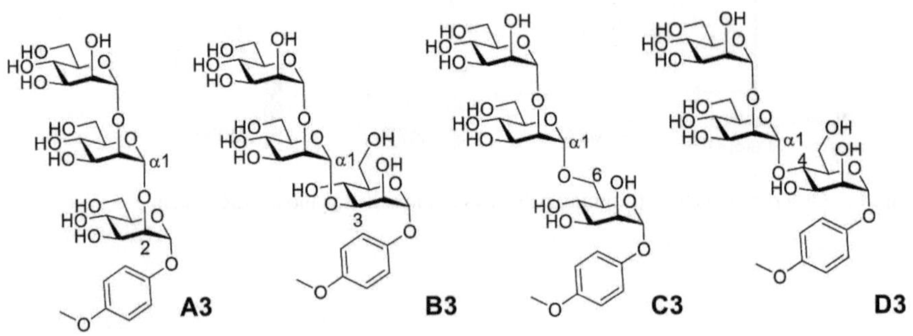


**Fig. 12** Chemical structures of **A3**, **B3**, and **C3** with mannosyl linkages of the A-, B-, and C-branch of M9 type glycan, and **D3** with unnatural mannosyl linkage

branch-responsive EDEM might be identified through further specificity analysis using purified or recombinant EDEMs and branch trisaccharides.

Moreover, competitive experiments with trisaccharides **A3**, **B3**, and **C3** and unnatural trisaccharide **D3** were performed. Interestingly, the hydrolysis activities of **A3** and **B3** were significantly improved in the presence of **D3** compared to those in its absence of **D3**. This improved mannosidase activity could be explained by **D3** functionalization as a pharmacological chaperone, which partially repaired the unfolded EDEMs. In contrast, the hydrolysis activity of **C3** with **D3** was almost the same as that without **D3**. Therefore, the unnatural trisaccharide **D3** is a promising lead compound for developing novel pharmacological chaperones.

## 6 Differences in Glycan Processing Profiles Between Healthy and Disease Models

In the above-mentioned sections, we introduced our findings on the functional analysis of specific glycan-related proteins involved in ERQC. However, because ERQC homeostasis is maintained by the balance of the action of various glycan-related proteins, it is also important to pay close attention to changes in the overall operating status of ERQC in response to fluctuations in the in vivo environment. In fact, the accumulation of misfolded (glyco)proteins causes ER stress, followed by misfolding diseases such as type 2 diabetes, neurodegenerative disorders, Alzheimer's disease, Parkinson's disease, Huntington's disease, amyotrophic lateral sclerosis, and prion diseases [78]. In this section, we demonstrated that the alternation in glycan processing in the ER can reflect disease states in type 2 diabetes and osteoporosis using a synthetic G1M9 glycan probe [79]. To observe the differences in glycan processing status between healthy and disease samples, we established a novel glycan-processing analysis system, described as the reconstructed glycan profile method. In this method, a structurally defined synthetic G1M9-type glycoprobe was incubated as an upstream glycan processing substrate with an ER fraction containing all glycan-related enzymes and lectins, followed by analysis of the reaction mixture to obtain the reconstructed glycan profiles over the reaction time. When comparing glycan profiles between ER fractions extracted from the livers of non-obese type 2 diabetes model rats, Goto-Kakizaki (GK) rats, and the related healthy model, Wistar (WT) rats, no differences were observed between the endogenous glycan profiles (Fig. 13a), clear differences were observed between the two reconstructed glycan profiles (Fig. 13b). Moreover, the reconstructed glycan profile from an osteoporosis model mouse (SAMP6) showed large differences from that of a normal aging model mouse (SAMR1) and a dementia model mouse (SAMP10) in terms of the detected amount of M8 isomers (Fig. 13c). The M8B isomer (secretion signal) was detected as a major M8 product in SAMR1 and SAMP10, whereas the M8A/C isomer (degradation signals) was detected as a major M8 product in SAMP6. This suggests that mannose-trimming activity is affected by disease conditions, such as the accumulation of misfolded glycoproteins. While the precise relationship between the onset of disease and alterations in glycan processing within the ER remains ambiguous, it is evident that an imbalance in ERQC is linked to diseases associated with ER stress.

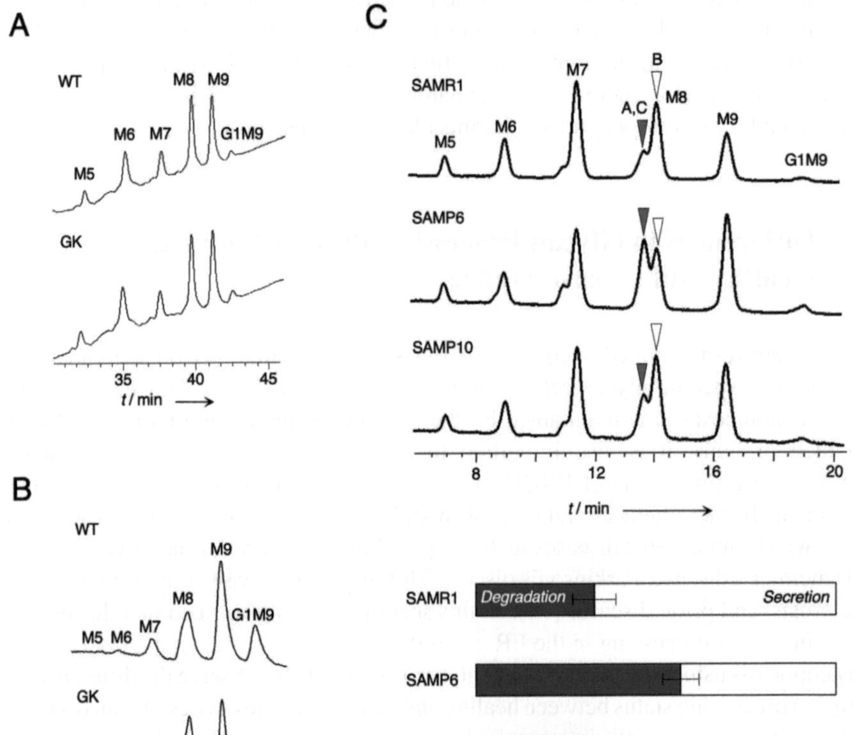


**Fig. 13** **A** Direct glycan profiles and **B** reconstructed glycan profiles of isolated ER lysate in healthy (WT) and non-obese type 2 diabetes rat models (GK). **C** Reconstructed glycan profiles of isolated ER lysates in osteoporosis and dementia (one of the neurodegenerative disorders) model mice, senescence-accelerated mouse prone 6 (SAMP6) and SAMP10, and the healthy model SAMR1. M8A/C and M8B showed glycan signals of degradation and secretion. These figures were adapted from [79].

## 7 Conclusion

Chemical approaches using synthetic high-mannose-type glycan probes have substantially advanced our understanding of the molecular mechanisms underlying glycoprotein folding, quality control, and ER trafficking. These studies clarified the substrate specificities of key glycan-processing enzymes and lectin chaperones, such as UGGT1, glucosidase II, CRT, and EDEMs, highlighting the importance of glycan branching patterns, aglycone hydrophobicity, and peptide context around glycosylation sites. Furthermore, inhibitor assays and the reconstructed glycan profile method revealed discrete mannose-trimming pathways and their alterations under disease conditions, providing molecular insights into ER stress-related disorders. These findings demonstrate the unique value of synthetic glycan probes in dissecting complex biological processes at the molecular level. Future research leveraging these chemical tools will enable the design of targeted ligands or inhibitors for modulating ER quality control, potentially leading to novel therapeutic strategies for diseases caused by protein misfolding and beyond.

## References

1. Varki A (2017) Biological role of glycans. Glycobiology 27:3–49. https://doi.org/10.1093/glycob/cww086
2. Ito Y, Kajihara Y, Takeda Y (2020) Chemical-synthesis-based approach to glycoprotein functions in the endoplasmic reticulum. Chem - Eur J 26:15461–15470. https://doi.org/10.1002/chem.202004158
3. Ninagawa S, George G, Mori K (2021) Mechanisms of productive folding and endoplasmic reticulum-associated degradation of glycoproteins and non-glycoproteins. Biochim Biophys Acta Gen Subj 1865:129812. https://doi.org/10.1016/j.bbagen.2020.129812
4. Kuribara T, Totani K (2020) Structural insights into *N*-linked glycan-mediated protein folding from chemical and biological perspectives. Curr Opin Struct Biol 68:41–47. https://doi.org/10.1016/j.sbi.2020.11.004
5. Aebi M. (2013) N-linked protein glycosylation in the ER. Biochim Biophys Acta—Mol Cell Res 1833:2430–2437. https://doi.org/10.1016/j.bbamcr.2013.04.001
6. Alaimo C, Catrein I, Morf L, Marolda CL, Callewaert N, Valvano MA, Feldman MF, Aebi M (2006) Two distinct but interchangeable mechanisms for flipping of lipid-linked oligosaccharides. EMBO J 25:967–976. https://doi.org/10.1038/sj.emboj.7601024
7. Raetz CRH, Whitfield C (2002) Lipopolysaccharide endotoxins. Annu Rev Biochem 71:635–700. https://doi.org/10.1146/annurev.biochem.71.110601.135414
8. Yan A, Guan Z, Raetz CRH (2007) An undecaprenyl phosphate-aminoarabinose flippase required for polymyxin resistance in Escherichia coli. J Biol Chem 282:36077–36089. https://doi.org/10.1074/jbc.M706172200
9. Ramírez AS, Locher KP (2023) Structural and mechanistic studies of the *N*-glycosylation machinery: from lipid-linked oligosaccharide biosynthesis to glycan transfer. Glycobiology 33:861–872. https://doi.org/10.1093/glycob/cwad053
10. Ware FE, Vassilakos A, Peterson PA, Jackson MR, Lehrman MA, Williams DB (1995) The molecular chaperone calnexin binds $Glc_1Man_9GlcNAc_2$ oligosaccharide as an initial step in recognizing unfolded glycoproteins. J Biol Chem 270:4697–4704. https://doi.org/10.1074/jbc.270.9.4697

11. Michalak M, Milner RE, Burns K, Opas M (1992) Calreticulin. Biochem J 28:681–692. https://doi.org/10.1042/bj2850681
12. Frickel E-M, Frei P, Bouvier M, Stafford WF, Helenius A, Glockshuber R, Ellgaard L (2004) ERp57 is a multifunctional thiol-disulfide oxidoreductase. J Biol Chem 279:18277–18287. https://doi.org/10.1074/jbc.M314089200
13. Nakao H, Seko A, Ito Y, Sakono M (2017) PDI family protein ERp29 recognizes P-domain of molecular chaperone calnexin. Biochem Biophys Res Comm 487:763–767. https://doi.org/10.1016/j.bbrc.2017.04.139
14. Kozlov G, Bastos-Aristizabal S, Määttänen P, Rosenauer A, Zheng F, Killikelly A, Trempe J-F, Thomas DY, Gehring K (2010) Structural basis of cyclophilin B binding by the calnexin/calreticulin P-domain. J Biol Chem 285:35551–35557. https://doi.org/10.1074/jbc.M110.160101
15. Trombetta ES, Simons JF, Helenius A (1996) Endoplasmic reticulum glucosidase II is composed of a catalytic subunit, conserved from yeast to mammals, and a tightly bound noncatalytic HDEL-containing subunit. J Biol Chem 271:27509–27516. https://doi.org/10.1074/jbc.271.44.27509
16. Pelletier MF, Marcil A, Sevigny G, Jakob CA, Tessier DC, Chevet E, Menard R, Bergeron JJM, Thomas DY (2000) The heterodimeric structure of glucosidase II is required for its activity, solubility, and localization in vivo. Glycobiology 10:815–827. https://doi.org/10.1093/glycob/10.8.815
17. George G, Ninagawa S, Yagi H, Saito T, Ishikawa T. Sakuma T, Yamamoto T, Imami K, Ishihama Y, Kato K, Okada T, Mori K (2020) EDEM2 stably disulfide-bonded to TXNDC11 catalyzes the first mannose trimming step in mammalian glycoprotein ERAD. eLife 9:e53455. https://doi.org/10.7554/eLife.53455
18. Yamamoto K (2014) Intracellular lectins are involved in quality control of glycoproteins. Proc Jpn Acad Ser B 90:67–82. https://doi.org/10.2183/pjab.90.67
19. D'Alessio C, Caramelo JJ, Parodi AJ (2010) UDP-Glc:glycoprotein glucosyltransferase-glucosidase II, the ying-yang of the ER quality control. Semin Cell Dev Biol 21:491–499. https://doi.org/10.1016/j.semcdb.2009.12.014
20. Trombetta ES, Helenius A (2000) Conformational requirements for glycoprotein reglucosylation in the endoplasmic reticulum. J Cell Biol 148:1123–1130. https://doi.org/10.1083/jcb.148.6.1123
21. George G, Ninagawa S, Yagi H, Furukawa J, Hashii N, Ishii-Watabe A, Deng Y, Matushita K, Ishikawa T, Mamahit YP, Maki Y, Kajihara Y, Kato K, Okada T, Mori K (2021) Purified EDEM3 or EDEM1 alone produces determinant oligosaccharide structures from M8B in mammalian glycoprotein ERAD. eLife 10:e70357. https://doi.org/10.7554/eLife.70357
22. Kuznetsov G, Chen LB, Nigam SK (1994) Several endoplasmic reticulum stress proteins, including ERp72, interact with thyroglobulin during its maturation. J Biol Chem 269:22990–22995. https://doi.org/10.1016/S0021-9258(17)31608-3
23. Trombetta ES, Parodi AJ (2005) Glycoprotein reglucosylation. Methods 35:328–337. https://doi.org/10.1016/j.ymeth.2004.10.004
24. Wu Y, Swulius MT, Moremen KW, Sifers RN (2003) Elucidation of the molecular logic by which misfolded α1-antitrypsin is preferentially selected for degradation. Proc Natl Acad Sci USA 100:8228–8234. https://doi.org/10.1073/pnas.1430537100
25. Hosokawa N, Tremblay LO, You Z, Herscovic A, Wada I, Nagata K (2003) Enhancement of endoplasmic reticulum (ER) degradation of misfolded null Hong Kong α1-antitrypsin by human ER mannosidase I. J Biol Chem 278:26287–26294. https://doi.org/10.1074/jbc.M303395200
26. Aikawa J, Matsuo I, Ito Y (2012) In vitro mannose trimming property of human ER α-1,2 mannosidase I. Glycoconj J 29:35–45. https://doi.org/10.1007/s10719-011-9362-1
27. Matsuo I, Totani K, Tatami A, Ito Y (2006) Comprehensive synthesis of ER related high-mannose-type sugar chains by convergent strategy. Tetrahedron 62:8262–8277. https://doi.org/10.1016/j.tet.2006.06.045

28. Matsuo I, Ito Y (2005) Systematic synthesis of ER related *N*-glycans using convergent strategy. Trends Glycosci Glycotechnol 17:85–95. https://doi.org/10.4052/tigg.17.85
29. Takeda Y, Totani K, Matsuo I, Ito Y (2009) Chemical approaches toward understanding glycan-mediated protein quality control. Curr Opin Chem Biol 13:582–591. https://doi.org/10.1016/j.cbpa.2009.09.011
30. Kuribara T, Hirose M, Tajima N, Totani K (2024) Chemical study of high-mannose-type glycan-related proteins to understand their functions at the molecular level. Trends Glycosci Glycotechnol 36:E94–E103. https://doi.org/10.4052/tigg.2039.7E
31. Grinna LS, Robbins PW (1980) Substrate specificities of rat liver microsomal glucosidases which process glycoproteins. J Biol Chem 255:2255–2258. https://doi.org/10.1016/S0021-9258(19)85881-7
32. Totani K, Ihara Y, Matsuo I, Ito Y (2006) Substrate specificity analysis of endoplasmic reticulum glucosidase II using synthetic high mannose-type glycans. J Biol Chem 281:31502–31508. https://doi.org/10.1016/S0021-9258(19)84063-2
33. Hase S, Ikenaka T, Matsushima Y (1987) Structure analyses of oligosaccharides by tagging of the reducing end sugars with a fluorescent compound. Biochem Biophys Res Comm 85:257–263. https://doi.org/10.1016/S0006-291X(78)80037-0
34. Bigge JC, Patel TP, Bruce JA, Goulding PN, Charles SM, Parekh RB (2002) Nonselective and efficient fluorescent labeling of glycans using 2-amino benzamide and anthranilic acid. Anal Biochem 230:229–238. https://doi.org/10.1006/abio.1995.1468
35. Parodi AJ (2000) Protein glucosylation and its role in protein folding. Annual Rev Biochem 69:69–93. https://doi.org/10.1146/annurev.biochem.69.1.69
36. Yoshida Y, Chiba T, Tokunaga F, Kawasaki H, Iwai K, Suzuki T, Ito Y, Matsuoka K, Yoshida M, Tanaka K, Tai T (2002) Nature 418:438–442. https://doi.org/10.1038/nature00890
37. Hagihara S, Totani K, Matsuo I, Ito Y (2005) Thermodynamic analysis of interactions between N-linked sugar chains and F-box protein Fbs1. J Med Chem 48:3126–3129. https://doi.org/10.1021/jm0489511
38. Caramelo JJ, Castro OA, Prat-Gay G, Parodi AJ (2004) The endoplasmic reticulum glucosyltransferase recognizes nearly native glycoprotein folding intermediates. J Biol Chem 279:46280–46285. https://doi.org/10.1074/jbc.M408404200
39. Caramelo JJ, Castro OA, Alonso LG, Prat-Gay G, Parodi AJ (2002) UDP-Glc:glycoprotein glucosyltransferase recognizes structured and solvent accessible hydrophobic patches in molten globule-like folding intermediates. Proc Natl Acad Sci USA 100:86–91. https://doi.org/10.1073/pnas.262661199
40. Sousa MC, Ferrero-Garcia MA, Parodi AJ (1991) Recognition of the oligosaccharide and protein moieties of glycoproteins by the UDP-Glc:glycoprotein glucosyltransferase. Biochemistry 31:97–105. https://doi.org/10.1021/bi00116a015
41. Totani K, Ihara Y, Matsuo I, Koshino H, Ito Y (2005) Synthetic substrates for an endoplasmic reticulum protein-folding sensor, UDP-glucose: glycoprotein glucosyltransferase. Angew Chem Int Ed 117:8164–8168. https://doi.org/10.1002/ange.200502723
42. Totani K, Ihara Y, Tsujimoto T, Matsuo I, Ito Y (2009) The recognition motif of the glycoprotein-folding sensor enzyme UDP-Glc:glycoprotein glucosyltransferase. Biochemistry 48:2933–2940. https://doi.org/10.1021/bi8020586
43. Totani K, Yamaya K, Hirano M, Ito Y (2017) Influence of aglycone structures on *N*-glycan processing reactions in the endoplasmic reticulum. Carbohydr Res 439:16–22. https://doi.org/10.1016/j.carres.2016.12.008
44. Kudo T, Hirano M, Ishihara T, Shimura S, Totani K (2014) Glycopeptide probes for understanding peptide specificity of the folding sensor enzyme UGGT. Bioorg Med Chem Lett 24:5563–5567. https://doi.org/10.1016/j.bmcl.2014.11.013
45. Trombetta ES (2003) The contribution of N-glycans and their processing in the endoplasmic reticulum to glycoprotein biosynthesis. Glycobiology 13:77R-91R. https://doi.org/10.1093/glycob/cwg075
46. Parodi AJ (1993) N-glycosylation in trypanosomatid protozoa. Glycobiology 3:193–199. https://doi.org/10.1093/glycob/3.3.193

47. Michalak M, Corbett EF, Masaeli N, Nakamura K, Opas M (1999) Calreticulin: one protein, one gene, many functions. Biochem J 344:281–292. https://doi.org/10.1042/bj3440281
48. Coppolino MG, Dedhar S (1998) Calreticulin. Int J Biochem Cell Biol 30:553–558. https://doi.org/10.1016/S1357-2725(97)00153-2
49. Krause KH, Michalak M (1997) Calreticulin. Cell 88:439–443. https://doi.org/10.1016/S0092-8674(00)81884-X
50. Michalak M, Groenendyk J, Szabo E, Gold LI, Opas M (2009) Calreticulin, a multi-process calcium-buffering chaperone of the endoplasmic reticulum. Biochem J 417:651–666. https://doi.org/10.1042/BJ20081847
51. Matsuo I, Wada M, Manabe S, Yamaguchi Y, Otake K, Kato K, Ito Y (2003) Synthesis of monoglucosylated high-mannose-type dodecasaccharide, a putative ligand for molecular chaperone, calnexin, and calreticurin. J Am Chem Soc 125:3402–3403. https://doi.org/10.1021/ja021288q
52. Ito Y, Hagihara S, Matsuo I, Totani K (2005) Structural approaches to the study of oligosaccharides in glycoprotein quality control. Curr Opin Struct Biol 15:481–489. https://doi.org/10.1016/j.sbi.2005.08.012
53. Arai MA, Matsuo I, Hagihara S, Totani K, Maruyama J, Kitamoto K, Ito Y (2005) Design and synthesis of oligosaccharides that interfere with glycoprotein quality-control systems. ChemBioChem 6:2281–2289. https://doi.org/10.1002/cbic.200500143
54. Kapoor M, Srinivas H, Kandiah E, Gemma E, Ellgaard L, Oscarson S, Helenius A, Surolia A (2003) Interactions of substrate with calreticulin, an endoplasmic reticulum chaperone. J Biol Chem 278:6194–6200. https://doi.org/10.1074/jbc.M209132200
55. Ito Y, Hagihara S, Arai MA, Matsuo I, Takatani M (2004) Synthesis of fluorine substituted oligosaccharide analogues of monoglucosylated glycan chain, a proposed ligand of lectin-chaperone calreticulin and calnexin. Glycoconj J 21:257–266. https://doi.org/10.1023/B:GLYC.0000045109.60425.2e
56. Hirano M, Adachi Y, Ito Y, Totani K (2015) Calreticulin discriminates the proximal region at the *N*-glycosylation site of $Glc_1Man_9GlcNAc_2$ ligand. Biochem Biophys Res Comm 466:350–355. https://doi.org/10.1016/j.bbrc.2015.09.026
57. Totani K, Matsuo I, Ihara Y, Ito Y (2006) High-mannose-type glycan modifications of dihydrofolate reductase using glycan–methotrexate conjugates. Bioorg Med Chem 14:5220–5229. https://doi.org/10.1016/j.bmc.2006.04.001
58. Kuribara T, Kojima T, Kobayashi Y, Hirose M, Shibayama K, Takeda Y, Totani K (2025) Development of a calreticulin-targeting glycan ligand based on a hybrid binding concept. Glycobiology 35:cwaf015. https://doi.org/10.1093/glycob/cwaf015
59. Saito Y, Ihara Y, Leach MR, Cohen-Doyle MF, Williams DB (1999) Calreticulin functions in vitro as a molecular chaperone for both glycosylated and non-glycosylated proteins. EMBO J 18:6718–6729. https://doi.org/10.1093/emboj/18.23.6718
60. Lederkremer GZ (2009) Glycoprotein folding, quality control and ER-associated degradation. Curr Opin Struct Biol 19:515–523. https://doi.org/10.1016/j.sbi.2009.06.004
61. Hosokawa N, Wada I, Hasegawa K, Yorihuzi T, Tremblay LO, Herscovics A, Nagata K (2001) A novel ER alpha-mannosidase-like protein accelerates ER-associated degradation. EMBO Rep 2:415–422. https://doi.org/10.1093/embo-reports/kve084
62. Olivari S, Cali T, Salo KE, Paganetti P, Ruddock LW, Molinari M (2006) EDEM1 regulates ER-associated degradation by accelerating demannosylation of folding-defective polypeptides and by inhibiting their covalent aggregation. Biochem Biophys Res Commun 349:1278–1284. https://doi.org/10.1016/j.bbrc.2006.08.186
63. Hosokawa N, Tremblay LO, Sleno B, Kamiya Y, Wada I, Nagata K, Kato K, Herscovics A (2010) EDEM1 accelerates the trimming of $\alpha$1,2-linked mannose on the C branch of *N*-glycans. Glycobiology 20:567–575. https://doi.org/10.1093/glycob/cwq001
64. Olivari S, Galli C, Alanen H, Ruddock L, Molinari M (2005) A novel stress-induced EDEM variant regulating endoplasmic reticulum-associated glycoprotein degradation. J Biol Chem 280:2423–2428. https://doi.org/10.1074/jbc.C400534200

65. Mast SW, Diekman K, Karaveg K, Davis A, Sifers RN, Moremen KW (2005) Human EDEM2, a novel homolog of family 47 glycosidases, is involved in ER-associated degradation of glycoproteins. Glycobiology 15:421–436. https://doi.org/10.1093/glycob/cwi014
66. Hirao K, Natsuka Y, Tamura T, Wada I, Morito D, Natsuka S, Romero P, Sleno B, Tremblay LO, Herscovics A, Nagata K, Hosokawa N (2006) EDEM3, a soluble EDEM homolog, enhances glycoprotein endoplasmic reticulum-associated degradation and mannose trimming. J Biol Chem 281:9650–9658. https://doi.org/10.1074/jbc.M512191200
67. Ninagawa S, Okuda T, Sumitomo Y, Kamiya Y, Kato K, Horimoto S, Ishikawa T, Takeda T, Sakuma T, Yamamoto T, Mori K (2014) EDEM2 initiates mammalian glycoprotein ERAD by catalyzing the first mannose trimming step. J Cell Biol 206:347–356. https://doi.org/10.1083/jcb.201404075
68. Shenkman M, Ron E, Yehuda R, Benyair R, Khalaila I, Lederkremer GZ (2018) Mannosidase activity of EDEM1 and EDEM2 depends on an unfolded state of their glycoprotein substrates. Commun Biol 1:172. https://doi.org/10.1038/s42003-018-0174-8
69. Pan S, Wang S, Utama B, Huang L, Bloc N, Estes MK, Moremen KW, Sifers RN (2011) Golgi localization of ERManI defines spatial separation of the mammalian glycoprotein quality control system. Mol Biol Cell 22:2810–2822. https://doi.org/10.1091/mbc.e11-02-0118
70. Rymen D, Peanne R, Millon MB, Race V, Sturiale L, Garozzo D, Mills P, Clayton CG, Asteggiano D, Quelhas A, Cansu E, Martins MC, Nassogne M, Goncalves-Rocha P, Topaloglu H, Jaeken J, Foulquier F, Matthijs G (2013) MAN1B1 deficiency: an unexpected CDG-II. PLoS Genet 9: Article e1003989. https://doi.org/10.1371/journal.pgen.1003989
71. Sun AH, Collette JR, Sifers RN (2020) The cytoplasmic tail of human mannosidase Man1b1 contributes to catalysis-independent quality control of misfolded alpha1-antitrypsin. Proc Natl Acad Sci USA 117:24825–24836. https://doi.org/10.1073/pnas.1919013117
72. Frenkel Z, Gregory W, Kornfeld S, Lederkremer GZ (2003) Endoplasmic reticulum-associated degradation of mammalian glycoproteins involves sugar chain trimming to $Man_{6-5}GlcNAc_2$. J Biol Chem 278:34119–34124. https://doi.org/10.1074/jbc.M305929200
73. Kuribara T, Hirano M, Speciale G, Williams SJ, Ito Y, Totani K (2017) Selective manipulation of discrete mannosidase activities in the endoplasmic reticulum by using reciprocally selective inhibitors. ChemBioChem 18:1027–1035. https://doi.org/10.1002/cbic.201700081
74. Caramelo JJ, Parodi AJ (2015) A sweet code for glycoprotein folding. FEBS Lett 589:3379–3387. https://doi.org/10.1016/j.febslet.2015.07.021
75. Benyair R, Ogen-Shtern N, Mazkereth N, Shai B, Ehrlich M, Lederkremer GZ (2014) Mammalian ER mannosidase I resides in quality control vesicles, where it encounters its glycoprotein substrates. Mol Biol Cell 26:172–184. https://doi.org/10.1091/mbc.E14-06-1152
76. Słomińska-Wojewdózka M, Sandvig K (2015) The Role of lectin-carbohydrate interactions in the regulation of ER-associated protein degradation. Molecules 20:9816–9846. https://doi.org/10.3390/molecules20069816
77. Nitta K, Kuribara T, Totani K. (2021) Synthetic trisaccharides reveal discrimination of endo-glycosidic linkages by exo-acting α-1,2 mannosidases in the endoplasmic reticulum. Org Biomol Chem 19:4137–4145. https://doi.org/10.1039/D1OB00428J
78. Kuribara T, Totani K (2022) Oligomannose-type glycan processing in the endoplasmic reticulum and its importance in misfolding diseases. Biology 11:199. https://doi.org/10.3390/biology11020199
79. Iwamoto S, Isoyama M, Hirano M, Yamaya K, Ito Y, Matsuo I, Totani K (2012) Reconstructed glycan profile for evaluation of operating status of the endoplasmic reticulum glycoprotein quality control. Glycobiology 23:121–131. https://doi.org/10.1093/glycob/cws130

# Pseudo-glycans as Molecular Tools for Chemical Glycobiology

**Makoto Yoritate and Go Hirai**

**Abstract** Glycans and glycoconjugates are integral to a wide range of biological functions, including energy storage, immune modulation, and molecular recognition. To elucidate the molecular mechanisms underlying these functions, a variety of chemical biology strategies have been developed. For instance, fluorescent probes have been employed to visualize the dynamics and subcellular localization of glycans, while photoaffinity probes have enabled the identification of glycan-binding proteins in complex biological systems. In parallel, *C*-glycoside analogs—resistant to enzymatic hydrolysis—have been developed as stable mimics of native glycans, preserving both structural features and biological activity. Given that specific functional groups within glycan structures often play decisive roles in their function, approaches that minimize the extent of chemical modification are critical for accurate functional analysis. Guided by this rationale, our group has developed a series of glycan analogs and photoaffinity probes that retain the native structure and physicochemical properties of the parent glycans. This chapter outlines the conceptual basis, development, and applications of these molecular tools in the context of chemical glycobiology.

**Keywords** *C*-glycoside · Photo-affinity probe · Photoredox · Glycoconjugate · Linkage-editing strategy

---

M. Yoritate (✉) · G. Hirai (✉)
Graduate School of Pharmaceutical Sciences, Kyushu University, Fukuoka, Japan
e-mail: m.yoritate@phar.kyushu-u.ac.jp

G. Hirai
e-mail: gohirai@phar.kyushu-u.ac.jp

 
K. Toshima and Y. Fujimoto (eds.), *Chemical Biology of Carbohydrates*,
https://doi.org/10.1007/978-981-95-5335-8_6

# 1 Photoaffinity Labeling in Chemical Glycobiology

## *1.1 Introduction*

Deciphering the molecular interaction networks that underpin biological phenomena remains a major challenge in life science research, especially for drug discovery, where identification of appropriate target biomolecules is critical [1]. Indeed, in chemical biology—a field that seeks to understand biological systems through the lens of chemistry—one of the central aims is to identify the molecular targets of bioactive compounds and to elucidate their mechanisms of action. At the same time, the development of methodologies and tools for reliably identifying these direct targets has also been an area of active research.

One such approach is affinity labeling, a method in which a reactive group incorporated into a bioactive molecule forms a covalent bond with a nearby functional group on the target molecule [2]. Although identifying targets from cell lysates poses challenges such as protein conformational changes or destabilization of protein complexes, affinity labeling offers the advantage of covalent bond formation, enabling target capture even at the cellular or, in some cases, whole-organism level.

A widely used subclass of this technique is photoaffinity labelling [3] in which a photo-reactive group (PRG) generates a reactive species upon UV irradiation, allowing covalent crosslinking to occur. This method has long been employed in chemical biology (Fig. 1). After covalent crosslinking of the bioactive molecule and its protein target, a detection tag is required to visualize or isolate the labeled protein. Molecules that carry both a photo-reactive group and a tag are referred to as photoaffinity probes. While traditional tags include radioactive isotopes and fluorophores, recent advances in bioorthogonal chemistry have enabled the use of compact functional handles such as alkynes and azides [4].

In parallel, affinity labeling using electrophilic reactive groups has also been extensively studied and successfully applied in various contexts [5]. In contrast, while the number of applications of photoaffinity labeling has steadily increased, the development of new photo-reactive groups has been limited. Although conventional photo-reactive groups are widely used and are well-established, there is still room for improvement in terms of stability, reactivity, and physicochemical properties.

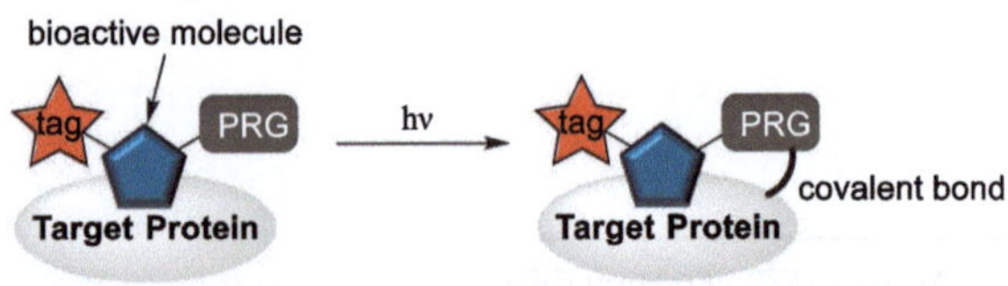


**Fig. 1** Photoaffinity labeling

## 1.2 Classification of Photo-Reactive Groups

An ideal photo-reactive group should remain stable at physiological pH prior to light activation, exert no effect on the biological activity of the parent molecule, and be excitable by light wavelengths that do not damage biological samples. It should also be compact in size to minimize steric perturbation. However, it is fair to say that no existing photo-reactive group fully satisfies all these criteria.

Photo-reactive groups commonly used in chemical biology can be broadly classified into two categories based on their reactivity. One major class includes groups that release molecular nitrogen ($N_2$) upon UV irradiation, such as aryl azides and aryl/aliphatic diazirines. These groups generate reactive intermediates irreversibly.

Aryl azides **1**, first reported by Knowles et al. in 1969 (Fig. 2), [6] typically require UV light around 300 nm for activation, and in some cases, prolonged irradiation is necessary. Upon photoexcitation, aryl azides generate nitrene species **3**. In the absence of substituents on the aryl ring (compound **2**), the nitrene tends to undergo rearrangement into chemical species such as benzazirines **4** or azepines **5**, which primarily react with nucleophilic amino acids. However, the introduction of appropriate substituents on the aromatic ring prolongs the lifetime of the nitrene species and enables them to react more broadly. Nitrenes can, in principle, insert into C–H bonds of amino acid side chains located in close proximity, allowing covalent crosslinking to a wide range of residues.

One well-known example is the fluorinated aryl azide **6**, developed by Platz and co-workers, which exhibits enhanced reactivity and stability. In addition, a variety of advanced aryl azide-based probes have been developed, including di-azide probes **7** that carry azide tags [7, 8] and turn-on fluorescent probes **8** that become strongly fluorescent only after labeling [9, 10].

In 1973, Knowles and Smith demonstrated that aryl diazirines **9** could serve as effective photo-reactive groups (Fig. 3) [11]. Upon UV irradiation, diazirines are converted into carbene species **10**, which, like nitrenes, can formally insert into carbon–hydrogen bonds of amino acids. However, diazirines also undergo rearrangement to form diazo compounds **11**, which, although capable of reacting with nucleophilic amino acids, are generally less reactive than carbenes.

A major advancement came in 1980, when Brunner and co-workers developed the trifluoromethyl-substituted diazirine **12**, in which the rearrangement to diazo species is suppressed, thereby enhancing the effective reactivity of the carbene intermediate [12]. This scaffold has since become one of the most widely used photo-reactive groups.

Aliphatic diazirines, in particular, are compact and exhibit low hydrophobicity, making them attractive for the design of minimally perturbing photoaffinity probes. More compact probes, including the tag moiety (e.g., compound **13**), have also been developed based on aliphatic diazirines [13]. However, in these systems, the corresponding diazo intermediates are often unstable and form nonspecific covalent bonds with nucleophilic amino acids, potentially leading to off-target labeling.

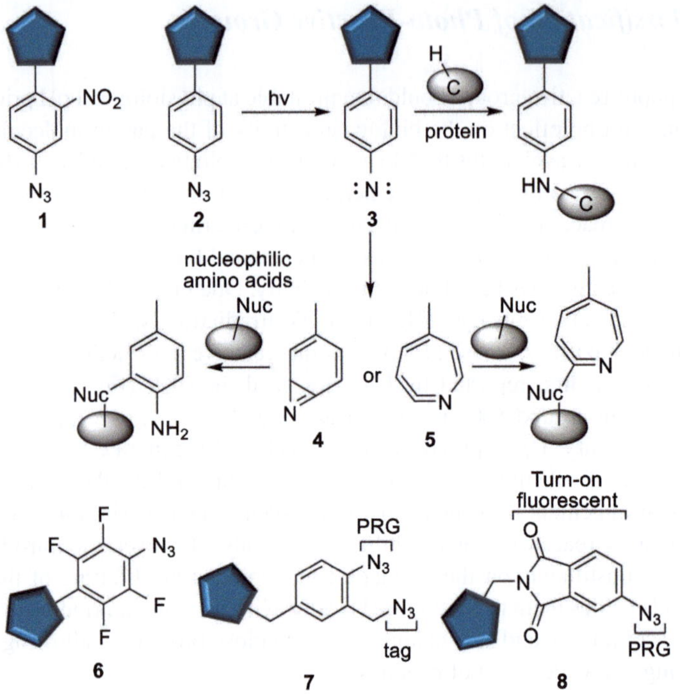


**Fig. 2** Structures and reactions of photo-reactive groups bearing aromatic azide

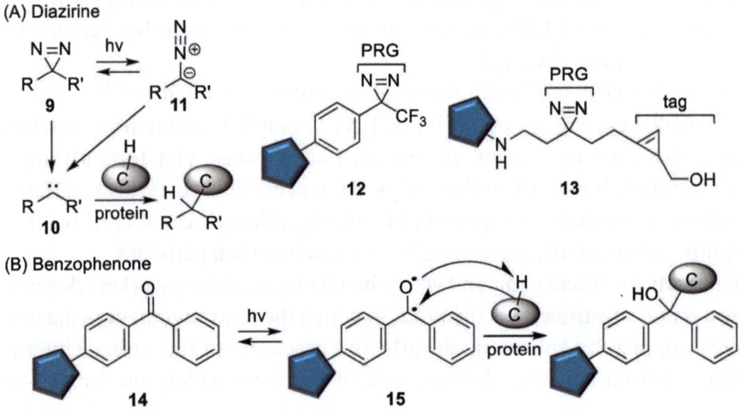


**Fig. 3** Structures and reactions of photo-reactive groups bearing diazirine and benzophenone

An alternative class of photo-reactive groups is represented by benzophenone **14**, which exploits the excited-state reactivity of the carbonyl group and was first reported by Galardy and Printz in 1973 [14]. The ketone group of benzophenone **14** is excited by light around 350 nm and generates a biradical-like reactive species **15**. After intersystem crossing from the singlet to the triplet excited state, it abstracts a hydrogen atom from a nearby carbon–hydrogen bond in the target molecule, forming two carbon-centered radicals, which then form a covalent bond. If there is no appropriately oriented C–H bond near the excited species, it returns to the ground state. Therefore, benzophenone **14** can be repeatedly excited, and it has the potential to achieve high labeling efficiency.

These photo-reactive groups exhibit good reactivity and, except for aryl azides (which may be reduced by thiols), are generally stable under physiological conditions. However, all of them are hydrophobic and relatively large, raising concerns that they could alter the function and physicochemical properties of the original bioactive molecule. Moreover, the intrinsic hydrophobicity of photo-reactive groups could promote nonspecific interactions with surrounding molecules, resulting in covalent modification of non-target species—an inherent drawback of photoaffinity labeling strategies. This issue becomes particularly prominent when the affinity between the probe and its target is low, necessitating the use of high probe concentrations. In such cases, multiple labeled protein bands are frequently observed in SDS-PAGE analyses.

Moreover, even if the photoaffinity probe retains the same activity as the original molecule, it may label not the directly interacting target protein but an adjacent protein, which could lead to misinterpretation of the results.

Nearly 50 years have passed since photo-reactive groups were first developed. During that time, many improvements in their reactivity and synthetic methods have been reported. However, attempts to discover new photo-reactive groups that are more compact and less hydrophobic are a quite recent development. In the following section, we describe some examples, including our own work.

## *1.3 Tetrazole-Based Photo-Reactive Group*

In 2008, Lin and co-workers reported a new bioconjugation method based on tetrazole derivatives [15]. Upon irradiation of a 2,5-disubstituted tetrazole **16** with ~ 300 nm light, a nitrile imine **17** is generated. Since this species undergoes a 1,3-dipolar cycloaddition with alkenes, the reaction was named the "photoclick reaction" and was initially considered a candidate for bioorthogonal chemistry (Fig. 4). In practice, however, nitrile imines **17** exhibit high electrophilicity and also react with nucleophiles, which imposes certain limitations on their use as truly bioorthogonal reactions. On the other hand, this electrophilic reactivity is advantageous for application as a photo-reactive group in affinity labeling.

In 2016, Yao et al. and Lin et al. independently reported the use of 2,5-disubstituted tetrazole derivatives as photo-reactive groups (compounds **18** and **19**, respectively)

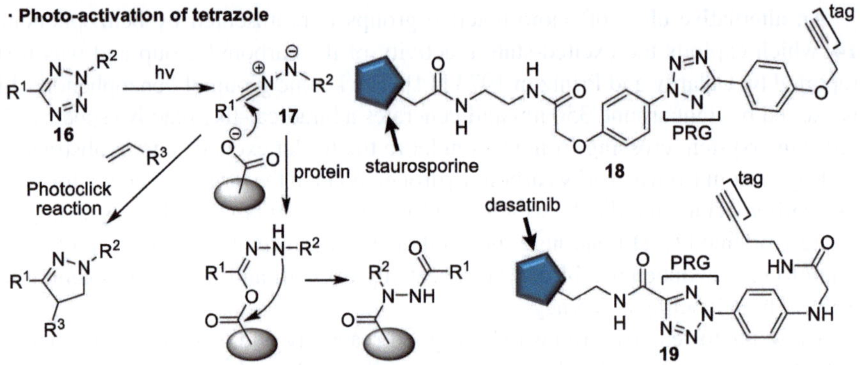


**Fig. 4** Photochemical reaction of 2,5-substituted tetrazole for selective labeling of carboxylic acid

[16, 17]. Here, staurosporine and dasatinib are compounds known to bind to the ATP-binding sites of their target kinases. Since nitrile imines are generated along with the release of molecular nitrogen, the reaction of this class of photo-reactive group is irreversible. According to Yao's report, nitrile imines react most efficiently with carboxylates, and the resulting adducts undergo 1,4-acyl migration to generate stable *N*,*N*′-diacylhydrazides. Lin and colleagues reported that photoaffinity labeling of Bruton's tyrosine kinase by the dasatinib-based probe **19** occurs at Glu488.

Whereas conventional photoaffinity reagents primarily form covalent bonds through insertion into C–H bonds, this new type of photo-reactive group generates an electrophilic species, offering a potentially complementary mechanism to existing systems.

## 1.4 A-Ketoamide-Based Photo-Reactive Probe

We considered that the conventional photo-reactive groups described above are poorly suited for probing glycan–protein interactions. Glycans are highly hydrophilic molecules that engage in selective molecular recognition via networks of hydroxyl groups and subtle C–H-mediated interactions. Yet, this very hydrophilicity contributes to the generally weak affinities observed in glycan–protein binding. Photoaffinity labeling represents a powerful strategy for capturing such transient interactions; however, major photo-reactive groups rely on short-lived intermediates, which often compromise labeling efficiency.

Benzophenone **14**, with its ability to repeatedly generate reactive species upon irradiation, appears ideal for glycan–protein interactions in principle. However, its pronounced hydrophobicity risks altering glycan properties and promoting extensive nonspecific interactions. Thus, we aimed to develop a photo-reactive group that, like benzophenone **14**, can be repeatedly excited, but is more hydrophilic and compact.

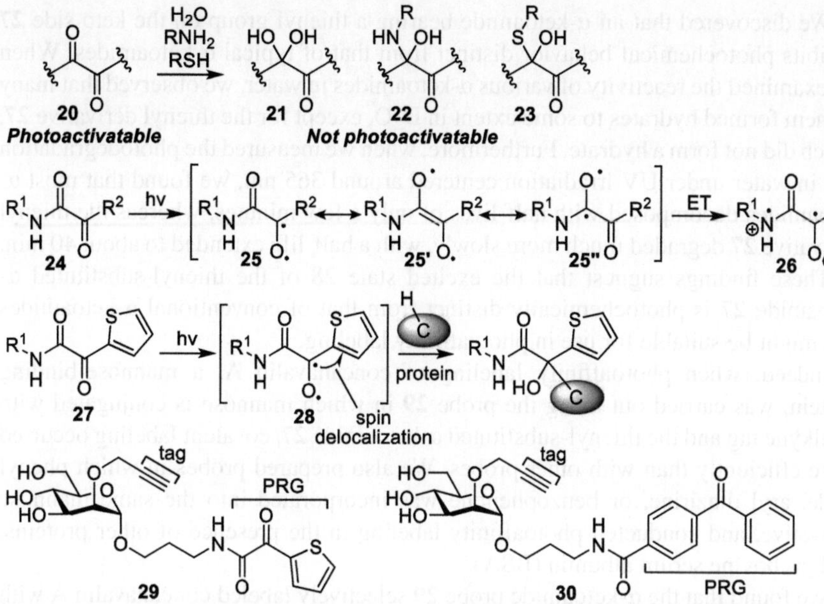


**Fig. 5** Ketoamide-based photoaffinity group

With this in mind, we focused on 1,2-dicarbonyl compounds **20** as candidates that can be excited by light around 350 nm (Fig. 5) [18]. The 1,2-dicarbonyl structure is polarized and does not necessarily require conjugation with aromatic rings for excitation, making it a promising candidate for this purpose. However, as we began our investigation, we immediately identified two major issues.

The first problem was the high electrophilicity of the 1,2-dicarbonyl structure. Since photoaffinity labeling experiments are typically conducted in complex mixtures containing various proteins, highly nucleophilic residues such as lysine and cysteine, as well as water, are always present. The 1,2-dicarbonyl structure potentially reacts with these nucleophiles, forming undesired species such as hydrates **21** or hemiacetals **22** and **23**, which are photoinactive and lead to nonspecific binding. Therefore, we concluded that only α-ketoamides **24**, which are thought to be the least electrophilic among 1,2-dicarbonyl compounds, would be practical.

However, a second issue emerged regarding the photo-reactivity of the α-ketoamide structure. In the case of α-ketoamide **24**, it has been proposed that the biradical-like excited state **25** generated by photoirradiation is extremely short-lived. Before undergoing intersystem crossing to the triplet state, it is rapidly converted into a zwitterionic biradical intermediate **26** via electron transfer from the amide moiety [19]. As a result, degradation or intramolecular cyclization occurs, depending on the substitution pattern. Thus, it was considered unlikely that α-ketoamide **24** would undergo intersystem crossing to a triplet state and react with nearby C–H bonds, like benzophenone **14**.

We discovered that an α-ketoamide bearing a thienyl group on the keto side **27** exhibits photochemical behavior distinct from that of typical α-ketoamides. When we examined the reactivity of various α-ketoamides in water, we observed that many of them formed hydrates to some extent in $D_2O$, except for the thienyl derivative **27**, which did not form a hydrate. Furthermore, when we measured the photodegradation rate in water under UV irradiation centered around 365 nm, we found that most α-ketoamides decomposed with half-lives of only a few minutes, whereas the thienyl derivative **27** degraded much more slowly, with a half-life extended to about 40 min.

These findings suggest that the excited state **28** of the thienyl-substituted α-ketoamide **27** is photochemically distinct from that of conventional α-ketoamides and might be suitable for use in photoaffinity labeling.

Indeed, when photoaffinity labeling of concanavalin A, a mannose-binding protein, was carried out using the probe **29** in which mannose is conjugated with an alkyne tag and the thienyl-substituted α-ketoamide **27**, covalent labeling occurred more efficiently than with other probes. We also prepared probes in which phenyl azide, aryl diazirine, or benzophenone was incorporated into the same mannose derivative, and conducted photoaffinity labeling in the presence of other proteins, such as bovine serum albumin (BSA).

We found that the α-ketoamide probe **29** selectively labeled concanavalin A with minimal nonspecific labeling of other proteins. In contrast, the other probes showed lower selectivity. In particular, the benzophenone-based probe **30** gave rise to a considerable amount of nonspecific binding.

The thienyl-substituted α-ketoamide **27** represents a novel photo-reactive group. The fact that it can selectively label concanavalin A in the presence of other proteins, even when the interaction with mannose has a dissociation constant in the millimolar range (~ $10^{-3}$ M), is a particularly noteworthy feature [20].

## *1.5 Development of Bifunctional Ketoamide Probe*

The previously developed thienyl-substituted α-ketoamide lacked a secondary functional handle for post-labeling introduction of reporter groups such as fluorescent dyes or biotin [21]. In the previous design, represented by probe **29**, the alkyne moiety for post-labeling detection was introduced into the mannose residue, potentially compromising its role in target protein recognition. To avoid this issue, we designed a series of probes (**31–33, Fig. 6**) that preserve the native mannose structure while incorporating both a photo-reactive group and a bioorthogonal tagging site by embedding the alkyne within the thienyl ring.

Probes **31–33** were synthesized by Sonogashira coupling of iodothienyl ketoesters **34**, followed by hydrolysis of the ester to furnish alkynylated α-keto acids **35**, which were then condensed with 1-(3-aminopropyl) mannose **36**. To assess the excited-state reactivity of the resulting probes, we determined their photodegradation rates under 365 nm irradiation using HPLC analysis. Compared to the non-substituted

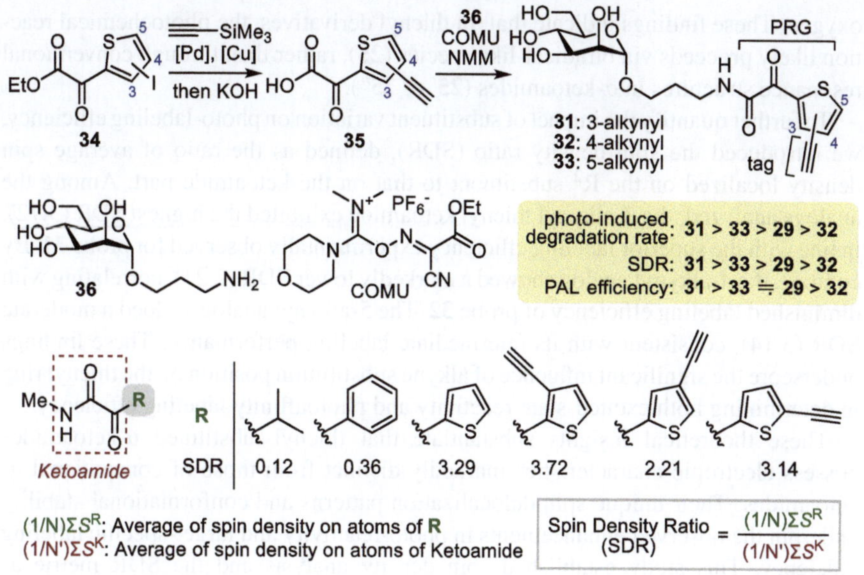


**Fig. 6** Bifunctional alkynylated α-ketoamide probes and computational analysis of spin density ratio of model α-ketoamides

analog **29**, the alkynylated thienyl ketoamides exhibited significantly shorter half-lives, indicative of increased photosensitivity. Notably, the alkyne position on the thienyl ring had a marked influence on degradation rate, with 3-alkynyl derivative **31** displaying the lowest stability (Fig. 6).

The binding affinities of the probes toward concanavalin A (ConA) in the dark were determined by isothermal titration calorimetry (ITC). While overall affinities remained within a similar range, probe **31** exhibited the highest affinity, followed sequentially by **29**, **32**, and **33**. To assess labeling efficiency, the probes were incubated with ConA, followed by conjugation with fluorophore via Cu(I)-catalyzed azide–alkyne cycloaddition (CuAAC), and analysis by SDS-PAGE. Given the spatial proximity of the alkyne moiety to the photoexcited carbonyl group, probe **31** was initially expected to display diminished photoaffinity labeling efficiency. This notion is consistent with its notably rapid photodegradation kinetics. However, the probe **31** afforded the most efficient labeling of ConA.

To elucidate the origin of the favorable properties observed in thienyl-substituted α-ketoamides, we performed density functional theory (DFT) calculations. Our analysis focused on the spin density distribution in the photoexcited triplet state ($T_1$), comparing α-ketoamides bearing phenyl or isopropyl groups with their thienyl-substituted counterparts. Consistent with prior reports, phenyl analogs exhibited spin density predominantly localized on the carbonyl and amide oxygen atoms. In contrast, thienyl-substituted analogs displayed notable spin delocalization across the thienyl ring, accompanied by a marked reduction in spin density on the amide-side oxygen. These findings indicate that, in thienyl derivatives, the photochemical reaction likely proceeds via biradical-like species (**25**), rather than through conventional resonance structures in α-ketoamides (**25′** or **25″**).

To further quantify the impact of substituent variation on photo-labeling efficiency, we introduced the spin density ratio (SDR), defined as the ratio of average spin density localized on the $R^2$ substituent to that on the ketoamide part. Among the analogs analyzed, the 3-alkynyl thienyl ketoamide exhibited the highest SDR (3.72), in line with the superior labeling efficiency experimentally observed for probe **31**. By contrast, the 4-alkynyl analog showed a markedly lower SDR (2.21), correlating with diminished labeling efficiency of probe **32**. The 5-alkynyl analog yielded a moderate SDR (3.14), consistent with its intermediate labeling performance. These findings underscore the significant influence of alkyne substitution position on the thienyl ring in determining both excited-state reactivity and photoaffinity labeling efficiency.

These theoretical insights substantiate that thienyl-substituted α-ketoamides possess electronic characteristics markedly distinct from those of conventional α-ketoamides. Their unique spin delocalization patterns and conformational stability underpin the observed enhancements in photo-reactivity and target-specific labeling efficiency. This study established spin density analysis and the SDR metric as powerful tools for the rational design of next-generation photo-reactive motifs. Indeed, the α-ketoamide framework offers significant potential to generate chemical probes capable of selectively interrogating complex biomolecular interactions.

Since our work, several photo-reactive groups have been introduced [22–25] and proximity-based labeling methods such as μMAP have been developed [26]. These advances have broadened the experimental toolbox in chemical glycobiology, and further innovations are expected to drive continued progress in the field.

## 2 *C*-Glycoside Analogs of Glycans and Glycoconjugates

### *2.1 Introduction*

Glycosidic linkages (acetal bonds) in oligosaccharides and glycoconjugates are enzymatically formed and cleaved by glycosyltransferases and glycoside hydrolases. In contrast, *C*-glycosidic linkages—where the glycosidic oxygen is replaced by a carbon atom—are both chemically and biologically stable. Similarly, *S*-glycosides and *Se*-glycosides, in which sulfur or selenium replaces oxygen, respectively, are regarded as analogs resistant to enzymatic degradation. These substitutions involve elements from the same period or group in the periodic table. While their two-dimensional structures appear similar, their three-dimensional geometries differ significantly. Comparative analysis of bond lengths and angles among dimethyl ether, methylene, sulfide, and selenide units reveals that sulfides and selenides deviate substantially from ethers and methylenes (Fig. 7a). Based on these observations, *C*-glycosides are considered the most faithful mimics of *O*-glycosides in terms of bond geometry.

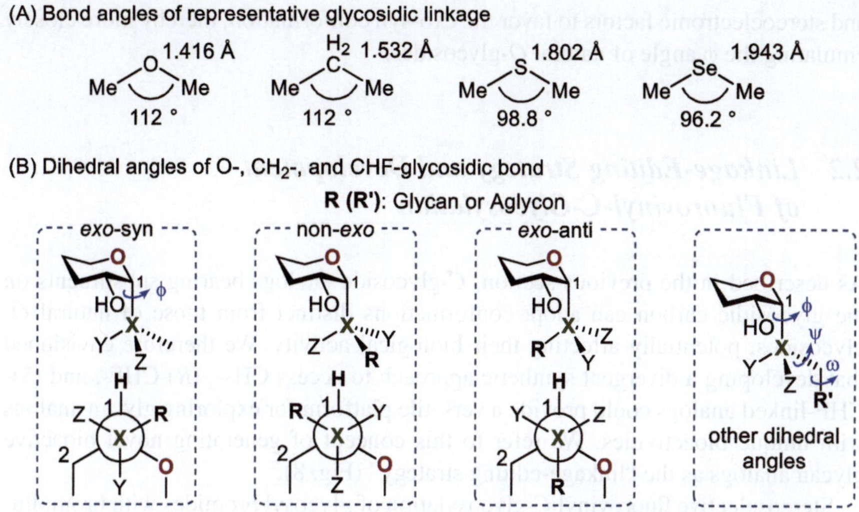


| | |
|---|---|
| native | X = O, Y = Z = LP |
| $CH_2$ | X = C, Y = Z = H |
| (*R*)-CHF | X = C, Y = H, Z = F |
| (*S*)-CHF | X = C, Y = F, Z = H |

| | Steric effect | Stereoelectronic factors: native | $CH_2$ | (*R*)-CHF | (*S*)-CHF |
|---|---|---|---|---|---|
| *exo*-syn | O | O | × | × | O |
| non-*exo* | △ | × | × | O | O |
| *exo*-anti | × | O | × | O | × |

**Fig. 7** Structural properties of various glycan linkages. **a** Bond angles of *O*-, $CH_2$, *S*-, and *Se*-glycosides. **b** Dihedral angles of *O*-, $CH_2$-, (*R*)-CHF-, and (*S*)-CHF-glycosides

In addition, each glycoside analog exhibits distinct dihedral angles (φ) across the glycosidic bond. Typical α-glycoside conformations are illustrated in Fig. 7b. The α-*O*-glycosidic linkage is stabilized in exo-syn and exo-anti conformations via the exo-anomeric effect, with the exo-syn form typically dominant due to favorable sterics. In contrast, $CH_2$-glycosides lack these stereoelectronic effects, and their conformational preferences are dictated purely by steric factors, leading to greater flexibility.

To faithfully replicate not only bond lengths and angles but also the dihedral angle (φ) characteristic of natural glycosides, an additional structural feature is required. We addressed this by exploiting the chemical tunability of the methylene moiety, designing CHF-glycosidic linkages in which one hydrogen atom is replaced with fluorine. The C–F antibonding orbital ($\sigma^*_{C\text{–}F}$) exhibits strong electron-accepting properties and can participate in stabilizing hyperconjugative interactions when aligned anti to $\sigma_{C\text{–}H}$ or $\sigma_{C\text{–}C}$ orbitals—a phenomenon known as the fluorine gauche effect [27, 28]. In (*R*)-CHF-glycosides, conflicting steric and stereoelectronic influences attenuate conformational preferences, reducing the energy gap between exo-syn and non-exo conformers. In contrast, (*S*)-CHF-glycosides align both steric and stereoelectronic factors to favor the exo-syn conformation, thereby more closely emulating the φ angle of natural *O*-glycosides.

## 2.2 Linkage-Editing Strategy and Development of Fluorovinyl-C-Glycosylation

As described in the previous section, *C*-glycoside analogs bearing substituents on the glycosidic carbon can adopt conformations distinct from those of natural *O*-glycosides, potentially affecting their biological activity. We therefore envisioned that developing a divergent synthetic approach to access $CH_2$-, (*R*)-CHF-, and (*S*)-CHF-linked analogs could provide a versatile platform for exploring glycan analogs with unique bioactivities. We refer to this concept of generating novel bioactive glycan analogs as the "linkage-editing strategy" (Fig. 8).

Stereoselective fluorovinyl *C*-glycosylation of glycosyl bromides with bromofluoroolefins (BFOs), followed by chemo- and stereoselective hydrogenation of the fluorovinyl group, enables divergent access to $CH_2$-, (*R*)-CHF-, and (*S*)-CHF-linked analogs.

The synthetic method is as follows (Fig. 9). When a 2,3-carbonate-protected glycosyl bromide (**42**) and BFO (**43**) were subjected to blue LED irradiation in the presence of an iridium photocatalyst, a nickel catalyst, and supersilanol as a reductant, the reaction proceeded smoothly to afford the α-fluorovinyl *C*-glycoside **44** with high stereoselectivity. The reaction commenced with generation of silyl radical via single electron oxidation of tris(trimethylsilyl)silanol, followed by radical Brook rearrangement. The silyl radical abstracts a bromine atom from the glycosyl bromide **42** via halogen-atom transfer (XAT), generating glycosyl radical intermediate **45** [29]. Radical capture by Ni(II) complex **46** derived from Ni(0) and **43**, followed by

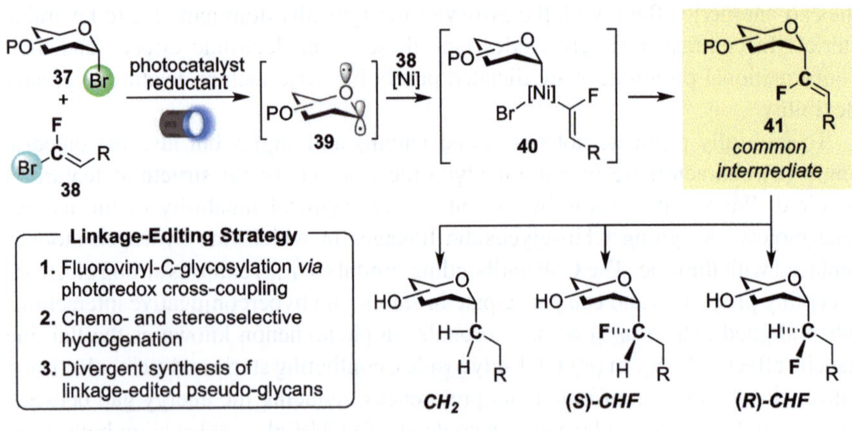


**Fig. 8** Divergent synthetic plan of linkage-edited pseudo-glycan analogs

reductive elimination of **47** gave the α-fluorovinyl-*C*-glycoside **44** [30]. The 2,3-carbonate protection on the glycosyl bromide was critical for both reactivity and stereoselectivity. This is likely because the glycosyl radical intermediate **45** adopts a $^{4}C_{1}$ conformation, in which the radical spin density is biased toward the α-face due to the anomeric effect, thereby promoting stereoselective coupling [31–33] (Fig. 9).

Optimization of the reaction conditions revealed that 2,4,6-collidine was the most generally effective base, and a broad range of solvents can be used, including acetone, acetonitrile, methyl-*tert*-butylether (MTBE), 1,2-dimethoxyethane (DME), toluene, or ethyl acetate.

The optimized reaction conditions exhibited a broad substrate scope, including glucose, galactose, and mannose-type donors with various protecting groups, and BFOs derived from glycans and lipids. For example, isomaltose (**48**, Glc-α(1 → 6)-Glc), α-galactosylceramide (**49**, α-GalCer), melibiosamine (**50**, Gal-α(1 → 6)-$GlcNH_2$), and (neo)trehalose (**51**, Glc-α(1 → 1)-Glc) were successfully synthesized in good yields and high selectivities.

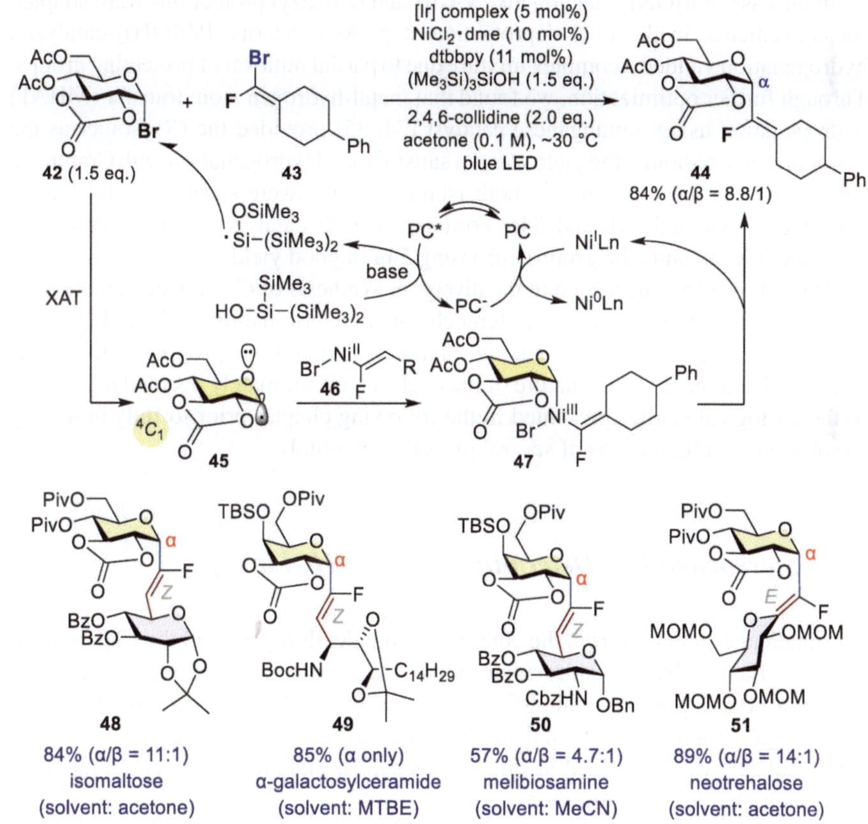


**Fig. 9** Photoredox coupling reaction: reaction mechanism and substrate scope

The 2′,3′-carbonate group of the isomaltose (IM)-type substrate **48** was removed to obtain the 2,3-diol **48′**, which was subjected to chemoselective and stereoselective hydrogenation (Fig. 10a). $Pd(OH)_2/C$-catalyzed hydrogenation afforded the (*S*)-CHF glycoside as the major product in 88% yield. In contrast, the Crabtree catalyst altered the selectivity due to the directing effect of the 2-hydroxyl group, yielding predominantly the (*R*)-CHF glycoside. Furthermore, hydrogenation using sulfur-poisoned platinum on carbon afforded the $CH_2$-glycoside in high yield via defluorinative hydrogenation. Finally, the removal of all protecting groups provided linkage-edited isomaltose analogs **52a**, **52b**, and **52c**.

Based on these findings, we investigated the hydrogenation of C-linked-α-galactosylceramide (α-GalCer) and melibiosamine ($MelNH_2$)-type substrates (Fig. 10b, c). Similar stereoselectivities were generally observed. In the case of the α-GalCer analog **49′**, Pearlman's catalyst afforded (*S*)-CHF-glycoside selectively and led to **53c**. In contrast, hydrogenation using Crabtree's catalyst afforded a 3.3:1 mixture of (*S*)-CHF- and (*R*)-CHF-isomers, from which the **53b** was synthesized. Pt/C-catalyzed hydrogenation followed by deprotection and acylation furnished **53a**.

In the case of $MelNH_2$ analog **50′**, *N*-Cbz and *O*-benzyl protections were adapted for convenience in the final deprotection step. As concerns, $Pd/(OH)_2$-catalyzed hydrogenation yielded a complex mixture due to partial removal of protecting groups. Through further optimization, we found that metal-hydrogen atom transfer (MHAT) hydrogenation using a manganese catalyst [34, 35] provided the (*S*)-isomer as the major product, although the yield was not satisfactory. Hydrogenation with Crabtree's catalyst gave a 3.3:1 mixture of both isomers, which were separated, and further deprotection yielded **54b** and **54c**. Fortunately, Pt/C-catalyzed hydrogenation did not affect the Bn and Cbz groups, affording **54a** in good yield.

Thus far, we had succeeded in the divergent synthesis of *C*-glycoside analogs via linkage editing. Nevertheless, the stereochemical determination of the CHF moiety presented a major bottleneck that substantially hindered our progress. Due to the prolonged and inconclusive nature of the analysis, we ultimately decided to advance to the biological studies presented in the following chapter prior to fully assigning the absolute configurations of several pivotal compounds.

## 2.3 Stereochemical Determination of CHF-Linkage

The challenge stemmed from the conformational flexibility around the CHF moiety, which is flanked by rotatable single bonds. As a result, conventional techniques such as *J*-coupling analysis and ROE spectroscopy failed to provide a solution at the initial stage. Moreover, many of the CHF-containing compounds, including synthetic intermediates, were obtained as amorphous solids, making X-ray crystallographic analysis impractical in most cases.

A breakthrough came with intermediate **55** in the synthesis of (*R*)-CHF-isomaltose (Fig. 11a). Upon derivatization with camphorsultam reagent **56**, the resulting compound (*R*)-CHF-**57** appeared entirely amorphous under polarized light and

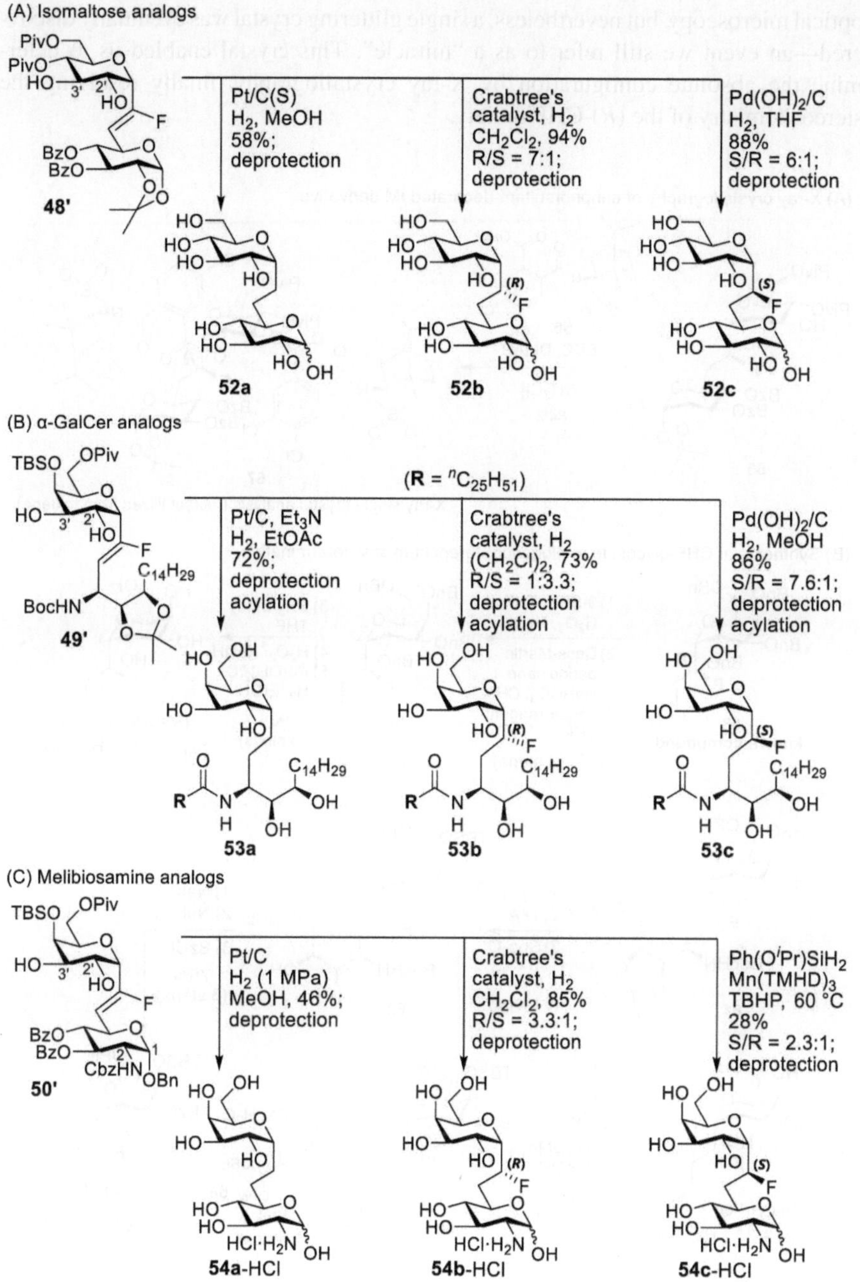


**Fig. 10** Synthesis of linkage-edited analogs of **a** isomaltose, **b** α-galactosylceramide, and **c** melibiosamine

optical microscopy, but nevertheless, a single glittering crystal was eventually discovered—an event we still refer to as a “miracle”. This crystal enabled us to determine the absolute configuration by X-ray crystallography, finally resolving the stereochemistry of the (*R*)-CHF moiety.

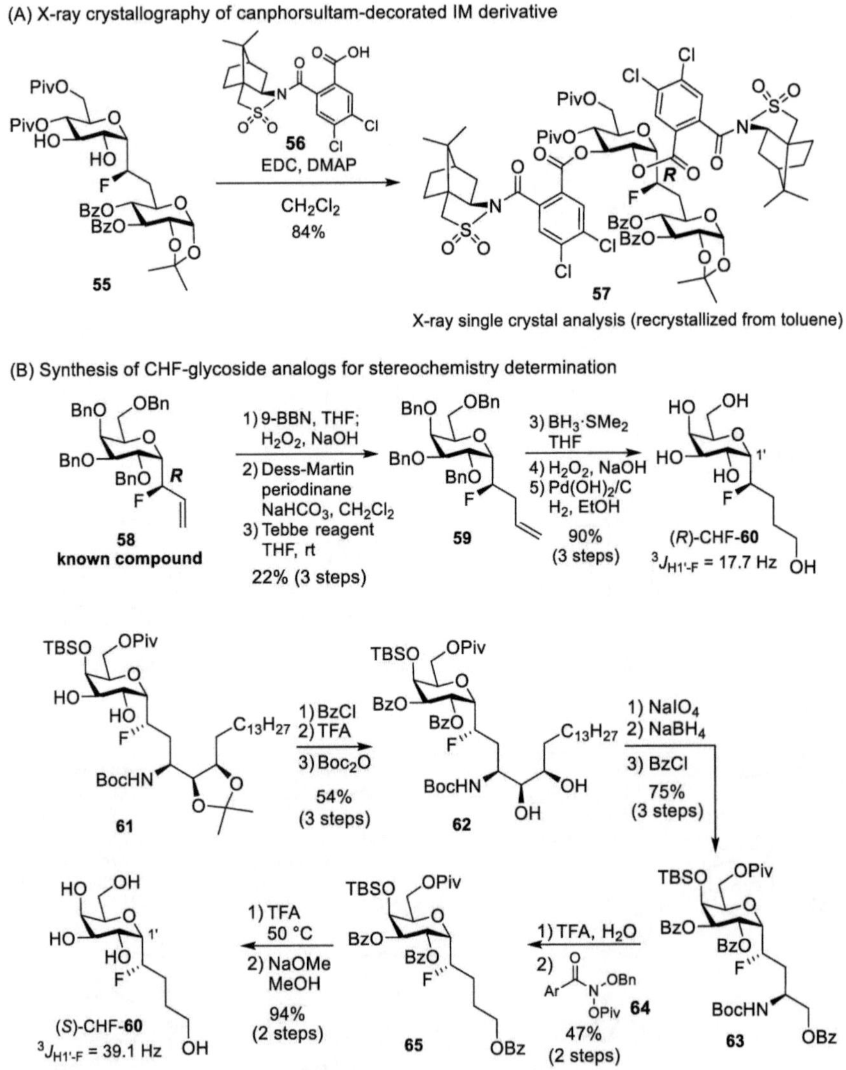


**Fig. 11** Determination of stereochemistry by **a** X-ray crystallography and **b** derivatization of known CHF-glycoside and synthetic intermediate

Unfortunately, a second miracle did not occur for (*S*)-CHF-**57**, and crystallization attempts for other CHF compounds also failed. Faced with these limitations, we turned to indirect strategies. Specifically, we planned to correlate our synthetic compounds with known CHF-glycosides through chemical transformation and NMR comparison (Fig. 11b). Starting from known (*R*)-CHF-glycoside **58**, [36] we performed a series of hydroboration–oxidation, olefination, and global deprotection steps to obtain compound **60**. In parallel, we converted an α-GalCer intermediate **61** into diol **62**, and the diol was cleaved with sodium periodate. After reduction of the resulting aldehyde and benzoylation of the alcohol, compound **63** was subjected to a deamination reaction using anomeric amide **64** to generate compound **65**. Finally, removal of all protecting groups furnished a CHF-glycoside **60**, but the resulting NMR spectrum did not match that of the (*R*)-CHF-**60** sample. From this, we concluded that the compound was the (*S*)-CHF isomer.

Of course, a mismatched NMR spectrum alone does not constitute definitive proof of opposite configuration. However, through these efforts, a variety of CHF-glycosides had been synthesized, and their NMR data collected. Careful analysis revealed a consistent trend: the $^3J_{H1'\text{-}F}$ coupling constant was typically 25–40 Hz for (*S*)-CHF-glycosides, while it remained lower, around 5–20 Hz, for the (*R*)-isomers. This trend is consistent with the qualitative conformational prediction (Fig. 7b) that (*S*)-CHF-glycosides would favor the anti-conformation between C1′-H and C-F sigma bonds. In addition, for several representative compounds, we conducted thorough conformational analysis using a combination of *J*-coupling analysis, ROE, and H–F HOESY experiments. These intensive NMR analyses provided support for correlating the $^3J_{H1'\text{-}F}$ values with the stereochemistry (see the original publications [37, 38] and their supporting information for detailed analysis and experimental procedures).

Thus, through careful analysis and cumulative evidence, we were ultimately able to assign the absolute configurations of all CHF-glycosides in this study with confidence.

## 2.4 Biological Evaluation of C-Linked Isomaltose Analogs

Isomaltose (IM), a disaccharide composed of Glc-α(1 → 6)-Glc, is known to act as a signaling molecule that induces amylase production in *Aspergillus nidulans* [39]. However, because native IM is also a substrate for glycoside hydrolases in the fungal system, its signaling effect is transient and rapidly diminished by enzymatic degradation. This complicates the mechanistic analysis of how IM triggers transcriptional pathways regulating amylase production. To address this limitation, *C*-glycoside analogs of IM were evaluated for their ability to function as enzymatically stable mimics of native IM.

The biological activity of C-linked isomaltose (IM) analogs **52a–c** was evaluated (Fig. 12). While native IM transiently induced amylase production, the activity was diminished in 24 h, likely due to enzymatic degradation. In contrast, the $CH_2$-linked

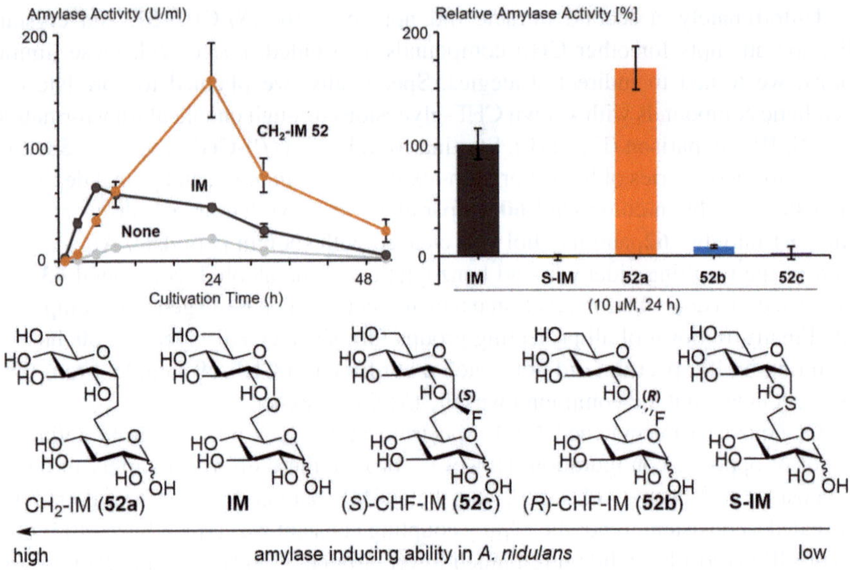


**Fig. 12** Biological evaluation of linkage-edited isomaltose analogs

analog **52a** induced more sustained and enhanced production, consistent with its resistance to glycosidic hydrolysis. The *S*-glycoside analog **S-IM** failed to elicit activity, underscoring the importance of glycosidic geometry for biological function.

Strikingly, CHF-linked analogs **52b** and **52c** showed negligible activity, despite differing from **52a** by only a single fluorine atom. These results suggest that minor stereoelectronic perturbations at the glycosidic linkage could abolish signaling function, possibly by affecting the conformational dynamics necessary for molecular recognition or transport.

Collectively, these findings demonstrate that $CH_2$-linked isomaltose functions as a stable and active mimic of native IM and emphasize the tight conformational requirements of glycan-mediated signaling in fungal systems.

## 3 Biological Evaluation of C-Linked α-Galactosylceramide Analogs

α-Galactosylceramide (α-GalCer) is a glycolipid known for its potent immunostimulatory properties [40]. Upon presentation by CD1d on antigen-presenting cells, α-GalCer is recognized by the T-cell receptor (TCR) on invariant natural killer T (iNKT) cells. This interaction leads to iNKT cell activation and induces the production of cytokines such as IFN-γ and IL-4, resulting in various downstream immunological responses, including anti-tumor and adjuvant effects. Owing to its well-characterized and specific biological activity, α-GalCer serves as an excellent model compound for investigating the effects of glycosidic linkage modifications on iNKT cell activation.

The immunostimulatory activity of C-linked α-GalCer analogs was evaluated using a combination of in vivo and in vitro assays. In an OVA immunization model, native α-GalCer enhanced the production of OVA-specific IgG and IgM antibodies, while the C-linked analogs **52a–c** exhibited reduced or negligible adjuvant activity [41]. Consistently, cytokine secretion from mouse splenocytes stimulated with these analogs mirrored this trend: the $CH_2$-linked analog **52a** induced weak cytokine production, whereas the CHF-linked analogs **52b** and **52c** failed to elicit detectable responses.

To further investigate iNKT cell activation by these analogs, a reporter assay was performed using NFAT-GFP Jurkat cells co-expressing mouse CD1d and the invariant NKT TCR, in which GFP fluorescence directly reflects TCR-mediated activation. As expected, native α-GalCer elicited robust GFP induction, confirming its potent agonist properties. In contrast, the $CH_2$-linked analog **52a** triggered only weak GFP expression, indicative of partial activation. Notably, both CHF-linked analogs, particularly (*R*)-CHF-α-GalCer **52b**, failed to induce any GFP signal across all concentrations tested, suggesting a complete loss of agonist activity (Fig. 13a).

Interestingly, in the case of co-treatment with native α-GalCer, the CHF-linked analogs exhibited antagonistic behavior (Fig. 13b). The (*S*)-CHF analog moderately suppressed GFP induction, whereas the (*R*)-CHF analog significantly reduced GFP expression in a dose-dependent manner. These findings raise the possibility that, although direct interaction of CHF-linked α-GalCer analogs with CD1d remains unverified, these analogs still occupy the CD1d binding groove and function as

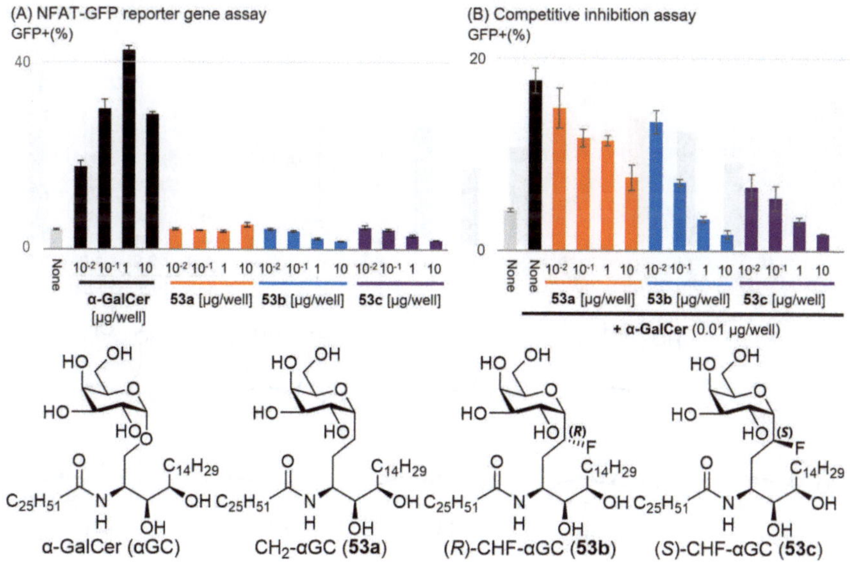


**Fig. 13** Biological evaluation of linkage-edited α-galactosylceramide analogs

competitive antagonists. This striking shift in function—driven exclusively by modification of the glycosidic linkage—emphasizes the pivotal influence of linkage geometry and local conformational features in shaping immune responses.

## 4 Biological Evaluation of C-Linked Melibiosamine Analogs

Melibiosamine ($MelNH_2$) is a non-natural disaccharide composed of galactose α(1 → 6)-linked glucosamine. It selectively inhibits the growth of human leukemia K562 cells with minimal toxicity toward normal human fibroblasts such as HUC-F2 [42]. This unique biological profile makes $MelNH_2$ a promising lead structure for anticancer glycomimetics and provides an attractive model system for evaluating the effects of linkage modification on cytotoxic activity.

The antiproliferative activities of O-linked $MelNH_2$ and C-linked analogs **54a–c** were evaluated in K562 human leukemia cells (Fig. 14). Cells were incubated in the presence of each analog at a final concentration of 5 mM. Cell proliferation was monitored over 72–96 h using a trypan blue exclusion assay. Among the analogs, the (*R*)-CHF analog **54b**-HCl exhibited the most potent activity, with an $IC_{50}$ approximately fivefold lower than that of O-linked $MelNH_2$. In contrast, both the $CH_2$ analog **54a**-HCl and the (*S*)-CHF analog **54c**-HCl showed slightly reduced or negligible activity compared to the original $MelNH_2$.

These findings again underscore the crucial role of the glycosidic linkage in modulating the biological function of glycans and glycolipids. Notably, (*S*)-CHF analogs

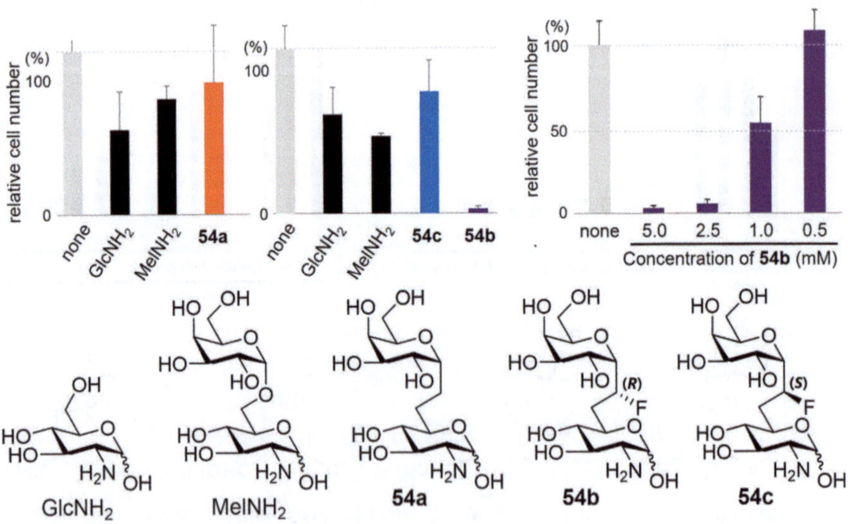


**Fig. 14** Biological evaluation of linkage-edited melibiosamine analogs

consistently exhibited reduced or even reversed bioactivity compared with their *O*-linked counterparts. This was unexpected, as our initial hypothesis was that the (*S*)-CHF linkage would mimic the φ dihedral angle of native *O*-glycosides via the fluorine gauche effect, thereby potentially preserving or even enhancing bioactivity (Fig. 7). In contrast, the $CH_2$- and (*R*)-CHF analogs, which exhibit different conformational distributions, unexpectedly showed comparable or enhanced activity in some cases.

Accurately predicting the impact of structural modifications on biological activity remains difficult. Ongoing mechanistic studies employing computational, chemical biology, and structural biology approaches aim to address this challenge.

## 5 Conclusion

In this chapter, we have described the development of minimally perturbing chemical tools for probing and modulating glycan functions in their native structural and physicochemical contexts. Thienyl-substituted α-ketoamide-based photoaffinity probes emerged as a novel class of compact, less hydrophobic photoreactive groups, offering superior selectivity and labeling performance. Detailed conformational analyses and spin-density calculations uncovered a mechanistic paradigm distinct from that of conventional photo-crosslinkers, establishing a new platform for selective molecular labeling in biologically complex environments.

Concurrently, we achieved the divergent synthesis of $CH_2$-, (*R*)-CHF-, and (*S*)-CHF-linked analogs of biologically relevant glycans and glycoconjugates. Biological evaluations of these analogs revealed that subtle variations in linkage structure and stereoelectronic factors have dramatic effects on biological activity, such as receptor binding, immune activation, or cellular selectivity. Notably, analogs bearing non-natural linkages often exhibited divergent functional profiles, underscoring the importance of going beyond simple structural mimicry.

These studies collectively highlight the importance of integrated molecular design tailored to the specific structural and physicochemical context of each target glycan. The ability to strategically design ideal molecular architectures—and to realize them through precise synthetic chemistry—is essential for advancing chemical biology research in a fundamental and sustainable way. As the field moves forward, this approach will continue to serve as a cornerstone for exploring the roles of glycans in complex biological systems and for developing novel molecular probes and therapeutic leads.

## References

1. Schenone M, Dančík V, Wagner BK, Clemons PA (2013) Target identification and mechanism of action in chemical biology and drug discovery. Nat Chem Biol 9:232–240. https://doi.org/10.1038/nchembio.1199

2. Takaoka Y, Ojida A, Hamachi I (2013) Protein organic chemistry and applications for labeling and engineering in live-cell systems. Angew Chem Int Ed Engl 52:4088–4106. https://doi.org/10.1002/anie.201207089
3. Sumranjit J, Chung SJ (2013) Recent advances in target characterization and identification by photoaffinity probes. Molecules 18:10425–10451. https://doi.org/10.3390/molecules180910425
4. Patterson DM, Nazarova LA, Prescher JA (2014) Finding the right (bioorthogonal) chemistry. ACS Chem Biol 9:592–605. https://doi.org/10.1021/cb400828a
5. Yamaura K, Kiyonaka S, Hamachi I (2017) Construction of protein-based biosensors using ligand-directed chemistry for detecting analyte binding. Methods Enzymol 589:253–280. https://doi.org/10.1016/bs.mie.2017.02.006
6. Fleet GWJ, Porter RR, Knowles JR (1969) Affinity labelling of antibodies with aryl nitrene as reactive group. Nature 224:511–512. https://doi.org/10.1038/224511a0
7. Kotzyba-Hibert F, Kapfer I, Goeldner M (1995) Recent trends in photoaffinity labeling. Angew Chem, Int Ed Engl 34:1296–1312. https://doi.org/10.1002/anie.199512961
8. Hosoya T, Hiramatsu T, Ikemoto T, Nakanishi M, Aoyama H, Hosoya A, Iwata T, Maruyama K, Endo M, Suzuki M (2004) Novel bifunctional probe for radioisotope-free photoaffinity labeling: compact structure comprised of photospecific ligand ligation and detectable tag anchoring units. Org Biomol Chem 2:637–641. https://doi.org/10.1039/b316221d
9. Young MJT, Platz MS (1989) Polyfluorinated aryl azides as photoaffinity labelling reagents; the room temperature CH insertion reactions of singlet pentafluorophenyl nitrene with alkanes. Tetrahedron Lett 30:2199–2202. https://doi.org/10.1016/s0040-4039(00)99647-3
10. Chiba K, Asanuma M, Ishikawa M, Hashimoto Y, Dodo K, Sodeoka M, Yamaguchi T (2017) Specific fluorescence labeling of target proteins by using a ligand–4-azidophthalimide conjugate. Chem Commun (Camb) 53:8751–8754. https://doi.org/10.1039/c7cc03252h
11. Smith RAG, Knowles JR (1973) Aryldiazirines. Potential reagents for photolabeling of biological receptor sites. J Am Chem Soc 95:5072–5073. https://doi.org/10.1021/ja00796a062
12. Brunner J, Senn H, Richards FM (1980) 3-Trifluoromethyl-3-phenyldiazirine. A new carbene generating group for photolabeling reagents. J Biol Chem 255:3313–3318. https://doi.org/10.1016/s0021-9258(19)85701-0
13. Pan S, Jang S-Y, Wang D, Liew SS, Li Z, Lee J-S, Yao SQ (2017) A suite of "Minimalist" photo-crosslinkers for live-cell imaging and chemical proteomics: case study with BRD4 inhibitors. Angew Chem Int Ed Engl 56:11816–11821. https://doi.org/10.1002/anie.201706076
14. Galardy RE, Craig LC, Printz MP (1973) Benzophenone triplet: a new photochemical probe of biological ligand-receptor interactions. Nat New Biol 242:127–128. https://doi.org/10.1038/newbio242127a0
15. Song W, Wang Y, Qu J, Madden MM, Lin Q (2008) A photoinducible 1,3-dipolar cycloaddition reaction for rapid, selective modification of tetrazole-containing proteins. Angew Chem Weinheim Bergstr Ger 120:2874–2877. https://doi.org/10.1002/ange.200705805
16. Li Z, Qian L, Li L, Bernhammer JC, Huynh HV, Lee J-S, Yao SQ (2016) Tetrazole photoclick chemistry: reinvestigating its suitability as a bioorthogonal reaction and potential applications. Angew Chem Int Ed Engl 55:2002–2006. https://doi.org/10.1002/anie.201508104
17. Herner A, Marjanovic J, Lewandowski TM, Marin V, Patterson M, Miesbauer L, Ready D, Williams J, Vasudevan A, Lin Q (2016) 2-aryl-5-carboxytetrazole as a new photoaffinity label for drug target identification. J Am Chem Soc 138:14609–14615. https://doi.org/10.1021/jacs.6b06645
18. Horowitz A, Meller R, Moortgat G (2001) The UV–VIS absorption cross sections of the α-dicarbonyl compounds: pyruvic acid, biacetyl and glyoxal. J Photochem Photobiol A—Chem 146:19–27. https://doi.org/10.1016/S1010-6030(01)00601-3
19. Chesta CA, Whitten DG (1992) Photocyclization of.alpha.-keto amides in homogeneous solution and aqueous cyclodextrin media. The role of zwitterions and diradicals in photoinduced electron transfer reactions. J Am Chem Soc 114:2188–2197. https://doi.org/10.1021/ja00032a038

20. Ota E, Usui K, Oonuma K, Koshino H, Nishiyama S, Hirai G, Sodeoka M (2018) Thienyl-substituted α-Ketoamide: a less hydrophobic reactive group for photo-affinity labeling. ACS Chem Biol 13:876–880. https://doi.org/10.1021/acschembio.7b00988
21. Fang K, Hashimoto M, Jockusch S, Turro NJ, Nakanishi K (1998) A bifunctional photoaffinity probe for ligand/receptor interaction studies. J Am Chem Soc 120:8543–8544. https://doi.org/10.1021/ja9817186
22. Guo A-D, Wei D, Nie H-J, Hu H, Peng C, Li S-T, Yan K-N, Zhou B-S, Feng L, Fang C, Tan M, Huang R, Chen X-H (2020) Light-induced primary amines and o-nitrobenzyl alcohols cyclization as a versatile photoclick reaction for modular conjugation. Nat Commun 11:5472. https://doi.org/10.1038/s41467-020-19274-y
23. Page ACS, Scholz SO, Keenan KN, Spradlin JN, Belcher BP, Brittain SM, Tallarico JA, McKenna JM, Schirle M, Nomura DK, Toste FD (2022) Photo-Brook rearrangement of acyl silanes as a strategy for photoaffinity probe design. Chem Sci 13:3851–3856. https://doi.org/10.1039/d2sc00426g
24. Bon C, Goretzki B, Flamme M, Shelton C, Davis H, Lima F, Garcia F, Brittain S, Brocklehurst CE (2024) Oxadiazolines as photoreleasable labels for drug target identification. J Am Chem Soc 146:26759–26765. https://doi.org/10.1021/jacs.4c06936
25. Garami K, Péczka N, Petri L, Imre T, Langó T, Szabó Z, Orgován Z, Szabó P, Keserü GM, Ábrányi-Balogh P (2025) Target agnostic photoaffinity labelling by sulfonylhydrazones. Angew Chem Int Ed Engl 64:e202408701. https://doi.org/10.1002/anie.202408701
26. Geri JB, Oakley JV, Reyes-Robles T, Wang T, McCarver SJ, White CH, Rodriguez-Rivera FP, Parker DL Jr, Hett EC, Fadeyi OO, Oslund RC, MacMillan DWC (2020) Microenvironment mapping via Dexter energy transfer on immune cells. Science 367:1091–1097. https://doi.org/10.1126/science.aay4106
27. Thiehoff C, Rey YP, Gilmour R (2017) The fluorine gauche effect: a brief history. Isr J Chem 57:92–100. https://doi.org/10.1002/ijch.201600038
28. Linclau B, Ardá A, Reichardt N-C, Sollogoub M, Unione L, Vincent SP, Jiménez-Barbero J (2020) Fluorinated carbohydrates as chemical probes for molecular recognition studies. Current status and perspectives. Chem Soc Rev 49:3863–3888. https://doi.org/10.1039/c9cs00099b
29. Zhang P, Le C "chip", MacMillan DWC (2016) Silyl radical activation of alkyl halides in metallaphotoredox catalysis: a unique pathway for cross-electrophile coupling. J Am Chem Soc 138:8084–8087. https://doi.org/10.1021/jacs.6b04818
30. Yuan M, Song Z, Badir SO, Molander GA, Gutierrez O (2020) On the nature of C(sp3)-C(sp2) bond formation in nickel-catalyzed tertiary radical cross-couplings: a case study of Ni/Photoredox catalytic cross-coupling of alkyl radicals and aryl halides. J Am Chem Soc 142:7225–7234. https://doi.org/10.1021/jacs.0c02355
31. Kiya N, Hidaka Y, Usui K, Hirai G (2019) Synthesis of $CH_2$-linked α(1,6)-disaccharide analogues by α-selective radical coupling C-glycosylation. Org Lett 21:1588–1592. https://doi.org/10.1021/acs.orglett.9b00133
32. Hidaka Y, Kiya N, Yoritate M, Usui K, Hirai G (2020) Synthesis of $CH_2$-linked α-galactosylceramide and its glucose analogues through glycosyl radical-mediated direct C-glycosylation. Chem Commun 56:4712–4715. https://doi.org/10.1039/d0cc00785d
33. Kudo N, Chiba S, Ono S, Nagatoishi M, Yoritate M, Karasawa S, Usui K, Hirai G (2024) Stereoselective synthesis of C-aryl-α-glycosides by reductive C(sp2)-C(sp3) cross-coupling reaction. Synlett 36:357–362. https://doi.org/10.1055/a-2352-4760
34. Iwasaki K, Wan KK, Oppedisano A, Crossley SWM, Shenvi RA (2014) Simple, chemoselective hydrogenation with thermodynamic stereocontrol. J Am Chem Soc 136:1300–1303. https://doi.org/10.1021/ja412342g
35. Crossley SWM, Obradors C, Martinez RM, Shenvi RA (2016) Mn-, Fe-, and Co-catalyzed radical hydrofunctionalizations of olefins. Chem Rev 116:8912–9000. https://doi.org/10.1021/acs.chemrev.6b00334
36. Altiti AS, Bachan S, Alrowhani W, Mootoo DR (2015) An organocatalytic strategy for the stereoselective synthesis of C-galactosides with fluorine at the pseudoanomeric carbon. Org Biomol Chem 13:10328–10335. https://doi.org/10.1039/c5ob01471a

37. Moriyama T, Yoritate M, Kato N, Saika A, Kusuhara W, Ono S, Nagatake T, Koshino H, Kiya N, Moritsuka N, Tanabe R, Hidaka Y, Usui K, Chiba S, Kudo N, Nakahashi R, Igawa K, Matoba H, Tomooka K, Ishikawa E, Takahashi S, Kunisawa J, Yamasaki S, Hirai G (2024) Linkage-editing pseudo-glycans: a reductive α-fluorovinyl-C-glycosylation strategy to create glycan analogs with altered biological activities. J Am Chem Soc 146:2237–2247. https://doi.org/10.1021/jacs.3c12581
38. Moritsuka N, Kiya N, Moriyama T, Koshino H, Yoritate M, Matoba H, Hirai G (2024) Linkage-editing of melibiosamine: synthesis and biological evaluation of $CH_2$- and CHF-linked analogs. J Org Chem 89:11909–11920. https://doi.org/10.1021/acs.joc.4c01143
39. Kato N, Murakoshi Y, Kato M, Kobayashi T, Tsukagoshi N (2002) Isomaltose formed by alpha-glucosidases triggers amylase induction in Aspergillus nidulans. Curr Genet 42:43–50. https://doi.org/10.1007/s00294-002-0325-8
40. Morita M, Motoki K, Akimoto K, Natori T, Sakai T, Sawa E, Yamaji K, Koezuka Y, Kobayashi E, Fukushima H (1995) Structure-activity relationship of alpha-galactosylceramides against B16-bearing mice. J Med Chem 38:2176–2187. https://doi.org/10.1021/jm00012a018
41. Ban Y, Dong W, Zhang L, Zhou T, Altiti AS, Ali K, Mootoo DR, Blaho VA, Hla T, Ren Y, Ma X (2019) Abrogation of endogenous glycolipid antigen presentation on myelin-laden macrophages by D-sphingosine ameliorates the pathogenesis of experimental autoimmune encephalomyelitis. Front Immunol 10:404. https://doi.org/10.3389/fimmu.2019.00404
42. Hosomi O, Misawa Y, Takeya A, Matahira Y, Sugahara K, Kubohara Y, Yamakura F, Kudo S (2009) Novel oligosaccharide has suppressive activity against human leukemia cell proliferation. Glycoconj J 26:189–198. https://doi.org/10.1007/s10719-008-9175-z

# Elucidating the Immune Functions of Glycans Through Synthesis-Based Approaches

**Koichi Fukase, Atsushi Shimoyama, and Yoshiyuki Manabe**

**Abstract** The immunological functions of glycans were investigated using well-defined synthetic glycans, with the aim of enabling precise immune modulation. Efficient methods for glycan synthesis were established, enabling identification of structural determinants of immunological activity with structurally defined synthetic glycans. Immune modulation was achieved using bioactive glycans and inhibitors of glycan biosynthetic enzymes. Furthermore, aspects of the emergent functions of glycans were elucidated by employing glycan conjugates, including self-adjuvanting vaccine constructs.

**Keywords** Glycosylation · Innate immunity · Adjuvant · Vaccine · *N*-glycan

## 1 Introduction

The cell surface is coated with diverse glycans—including glycoproteins, glycolipids, and glycosaminoglycans—that play essential roles in cell–cell recognition and adhesion, in discrimination of self from non-self and altered self (e.g., cancer cells or virus-infected cells), and in a wide range of biological processes such as immune responses, infection, and inflammation. We have investigated the immunological functions of glycans—beyond the canonical roles of individual motifs—with particular emphasis on emergent functions arising from multivalent presentation and contextual effects within glycoconjugates.

K. Fukase (✉) · A. Shimoyama · Y. Manabe
Department of Chemistry, Graduate School of Science, The University of Osaka, Toyonaka, Osaka, Japan
e-mail: koichi@chem.sci.osaka-u.ac.jp

A. Shimoyama
e-mail: ashimo@chem.sci.osaka-u.ac.jp

Y. Manabe
e-mail: manabey12@chem.sci.osaka-u.ac.jp



K. Toshima and Y. Fujimoto (eds.), *Chemical Biology of Carbohydrates*,
https://doi.org/10.1007/978-981-95-5335-8_7

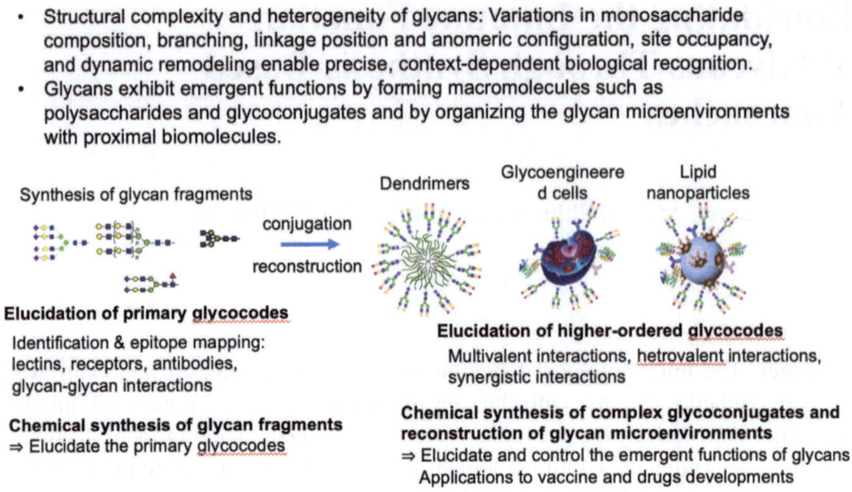


**Fig. 1** Elucidation of glycan functions

Kasai and Hirabayashi [1] pioneered the glycocode concept, providing a foundational framework for interpreting glycan information. Building on this foundation, we define the primary glycocode as the glycan motif recognized by glycan-recognizing molecules—including lectins, glycosyltransferases, and glycosidases [1–3].

Polysaccharides often carry multiple binding sites that constitute primary glycocodes, and multivalent engagement of these sites gives rise to higher-order activities. In some cases, the functional entity is the glycoconjugate as a whole; the native activity resides in the integrated glycoprotein, glycolipid, or proteoglycan, whereas the glycan moiety alone is insufficient (Fig. 1).

Accordingly, we define as a "higher-order glycocode" any architecture in which multiple primary glycocodes are presented within the same molecule or as a cluster and, through multivalent interactions and cooperation with other biomolecules, emergently yields affinity, selectivity, and functions that exceed simple additivity. Representative examples include (i) multivalent binding of polysaccharides [4–7], (ii) heterovalent (or heteromultivalent) recognition in which glycans act together with other elements (peptides, lipids, aromatic residues, etc.) [8, 9], and (iii) recognition of glycolipid clusters on membranes [10–14].

For example, in C-type lectin (-like) receptors such as DC-SIGN, Langerin, MGL, Dectin-1, CLEC-2, and NKG2D, contributions from distal secondary binding sites have been reported in addition to the canonical carbohydrate recognition domain (CRD). Such composite recognition enhances ligand selectivity and permits fine-tuning of biological responses [15, 16].

Functional studies of glycans are challenging due to their structural complexity and heterogeneity, as well as the persistent risk that natural preparations retain trace bioactive contaminants despite exhaustive purification. Therefore, we have investigated the developments of efficient methods for glycan synthesis and applied them to the preparation of glycans relevant to immune regulation. Using these synthetic glycans, together with peptide/protein assemblies and dendrimers displaying defined glycans, we identified minimal immunostimulatory motifs and elucidated emergent glycan functions in immune modulation.

## 1.1 Molecular Basis of the Immunoregulatory Functions of N-Glycans

Asparagine-linked glycans (*N*-glycans) are structurally diverse and play key roles in immune regulation. For example, core fucosylation of the Fc *N*-glycan suppresses antibody-dependent cellular cytotoxicity (ADCC) in therapeutic antibodies [17], whereas core fucosylation of the TGF-β receptor promotes TGF-β signaling, thereby suppressing immune responses [18]. Sialic acids on *N*-glycans are recognized by Siglecs and generally dampen immunity. Core fucosylation also influences numerous biological processes—for instance, it can inhibit antitumor responses by increasing PD-1 expression on T cells [19], and core fucosylation of T-cell glycans is upregulated under inflammatory conditions [20]. However, endogenous mammalian receptors that recognize core fucose had not been identified until recently.

We discovered that Dectin-1, a lectin known to recognize β-glucans, recognizes the core fucose on IgG [21]. Dectin-1 engages both an aromatic amino acid adjacent to the *N*-glycosylated Asn site and the core fucose, an interaction we termed heterovalent interaction; this corresponds to a secondary glycocode (Fig. 2). Dectin-1 thus represents the first lectin that recognizes a specific *N*-glycan on antibodies in a heterovalent and selective manner.

Recently, Seeling et al. reported that Dectin-1 interacts with anti-inflammatory IgG (IVIg) and FcγRIIb, contributing to suppression of inflammation [22]. Although

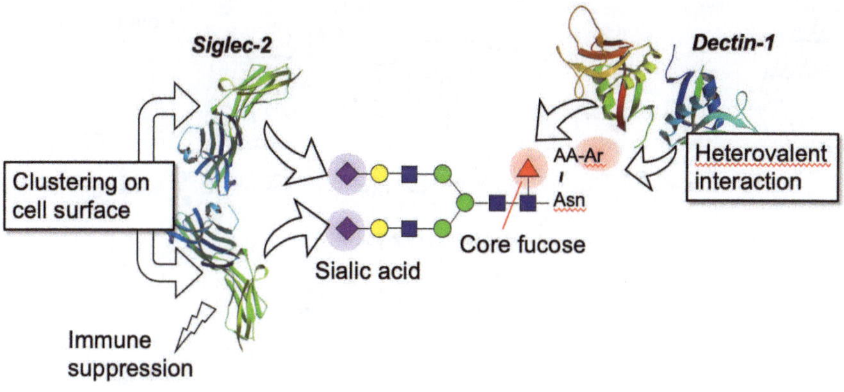


**Fig. 2** Multivalent interactions involving *N*-glycans

they did not discuss heterovalent recognition of core-fucosylated IgG, their findings support the idea that core fucose immunosuppresses IgG function. Relatedly, heteromultivalent binding that involves a carbohydrate-recognition site together with a distal pocket in C-type lectins has been reported by Rademacher and colleagues [15, 16].

Core fucosylation is catalyzed by α1,6-fucosyltransferase (FUT8). Using high-throughput screening (HTS) of Osaka University's in-house chemical library, we identified a FUT8-inhibitory pharmacophore, and subsequent structure optimization yielded an inhibitor **1** with $K_D = 49$ nM (Fig. 3) [23]. Compound **1** binds to FUT8 only in the presence of GDP. Mechanistic studies suggested the in-enzyme formation of a naphthoquinone methide derivative that reacts with FUT8. A similar covalent inhibitory mode, involving generation of a quinone-methide intermediate followed by nucleophilic addition to form a covalent adduct, has been documented previously [24, 25]; notably, reversible covalent inhibition has also been reported [25].

Furthermore, prodrug derivatization improved stability and, in cell-based assays, prodrug **2** suppressed core-fucose expression and the ensuing epidermal growth factor receptor (EGFR) signaling. Another FUT8 inhibitor also attenuated inflammatory signaling in T cells. Although preliminary, oral activity was observed, opening a path toward the development of therapeutics that target core fucosylation [23].

We also carried out the precision synthesis of sialic acid–containing glycans and employed them for the functional analysis of sialic acid–recognizing proteins.

In general, when NHAc groups are present on a glycosyl donor and/or acceptor, intermolecular hydrogen bonding of NHAc group markedly diminishes glycosylation reactivity. To address this issue, we developed a "diacetyl strategy" for the synthesis of NHAc-bearing glycans, in which the NHAc is temporarily converted to an *N*,*N*-diacetyl imide ($NAc_2$) [26].

As an application to the synthesis of sialyl glycans, we prepared disialylated tetrasaccharide **8**, a structural motif found in the *N*-glycans of human factor X and

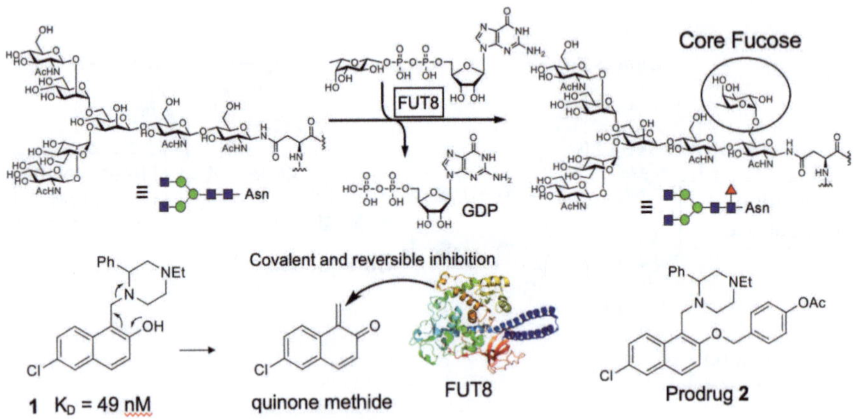


**Fig. 3** Development of inhibitors of FUT8 (α1,6-Fucosyltransferase)

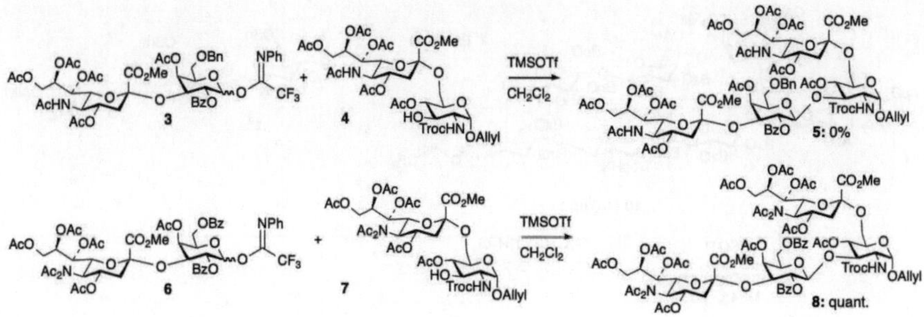

**Fig. 4** Synthesis of disialylated tetrasaccharide **8**

fetuin (Fig. 4). The glycosylation of sialyl disaccharides **3** and **4** exhibited very low reactivity, and the desired product was not obtained. In contrast, fragments **6** and **7**, in which both sialic acid residues were protected as $NAc_2$, showed markedly improved reactivity, furnishing tetrasaccharide **8** in quantitative yield [27].

Using this diacetyl strategy, we further synthesized a range of sialylated glycans, including a tetraantennary *N*-glycan **15** and biantennary *N*-glycans in which one acetyl group was selectively deuterated [28]. In the synthesis of tetraantennary N-glycan **15** (Scheme 1), glycosylation between heptasaccharide donor **10** and trisaccharide acceptor **11** in $Et_2O/CH_2Cl_2 = 1/1$ proceeded with complete α-selectivity; subsequent benzylidene cleavage afforded decasaccharide **12** in 33% yield (BRSM 49%) over two steps. The following glycosylation of **12** with **13**, conducted in a higher-ether solvent mixture ($Et_2O/CH_2Cl_2 = 5/1$) at 0 °C with TBDPSOTf as promoter, gave compound **14** in 36% yield. Global deprotection of **14** then furnished the fully sialylated tetraantennary *N*-glycan **15**.

Using asymmetrically deuterated *N*-glycans, we found that H1N1 neuraminidase preferentially cleaves the terminal sialic acid on the α1, 3-branched arm rather than on the α1,6-branched arm. Treatment of a tetraantennary *N*-glycan with H1N1 neuraminidase revealed no multivalency effect, highlighting a pronounced difference from the multivalent recognition by lectins [28].

We further systematically analyzed the interactions of sialylated glycans with various sialic-acid-recognizing proteins [29–35]. In particular, we showed that Siglec-2 (CD22) cooperatively engages two sialic acids located on the respective arms of a biantennary *N*-glycan, thereby promoting local accumulation (cis-clustering) [29]. These findings indicate that *N*-glycans and their assemblies can display emergent functions driven by multivalent recognition [29, 30].

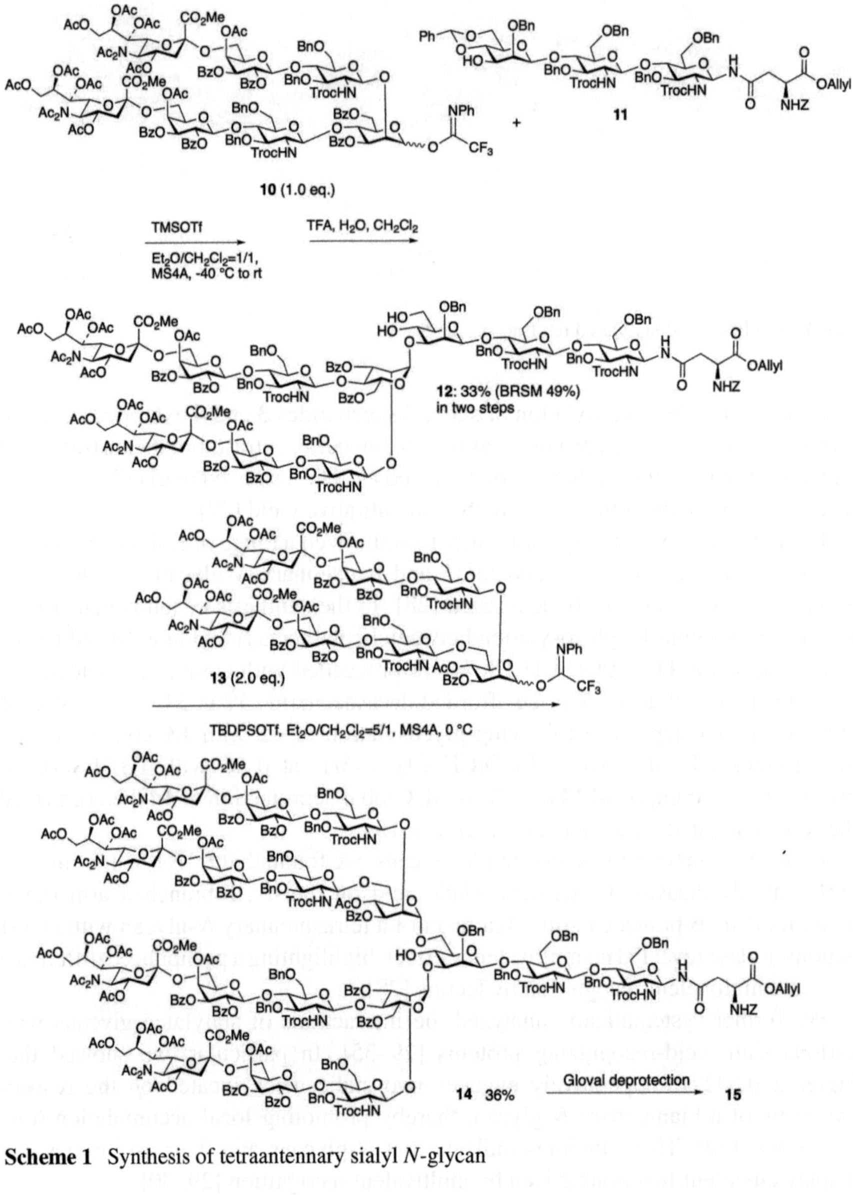


**Scheme 1** Synthesis of tetraantennary sialyl *N*-glycan

## 2 Bacterial Cell-Surface Glycoconjugates that Activate Host Defense System

The immunostimulatory activities of bacterial glycoconjugates—such as peptidoglycan and lipopolysaccharide (LPS)—have long been known and are now understood to arise from activation of the innate immune system. Because innate immunity not only defends against microbes but also intersects with allergy, autoimmunity, chronic inflammation, and cancer, it has attracted considerable interest. At the School of Science, Osaka University, the syntheses of muramyl dipeptide (MDP) and lipid A—partial structures of peptidoglycan and LPS, respectively—were accomplished, establishing these motifs as the active principles [36, 37]. Building on this foundation, we elucidated the mechanisms by which these immunostimulatory glycoconjugates are recognized and act, thereby providing a rigorous molecular basis for their recognition by innate immunity.

### *2.1 Synthesis of Peptidoglycan Fragments and Mechanisms of Bioactivity*

Although MDP was identified in the mid-1970s as the minimal active motif of peptidoglycan, its mechanism of action remained unclear. We synthesized libraries of peptidoglycan fragments, including the first chemical synthesis of the repeating glycan strands of peptidoglycan [38–44]. In collaboration with Inohara and Núñez, we demonstrated that the cytosolic proteins NOD1 and NOD2 are receptors for peptidoglycan, with γ-D-Glu-meso-DAP (iE-DAP) and MDP serving as their respective minimal ligands (Fig. 5) [45, 46]; these constitute primary glycocodes. Bacteria release NOD1/2 ligands into their environment, and stimulation by these ligands is thought to activate innate immunity. We purified NOD1-activating factors from *Escherichia coli* culture supernatants and identified several peptidoglycan fragments [47]. In parallel, we accomplished the first total synthesis of tracheal cytotoxin (TCT)—a disaccharide tetrapeptide containing anhydro muramic acid—and related structures (Fig. 5) [41]. TCT showed low NOD1 activity, suggesting it is not a true toxin. By contrast, the tripeptide lacking the C-terminal D-Ala (derived from TCT) proved to be the major NOD1 ligand in *E. coli* supernatants, exhibiting potent NOD1 activity. The molecular recognition of various proteins including peptidoglycan-recognition proteins (PGRPs) in insects and mammals was clarified by using our synthetic compounds (Fig. 5) [48–54].

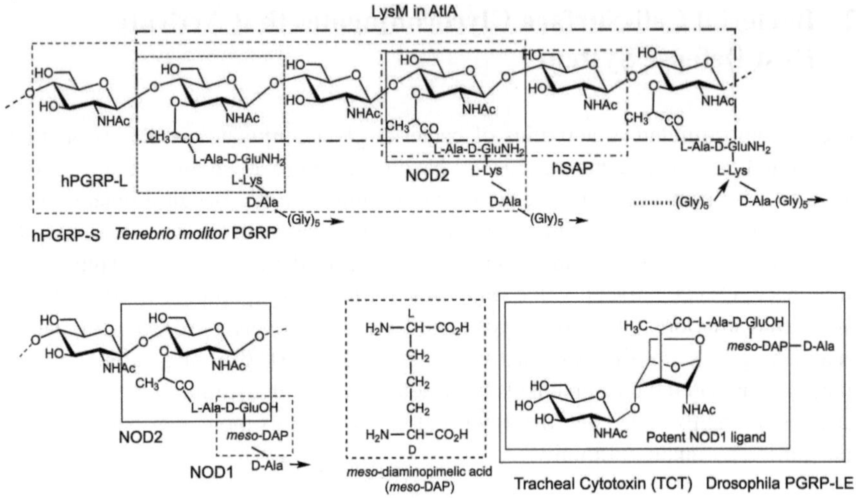


**Fig. 5** Peptidoglycan glycocodes

## 2.2 *Synthesis of LPS and Mechanisms of Bioactivity*

Endotoxin is a toxin of Gram-negative bacteria that provokes a wide range of host responses involving immunity and inflammation—including lethal (endotoxic) shock, pyrogenicity, complement activation, leukocyte activation, vascular endothelial injury, disseminated intravascular coagulation (DIC), enhanced antibody production, and promotion of phagocytosis. The chemical entity of endotoxin is lipopolysaccharide (LPS), a component of the bacterial outer membrane, whose structure comprises the bioactive principle lipid A linked to a polysaccharide region via the acidic sugar Kdo (3-deoxy-D-manno-oct-2-ulosonic acid) [36, 37, 55, 56] (Fig. 6). In canonical *E. coli* LPS, lipid A moiety is recognized by the TLR4/MD2 receptor, activating two signaling pathways; their combined action stimulates both innate and, secondarily, adaptive immunity (e.g., antibody production), but also elicits toxicities such as lethal inflammation. Through synthesis of lipid A and LPS partial structures, we contributed to defining the mechanisms underlying LPS/lipid A activity.

Although chemical syntheses of *E. coli* lipid A **16** and its biosynthetic precursor lipid IVa **17** were reported in 1983–1985, yields and efficiencies were unsatisfactory, and purification of the final products remained problematic. We improved glycosylation, protecting-group strategies, and overall synthetic design, and liquid–liquid partition chromatography for purification, thereby establishing a far more efficient route to lipid A [36, 37, 55, 56]. This enabled sufficient supply for biological studies and helped establish that lipid A signals via TLR4, with MD2 being essential.

We synthesized radiolabeled lipid A derivatives [57], which Miyake and colleagues used to directly analyze interactions with the TLR4/MD2 complex [58];

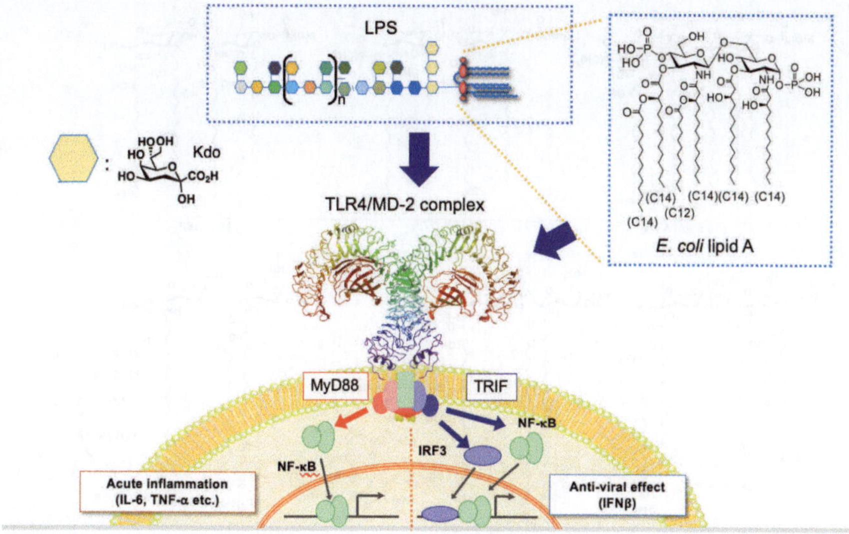


**Fig. 6** Activation of innate immunity by lipopolysaccharide (LPS)

we also showed that receptor dimerization upon stimulation underlies signal transduction [59, 60]. Subsequently, Ohto and Satow obtained co-crystals of human MD-2 with lipid IVa—an antagonist in humans—and determined their three-dimensional structures by X-ray crystallography [61]. Lee and colleagues reported X-ray crystal structures of the human TLR4/MD2 complex bound to the antagonist Eritoran and to the agonist LPS [62, 63], and Ohto and co-workers determined X-ray structures of the mouse TLR4/MD2 complex with LPS as well as lipid IVa (an agonist in mice, an agonist in humans) [64], thereby elucidating structural determinants critical for TLR4/MD2 activation. In parallel, we synthesized a variety of related analogues and evaluated their biological activities, thereby elucidating structural features essential for lipid A bioactivity—specifically, that the lipid moieties and acidic functional groups are critical for TLR4/MD2 recognition.

In parallel, we synthesized a variety of related analogues and evaluated their biological activities, thereby elucidating structural features essential for lipid A bioactivity—specifically, that the lipid moieties and acidic functional groups are critical for TLR4/MD2 recognition [36, 37, 65, 66] (Fig. 7). We showed that monophosphoryl lipid A **19** (MPL; dephosphorylated at the 1-position) exhibits moderate adjuvant activity [67], and Miyake and colleagues demonstrated that this correlates with reduced TLR4/MD2 dimerization [68]. GlaxoSmithKline developed 3D-MPL from Salmonella LPS and found moderate adjuvanticity with suppression of NF-κB (antibacterial) signaling while retaining antiviral pathways, leading to clinical implementation as a vaccine adjuvant [69, 70].

To ensure usefulness as an adjuvant, it is essential to attenuate the intrinsic toxicity of lipid A. Guided by a chemical-ecology perspective, we therefore pursued the development of low-toxicity adjuvants based on lipid A structures from commensal

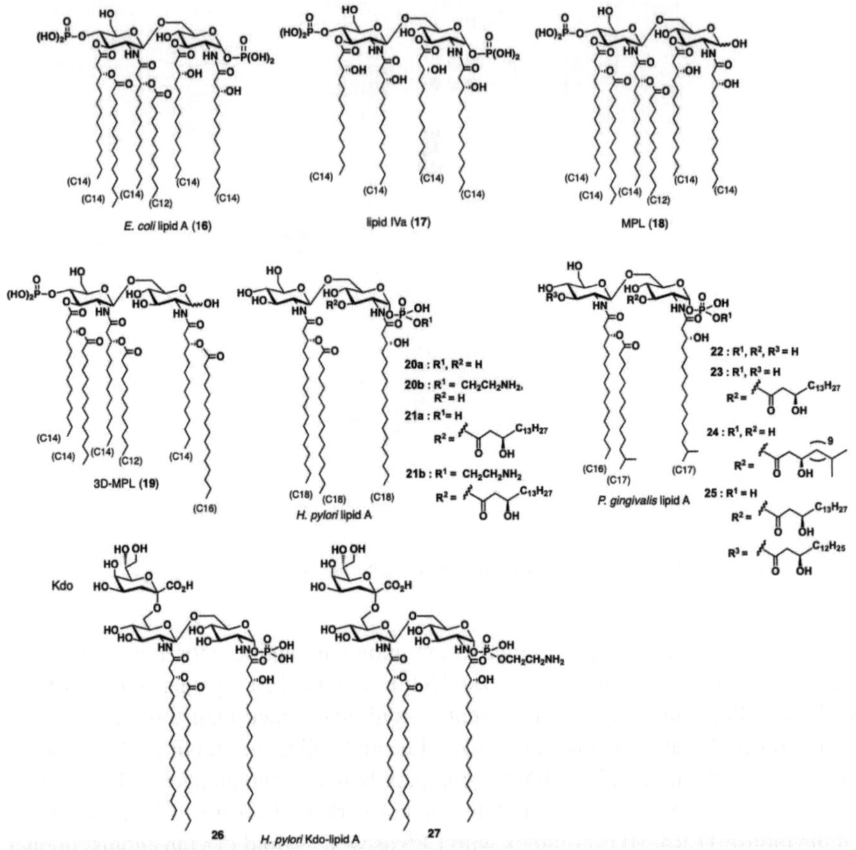

**Fig. 7** Various types of lipid A and derivatives

and parasitic bacteria. Their LPSs, shaped by coevolution, possess distinct chemical structures that avoid strong pro-inflammatory toxicity.

Parasitic bacteria such as *Helicobacter pylori* (gastric ulcer) and *Porphyromonas gingivalis* (periodontitis) possess several lipid A types that differ from the *E. coli* prototype (Fig. 7), featuring longer but fewer acyl chains and monophosphoryl structures sometimes modified with ethanolamine. We synthesized LPS partial structures (Fig. 7) from these bacteria using a diversity-oriented, convergent strategy (Scheme 2) in which a common intermediate **30** was assembled and acyl chains were introduced stepwise [71–74]. For the α-selective glycosylation of the 3-deoxy sugar Kdo, we improved upon our glycosyl fluoride protocol by employing a conformationally constrained ("boat-locked") imidate donor **31**, achieving high stereoselectivity under mild conditions. Implementation in a microflow setup then furnished trisaccharide **32** with high selectivity and efficiency [72].

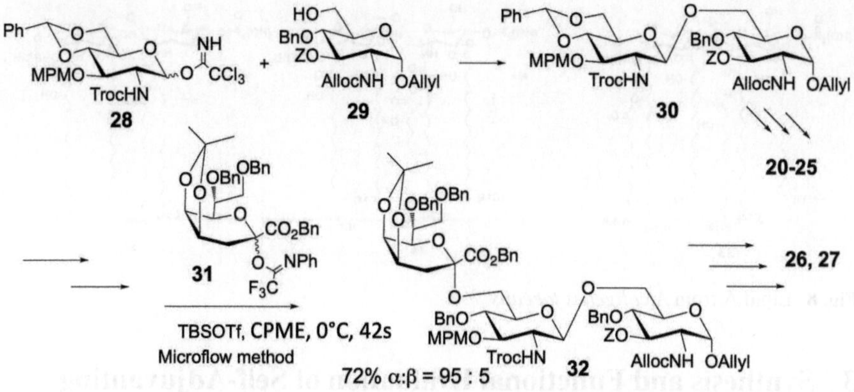


**Scheme 2** Synthesis of lipid A and Kdo-lipid A from parasitic bacteria

These compounds lacked strong pro-inflammatory activity; *H. pylori* lipid A **20a**, **21a**, Kdo-lipid A **26** and **27**, and *P. gingivalis* lipid A (tri- or tetra-acylated) **22**–**24**, acted as a TLR4–MD2 antagonist, whereas *H. pylori* lipid A **20b**, **21b** (ethanol amine type), and pepta-acylated *P. gingivalis* lipid A **25** were very weak agonists. All compounds induced IL-18 and/or IL-12 linked to chronic inflammation. Contrary to the conventional view that lipid A alone is the active principle, we found that in parasitic bacteria Kdo-lipid A **26** and **27** functions as the primary glycocode conveying essential properties. This indicates that parasites exploit LPS to evade acute, bactericidal inflammation, while promoting chronic inflammation, establishing LPS as a key molecule in the chemical ecology of host–microbe immune control.

Kunisawa and Kiyono showed that A. faecalis is a commensal resident of intestinal Peyer's patches [75], and we demonstrated that its LPS exerts low-inflammatory adjuvant activity [76]. The extracted LPS fraction was non-toxic yet activated immunity moderately in a TLR4-dependent manner and enhanced antibody production as effectively as *E. coli* LPS[76]. In collaboration with Kunisawa, Kiyono, and Molinaro, we found that *A. faecalis* produces a nonasaccharide lipooligosaccharide (LOS) whose lipid A is a mixture **33**–**35** with varying numbers of acyl chains (Fig. 8) [77]. Chemical synthesis of lipid A **33**–**35** revealed that only the hexa-acylated lipid A **33** exhibited adjuvant activity; the activity of **33** matched that of *A. faecalis* LOS, identifying **33** as the active center (primary glycocode). In vivo mouse studies showed that **33** and *A. faecalis* LOS are low-inflammatory adjuvants, enhancing antigen-specific IgA and IgG, and that vaccines formulated with **33** significantly protected against bacterial infection [78–81]. Because IgA maintains intestinal immune homeostasis, *A. faecalis* LOS and its lipid A constitute important immunoregulatory factors. By repurposing lipid A—long regarded as the toxic principle of endotoxin—as an immune-modulatory factor, we contributed to a major paradigm shift. Compound **33** has been commercialized as a reagent and is expected to advance toward practical vaccine-adjuvant use.

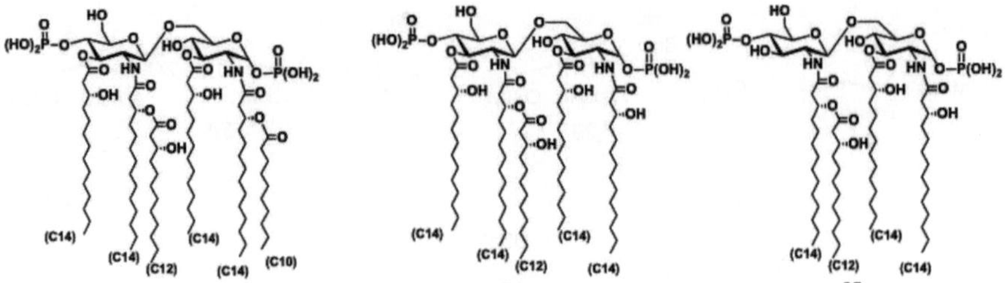

33 34 35

**Fig. 8** Lipid A from *Alcaligenes faecalis*

## 3 Synthesis and Functional Evaluation of Self-Adjuvanting Vaccines

Self-adjuvanting vaccines, in which antigens are covalently linked to adjuvants to elicit robust antigen-specific responses, are attracting attention as next-generation vaccines [82]; however, glycan-antigen-based versions remained unexplored. Using trisialyl-Tn (TriSTn)—a trimer of the cancer antigen sialyl-Tn (STn)—as the antigen (to exploit cluster effects for tumor selectivity), and $Pam_3CSK_4$ as a TLR2-ligand adjuvant, we synthesized ternary conjugates incorporating a T-cell epitope (Fig. 9) [83].

This vaccine strongly induced antibodies against TriSTn, and the antibodies specifically recognized tumor cells. In contrast, a non-conjugated mixture of $Pam_3CSK_4$, TriSTn, and the T-cell epitope failed to induce antibodies, underscoring the importance of covalent conjugation [83, 84].

This conjugation-centric strategy proved effective when applied to CH401 (a HER2-derived peptide antigen) and to the SARS-CoV-2 receptor-binding domain (RBD) [85, 86]. Thus, higher-order functionalization of glycans via conjugation concretely realizes the emergent functions of glycans in vivo and provides a powerful method for controlling biological functions.

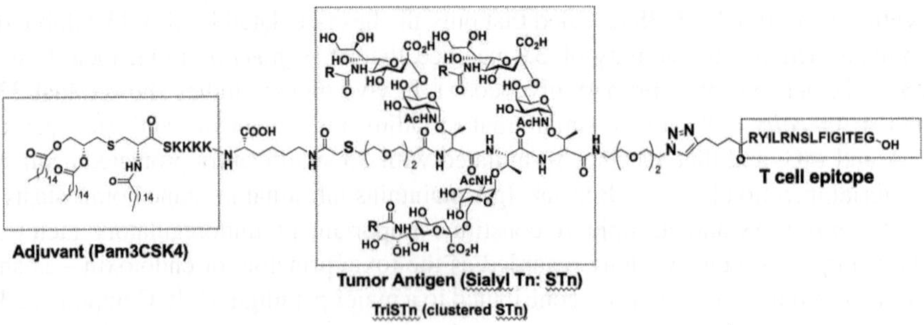


**Fig. 9** Structure of a self-adjuvanting vaccine

## 4 Conclusion

In summary, through an integrated strategy combining chemical synthesis with functional studies, we synthesized a wide range of bioactive glycans and glycoconjugates and chemically elucidated the molecular bases of glycan-mediated immune functions.

## References


1. Kasai K, Hirabayashi J (1996) Galectins: a family of animal lectins that decipher glycocodes. J Biochem 119:1–8. https://doi.org/10.1093/oxfordjournals.jbchem.a021192
2. Hebert DN, Garman SC, Molinari M (2005) The glycan code of the endoplasmic reticulum: asparagine-linked carbohydrates as protein maturation and quality-control tags. Trends Cell Biol 15:364–370. https://doi.org/10.1016/j.tcb.2005.05.007
3. Gabius HJ, Kayser K (2014) Introduction to glycopathology: the concept, the tools and the perspectives. Diagn Pathol 9:4. https://doi.org/10.1186/1746-1596-9-4
4. Mammen M, Choi S-K, Whitesides GM (1998) Polyvalent interactions in biological systems: implications for design and use of multivalent ligands and inhibitors. Angew Chem Int Ed Engl 37:2754–2794. https://doi.org/10.1002/(SICI)1521-3773(19981102)37:20%3c2754::AID-ANIE2754%3e3.0.CO;2-3
5. Imberty A, Varrot A (2008) Microbial recognition of human cell surface glycoconjugates. Curr Opin Struct Biol 18:567–576. https://doi.org/10.1016/j.sbi.2008.08.001
6. Roy R, Murphy PV, Gabius HJ (2016) Multivalent carbohydrate-lectin interactions: how synthetic chemistry enables insights into nanometric recognition. Molecules 21:629. https://doi.org/10.3390/molecules21050629
7. Quintana JI, Cañada FJ, Jiménez-Barbero J, Ardá A (2023) Exploring multivalent carbohydrate–protein interactions by NMR. Chem Soc Rev 52:1591–1613. https://doi.org/10.1039/D2CS00983H
8. Müller C, Despras G, Lindhorst TK (2016) Organizing multivalency in carbohydrate recognition. Chem Soc Rev 45:3275–3302. https://doi.org/10.1039/C6CS00165C
9. Jiménez Blanco JL, Ortiz Mellet C, García Fernández JM (2013) Multivalency in heterogeneous glycoenvironments: hetero-glycoclusters, -glycopolymers and -glycoassemblies. Chem Soc Rev 42:4518–4531. https://doi.org/10.1039/C2CS35219B
10. Simons K, Toomre D (2000) Lipid rafts and signal transduction. Nat Rev Mol Cell Biol 1:31–39. https://doi.org/10.1038/35036052
11. Lingwood D, Simons K (2010) Lipid rafts as a membrane-organizing principle. Science 327:46–50. https://doi.org/10.1126/science.1174621
12. Ohtsubo K, Marth JD (2006) Glycosylation in cellular mechanisms of health and disease. Cell 126:855–867. https://doi.org/10.1016/j.cell.2006.08.019
13. Johannes L, Wunder C, Shafaq-Zadah M (2016) Glycolipids and lectins in endocytic uptake processes. J Mol Biol 428:4792–4818. https://doi.org/10.1016/j.jmb.2016.10.027
14. Siukstaite L, Imberty A, Römer W (2021) Structural diversities of lectins binding to the glycosphingolipid Gb3. Front Mol Biosci 8:704685. https://doi.org/10.3389/fmolb.2021.704685
15. Wawrzinek R, Wamhoff E-C, Lefebre J, Rentzsch M, Bachem G, Domeniconi G, Schulze J, Fuchsberger FF, Zhang H, Modenutti C, Schnirch L, Marti MA, Schwardt O, Bräutigam M, Guberman M, Hauck D, Seeberger PH, Seitz O, Titz A, Ernst B, Rademacher C (2021) A remote secondary binding pocket promotes heteromultivalent targeting of DC-SIGN. J Am Chem Soc 143:18977–18988. https://doi.org/10.1021/jacs.1c07235

16. Lefèbre J, Falk T, Ning Y, Rademacher C (2024) Secondary sites of the C-type lectin-like fold. Chem Eur J 30:e202400660. https://doi.org/10.1002/chem.202400660
17. Shinkawa T, Nakamura K, Yamane N, Shoji-Hosaka E, Kanda Y, Sakurada M, Uchida K, Anazawa H, Satoh M, Yamasaki M, Hanai N, Shitara K (2003) The absence of fucose but not the presence of galactose or bisecting N-acetylglucosamine of human IgG1 complex-type oligosaccharides shows the critical role of enhancing antibody-dependent cellular cytotoxicity. J Biol Chem 278:3466–3473. https://doi.org/10.1074/jbc.M210665200
18. Wang X, Inoue S, Gu J, Miyoshi E, Noda K, Li W, Mizuno-Horikawa Y, Nakano M, Asahi M, Takahashi M, Uozumi N, Ihara S, Lee SH, Ikeda Y, Yamaguchi Y, Aze Y, Tomiyama Y, Fujii J, Suzuki K, Kondo A, Shapiro SD, López-Otín C, Kuwaki T, Okabe M, Honke K, Taniguchi N (2005) Dysregulation of TGF-β1 receptor activation leads to abnormal lung development and emphysema-like phenotype in core fucose-deficient mice. Proc Natl Acad Sci USA 102:15791–15796. https://doi.org/10.1073/pnas.0507375102
19. Okada M, Chikuma S, Kondo T, Hibino S, Machiyama H, Yokosuka T, Nakano M, Yoshimura A (2017) Blockage of core fucosylation reduces cell-surface expression of PD-1 and promotes anti-tumor immune responses of T cells. Cell Rep 20:1017–1028. https://doi.org/10.1016/j.celrep.2017.07.027
20. Fujii H, Shinzaki S, Iijima H, Wakamatsu K, Iwamoto C, Sobajima T, Kuwahara R, Hiyama S, Hayashi Y, Takamatsu S, Uozumi N, Kamada Y, Tsujii M, Taniguchi N, Takehara T, Miyoshi E (2016) Core fucosylation on T cells, required for activation of T-cell receptor signaling and induction of colitis in mice, is increased in patients with inflammatory bowel disease. Gastroenterology 150:1620–1632. https://doi.org/10.1053/j.gastro.2016.03.002
21. Manabe Y, Marchetti R, Takakura Y, Nagasaki M, Nihei W, Takebe T, Tanaka K, Kabayama K, Chiodo F, Hanashima S, Kamada Y, Miyoshi E, Dulal HP, Yamaguchi Y, Adachi Y, Ohno N, Tanaka H, Silipo A, Fukase K, Molinaro A (2019) The core fucose on an IgG antibody is an endogenous ligand of Dectin-1. Angew Chem Int Ed 58:18697–18702. https://doi.org/10.1002/anie.201911875
22. Seeling M, Pöhnl M, Kara S, Horstmann N, Riemer C, Wöhner M, Liang C, Brückner C, Eiring P, Werner A, Biburger M, Altmann L, Schneider M, Amon L, Lehmann CHK, Lee S, Kunz M, Dudziak D, Schett G, Bäuerle T, Lux A, Tuckermann J, Vögtle T, Nieswandt B, Sauer M, Böckmann RA, Nimmerjahn F (2023) Immunoglobulin G-dependent inhibition of inflammatory bone remodeling requires pattern recognition receptor Dectin-1. Immunity 56:1046-1063.e7. https://doi.org/10.1016/j.immuni.2023.02.019
23. Manabe Y, Takebe T, Kasahara S, Hizume K, Kabayama K, Kamada Y, Asakura A, Shinzaki S, Takamatsu S, Miyoshi E, García-García A, Vakhrushev SY, Hurtado-Guerrero R, Fukase K (2024) Development of a FUT8 Inhibitor with cellular inhibitory properties. Angew Chem Int Ed 63:e202414682. https://doi.org/10.1002/anie.202414682
24. McLean LR, Zhang Y, Li H, Li Z, Lukasczyk U, Choi YM, Han Z, Prisco J, Fordham J, Tsay JT, Reiling S, Vaz RJ, Li Y (2009) Discovery of covalent inhibitors for MIF tautomerase via cocrystal structures with phantom hits from virtual screening. Bioorg Med Chem Lett 19:6717–6720. https://doi.org/10.1016/j.bmcl.2009.09.106
25. Madonna S, Béclin C, Laras Y, Moret V, Marcowycz A, Lamoral-Theys D, Dubois J, Barthelemy-Requin M, Lenglet G, Depauw S, Cresteil T, Aubert G, Monnier V, Kiss R, David-Cordonnier M-H, Kraus J-L (2010) Structure–activity relationships and mechanism of action of antitumor bis 8-hydroxyquinoline substituted benzylamines. Eur J Med Chem 45:623–638. https://doi.org/10.1016/j.ejmech.2009.11.006
26. Fukase K, Manabe Y, Shimoyama A (2023) Diacetyl strategy for synthesis of NHAc-containing glycans: enhancing glycosylation reactivity via diacetyl imide protection. Front Chem 11:1319883. https://doi.org/10.3389/fchem.2023.1319883
27. Zhou J, Manabe Y, Tanaka K, Fukase K (2016) Efficient synthesis of the disialylated tetrasaccharide motif in N-glycans through an amide-protection strategy. Chem Asian J 11:1436–1440. https://doi.org/10.1002/asia.201600139
28. Shirakawa A, Manabe Y, Marchetti R, Yano K, Masui S, Silipo A, Molinaro A, Fukase K (2021) Chemical synthesis of sialyl N-glycans and analysis of their recognition by neuraminidase. Angew Chem Int Ed 60:24686–24693. https://doi.org/10.1002/anie.202111035

29. Di Carluccio C, Crisman E, Manabe Y, Forgione RE, Lacetera A, Amato J, Pagano B, Randazzo A, Zampella A, Lanzetta R, Fukase K, Molinaro A, Crocker PR, Martín-Santamaría S, Marchetti R, Silipo A (2020) Characterisation of the dynamic interactions between complex N-glycans and human CD22. ChemBioChem 21:129–140. https://doi.org/10.1002/cbic.201900295
30. Handa-Narumi M, Yoshimura T, Konishi H, Fukata Y, Manabe Y, Tanaka K, Bao G, Kiyama H, Fukase K, Ikenaka K (2018) Branched sialylated N-glycans are accumulated in brain synaptosomes and interact with Siglec-H. Cell Struct Funct 43:141–152. https://doi.org/10.1247/csf.18009
31. Di Carluccio C, Cerofolini L, Moreira M, Rosu F, Padilla-Cortés L, Gheorghita GR, Xu Z, Santra A, Yu H, Yokoyama S, Gray TE, St. Laurent CD, Manabe Y, Chen X, Fukase K, Macauley MS, Molinaro A, Li T, Bensing BA, Marchetti R, Gabelica V, Fragai M, Silipo A (2024) Molecular insights into O-linked sialoglycans recognition by the Siglec-like SLBR-N (SLBRUB10712) of Streptococcus gordonii. ACS Central Sci 10:447–459. https://doi.org/10.1021/acscentsci.3c01598
32. Di Carluccio C, Forgione RE, Bosso A, Yokoyama S, Manabe Y, Pizzo E, Molinaro A, Fukase K, Fragai M, Bensing BA, Marchetti R, Silipo A (2021) Molecular recognition of sialoglycans by streptococcal Siglec-like adhesins: toward the shape of specific inhibitors. RSC Chem Biol 2:1618–1630. https://doi.org/10.1039/D1CB00173F
33. Di Carluccio C, Forgione RE, Montefiori M, Civera M, Sattin S, Smaldone G, Fukase K, Manabe Y, Crocker PR, Molinaro A, Marchetti R, Silipo A (2021) Behavior of glycolylated sialoglycans in the binding pockets of murine and human CD22. iScience 24:101998. https://doi.org/10.1016/j.isci.2020.101998
34. Forgione RE, Di Carluccio C, Guzmán-Caldentey J, Gaglione R, Battista F, Chiodo F, Manabe Y, Arciello A, Del Vecchio P, Fukase K, Molinaro A, Martín-Santamaría S, Crocker PR, Marchetti R, Silipo A (2020) Unveiling molecular recognition of sialoglycans by human Siglec-10. iScience 23:101231. https://doi.org/10.1016/j.isci.2020.101231
35. Forgione RE, Di Carluccio C, Kubota M, Manabe Y, Fukase K, Molinaro A, Hashiguchi T, Marchetti R, Silipo A (2020) Structural basis for glycan-receptor binding by mumps virus hemagglutinin-neuraminidase. Sci Rep 10:1589. https://doi.org/10.1038/s41598-020-58559-6
36. Kusumoto S, Fukase K (2006) Synthesis of endotoxic principle of bacterial lipopolysaccharide and its recognition by the innate immune systems of hosts. Chem Rec 6:333–343. https://doi.org/10.1002/tcr.20098
37. Kusumoto S, Fukase K, Shiba T (2010) Key structures of bacterial peptidoglycan and lipopolysaccharide triggering the innate immune system of higher animals: chemical synthesis and functional studies. Proc Jpn Acad Ser B Phys Biol Sci 86:322–337. https://doi.org/10.2183/pjab.86.322
38. Inamura S, Fukase K, Kusumoto S (2001) Synthetic study of peptidoglycan partial structures: synthesis of tetrasaccharide and octasaccharide fragments. Tetrahedron Lett 42:7613–7616. https://doi.org/10.1016/S0040-4039(01)01619-7
39. Inamura S, Fujimoto Y, Kawasaki A, Shiokawa Z, Woelk E, Heine H, Lindner B, Inohara N, Kusumoto S, Fukase K (2006) Synthesis of peptidoglycan fragments and evaluation of their biological activity. Org Biomol Chem 4:232–242. https://doi.org/10.1039/B511866B
40. Fujimoto Y, Inamura S, Kawasaki A, Shiokawa Z, Shimoyama A, Hashimoto T, Kusumoto S, Fukase K (2007) Chemical synthesis of peptidoglycan fragments for elucidation of the immunostimulating mechanism. J Endotoxin Res 13(3):189–196. https://doi.org/10.1177/0968051907080739
41. Kawasaki A, Karasudani Y, Otsuka Y, Hasegawa M, Inohara N, Fujimoto Y, Fukase K (2008) Synthesis of diaminopimelic acid containing peptidoglycan fragments and tracheal cytotoxin (TCT) and investigation of their biological functions. Chem Eur J 14(33):10318–10330. https://doi.org/10.1002/chem.200801121
42. Wang N, Huang C-Y, Hasegawa M, Inohara N, Fujimoto Y, Fukase K (2013) Glycan sequence-dependent Nod2 activation investigated by using a chemically synthesized bacterial peptidoglycan fragment library. ChemBioChem 14:482–488. https://doi.org/10.1002/cbic.201200655

43. Wang Q, Matsuo Y, Pradipta AR, Inohara N, Fujimoto Y, Fukase K (2016) Synthesis of characteristic Mycobacterium peptidoglycan (PGN) fragments utilizing with chemoenzymatic preparation of meso-diaminopimelic acid (DAP), and their modulation of innate immune responses. Org Biomol Chem 14:1013–1023. https://doi.org/10.1039/C5OB02145F
44. Kadonaga Y, Wang N, Shimoyama A, Fujimoto Y, Fukase K (2025) Solid-phase oligosaccharide synthesis with highly complexed peptidoglycan fragments. Molecules 30:2787. https://doi.org/10.3390/molecules30132787
45. Inohara N, Ogura Y, Fontalba A, Gutierrez O, Pons F, Crespo J, Fukase K, Inamura S, Kusumoto S, Hashimoto M, Foster SJ, Moran AP, Fernandez-Luna JL, Nuñez G (2003) Host recognition of bacterial muramyl dipeptide mediated through NOD2. Implications for Crohn's disease. J Biol Chem 278(8):5509–5512. https://doi.org/10.1074/jbc.C200673200
46. Chamaillard M, Hashimoto M, Horie Y, Masumoto J, Qiu S, Saab L, Ogura Y, Kawasaki A, Fukase K, Kusumoto S, Valvano MA, Foster SJ, Mak TW, Nuñez G, Inohara N (2003) An essential role for NOD1 in host recognition of bacterial peptidoglycan containing diaminopimelic acid. Nat Immunol 4:702–707. https://doi.org/10.1038/ni945
47. Pradipta AR, Fujimoto Y, Hasegawa M, Inohara N, Fukase K (2010) Characterization of natural human nucleotide-binding oligomerization domain protein 1 (Nod1) ligands from bacterial culture supernatant for elucidation of immune modulators in the environment. J Biol Chem 285:23607–23613. https://doi.org/10.1074/jbc.M110.137893
48. Wang ZM, Li X, Cocklin RR, Wang M, Wang M, Fukase K, Inamura S, Kusumoto S, Gupta D, Dziarski R (2003) Human peptidoglycan recognition protein-L is an N-acetylmuramoyl-L-alanine amidase. J Biol Chem 278:49044–49052. https://doi.org/10.1074/jbc.M307758200
49. Kaneko T, Goldman WE, Mellroth P, Steiner H, Fukase K, Kusumoto S, Harley W, Fox A, Golenbock D, Silverman N (2004) Monomeric and polymeric Gram-negative peptidoglycan but not purified LPS stimulate the Drosophila IMD pathway. Immunity 20:637–649. https://doi.org/10.1016/S1074-7613(04)00104-9
50. Cho JH, Fraser IP, Fukase K, Kusumoto S, Fujimoto Y, Stahl GL, Ezekowitz RAB (2005) Human peptidoglycan recognition protein S is an effector of neutrophil-mediated innate immunity. Blood 106:2551–2558. https://doi.org/10.1182/blood-2005-02-0530
51. Park J-W, Je B-R, Piao S, Inamura S, Fujimoto Y, Fukase K, Kusumoto S, Söderhäll K, Ha N-C, Lee B-L (2006) A synthetic peptidoglycan fragment as a competitive inhibitor of the melanization cascade. J Biol Chem 281:7747–7755. https://doi.org/10.1074/jbc.M510058200
52. Yano T, Mita S, Ohmori H, Oshima Y, Fujimoto Y, Ueda R, Takada H, Goldman WE, Fukase K, Silverman N, Yoshimori T, Kurata S (2008) Autophagic control of listeria through intracellular innate immune recognition in drosophila. Nat Immunol 9:908–916. https://doi.org/10.1038/ni.1634
53. An J-H, Kurokawa K, Jung D-J, Kim M-J, Kim C-H, Fujimoto Y, Fukase K, Coggeshall KM, Lee B-L (2013) Human SAP is a novel peptidoglycan recognition protein that induces complement-independent phagocytosis of Staphylococcus aureus. J Immunol 191:3319–3327. https://doi.org/10.4049/jimmunol.1300940
54. Mesnage S, Dellarole M, Baxter NJ, Rouget J-B, Dimitrov JD, Wang N, Fujimoto Y, Hounslow AM, Lacroix-Desmazes S, Fukase K, Foster SJ, Williamson MP (2014) Molecular basis for bacterial peptidoglycan recognition by LysM domains. Nat Commun 5:4269. https://doi.org/10.1038/ncomms5269
55. Shimoyama A, Fukase K (2023) Bacterial glycolipid lipid As and their potential as adjuvants. In: Adamo R, Lay L (ed) Carbohydrate-based therapeutics. Wiley, pp 111–130, Chapter 4. https://doi.org/10.1002/9783527831326.ch4
56. Shimoyama A, Fukase K (2023) Chemical synthesis and immunomodulatory functions of bacterial lipid As. Methods Mol Biol 2613:33–53. https://doi.org/10.1007/978-1-0716-2910-9_4
57. Fukase K, Kirikae T, Kirikae F, Liu W-C, Oikawa M, Suda Y, Kurosawa M, Fukase Y, Yoshizaki H, Kusumoto S (2001) Synthesis of [3H]-labeled bioactive lipid A analogs and their use for detection of lipid A-binding proteins on murine macrophages. Bull Chem Soc Jpn 74:2189–2197. https://doi.org/10.1246/bcsj.74.2189

58. Saitoh S, Akashi S, Yamada T, Tanimura N, Kobayashi M, Konno K, Matsumoto F, Fukase K, Kusumoto S, Nagai Y, Kusumoto Y, Kosugi A, Miyake K (2004) Lipid A antagonist, lipid IVa, is distinct from lipid A in interaction with Toll-like receptor 4 (TLR4)-MD-2 and ligand-induced TLR4 oligomerization. Int Immunol 16:961–969. https://doi.org/10.1093/intimm/dxh097
59. Akashi S, Nagai Y, Ogata H, Oikawa M, Fukase K, Kusumoto S, Kawasaki K, Nishijima M, Hayashi S, Kimoto M, Miyake K (2001) Human MD-2 confers on mouse Toll-like receptor 4 species-specific lipopolysaccharide recognition. Int Immunol 13:1595–1599. https://doi.org/10.1093/intimm/13.12.1595
60. Akashi S, Saitoh S, Wakabayashi Y, Kikuchi T, Takamura N, Nagai Y, Kusumoto Y, Fukase K, Kusumoto S, Adachi Y, Kosugi A, Miyake K (2003) Lipopolysaccharide interaction with cell surface Toll-like receptor 4-MD-2: higher affinity than that with MD-2 or CD14. J Exp Med 198:1035–1042. https://doi.org/10.1084/jem.20031076
61. Ohto U, Fukase K, Miyake K, Satow Y (2007) Crystal structures of human MD-2 and its complex with antiendotoxic lipid IVa. Science 316:1632–1634. https://doi.org/10.1126/science.1139111
62. Kim HM, Park BS, Kim J-I, Kim SE, Lee J, Oh S-C, Enkhbayar P, Matsushima N, Lee H, Yoo OJ, Lee J-O (2007) Crystal structure of the TLR4–MD-2 complex with bound endotoxin antagonist Eritoran. Cell 130:906–917. https://doi.org/10.1016/j.cell.2007.08.002
63. Park BS, Song DH, Kim HM, Choi B-S, Lee H, Lee J-O (2009) The structural basis of lipopolysaccharide recognition by the TLR4–MD-2 complex. Nature 458:1191–1195. https://doi.org/10.1038/nature07830
64. Ohto U, Fukase K, Miyake K, Shimizu T (2012) Structural basis of species-specific endotoxin sensing by innate immune receptor TLR4/MD-2. Proc Natl Acad Sci USA 109:7421–7426. https://doi.org/10.1073/pnas.1201193109
65. Fukase Y, Fujimoto Y, Adachi Y, Suda Y, Kusumoto S, Fukase K (2008) Synthesis of Rubrivivax gelatinosus lipid A and analogues for investigation of the structural basis for immunostimulating and inhibitory activities. Bull Chem Soc Jpn 81:796–819. https://doi.org/10.1246/bcsj.81.796
66. Fukase K, Ueno A, Fukase Y, Oikawa M, Suda Y, Kusumoto S (2003) Synthesis and biological activities of lipid A analogs possessing β-glycosidic linkage at 1-position. Bull Chem Soc Jpn 76:485–500. https://doi.org/10.1246/bcsj.76.485
67. Yoshizaki H, Fukuda N, Sato K, Oikawa M, Fukase K, Suda Y, Kusumoto S (2001) First total synthesis of the Re-type lipopolysaccharide. Angew Chem Int Ed 40:1475–1480. https://doi.org/10.1002/1521-3773(20010417)40:8%3C1475::AID-ANIE1475%3E3.0.CO;2-V
68. Tanimura N, Saitoh S-I, Ohto U, Akashi-Takamura S, Fujimoto Y, Fukase K, Shimizu T, Miyake K (2014) The attenuated inflammation of MPL is due to the lack of CD14-dependent tight dimerization of the TLR4/MD-2 complex at the plasma membrane. Int Immunol 26:307–314. https://doi.org/10.1093/intimm/dxt071
69. Mata-Haro V, Cekic C, Martin M, Chilton PM, Casella CR, Mitchell TC (2007) The vaccine adjuvant monophosphoryl lipid A as a TRIF-biased agonist of TLR4. Science 316:1628–1632. https://doi.org/10.1126/science.1138963
70. Casella CR, Mitchell TC (2008) Putting endotoxin to work for us: monophosphoryl lipid A as a safe and effective vaccine adjuvant. Cell Mol Life Sci 65:3231–3240. https://doi.org/10.1007/s00018-008-8228-6
71. Fujimoto Y, Iwata M, Imakita N, Shimoyama A, Suda Y, Kusumoto S, Fukase K (2007) Synthesis of immunoregulatory Helicobacter pylori lipopolysaccharide partial structures. Tetrahedron Lett 48:6577–6581. https://doi.org/10.1016/j.tetlet.2007.07.036
72. Shimoyama A, Saeki A, Tanimura N, Tsutsui H, Miyake K, Suda Y, Fujimoto Y, Fukase K (2011) Chemical synthesis of Helicobacter pylori lipopolysaccharide partial structures and their selective proinflammatory responses. Chem Eur J 17:14464–14474. https://doi.org/10.1002/chem.201003581
73. Fujimoto Y, Shimoyama A, Suda Y, Fukase K (2012) Synthesis and immunomodulatory activities of Helicobacter pylori lipophilic terminus of lipopolysaccharide including lipid A. Carbohydr Res 356:37–43. https://doi.org/10.1016/j.carres.2012.03.013

74. Fujimoto Y, Shimoyama A, Saeki A, Kitayama N, Kasamatsu C, Tsutsui H, Fukase K (2013) Innate immunomodulation by lipophilic termini of lipopolysaccharide; synthesis of lipid As from Porphyromonas gingivalis and other bacteria and their immunomodulative responses. Mol BioSyst 9:987–996. https://doi.org/10.1039/C3MB25477A
75. Obata T, Goto Y, Kunisawa J, Sato S, Sakamoto M, Setoyama H, Matsuki T, Nonaka K, Shibata N, Gohda M, Kagiyama Y, Nochi T, Yuki Y, Fukuyama Y, Mukai A, Shinzaki S, Fujihashi K, Sasakawa C, Iijima H, Goto M, Umesaki Y, Benno Y, Kiyono H (2010) Indigenous opportunistic bacteria inhabit mammalian gut-associated lymphoid tissues and share a mucosal antibody-mediated symbiosis. Proc Natl Acad Sci USA 107:7419–7424. https://doi.org/10.1073/pnas.1001061107
76. Shibata N, Kunisawa J, Hosomi K, Fujimoto Y, Mizote K, Kitayama N, Shimoyama A, Mimuro H, Sato S, Kishishita N, Ishii KJ, Fukase K, Kiyono H (2018) Lymphoid tissue–resident Alcaligenes LPS induces IgA production without excessive inflammatory responses via weak TLR4 agonist activity. Mucosal Immunol 11:693–702. https://doi.org/10.1038/mi.2017.103
77. Shimoyama A, De Lorenzo F, Yamaura H, Mizote K, Palmigiano A, Pither MD, Speciale I, Uto T, Masui S, Sturiale L, Garozzo D, Hosomi K, Shibata N, Kabayama K, Fujimoto Y, Silipo A, Kunisawa J, Kiyono H, Molinaro A, Fukase K (2021) Lipopolysaccharide from gut-associated lymphoid tissue–resident Alcaligenes faecalis: complete structure determination and chemical synthesis of its lipid A. Angew Chem Int Ed 60:10023–10031. https://doi.org/10.1002/anie.202012374
78. Yoshii K, Hosomi K, Shimoyama A, Wang Y, Yamaura H, Nagatake T, Suzuki H, Lan H, Kiyono H, Fukase K, Kunisawa J (2020) Chemically synthesized Alcaligenes lipid A shows a potent and safe nasal vaccine adjuvant activity for the induction of Streptococcus pneumoniae–specific IgA and Th17-mediated protective immunity. Microorganisms 8:1102. https://doi.org/10.3390/microorganisms8081102
79. Liu Z, Hosomi K, Shimoyama A, Yoshii K, Sun X, Lan H, Wang Y, Yamaura H, Kenneth D, Saika A, Nagatake T, Kiyono H, Fukase K, Kunisawa J (2021) Chemically synthesized Alcaligenes lipid A as an adjuvant to augment immune responses to Haemophilus influenzae type B conjugate vaccine. Front Pharmacol 12:763657. https://doi.org/10.3389/fphar.2021.763657
80. Sun X, Hosomi K, Shimoyama A, Yoshii K, Lan H, Wang Y, Yamaura H, Nagatake T, Ishii KJ, Akira S, Kiyono H, Fukase K, Kunisawa J (2023) TLR4 agonist activity of Alcaligenes lipid A utilizes MyD88 and TRIF signaling pathways for efficient antigen presentation and T cell differentiation by dendritic cells. Int Immunopharmacol 117:109852. https://doi.org/10.1016/j.intimp.2023.109852
81. Sun X, Hosomi K, Shimoyama A, Yoshii K, Saika A, Yamaura H, Nagatake T, Kiyono H, Fukase K, Kunisawa J (2024) Alcaligenes lipid A induces superior IgA antibody production via recruiting cDC2s in the nose. Int Immunol 36:33–43. https://doi.org/10.1093/intimm/dxad045
82. Manabe Y, Fukase K (2023) Innovative vaccine strategy: self-adjuvanting conjugate vaccines. Methods Mol Biol 2613:55–72. https://doi.org/10.1007/978-1-0716-2910-9_5
83. Chang TC, Manabe Y, Fujimoto Y, Ohshima S, Kametani Y, Kabayama K, Nimura Y, Lin CC, Fukase K (2018) Syntheses and immunological evaluation of self-adjuvanting clustered N-acetyl and N-propionyl sialyl-Tn combined with a T-helper cell epitope as antitumor vaccine candidates. Angew Chem Int Ed Engl 57(27):8219–8224. https://doi.org/10.1002/anie.201804437
84. Chang TC, Manabe Y, Ito K, Yamamoto R, Kabayama K, Ohshima S, Kametani Y, Fujimoto Y, Lin CC, Fukase K (2022) Precise immunological evaluation rationalizes the design of a self-adjuvanting vaccine composed of glycan antigen, TLR1/2 ligand, and T-helper cell epitope. RSC Adv 12(29):18985–18993. https://doi.org/10.1039/d2ra03286d
85. Aiga T, Manabe Y, Ito K, Chang TC, Kabayama K, Ohshima S, Kametani Y, Miura A, Furukawa H, Inaba H, Matsuura K, Fukase K (2020) Immunological evaluation of co-assembling a lipidated peptide antigen and lipophilic adjuvants: self-adjuvanting anti-breast-cancer vaccine

candidates. Angew Chem Int Ed Engl 59(40):17705–17711. https://doi.org/10.1002/anie.202007999

86. Manabe Y, Gárate-Reyes B, Ito K, Hurtado-Guerrero R, Kabayama K, Fukase K (2024) Synthesis and immunological evaluation of TLR1/2 ligand-conjugated RBDs as self-adjuvanting vaccine candidates against SARS-CoV-2. Chem Commun 60:3946–3949. https://doi.org/10.1039/D4CC00462K

# Chemical Approaches to Explore the Function of Bacterial Glycolipid MPIase

**Kohki Fujikawa, Tsukiho Osawa, Ken-ichi Nishiyama, and Keiko Shimamoto**

**Abstract** Membrane Protein Integrase is a glycolipid found in the inner membrane of *Escherichia coli*, playing an essential role in the integration of membrane proteins and the translocation of secretory proteins. Structurally, MPIase consists of 9–11 repeating trisaccharide units composed of *N*-acetyl-D-glucosamine, *N*-acetyl-D-mannosamineuronic acid, and 4-*N*-acetyl-D-fucosamine. The glycan chain is anchored to the membrane via a pyrophosphate-linked diacylglycerol moiety. Due to the limited availability and structural heterogeneity of natural MPIase, we undertook a systematic synthesis of MPIase analogs to elucidate its mechanism of action. Structure–activity relationship studies revealed that specific functional groups and glycan chain length critically influence membrane protein integration activity. Furthermore, we observed synergistic effects between MPIase analogs and the membrane chaperone YidC, along with chaperone-like activity inherent to the phosphorylated glycan portion. These findings provide insight into a translocon-independent pathway for membrane protein integration in *E. coli*, in which MPIase recognizes highly hydrophobic nascent proteins, prevents their aggregation, facilitates their recruitment to the membrane, and collaborates with YidC to promote successful integration, while concurrently regenerating its own activity.

**Keywords** Glycolipid · Glycopyrophospholipid · Chaperon like activity · Membrane protein integration · *Escherichia coli*

K. Fujikawa (✉) · T. Osawa · K. Shimamoto (✉)
Bioorganic Research Institute, Suntory Foundation for Life Sciences, Seika-cho, Soraku-gun, Kyoto, Japan
e-mail: fujikawa@sunbor.or.jp

K. Shimamoto
e-mail: shimamot@sunbor.or.jp

K. Nishiyama
Department of Biological Chemistry and Food Sciences, Faculty of Agriculture, Iwate University, Morioka, Iwate, Japan

K. Shimamoto
Department of Chemistry, Graduate School of Science, Osaka University, Toyonaka, Osaka, Japan

 
K. Toshima and Y. Fujimoto (eds.), *Chemical Biology of Carbohydrates*,
https://doi.org/10.1007/978-981-95-5335-8_8

# 1 Introduction

Membrane proteins, which function as receptors, transporters, ion channels, and enzymes, are ubiquitous components of biological membranes in all organisms. The membrane integration of these proteins is mediated by channel complexes called translocons, which are conserved across the domains of life from prokaryotes such as *Escherichia coli* (*E. coli*) to eukaryotes including plants and mammals. This conservation underscores a universal mechanism of membrane protein integration [1, 2]. In *E. coli*, membrane protein integration occurs via two distinct pathways: a translocon-dependent pathway [3, 4] and a translocon-independent pathway [5, 6] (Fig. 1a). In 2006, Nishiyama et al. discovered Membrane Protein Integrase (MPIase) as an essential factor for both membrane protein integration pathways in *E. coli* [7–9]. Subsequently, we determined that MPIase is a novel glycolipid lacking any proteinaceous components and is composed of repeating units of three amino sugars—*N*-acetyl-D-glucosamine (GlcNAc), *N*-acetyl-D-mannosaminuronic acid (ManNAcA), and 4-*N*-acetyl-D-fucosamine (Fuc4NAc)—repeated approximately 9–11 times in the polysaccharide chain. This glycan chain is linked via a pyrophosphate moiety to a diacylglycerol anchor (Fig. 1b). Furthermore, approximately one-third of the GlcNAc residues at the C6 hydroxyl groups are acetylated. To date, no evidence has suggested the involvement of glycolipids in the process of membrane protein integration. In this context, the discovery of MPIase, an enzyme-like glycolipid, represents a conceptual breakthrough. Interestingly, the structure of MPIase closely resembles that of enterobacterial common antigen (ECA). While the glycan backbone of ECA shares the same trisaccharide repeat unit as MPIase, ECA contains a significantly greater number of repeats (trisaccharide × 18–55) compared to MPIase (trisaccharide × 9–11) and features a monophospholipid anchor instead of a pyrophospholipid moiety [10]. Notably, despite sharing an identical glycan sequence with MPIase, ECA lacks membrane protein integration activity [9]. This suggests that subtle structural differences can have a significant impact on biological function.

Using natural MPIase derived from *E. coli*, our studies demonstrated that the glycan moiety exhibits chaperone-like activity by interacting with membrane proteins to prevent their aggregation [9]. Additionally, MPIase possesses enzyme-like activity, enabling the integration of membrane proteins in amounts exceeding its own molar equivalent [8]. However, uncovering the functional mechanisms of MPIase poses significant challenges due to several factors: (i) MPIase constitutes only ~ 0.5% of the *E. coli* membrane, rendering it a minor component; (ii) the inherent heterogeneity in glycan chain length, acetyl (Ac) group number and positional variation, as well as fatty acid composition; and (iii) the biosynthetic pathway remains poorly understood, with MPIase-deficient strains being lethal, thereby precluding the use of molecular biological approaches alone. To overcome these challenges, we employed a chemical synthesis to produce homogeneous MPIase analogs with precisely controlled glycan chain lengths and functional group modifications in sufficient quantities. This approach has facilitated detailed investigations into the functional mechanisms of MPIase at the molecular and functional group level.

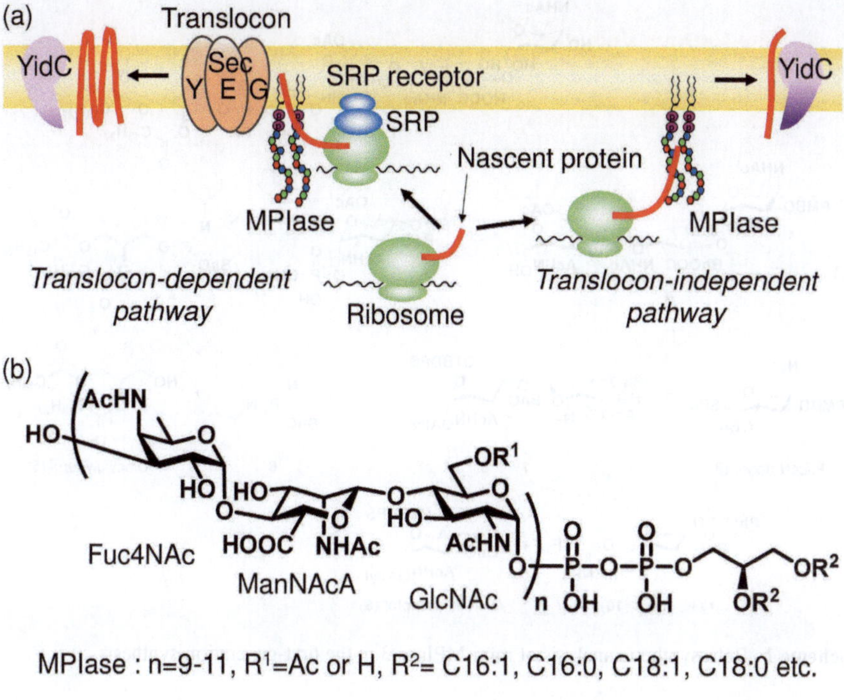


**Fig. 1** **a** Proposed model illustrating membrane protein integration via translocon-dependent and translocon-independent pathways in *E. coli.* In the translocon-dependent pathway, nascent proteins synthesizing from the ribosome are targeted to the membrane by the signal recognition particle (SRP) and its receptor, integrated into the membrane through the SecYEG translocon, and further positioned by the membrane insertase/chaperone YidC. YidC may act directly as an insertase and/or assist in the lateral integration of transmembrane domains into the lipid bilayer. MPIase is shown to be essential for both pathways, however the co-localization of MPIase with SecYEG and/or YidC remains speculative. **b** Structural representation of MPIase. The 6-hydroxyl group of the GlcNAc residue ($R^1$) is partially *O*-acetylated. The fatty acid composition of the diacylglycerol (DAG) moiety ($R^2$) varies depending on the growth conditions of *E. coli*

## 2 First-Generation Synthesis of MPIase Analogs

In the chemical synthesis of MPIase, several key challenges had to be addressed: (i) the synthesis of bacteria-specific constituent sugars such as Fuc4NAc and ManNAcA; (ii) the construction of β-linked mannosaminyl bonds; (iii) the introduction of 6-OAc groups on GlcNAc residues; (iv) the incorporation of lipid moieties via a labile pyrophosphate linkage; and (v) the formation of α-linked glucosaminyl bonds. As an initial step, we focused on the chemical synthesis of a trisaccharyl pyrophospholipid named mini-MPIase-3 (**1**), which constitutes the core structure of MPIase, and devised the synthetic strategy shown in Scheme 1 [11].

Initially, the monosaccharide building blocks were carefully designed and synthesized as follows. The GlcNAc acceptor **9** was equipped with a *tert*-butyldiphenylsilyl

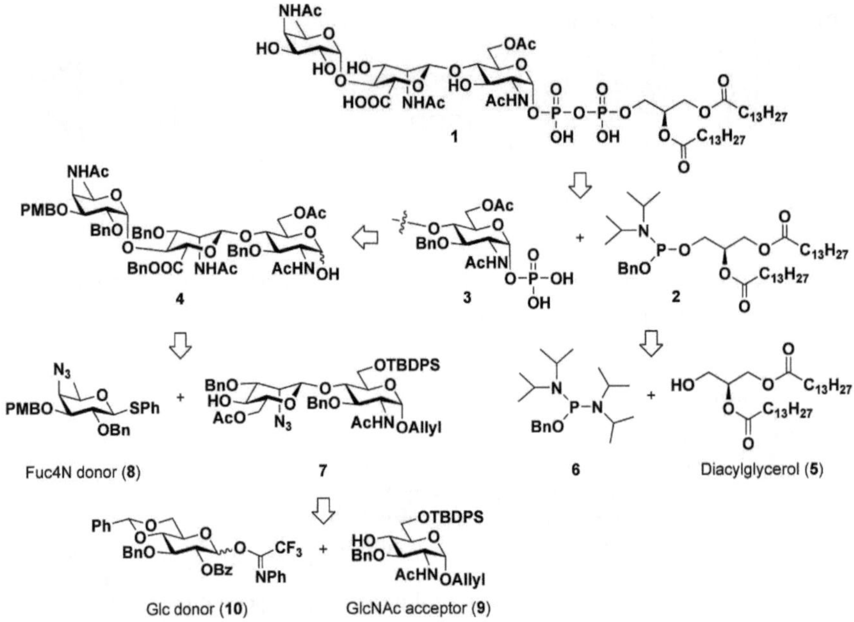


**Scheme 1** Retrosynthetic analysis of mini-MPIase-3 in the first-generation synthesis

(TBDPS) group at the 6-position and an allyl group at the anomeric position, allowing for their selective removal during subsequent introduction of Ac and phosphate groups (Scheme 1). The β-mannosaminyl linkage was constructed by first forming a β-linked disaccharide through neighboring group participation of the 2-O-benzoyl (Bz) group in the Glc donor **10**, followed by steric inversion of the hydroxyl group at C2 of Glc to yield intermediate **7**. The Fuc4N donor **8** was synthesized from Glc via reduction of the hydroxyl group at C6 and steric inversion at C4. A *p*-methoxybenzyl (PMB) group was introduced at the C3 hydroxyl of Fuc4N to allow for selective removal during future glycan elongation. After coupling the disaccharide acceptor **7** with the Fuc4N donor **8**, further transformations such as conversion to ManNAcA and installation of the GlcNAc 6-OAc group were planned. To protect the trisaccharide intermediate **4**, benzyl (Bn)-based protecting groups were employed, enabling global deprotection under neutral hydrogenolysis conditions to avoid base-sensitive hydrolysis of the 6-OAc group on GlcNAc, as well as the DAG and pyrophosphate linkages. Notably, this strategy has been employed for the synthesis of glycan substructures of MPIase during its structural elucidation, in order to enable NMR comparison with the natural product [9].

The synthesis of glycan substructures of ECA using a similar strategy has been reported by the group of G. J. Boons [12]. The synthesis of the ECA trisaccharide unit has also been reported independently by H. Paulsen's group [13]. However, in both cases, only the glycan moiety was synthesized, and the lipid portion was not incorporated. Given the lack of activity in ECA, it is inferred that the pyrophospholipid domain plays a pivotal role in MPIase function. Consequently, the construction of the glycan-pyrophospholipid structure represented one of the most challenging aspects of our synthetic efforts. To address this, we strategically planned the introduction of the labile pyrophospholipid moiety in the late stages of the synthesis, coupling the trisaccharyl phosphate intermediate **3** with the diacylglycerol phosphoramidite **2** [11, 14, 15]. For the fatty acid chains in DAG **5**, myristic acid (C14:0) was selected due to its synthetic accessibility and its phase transition temperature, which is comparable to that of natural membrane lipids.

The synthesis of compound **1** was carried out as outlined in Scheme 2. The condensation of the GlcNAc acceptor **9** with the Glc donor **10** afforded the desired β-linked disaccharide **11**. Subsequent removal of the 2-OBz group and steric inversion at C2 via intermediate **12** yielded the 2-azido derivative **13**. After removal of the benzylidene acetal protecting group, regioselective acetylation at the C6 hydroxyl group furnished the disaccharide acceptor **7**. Glycosylation with the Fuc4N donor **8** afforded the target trisaccharide **14** in 86% yield ($\alpha/\beta = 7/1$). Notably, the reactivity of **7** was unexpectedly low, requiring the use of 5.0 equivalents of **8**. Subsequently, the 6-OAc group of the ManN residue was removed and oxidized to afford the corresponding uronic acid structure **15**. The TBDPS group of the GlcNAc residue was converted to an Ac group, providing the 6-OAc GlcNAc derivative **16**. Reduction of the two azido groups was optimized using various conditions, including the Staudinger reaction, and the method reported by Boons's group employing $SnCl_2$, PhSH, and $Et_3N$ proved to be the most effective [12, 16]. Subsequent acetylation furnished the acetamido derivative **17**. The allyl glycoside at the reducing end was selectively removed to afford the trisaccharyl hemiacetal **4**. Reaction of **4** with commercially available diallyl phosphoramidite **18** afforded the corresponding phosphate ester **19**. Unintentionally, only the α-anomer was obtained after purification on silica gel. This presumably because the unstable β-anomer decomposed during the reaction or purification, likely due to the influence of the acetamido group at C2 of GlcNAc [17, 18]. Removal of the allyl esters yielded the trisaccharyl phosphate **3**. Coupling of **3** with the DAG compound **2**, followed by oxidation, furnished the protected trisaccharyl pyrophospholipid **20**. Global deprotection of the Bn-based protecting groups via hydrogenolysis gave **1** bearing the 6-OAc group on the GlcNAc residue.

The synthesized **1** was reconstituted into liposomes mimicking the *E. coli* membrane composition to evaluate its membrane protein integration activity [11]. Remarkably, even though the glycan chain length of **1** was approximately one-tenth that of natural MPIase, significant membrane protein integration activity was observed. This finding provided compelling evidence that the glycolipid is responsible for the biological activity (Fig. 4). Having established the utility of synthetic analogs, we subsequently embarked on the synthesis of various MPIase derivatives to facilitate structure–activity relationship studies (Fig. 2) [19]. These included MPIase analogs bearing glycan chains of varying lengths (trisaccharide, hexasaccharide, and nonasaccharide) as well as derivatives modified at functional groups predicted to interact with substrate proteins based on docking simulations (DS) [20], such as the 6-OAc group of GlcNAc, the carboxy group of ManNAcA, and the pyrophosphate

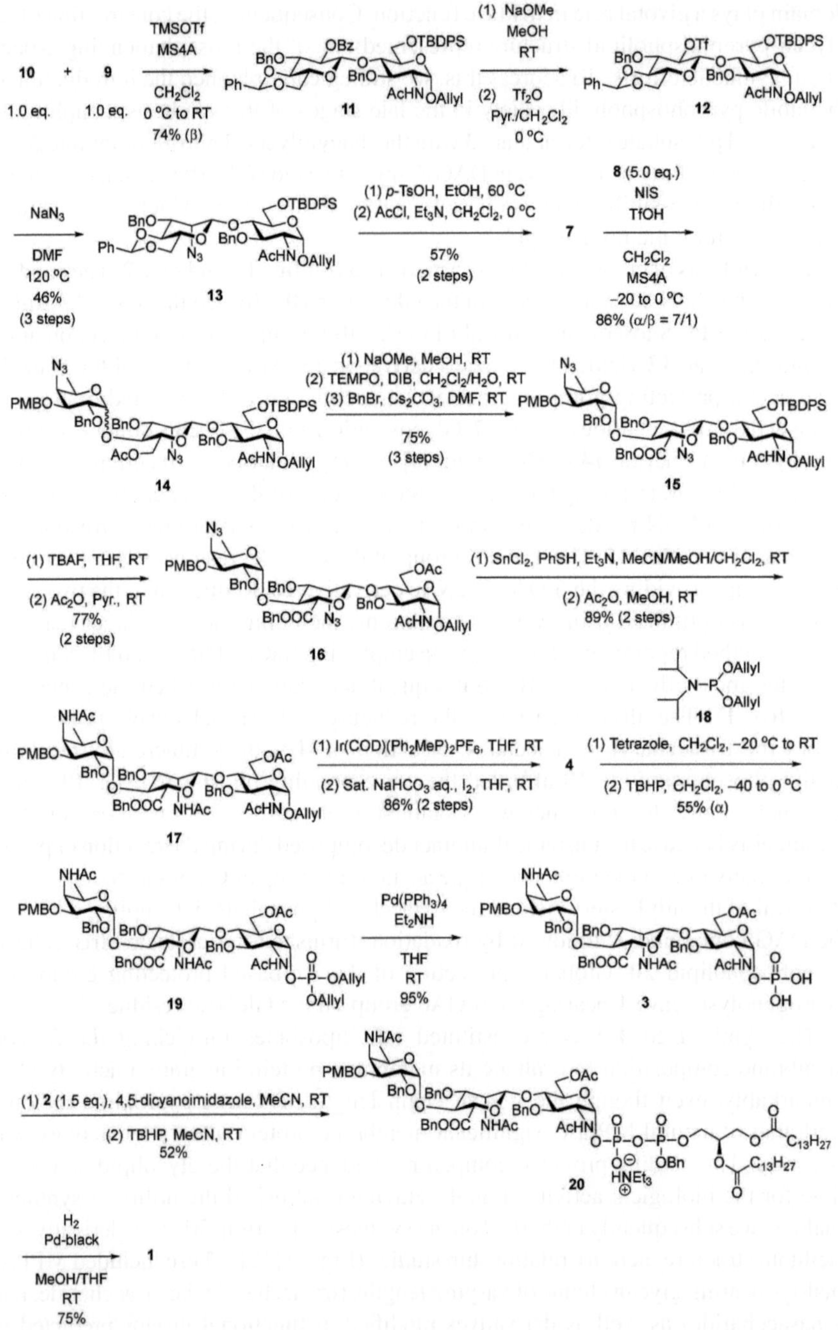


**Scheme 2** Synthesis of mini-MPIase-3 (**1**). TMSOTf: trimethylsilyl trifluoromethanesulfonate, MS: molecular sieves, $Tf_2O$: trifluoromethanesulfonic anhydride, *p*-TsOH: *p*-toluenesulfonic acid, $Et_3N$: triethylamine, NIS: *N*-iodosuccinimide, TfOH: trifluoromethanesulfonic acid, TEMPO: 2,2,6,6-tetramethylpiperidine 1-oxyl, DIB: (diacetoxyiodo)benzene, TBAF: tetrabutylammonium fluoride, $[Ir(COD)(PPh_2Me)_2]PF_6$: (1,5-cyclooctadiene)bis(methyldiphenylphosphine)iridium (I), TBHP: *tert*-butyl hydroperoxide, $Pd(PPh_3)_4$: tetrakis(triphenylphosphine)palladium (0)

moiety. In total, more than ten MPIase analogs were synthesized to enable detailed mechanistic investigations.

# 3 Second-Generation Synthesis of MPIase Analogs

Achieving α-glucosaminylation at the trisaccharide linkage represented the most critical challenge for constructing glycan-extended derivatives (Scheme 3). For this purpose, we developed the second generation synthesis employing GlcN unit **43** bearing a non-participating azido group at the C2 position and bulky protecting groups, including a *p*-bromobenzyl (*p*-BrBn) group at C3 and a TBDPS group at C6 positions based on the report by S.-C. Hung's group [21, 22]. Notably, the TBDPS group introduced at the C6 position of GlcNAc allows for selective deprotection, making it a valuable handle for the synthesis of C6-modified analogs. Given that MPIase consists of repeating trisaccharide units, glycan chain elongation was designed to proceed in increments of trisaccharides. To facilitate efficient synthesis, we designed a trisaccharide intermediate **40** incorporating orthogonal protecting groups: an allyl group at the non-reducing end and a *p*-methoxyphenyl (MP) group at the reducing end. This protection strategy allowed conversion of the trisaccharide into either trisaccharide donors (**36**, **37**, **38**, and **39**) or a trisaccharide acceptor **35**. For the construction of MPIase-type pyrophospholipids (**1**, **21–27**), we employed a similar strategy to that of the first-generation synthesis, in which the glycan monophosphate intermediates were coupled with the DAG-containing phosphoramidite reagent **2**. In contrast, for the assembly of ECA-type monophospholipids (**28** and **29**), the glycosyl hemiacetal intermediates were directly reacted with reagent **2** to install the lipid moiety.

## 3.1 Synthesis of Trisaccharide Intermediate

As depicted in Scheme 4, the synthesis of second-generation trisaccharide intermediate was designed to accommodate glycan chain elongation by employing a GlcNAc unit tailored for α-glucosaminylation [21, 22]. The construction of the disaccharide acceptor **41** starting from the GlcN acceptor **43** and Glc donor **10** proceeded smoothly, employing a same strategy to that used in the first-generation synthesis. Glycosylation of **41** with the Fuc4N donor **42** in diethyl ether at − 20 °C afforded the target trisaccharide in 75% yield with an α/β ratio of 7:1. In the second-generation synthesis, an azido group was introduced at C2 of the GlcN residue for α-glucosaminylation, which also significantly improved the reactivity of **41** compared to the first-generation approach. As a result, the reaction proceeded efficiently using only 1.2 equivalents of **42** with high yield. It is known that the acetamido group can sequester the acid catalyst and increase steric hindrance due to intermolecular hydrogen-bonding networks [23–25]. The elimination of such inhibitory effects would improve the reactivity. The

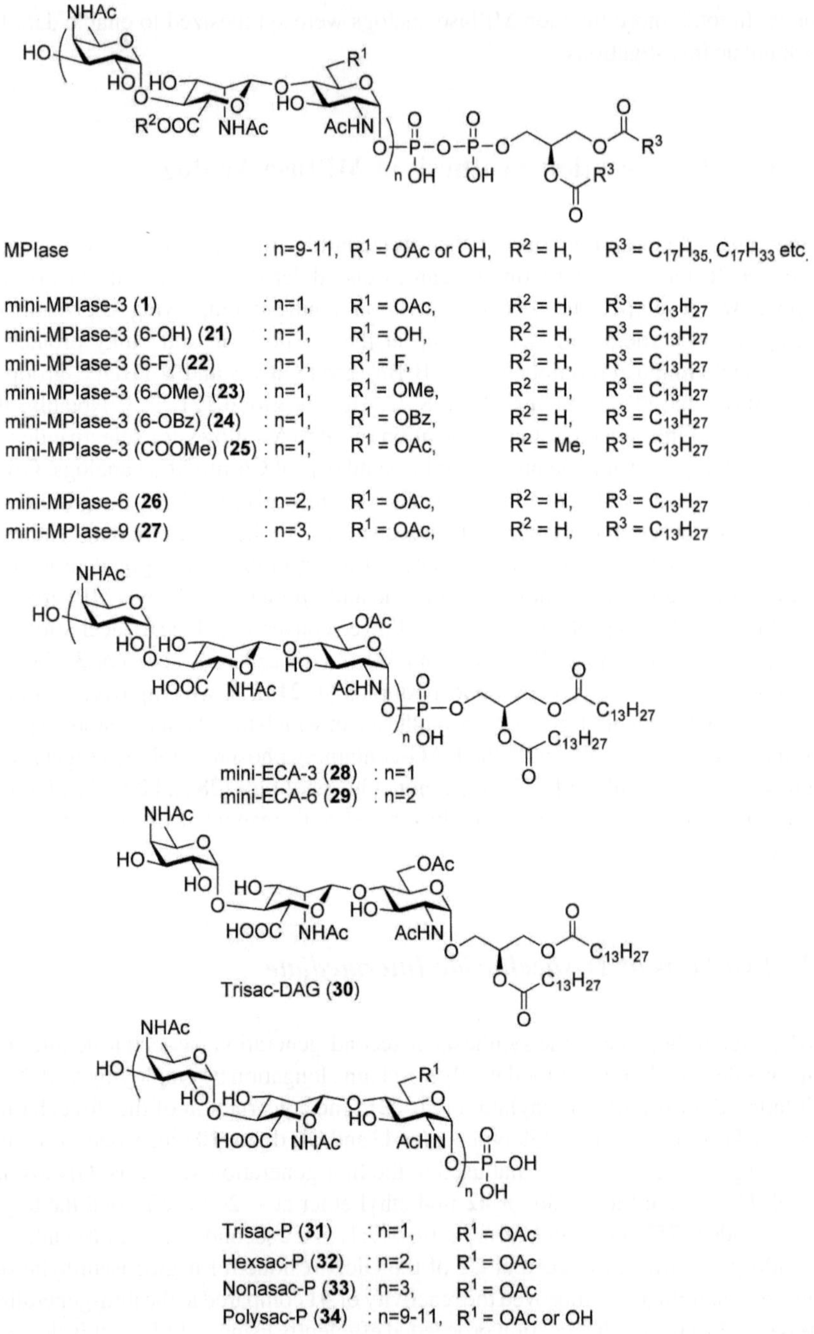


**Fig. 2** Structures of MPIase and its synthetic analogs

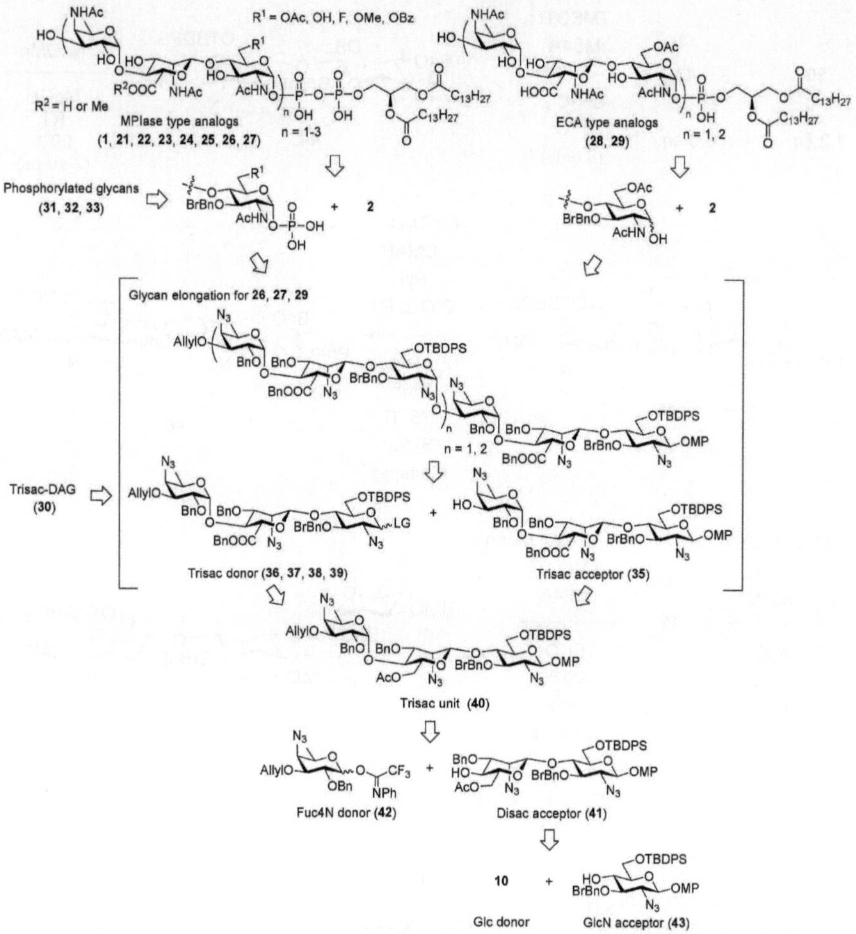


**Scheme 3** Retrosynthetic analysis of MPIase analog in the second-generation synthesis

desired α-anomer of the trisaccharide was isolated by silica gel chromatography, and oxidation of the primary hydroxyl group at C6 of the ManNAcA residue afforded the trisaccharide intermediate **40**.

## 3.2 *Synthesis of Trisaccharide Acceptors and Trisaccharide Donors*

Subsequently, **40** was converted into the corresponding trisaccharide acceptor **35** and trisaccharide donors (**36**–**39**) as outlined in Scheme 5. Removal of the allyl group at the non-reducing end of **40** using palladium chloride afforded **35** in 98% yield.

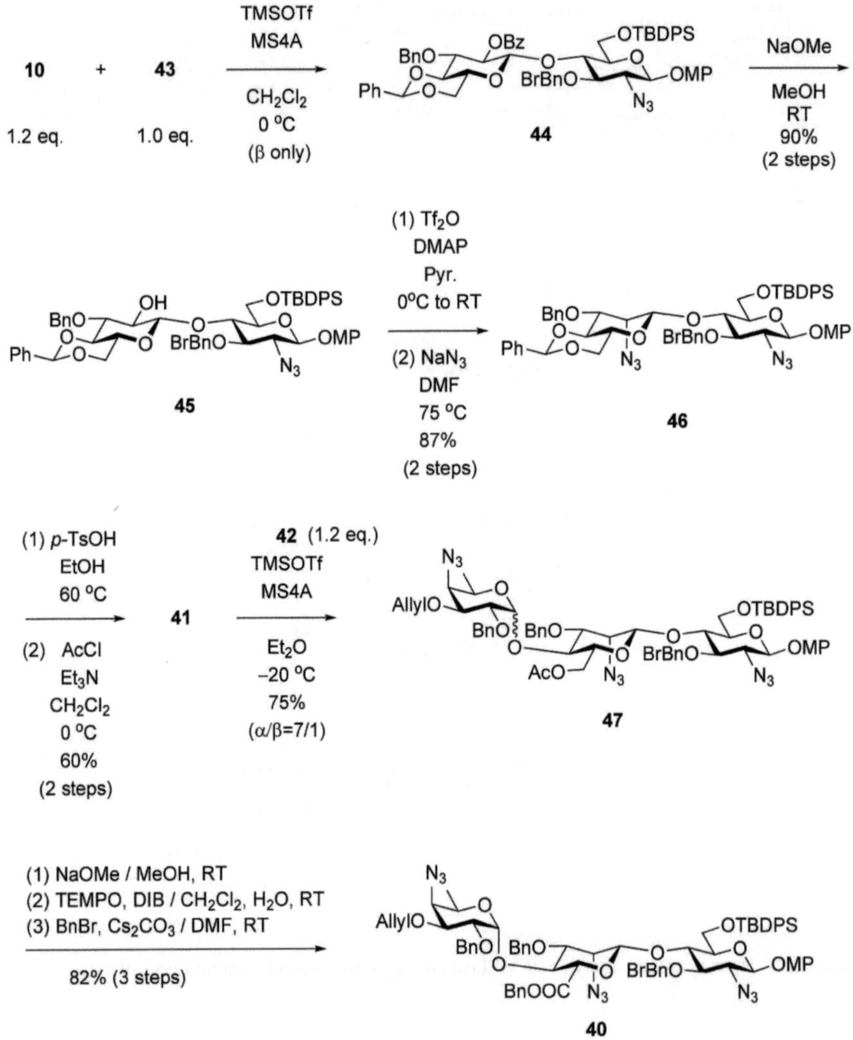


**Scheme 4** Synthesis of trisaccharide intermediate. DMAP: 4-dimethylaminopyridine

For the synthesis of the trisaccharide donors, the MP group at the reducing end of **40** was cleaved to generate the trisaccharyl hemiacetal **48**. Introduction of various leaving groups to **48** enabled the preparation of four distinct donors: a fluoride (**36**), *o*-alkynylbenzoate groups with *n*-butyl (**37**) or cyclopropyl (**38**) substituents, and an *N*-phenyl imidate group (**39**).

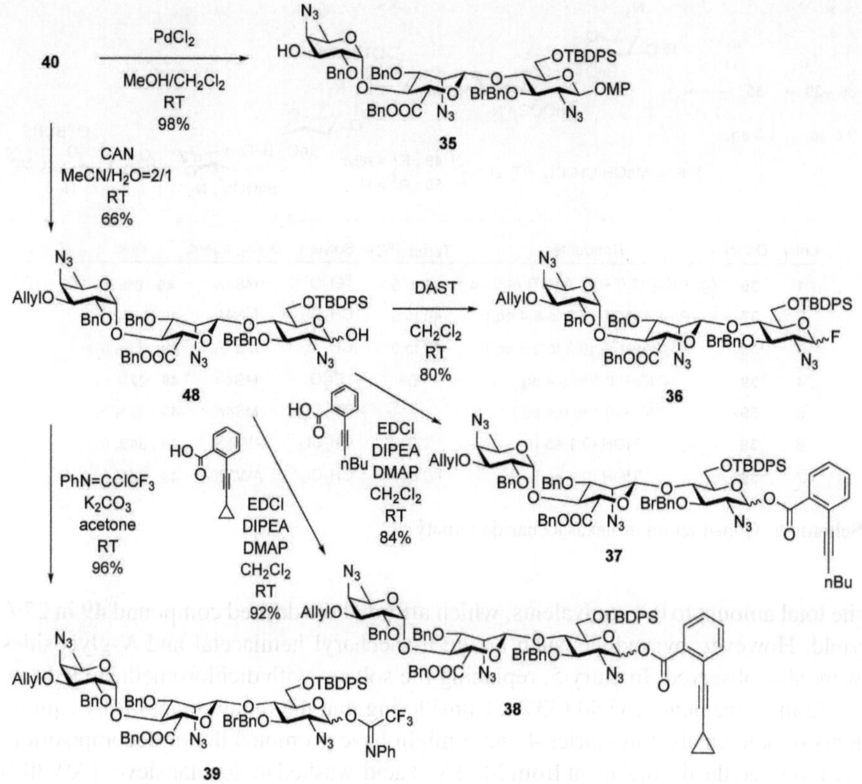


**Scheme 5** Synthesis of trisaccharide acceptors and trisaccharide donors. CAN: ceric ammonium nitrate, DAST: (diethylamino)sulfur trifluoride, EDCI: 1-(3-dimethylaminopropyl)-3-ethylcarbodiimide hydrochloride, DIPEA: *N, N*-diisopropylethylamine

## 3.3 Synthesis of Hexasaccharide

The synthesis of hexasaccharide was investigated using the trisaccharide donors (**36**–**39**, 2.0 eq.) and the trisaccharide acceptor (**35**, 1.0 eq.), as outlined in Scheme 6. In entry 1, glycosylation of the trisaccharyl fluoride donor **36** with the acceptor **35** in diethyl ether at 0 °C using a combination of hafnium and silver triflate catalysts [26] resulted in partial decomposition of the donor, and the desired hexasaccharide **49** was not obtained. In entries 2 and 3, *o*-alkynylbenzoate donors bearing *n*-butyl (**37**) and cyclopropyl (**38**) substituents, as reported by B. Yu's group [27, 28], were employed in dichloromethane at − 40 to 0 °C under gold catalysis. The desired product **49** was obtained in moderate yields (46% for *n*-butyl, 63% for cyclopropyl). Hemiacetal forms of the donors were identified as byproducts. In entry 4, glycosylation using the *N*-phenyl imidate donor **39**, following B. Yu's protocol [29], was carried out in diethyl ether at 0 °C with 0.1 equivalents of triflic acid as the promoter. Since the reaction did not reach completion, an additional portion of triflic acid was added to bring

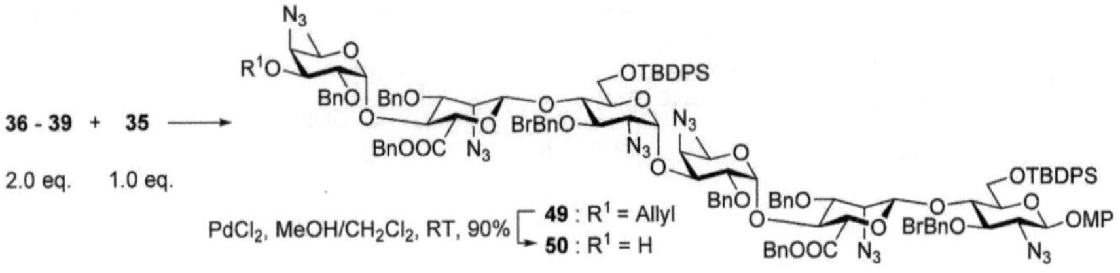


| Entry | Donor | Reagents | Temp. (°C) | Solvent | Drying agent | Yield |
|---|---|---|---|---|---|---|
| 1 | **36** | $Cp_2HfCl_2$ (2.0 eq.), AgOTf (4.0 eq.) | 0 to 50 | $Et_2O$ | MS4A | **49** : 0% (α) |
| 2 | **37** | $Ph_3PAuNTf_2$ (0.2 to 0.4 eq.) | –40 to 0 | $CH_2Cl_2$ | MS4A | **49** : 46% (α) |
| 3 | **38** | $Ph_3PAuNTf_2$ (0.2 to 0.4 eq.) | –40 to 0 | $CH_2Cl_2$ | MS4A | **49** : 63% (α) |
| 4 | **39** | TfOH (0.1 to 0.4 eq.) | 0 | $Et_2O$ | MS4A | **49** : 27% (α) |
| 5 | **39** | TfOH (0.1 to 0.4 eq.) | 0 | $CH_2Cl_2$ | MS4A | **49** : 33% (α) |
| 6 | **39** | TfOH (0.1 eq.) | 0 | $CH_2Cl_2$ | AW300 | **49** : 34% (α) |
| 7 | **39** | TfOH (0.1 eq.) | –40 to 0 | $CH_2Cl_2$ | AW300 | **49** : 72% (α) |

**Scheme 6** Construction of hexasaccharide moiety

the total amount to 0.4 equivalents, which afforded the desired compound **49** in 27% yield. However, byproducts such as the trisaccharyl hemiacetal and *N*-glycosides were also observed. In entry 5, replacing the solvent with dichloromethane led to a slight improvement in yield (33%). Considering that the requirement of 0.4 equivalents of acid catalyst in entries 4 and 5 might have promoted donor decomposition, we modified the drying agent from MS4A to acid-washed molecular sieves (AW300) in entry 6, which allowed us to reduce the acid catalyst loading to 0.1 equivalents without changing the yield (34%). In entry 7, initiating the reaction at –40 °C successfully suppressed side reactions and improved the yield to 72%. The newly formed glucosaminyl linkage was analyzed by measurement of the $^1J_{CH}$ coupling constant [30], which confirmed exclusive formation of the α-anomer in all reactions, thereby substantiating the efficacy of the GlcNAc design strategy proposed by S.-C. Hung's group. Finally, removal of the allyl group at the non-reducing end of **49** furnished the hexasaccharide acceptor **50**.

## 3.4 Synthesis of Nonasaccharide

Next, the synthesis of a nonasaccharide was investigated using the trisaccharide donor (**39**, 4.0 eq.) and the hexasaccharide acceptor (**50**, 1.0 eq.) (Scheme 7). In entry 1, the optimized condition for hexasaccharide formation (Scheme 6, entry 7) was applied with 4.0 equivalents of **39**. However, byproducts such as the trisaccharyl hemiacetal and *N*-glycosides were observed, and the yield of the target nonasaccharide **51** was limited to 35%. To suppress these side reactions, the reaction was initiated at a lower

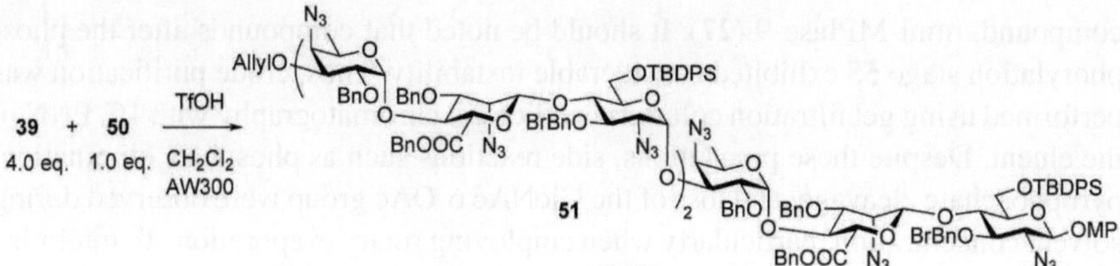


| Entry | Temp. (°C) | Yield |
|---|---|---|
| 1 | −40 to 0 | **51** : 35% (α) |
| 2 | −60 to −40 to −20 to 0 | **51** : 53% (α) |

**Scheme 7** Construction of nonasaccharide moiety

temperature (– 60 °C) in entry 2 and then gradually warmed to ambient temperature. This modification effectively minimized byproduct formation and improved the yield to 53%. Analysis of the newly formed glucosaminyl linkage, as determined by measurement of the $^1J_{CH}$ coupling constant [30], confirmed that glycosylation proceeded with complete α-selectivity.

## 3.5 *Synthesis of Nonasaccharyl Pyrophospholipids*

The obtained **51** was further elaborated through protecting group manipulations and lipid installation to afford the nonasaccharyl pyrophospholipid (mini-MPIase-9, **27**) (Scheme 8). The TBDPS group introduced for α-glucosaminylation of **51** was removed and replaced with an Ac group to yield **52**. The reduction of azido groups was achieved using the $SnCl_2$/thiol/$Et_3N$ protocol, applied in the first-generation synthesis of MPIase analogs. To minimize the strong odor of benzenethiol, it was substituted with *tert*-butyl benzenethiol, a higher-boiling thiol [31, 32]. Acetylation of the resulting amino groups furnished compound **53**. Subsequently, the MP group at the reducing end of **53** was cleaved using CAN to generate the nonasaccharyl hemiacetal **54**. Phosphorylation of **54** with the commercially available phosphoramidite reagent **18** bearing an allyl group, followed by oxidation and purification by silica gel chromatography, afforded exclusively the desired α-anomer of the nonasaccharide phosphate **55**. The high α-selectivity is likely attributable to the predominance of the α-anomer in the starting hemiacetal **54** and the instability of the β-phosphate, which presumably decomposed during purification [17, 18]. The allyl group of **55** was removed using palladium chloride to furnish intermediate **56**, which was coupled with the DAG-containing phosphoramidite reagent **2** and oxidized to yield the protected nonasaccharyl pyrophospholipid **57**. Owing to the pronounced instability of this intermediate, only minimal purification by gel filtration was undertaken prior to the final deprotection step. Global deprotection of Bn-based protecting groups by catalytic hydrogenation, followed by gel filtration (MeOH) and preparative HPLC (C18 column, $AcONH_4$ aqueous/acetone eluent), furnished the target compound, mini-MPIase-9 (**27**). It should be noted that compounds after the phosphorylation stage **55** exhibited considerable instability. Thus, crude purification was performed using gel filtration columns or silica gel chromatography with 1% $Et_3N$ in the eluent. Despite these precautions, side reactions such as phosphate elimination, pyrophosphate cleavage, and loss of the GlcNAc 6-OAc group were observed during solvent concentration, particularly when employing rotary evaporation. To minimize such decomposition, solvent removal by freeze-drying from $H_2O$/MeCN or $H_2O$/*tert*-butanol mixtures would be an effective approach [33].

This synthetic strategy was successfully applied to other MPIase analogs. For example, in the case of GlcNAc C6 modified derivatives, as illustrated in Scheme 9, selective deprotection of the TBDPS group in trisaccharide intermediate **40** afforded the corresponding hydroxyl compound (**58**). Subsequent treatment of **58** with the Meerwein reagent, DAST, or benzoic anhydride led to the formation of the C6 methylated (**59**), fluorinated (**60**), and benzoylated (**61**) derivatives, respectively. Thus, by utilizing the well-designed **40** as a key compound, a series of MPIase analogs were systematically synthesized.

# 4 Structure–Activity Relationship Study Using MPIase Analogs

## *4.1 Integration Activity Assay*

The synthesized compounds were reconstituted into liposomes prepared from *E. coli* membrane lipids, comprising phosphatidylethanolamine (PE), phosphatidylglycerol (PG), and cardiolipin (CL). Additionally, DAG was added to inhibit unfavorable spontaneous integrations (Fig. 3) [7, 8, 34, 35]. Membrane protein integration was evaluated by measuring the incorporation efficiency of the substrate protein produced in a cell-free system into the liposomal membrane. We used a modified phage coat protein (3L-Pf3) as a substrate since it has been used as a model protein for the translocon-independent integration pathway. In this assay, the *N*-terminal methionine of the protein was radioisotope-labeled with $^{35}S$ to enable quantitative evaluation of membrane integration.

## *4.2 Integration Activity of Liposomes Reconstituted with MPIase Analogs*

As previously described, even **1**, which possesses a glycan chain approximately one-tenth the length of natural MPIase, exhibited significant membrane integration activity of the model protein (3L-Pf3) into liposomal membranes (Fig. 4) [11, 19]. In contrast, the activity of the 6-hydroxy derivative **21**, lacking the GlcNAc 6-OAc

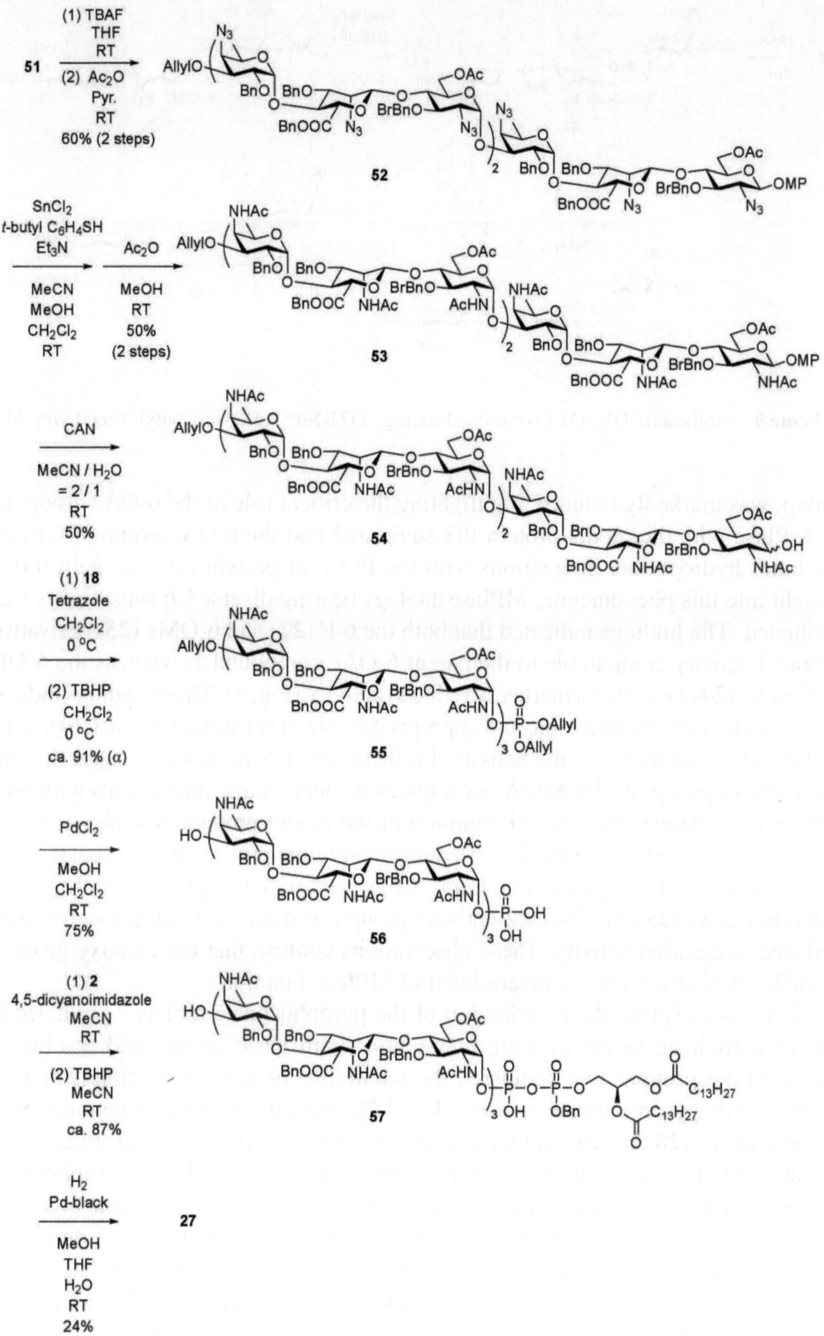


**Scheme 8** Synthesis of mini-MPIase-9 (**27**). TBAF: tetrabutylammonium fluoride

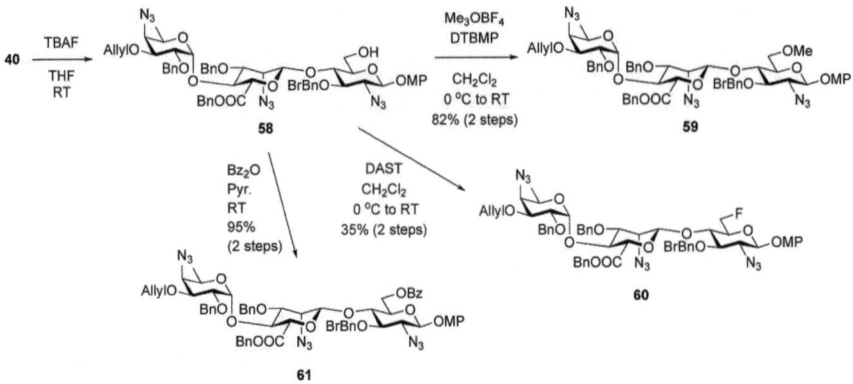


**Scheme 9** Synthesis of GlcNAc C6 modified analogs. DTBMP: 2,6-di-*tert*-butyl-4-methylpyridine

group, was markedly reduced, highlighting the critical role of the 6-OAc group. DS of MPIase with the model protein Pf3 suggested that the 6-OAc group of GlcNAc mediates hydrophobic interactions with the Pf3 coat protein [20]. To gain further insight into this phenomenon, MPIase analogs bearing diverse C6 substituents were evaluated. The findings indicated that both the 6-F (**22**) and 6-OMe (**23**) derivatives retained activity comparable to the parent 6-OAc compound **1**, whereas the 6-OBz derivative (**24**) exhibited a markedly reduced activity (Fig. 4). These findings indicate that a moderately hydrophobic and appropriately sized substituent at the C6 position of GlcNAc contributes to the activity. Furthermore, DS results also suggested that the carboxy group of ManNAcA participates in electrostatic interactions with basic amino acid residues (lysine and arginine) of the model protein. Notably, as is well known as the positive inside rule [36], many membrane proteins possess basic amino acid residues in their cytoplasmic loops. Consistent with this prediction, the methyl ester derivative (**25**), in which the carboxy group was masked, exhibited significantly reduced integration activity. These observations confirm that the carboxy group of ManNAcA is also a critical determinant of MPIase function.

To further explore the contribution of the pyrophosphate moiety—predicted by DS to participate in electrostatic interactions with basic amino acid residues of the substrate protein—we compared the membrane protein integration activity of MPIase-type pyrophosphate analogs (**1** and **26**) with that of ECA-type monophosphate analogs (**28** and **29**) and the phosphate-deficient derivative (**30**) (Fig. 5). The results demonstrated that both the trisaccharyl and hexasaccharyl pyrophosphate analogs exhibited high integration activity, whereas a progressive decline in activity was observed for the monophosphate analogs and the phosphate-free derivative. These findings clearly indicate that the presence of two phosphate groups in the linker moiety is a critical determinant for the integration activity of MPIase.

Next, the relationship between glycan chain length and integration activity was evaluated (Fig. 6). The results demonstrated that integration activity progressively increased with glycan chain elongation, as evidenced by comparisons among the

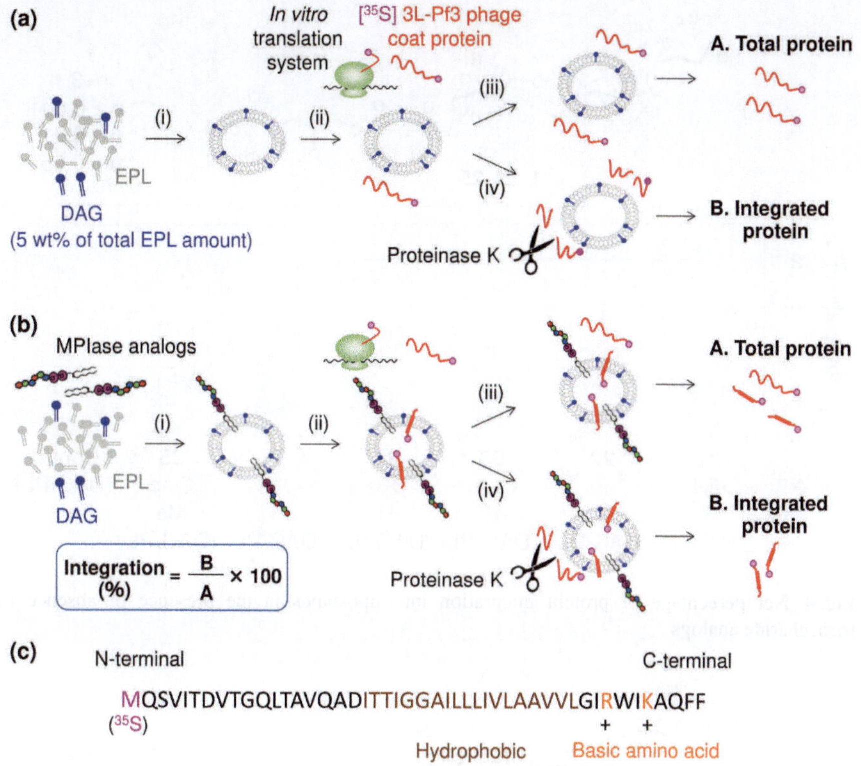


**Fig. 3** Overview of membrane protein integration assay. **a** Control liposome assay protocol. (i) Control liposomes are composed of DAG and *E. coli* phospholipids (EPL). (ii) In the presence of these liposomes, radioisotope-labeled 3L-Pf3 coat proteins are synthesized via a cell-free translation system. (iii) The resulting reaction mixture is split into two portions. One portion is analyzed by SDS-PAGE and visualized by autoradiography to determine the total amount of synthesized protein. (iv) The second portion is subjected to protease treatment to digest proteins located outside the liposomal membrane. **b** Assay using liposomes incorporating MPIase analogs. When functional MPIase analogs are incorporated into the liposomes, integrated protein (B) becomes detectable, indicating successful integration of the model protein into the liposomal membrane. **c** Amino acid sequence of the 3L-Pf3 coat protein employed in the assay

trisaccharide analogs (**1** and **21**), hexasaccharide (**26**), nonasaccharide (**27**), and natural MPIase (about 30 saccharide units).

To investigate the necessity of membrane anchoring by the lipid moiety (DAG) of MPIase, we performed an experiment in which the glycan monophosphate derivative **34**, lacking the lipid portion, was added to the membrane protein integration assay solution (Fig. 7a). Notably, even at a high concentration (1 mg/mL), the addition of **34** failed to promote integration of the model protein into liposomes. This observation indicates that MPIase requires membrane anchoring through its lipid moiety to facilitate the integration of proteins into membranes. In contrast, when **1** was reconstituted into liposomes at 5 w/w% phospholipid content, we observed that

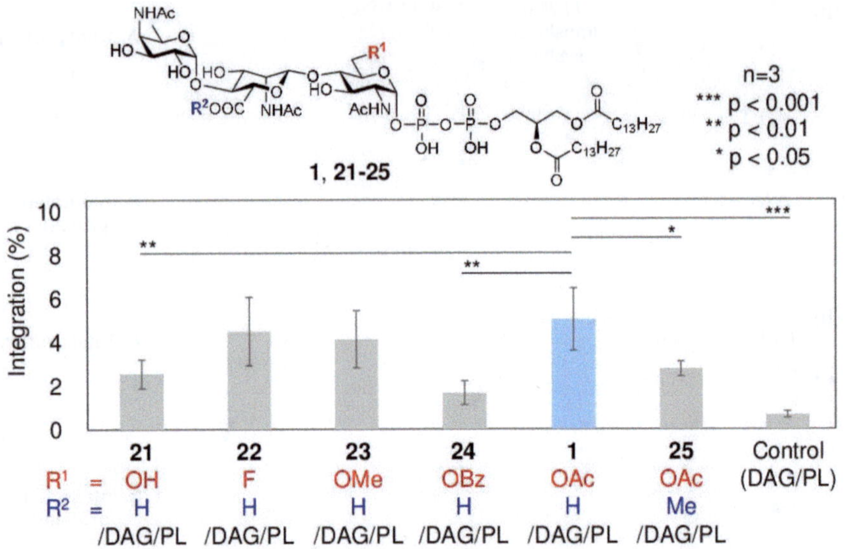


**Fig. 4** Net percentage of protein integration into liposomes in the presence or absence of trisaccharide analogs

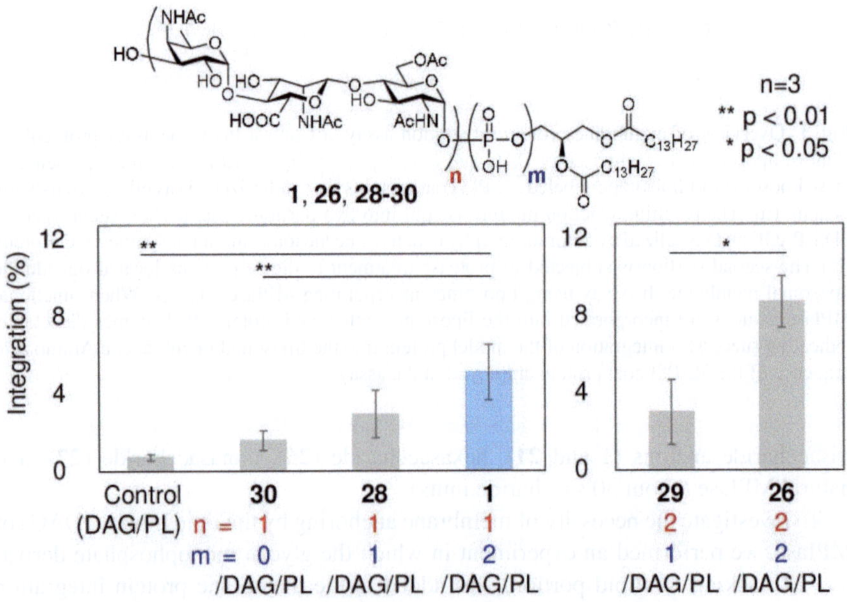


**Fig. 5** Net percentages of protein integration by pyrophosphate analogs

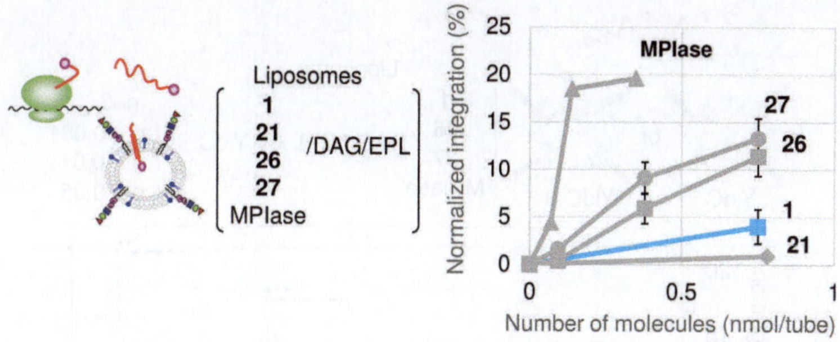


**Fig. 6** Normalized integration of synthetic analogs (**1, 21, 26,** and **27**) and natural MPIase. Values were calculated by subtracting background integration (control liposomes without MPIase) from the net integration percentages in each assay

adding the longer-chain glycan monophosphates (**33** and **34**) into the reaction solution enhanced the integration efficiency of **1** while the shorter ones (**31** and **32**) did not. We previously reported that **34** exhibits chaperone-like activity by preventing the aggregation of substrate proteins [9]. These findings suggest that long-chain glycan monophosphates help maintain the membrane-insertable conformation of the model protein, thereby enhancing the efficiency of membrane integration mediated by **1**.

We next examined the membrane integration activity under conditions where MPIase and the membrane chaperone YidC were co-reconstituted into liposomes (Fig. 8). YidC is recognized as a membrane chaperone that facilitates the translocation of substrate proteins by attracting them via electrostatic interactions [37–39].

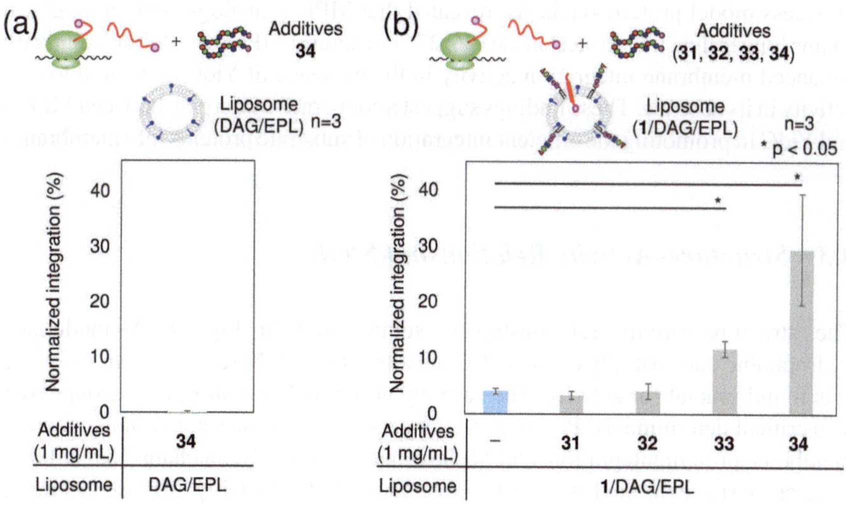


**Fig. 7** **a** Normalized integration values upon addition of anchorless analog (**34**) in the presence of control liposome. **b** Normalized integration values upon addition of anchorless analogs (**31, 32, 33,** and **34**) in the presence of **1** containing liposome

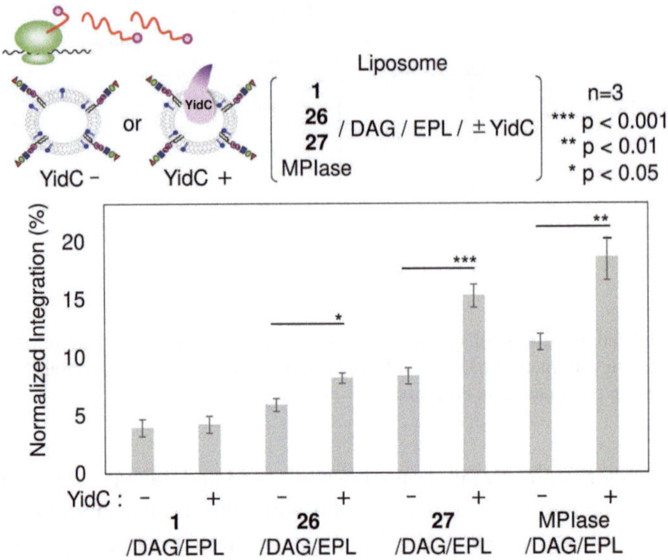


**Fig. 8** Normalized integration values of MPIase analogs co-reconstituted with or without YidC under substrate-excess conditions. Data represent means from at least three independent experiments

Notably, YidC alone is insufficient to mediate membrane protein integration [40], leading us to hypothesize that its coexistence with MPIase is essential for supporting this process. Consistent with this hypothesis, co-reconstitution of various MPIase analogs (**1**, **26**, **27**, and natural MPIase) with YidC into liposomes, under conditions of excess model protein synthesis, revealed that MPIase analogs containing glycan chains longer than a hexasaccharide (**26**, **27**, and natural MPIase) exhibited markedly enhanced membrane integration activity in the presence of YidC compared to their activity in its absence. These findings suggest a cooperative interplay between MPIase and YidC in promoting the efficient integration of substrate proteins into membranes.

## 4.3 Structure–Activity Relationship Study

The structure–activity relationship is summarized in Fig. 9. A moderately hydrophobic and spatially compact C6 substituent on GlcNAc, such as an Ac group, was found crucial for activity. The carboxy group and two phosphate groups were also critical determinants. Progressive elongation of the trisaccharide unit enhanced membrane protein integration, and MPIase analogs with glycan chains longer than a hexasaccharide exhibited synergistic effects with YidC. Notably, MPIase lacking its lipid moiety (DAG) failed to promote substrate integration, highlighting the importance of membrane anchoring. In contrast, long phosphorylated glycans displayed chaperone-like activity by preventing substrate aggregation, which in turn improved

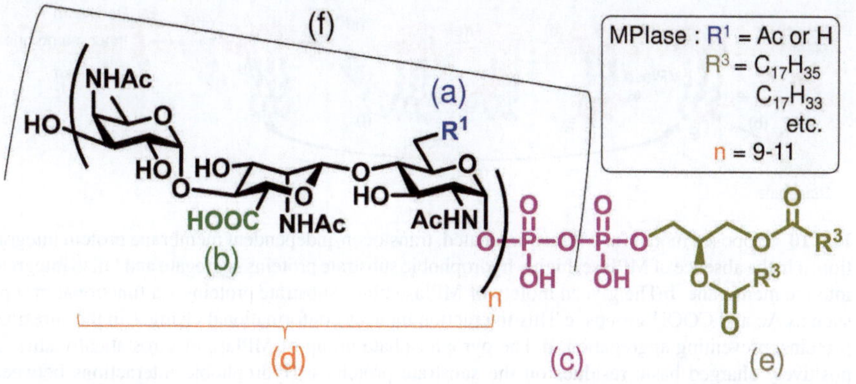


**Fig. 9** Structural determinants of MPIase function. **a** A compact and hydrophobic substituent at the C6 position of GlcNAc ($R^1$) is required to preserve integration activity. **b** The carboxy group of ManNAcA is important for functional activity. **c** The pyrophosphate moiety is crucial, as reduced phosphorylation abolishes integration function. **d** Integration activity correlates with the number of repeating trisaccharide units. A single trisaccharide ($n = 1$) enables basal integration, whereas a longer glycan chain ($n \geq 2$) is necessary for YidC-dependent enhancement. **e** Membrane anchoring through the lipid moiety is indispensable for effective membrane protein integration. **f** Anti-aggregation activity requires both phosphate groups and an extended glycan chain ($n \geq 3$)

the membrane integration efficiency of compound **1** in liposomes. To elucidate the molecular mechanism of MPIase, various physicochemical approaches—including circular dichroism (CD), surface plasmon resonance (SPR), saturation transfer difference (STD)-NMR, and solid-state NMR—revealed key properties of MPIase, such as its chaperone-like activity in preventing protein aggregation, its direct interactions with substrate proteins, and its ability to increase membrane mobility [20, 41–43].

## *4.4 Mechanism for Translocon-Independent Integration Involving MPIase*

Based on these structure–activity relationship studies, we propose a mechanistic model for the translocon-independent membrane protein integration in the *E. coli* inner membrane (Fig. 10) [43–45]. In this model, the glycan moiety of MPIase captures substrate proteins via its functional groups and induces conformational changes that prevent aggregation. The strongly negative charge of the pyrophosphate moiety subsequently attracts substrate proteins to the membrane surface, facilitating their integration into membrane regions where MPIase enhances lipid mobility. Finally, the substrate proteins are handed over to YidC, enabling MPIase to regenerate its integration activity.

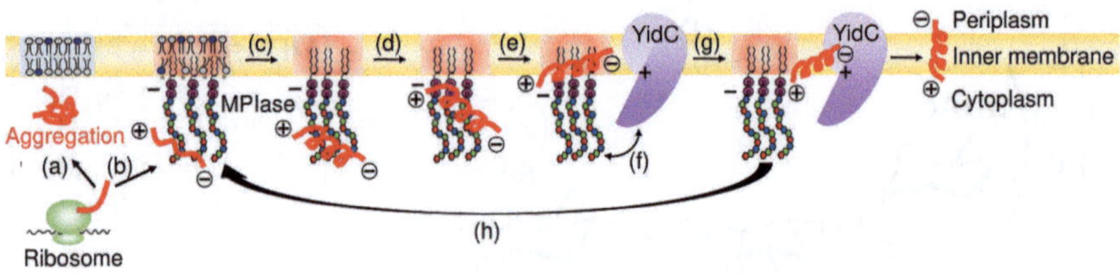


**Fig. 10** Proposed model for MPIase-mediated, translocon-independent membrane protein integration. **a** In the absence of MPIase, highly hydrophobic substrate proteins aggregate and fail to integrate into the membrane. **b** The glycan moiety of MPIase binds substrate proteins via functional groups such as Ac and COOH groups. **c** This interaction induces conformational changes in the substrate proteins, preventing aggregation. **d** The pyrophosphate group of MPIase electrostatically attracts positively charged basic residues on the substrate protein. **e** Hydrophobic interactions between the substrate protein and the membrane surface drive membrane integration. **f** YidC further facilitates the integration process. Longer glycan chains may promote cooperative interactions between MPIase and YidC and/or contribute to microdomain formation where both components colocalize. **g** Substrate proteins are transferred to YidC through electrostatic interactions and are fully incorporated into the membrane. **h** Free MPIase is regenerated for subsequent integration cycles. In addition, MPIase enhances membrane mobility (indicated in red), which may further support the integration process

## 5 Conclusion

In our efforts to elucidate the function of MPIase, a unique glycolipid residing in the *E. coli* inner membrane, we conducted structure–activity relationship studies employing a synthetic chemistry-based approach. These investigations led to the identification of the minimal active structure of MPIase, clarification of the contributions of its key functional groups, demonstration of activity enhancement through glycan chain elongation, and the discovery of a synergistic effect with the membrane chaperone YidC. Collectively, these findings enabled us to propose a mechanistic model for membrane protein integration in *E. coli*. The synthetic chemistry approach represents a powerful strategy for the functional dissection of bioactive glycans at the molecular and functional group levels, particularly in cases where natural products are scarce or structurally heterogeneous. Since glycan biosynthesis involves diverse enzymatic reactions, the absence of information on the responsible genes and enzymes, or the lethality of gene knockouts, often limits the utility of molecular biological methods for obtaining modified glycans. Moreover, the development of functionalized MPIase analogs bearing fluorescent or photo-reactive groups is anticipated to provide valuable tools for addressing unresolved questions. Visualizing the dynamics of MPIase within membranes and elucidating its cooperative interactions with membrane transport proteins, such as the Sec translocon and YidC, would offer critical insights into the essential roles of MPIase in both translocon-dependent membrane integration and the translocation of diverse proteins. We hope that this chemical approach will shed new light on the biological functions of glycolipids.

## References

1. Kuhn A, Koch H-G, Dalbey RE (2017) Targeting and insertion of membrane proteins. EcoSal Plus 7. https://doi.org/10.1128/ecosalplus.ESP-0012-2016
2. Oswald J, Njenga R, Natriashvili A, Sarmah P, Koch H-G (2021) The dynamic SecYEG translocon. Front Mol Biosci 8:1–30. https://doi.org/10.3389/fmolb.2021.664241
3. Tsirigotaki A, De Geyter J, Sostaric N, Economou A, Karamanou S (2017) Protein export through the bacterial Sec pathway. Nat Rev Microbiol 15:21–36. https://doi.org/10.1038/nrmicro.2016.161
4. Smets D, Loos MS, Karamanou S, Economou A (2019) Protein transport across the bacterial plasma membrane by the Sec pathway. Protein J 38:262–273. https://doi.org/10.1007/s10930-019-09841-8
5. Kiefer D, Kuhn A (1999) Hydrophobic forces drive spontaneous membrane insertion of the bacteriophage Pf3 coat protein without topological control. EMBO J 18:6299–6306. https://doi.org/10.1093/emboj/18.22.6299
6. Chen M, Samuelson JC, Jiang F, Muller M, Kuhn A, Dalbey RE (2002) Direct interaction of YidC with the Sec-independent Pf3 coat protein during its membrane protein insertion. J Biol Chem 277:7670–7675. https://doi.org/10.1074/jbc.M110644200
7. Nishiyama K, Ikegami A, Moser M, Schiltz E, Tokuda H, Muller M (2006) A derivative of lipid A is involved in signal recognition particle/SecYEG-dependent and -independent membrane integrations. J Biol Chem 281:35667–35676. https://doi.org/10.1074/jbc.M608228200
8. Nishiyama K, Maeda M, Abe M, Kanamori T, Shimamoto K, Kusumoto S, Ueda T, Tokuda H (2010) A novel complete reconstitution system for membrane integration of the simplest membrane protein. Biochem Biophys Res Commun 394:733–736. https://doi.org/10.1016/j.bbrc.2010.03.061
9. Nishiyama K, Maeda M, Yanagisawa K, Nagase R, Komura H, Iwashita T, Yamagaki T, Kusumoto S, Tokuda H, Shimamoto K (2012) MPIase is a glycolipozyme essential for membrane protein integration. Nat Commun 3:1260. https://doi.org/10.1038/ncomms2267
10. Mitchell AM, Srikumar T, Silhavy TJ (2018) Cyclic enterobacterial common antigen maintains the outer membrane permeability barrier of *Escherichia coli* in a manner controlled by YhdP. mBio. 9. https://doi.org/10.1128/mBio.01321-18
11. Fujikawa K, Suzuki S, Nagase R, Ikeda S, Mori S, Nomura K, Nishiyama K, Shimamoto K (2018) Syntheses and activities of the functional structures of a glycolipid essential for membrane protein integration. ACS Chem Biol 13:2719–2727. https://doi.org/10.1021/acschembio.8b00654
12. Liu L, Zha J, DiGiandomenico A, McAllister D, Stover CK, Wang Q, Boons GJ (2015) Synthetic enterobacterial common antigen (ECA) for the development of a universal immunotherapy for drug-resistant enterobacteriaceae. Angew Chem Int Ed Engl 54:10953–10957. https://doi.org/10.1002/anie.201505420
13. Paulsen H, Lorentzen JP (1985) Synthesis of the immunologically essential saccharide sequence of the "enterobacterial common antigen." Angew Chem Int Ed Engl 24:773–775. https://doi.org/10.1002/anie.198507732
14. Nguyen HQ, Davis RA, Gervay-Hague J (2014) Synthesis and structural characterization of three unique helicobacter pylori alpha-cholesteryl phosphatidyl glucosides. Angew Chem Int Ed Engl 53:13400–13403. https://doi.org/10.1002/anie.201406529
15. Gold H, Delft PV, Meeuwenoord N, Codee JDC, Filippov DV, Eggink G, Overkleeft HS, Marel GAvd (2008) Synthesis of sugar nucleotides by application of phosphoramidites. J Org Chem 73:9458–9460. https://doi.org/10.1021/jo802021t
16. Bartra M, Romea P, Urpi F, Vilarrasa J (1990) A fast procedure for the reduction of azides and nitro compounds based on the reducing ability of $Sn(SR)_3$- species. Tetrahedron 46:587–594. https://doi.org/10.1016/S0040-4020(01)85439-9
17. Sabesan S, Neira S (1992) Synthesis of glycosyl phosphates and azides. Carbohydr Res 223:169–185. https://doi.org/10.1016/0008-6215(92)80015-S

18. Sim MM, Kondo H, Wong C-H (1993) Synthesis and use of glycosyl phosphites: an effective route to glycosyl phosphates, sugar nucleotides, and glycosides. J Am Chem Soc 115:2260–2267. https://doi.org/10.1021/ja00059a023
19. Fujikawa K, Han Y, Osawa T, Mori S, Nomura K, Muramoto M, Nishiyama K, Shimamoto K (2023) Structural requirements of a glycolipid MPIase for membrane protein integration. Chem Eur J 29:e202300437. https://doi.org/10.1002/chem.202300437
20. Mori S, Nomura K, Fujikawa K, Osawa T, Shionyu M, Yoda T, Shirai T, Tsuda S, Yoshizawa-Kumagaye K, Masuda S, Nishio H, Yoshiya T, Suzuki S, Muramoto M, Nishiyama K, Shimamoto K (2022) Intermolecular interactions between a membrane protein and a glycolipid essential for membrane protein integration. ACS Chem Biol 17:609–618. https://doi.org/10.1021/acschembio.1c00882
21. Zulueta MML, Lin SY, Lin YT, Huang CJ, Wang CC, Ku CC, Shi Z, Chyan CL, Irene D, Lim LH, Tsai TI, Hu YP, Arco SD, Wong CH, Hung SC (2012) α-Glycosylation by D-glucosamine-derived donors: synthesis of heparosan and heparin analogues that interact with mycobacterial heparin-binding hemagglutinin. J Am Chem Soc 134:8988–8995. https://doi.org/10.1021/ja302640p
22. Zulueta MML, Lin SY, Hung SC (2013) Chemical synthesis of oligosaccharides based on heparin and heparan sulfate. Trends Glycosci Glycotechnol. 25:141–158. https://doi.org/10.4052/tigg.25.141
23. Crich D, Dudkin V (2001) Why are the hydroxy groups of partially protected N-acetylglucosamine derivatives such poor glycosyl acceptors, and what can be done about It? A comparative study of the reactivity of N-acetyl-, N-phthalimido-, and 2-azido-2-deoxy-glucosamine derivatives in glycosylation. 2-picolinyl ethers as reactivity-enhancing replacements for benzyl ethers. J Am Chem Soc 123:6819–6825. https://doi.org/10.1021/ja010086b
24. Kononov LO, Malysheva NN, Kononova EG, Orlova AV (2008) Intermolecular hydrogen-bonding pattern of a glycosyl donor: the key to understanding the outcome of sialylation. Eur J Org Chem 2008:3251–3255. https://doi.org/10.1002/ejoc.200800324
25. Nagasaki M, Manabe Y, Minamoto N, Tanaka K, Silipo A, Molinaro A, Fukase K (2016) Chemical synthesis of a complex-type N-glycan containing a core fucose. J Org Chem 81:10600–10616. https://doi.org/10.1021/acs.joc.6b02106
26. Suzuki K, Maeta H, Matsumoto T (1989) An improved procedure for metallocene-promoted glycosidation. Enhanced reactivity by employing 1:2-ratio of Cp2HfCl2-AgClO4. Tetrahedron Lett 30:4853–4856. https://doi.org/10.1016/S0040-4039(01)80526-8
27. Li Y, Yang X, Liu Y, Zhu C, Yang Y, Yu B (2010) Gold(I)-catalyzed glycosylation with glycosyl ortho-alkynylbenzoates as donors: general scope and application in the synthesis of a cyclic triterpene saponin. Chem Eur J 16:1871–1882. https://doi.org/10.1002/chem.200902548
28. Yu B (2018) Gold(I)-catalyzed glycosylation with glycosyl o-alkynylbenzoates as donors. Acc Chem Res 51:507–516. https://doi.org/10.1021/acs.accounts.7b00573
29. Yu B, Tao H (2001) Glycosyl trifluoroacetimidates. part 1: preparation and application as new glycosyl donors. Tetrahedron Lett 42:2405–2407. https://doi.org/10.1016/S0040-4039(01)00157-5
30. Podlasek CA, Wu J, Stripe WA, Bondo PB, Serianni AS (1995) [13C]-enriched methyl aldopyranosides: structural interpretations of 13C spin-coupling constants and chemical shifts. J Am Chem Soc 117:8635–8644. https://doi.org/10.1021/ja00138a020
31. Dohi H, Nishida Y (2014) Odorless access to thioglycosides for oligosaccharide synthesis: their design and advanced procedures for thioglycosidation. Trends Glycosci Glycotechnol. 26:119–130. https://doi.org/10.4052/tigg.26.119
32. Asano S, Tanaka HN, Imamura A, Ishida H, Ando H (2019) P-tert-butyl groups improve the utility of aromatic protecting groups in carbohydrate synthesis. Org Lett 21:4197–4200. https://doi.org/10.1021/acs.orglett.9b01372
33. Osawa T, Nomura K, Shimamoto K, Fujikawa K (2025) Synthesis of a fluorescent analog for exploring the functions of the bacterial glycopyrophospholipid MPIase. Carbohydr Res 553:109483. https://doi.org/10.1016/j.carres.2025.109483

34. Kawashima Y, Miyazaki E, Muller M, Tokuda H, Nishiyama K (2008) Diacylglycerol specifically blocks spontaneous integration of membrane proteins and allows detection of a factor-assisted integration. J Biol Chem 283:24489–24496. https://doi.org/10.1074/jbc.M801812200
35. Koch H-G, Hengelage T, Neumann-Haefelin C, MacFarlane J, Hoffschulte HK, Schimz K-L, Mechler B, Muller M (1999) In vitro studies with purified components reveal signal recognition particle (SRP) and SecA/SecB as constituents of two independent protein-targeting pathways of *escherichia coli*. Mol Biol Cell 10:2163–2173. https://doi.org/10.1091/mbc.10.7.2163
36. Bogdanov M, Xie J, Dowhan W (2009) Lipid-protein interactions drive membrane protein topogenesis in accordance with the positive inside rule. J Biol Chem 284:9637–9641. https://doi.org/10.1074/jbc.R800081200
37. Scotti PA, Urbanus ML, Brunner J, L.de Gier J-W, Heijne G, Does C, Driessen AJM, Oudega B, Luirink J (2000) YidC, the *escherichia coli* homologue of mitochondrial oxa1p, is a component of the Sec translocase. EMBO J 19:542–549. https://doi.org/10.1093/emboj/19.4.542
38. Kumazaki K, Chiba S, Takemoto M, Furukawa A, Nishiyama K, Sugano Y, Mori T, Dohmae N, Hirata K, Nakada-Nakura Y, Maturana AD, Tanaka Y, Mori H, Sugita Y, Arisaka F, Ito K, Ishitani R, Tsukazaki T, Nureki O (2014) Structural basis of Sec-independent membrane protein insertion by YidC. Nature 509:516–520. https://doi.org/10.1038/nature13167
39. Kumazaki K, Kishimoto T, Furukawa A, Mori H, Tanaka Y, Dohmae N, Ishitani R, Tsukazaki T, Nureki O (2014) Crystal structure of *escherichia coli* YidC, a membrane protein chaperone and insertase. Sci Rep 4:7299. https://doi.org/10.1038/srep07299
40. Sasaki M, Nishikawa H, Suzuki S, Moser M, Huber M, Sawasato K, Matsubayashi HT, Kumazaki K, Tsukazaki T, Kuruma Y, Nureki O, Ueda T, Nishiyama K (2019) The bacterial protein YidC accelerates MPIase-dependent integration of membrane proteins. J Biol Chem 294:18898–18908. https://doi.org/10.1074/jbc.RA119.011248
41. Nomura K, Yamaguchi T, Mori S, Fujikawa K, Nishiyama K, Shimanouchi T, Tanimoto Y, Morigaki K, Shimamoto K (2019) Alteration of membrane physicochemical properties by two factors for membrane protein integration. Biophys J 117:99–110. https://doi.org/10.1016/j.bpj.2019.05.014
42. Nomura K, Mori S, Fujikawa K, Osawa T, Tsuda S, Yoshizawa-Kumagaye K, Masuda S, Nishio H, Yoshiya T, Yoda T, Shionyu M, Shirai T, Nishiyama K, Shimamoto K (2022) Role of a bacterial glycolipid in Sec-independent membrane protein insertion. Sci Rep 12:12231. https://doi.org/10.1038/s41598-022-16304-1
43. Mori S, Shionyu M, Shimamoto K, Nomura K (2024) Bacterial glycolipid acting on protein transport across membranes. ChemBioChem 25:e202300808. https://doi.org/10.1002/cbic.202300808
44. Shimamoto K, Fujikawa K, Osawa T, Mori S, Nomura K, Nishiyama K (2024) Key contributions of a glycolipid to membrane protein integration. Proc Jpn Acad, Ser B 100:387–413. https://doi.org/10.2183/pjab.100.026
45. Fujikawa K, Mori S, Nishiyama K, Shimamoto K (2022) A bacterial glycolipid essential for membrane protein integration. Adv Carbohydr Chem Biochem 81:95–129. https://doi.org/10.1016/bs.accb.2022.09.005

# Glycolipid-Recognizing C-Type Lectin Receptors: Ligand Structures and Molecular Probes for Biofunctional Analysis

**Takanori Matsumaru and Yukari Fujimoto**

**Abstract** C-type lectin receptors (CLRs) play pivotal roles in recognizing specific structures of carbohydrate-based ligands and in regulating innate immune responses. Several CLRs are known to recognize glycolipid ligands derived from microbial or endogenous sources. Among these glycolipid-recognizing receptors, this review focuses on Mincle and DCAR, summarizing the structural features and immunological activities of representative or recently developed synthetic ligands, with particular emphasis on their application to the design of functional molecular probes. These ligand-based probes have proven useful not only for elucidating receptor–ligand interactions, but also for visualizing the dynamics of the receptors in live cells. Notably, we have established fluorescence-labeled molecular probes derived from synthetic glycolipids, which have enabled the analysis of intracellular trafficking of both Mincle and its ligands. Such molecular tools offer valuable insights into the mechanisms of CLR-mediated immune signaling and contribute to the development of innovative therapeutic and diagnostic strategies.

**Keywords** C-type lectin receptors (CLRs) · Mincle · DCAR · Glycolipids · Chemical probe · Live-cell imaging

## 1 Introduction

C-type lectin receptors (CLRs) constitute a family of transmembrane lectin receptors characterized by the presence of calcium-dependent carbohydrate recognition domains [1]. As key pattern recognition receptors (PRRs) in the innate immune

---

T. Matsumaru · Y. Fujimoto (✉)
Department of Chemistry, Faculty of Science and Technology, Keio University, Yokohama, Kanagawa, Japan
e-mail: fujimotoy@chem.keio.ac.jp

Y. Fujimoto
Human Biology Microbiome Quantum Research Center (Bio2Q), Keio University, Shinjuku-ku, Tokyo, Japan

 
K. Toshima and Y. Fujimoto (eds.), *Chemical Biology of Carbohydrates*,
https://doi.org/10.1007/978-981-95-5335-8_9

system, CLRs are predominantly expressed on myeloid immune cells such as macrophages and dendritic cells, where they contribute to immune regulation by recognizing both pathogen-associated molecular patterns (PAMPs) and endogenous damage-associated molecular patterns (DAMPs) [2]. CLR-mediated signaling is initiated upon ligand binding, leading to signal transduction through immunoreceptor tyrosine-based activation motifs (ITAMs) or inhibitory motifs (ITIMs), located either within the receptors themselves or in associated adaptor proteins (e.g., FcRγ) [2, 3]. Upon ligand engagement, CLRs activate downstream signaling pathways involving molecules such as Syk and CARD9, which trigger various immune responses, including the production of pro-inflammatory cytokines and the enhancement of antigen presentation [4, 5]. Through these mechanisms, CLRs play a critical role in bridging innate and adaptive immunity.

CLRs recognize a wide spectrum of ligands, most notably polysaccharides and glycoproteins [6, 7]. Glycolipids have recently garnered attention as the ligands for specific CLRs to modulate various immune responses. These glycolipid ligands include not only structural components of microbial pathogens but also endogenous immunomodulatory molecules. The specificity of CLR–glycolipid recognition is largely determined by calcium-binding sites and adjacent amino acid residues.

Recent research on CLR ligands has advanced the understanding of their structure–activity relationships, immunomodulatory functions, and the development of molecular probes for further analysis of these receptors and their ligands. This review focuses on glycolipid-recognizing CLRs, particularly Mincle and DCAR, with an emphasis on their ligands and ligand-based molecular probes. In particular, the structural features of their ligands are summarized, along with recent advances in the design and application of synthetic molecular probes, including fluorescently labeled glycolipids and selective ligands. These chemical tools are emerging as powerful platforms for biofunctional studies, receptor imaging, and drug discovery targeting CLR-mediated immune responses.

## 2 Mincle Ligands and the Molecular Probes

This section summarizes the key features of the C-type lectin receptor Mincle and representative glycolipid ligands reported in recent years, with a focus on their structural features and immunological activities. Particular attention is also given to the development of molecular probes derived from these ligands, which serve as powerful tools for biofunctional analysis and receptor visualization.

### *2.1 Mincle*

Mincle (macrophage-inducible C-type lectin, CLEC4E) is a representative C-type lectin receptor (CLR) involved in glycolipid recognition within the innate immune system. It plays a pivotal role in sensing lipid-containing microbial components as well as endogenous danger signals [8–12]. Mincle is predominantly expressed on myeloid immune cells, such as macrophages, and functions through a signaling complex formed with the common Fc receptor γ-chain (FcRγ), which contains an immunoreceptor tyrosine-based activation motif (ITAM) [13]. Upon ligand engagement, Mincle activates the Syk–CARD9 signaling cascade, leading to the production of pro-inflammatory cytokines, stimulation of antigen-presenting cell functions and promotion of Th1/Th17-type immune polarization [14].

Mincle is capable of recognizing a wide range of complex lipids, including various glycolipids and acylglycerol-based structures [9–12, 15] One well-characterized ligand is trehalose-6,6'-dimycolate (TDM) from *Mycobacterium tuberculosis*, which is further discussed in the following sections of this review. Notably, Mincle exhibits structural selectivity not only for the glycan moiety but also for the hydrophobic regions of ligands, such as lipid chain length, degree of branching, and saturation. Structural studies have provided significant insights into the molecular basis of ligand recognition by Mincle. As shown in Fig. 1, in the binding model with trehalose monoester (C26), calcium ions at the binding site, which are coordinated by neighboring amino acid residues including Glu169, Pro170, and Asn171 of the EPN motif, play a critical role in mediating interactions [16] with the glycan moiety of glycolipids, particularly trehalose [17–19]. Arg183 has also been reported to contribute to glycan recognition. Regarding the lipid moiety of the ligand, it is suggested that hydrophobic interactions occur near the calcium ions at the surface of Mincle, where several hydrophobic amino acid residues are located. In addition, surface-exposed polar residues within this hydrophobic region, such as Asp178 and Lys106, have been suggested as potential sites for enhancing or selectively modulating ligand interactions, as described in Sects. 2.2.2 and 2.2.4. Furthermore, recent structural investigations, including NMR-based interaction analyses, have been actively conducted to better understand Mincle's specificity toward the lipid moiety of ligands (Fig. 1) [20, 21].

These insights into structure–activity relationships (SARs) have enabled the rational design of selective Mincle agonists and have been applied to the development of functional molecular probes. These probes serve as powerful tools for elucidating receptor–ligand interactions, visualizing receptor dynamics through live-cell imaging, and evaluating immunomodulatory functions.

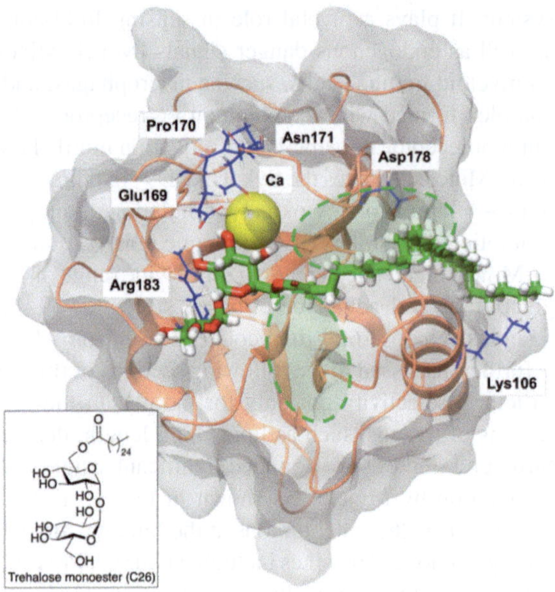


**Fig. 1** Human Mincle in complex with the model ligand trehalose monoester (C26), illustrating ligand recognition. The green dotted-line-enclosed areas represent additional hydrophobic regions predicted by NMR analysis [20]

## 2.2 *Various Mincle Ligands and Their Derivatives*

### 2.2.1 Trehalose Diesters (TDEs): Trehalose Dimycolate (TDM) and Their Analogs

Trehalose-6,6'-dimycolate (TDM), known as the "cord factor," represents a major class of glycolipids enriched in the cell envelope of *Mycobacterium* species, particularly *Mycobacterium tuberculosis* (Fig. 2) [22, 23]. TDMs are composed of an α,α'-trehalose scaffold bearing ester-linked mycolic acids at the C6 and C6' positions. The attached mycolic acids are highly diverse in structure, varying considerably in carbon chain length, degree of branching, and the presence of distinct functional groups [24, 25]. The field of Mincle ligand recognition underwent a major breakthrough in 2009. Yamasaki et al. identified the C-type lectin receptor Mincle as a direct pattern recognition receptor for both TDM and its simplified synthetic analogs, trehalose dibehenate (TDB) [8]. They further showed that bone marrow-derived macrophages (BMMs) from Mincle-deficient mice failed to produce tumor necrosis factor (TNF) and macrophage inflammatory protein 2 (MIP-2) upon stimulation with TDM or TDB.

However, detailed structure–activity relationship (SAR) analysis has been hampered by the structural complexity and heterogeneity of naturally derived TDMs (the TDM structure depicted in the figure is based on the chemical structure described in Reference 8). Baird et al. developed a total synthesis strategy to construct structurally defined mycolic acids, which were subsequently esterified at the 6- and 6'-positions of α,α'-trehalose to yield a library of synthetic TDMs featuring distinct

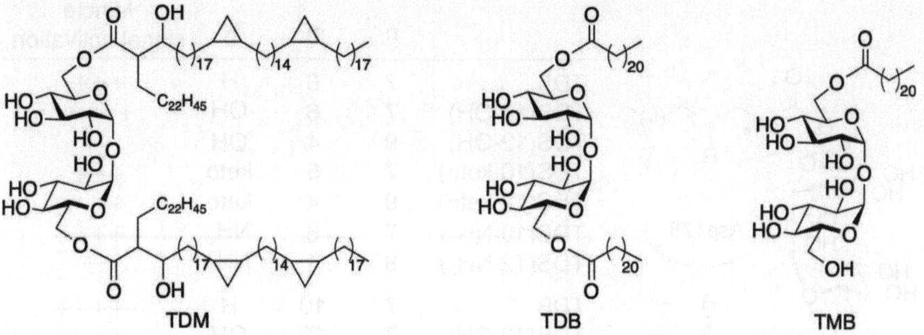


**Fig. 2** The structures of trehalose-6,6'-dimycolate (TDM) and its analogs, trehalose dibehenate (TDB), and trehalose monobehenate (TMB)

functional groups [26–28]. In addition, several trehalose diesters bearing branched [29–31] or linear fatty acids [32] and trehalose monoesters [33, 34] have also been reported to function as Mincle agonists.

### 2.2.2 Trehalose Diesters (TDEs) with Polar-Functionalized Lipid Moieties for Enhanced Biofunctionality

To develop trehalose diester derivatives with improved physicochemical properties, particularly enhanced solubility, while retaining or increasing their Mincle agonistic activity, our group introduced polar functional groups into the lipid moieties of trehalose diesters (TDEs) [35]. Modification of the fatty acid moiety was guided by the binding model of human Mincle and trehalose monoester (Fig. 1), which indicated the potential involvement of a polar amino acid residue (Asp178) in the vicinity of the ligand's lipid chain, specifically around the C10 and C12 positions, thus providing a rational basis for this design strategy. To this end, we synthesized TDE derivatives by esterifying trehalose with C18 or C22 fatty acids bearing hydroxy, oxo, or amino groups at the 10- or 12-position of the lipid chain (Fig. 3). Among these compounds, the derivatives bearing a hydroxy group at the 10-position, such as TDS(10-OH) and TDB(10-OH), exhibited Mincle-mediated signaling activity in NFAT-GFP reporter cells that was comparable or superior to that of conventional TDS and TDM. Notably, TDS(10-OH) demonstrated over an eightfold increase in solubility in DMSO compared to unmodified TDS, which is advantageous for use in biological assays and structural studies. These findings indicated that Mincle can tolerate polar functional groups introduced at specific positions in the lipid chain. This observation demonstrated the potential of structural modification of TDEs as a strategy to modulate their physicochemical properties and to facilitate the development of novel molecular probes. On the other hand, while strong activity had been anticipated for the amine-type compound based on its potential interaction with the side-chain carboxyl group of Asp178, the observed activity was comparable to or

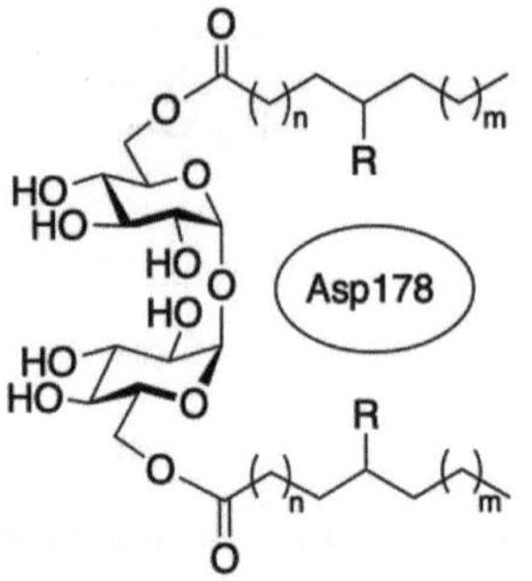


| | n | m | R | Mincle signal activation |
|---|---|---|---|---|
| TDS | 7 | 6 | H | +++ |
| TDS(10-OH) | 7 | 6 | OH | ++++ |
| TDS(12-OH) | 9 | 4 | OH | ++ |
| TDS(10-keto) | 7 | 6 | keto | +++ |
| TDS(12-keto) | 9 | 4 | keto | +++ |
| TDS(10-$NH_2$) | 7 | 6 | $NH_2$ | +++ |
| TDS(12-$NH_2$) | 9 | 4 | $NH_2$ | ++ |
| TDB | 7 | 10 | H | +++++ |
| TDB(10-OH) | 7 | 10 | OH | +++++ |
| TDB(10-keto) | 7 | 10 | keto | +++ |
| TDB(10-$NH_2$) | 7 | 10 | $NH_2$ | +++ |

**Fig. 3** The structures and human Mincle-mediated signaling activity of TDEs modified with polar functional groups in the lipid moiety

lower than that of TDS and TDB. This study has provided the first systematic demonstration that functional group modification of the lipid moiety of TDEs can influence their interaction with Mincle, thereby offering valuable insights into the rational design of Mincle-targeted immunomodulatory agents.

### 2.2.3 β-Mannosyloxymannitol Glycolipid "44-2"

A unique glycolipid structure, "44-2" from *Malassezia pachydermatis* was also reported as a Mincle ligand [36]. It features a rare L-mannitol core esterified at the 1-, 3-, and 4-positions with three mannosylated fatty acids, specifically one residue of 10-*O*-(β-D-mannopyranosyl-[1→2]-β-D-mannopyranosyl)-10-hydroxyoctadecanoic acid and two residues of 10-*O*-β-D-mannopyranosyl-10-hydroxyoctadecanoic acid (Fig. 4), in which the position of the 10-hydroxy group in the fatty acyl chains remains unidentified. Functionally, natural "44-2" strongly activated mouse Mincle-expressing NFAT-GFP reporter cells, exhibiting activity comparable to that of TDM. Direct binding of "44-2" to a Mincle-Ig fusion protein was also confirmed by ELISA. Moreover, "44-2" induced robust TNF production in wild-type bone marrow-derived dendritic cells (BMDCs).

The total synthesis and immunological evaluation of "44-2" were carried out independently by Williams et al. [37] and Fujimoto et al. [38]. In our approach, we designed and achieved the total synthesis of "44-2" and its derivatives to conduct a comprehensive structure–activity relationship (SAR) study of β-1,2-mannosyloxymannitol glycolipids and their analogs. This study aimed to elucidate the molecular features essential for Mincle activation, focusing on unique, complex, medium-sized glycolipids, "44-2"-derived structures, capable of interacting with a broad region of the receptor, building also on insights from trehalose diesters (TDEs) with polar-functionalized lipid moieties [35]. To examine the role of stereochemistry

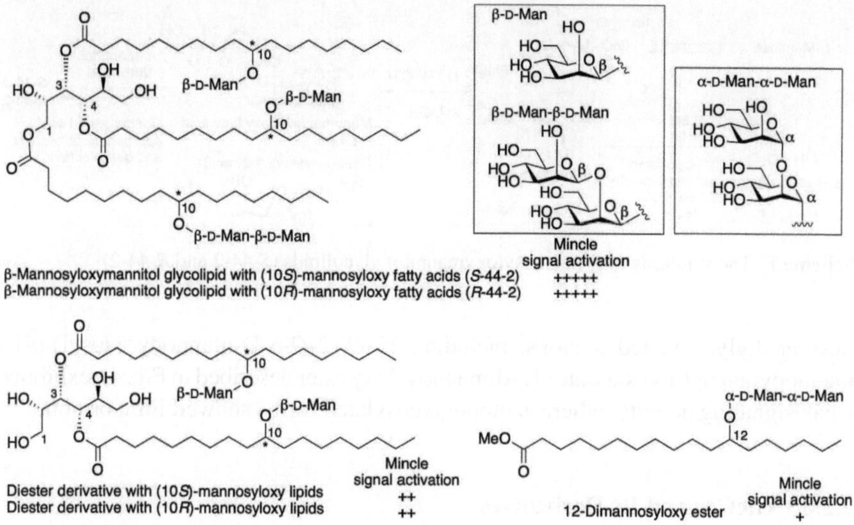


**Fig. 4** The chemical structures of β-mannosyloxymannitol glycolipids (*S*-44-2 and *R*-44-2) and its analogs. In Mincle signal activation, "+++++" was shown to exhibit activity comparable to that of TDB. Mincle signaling activation was shown in relative activity against TDB (see Fig. 3)

at the 10-position in Mincle-mediated signaling, we synthesized mannosyloxymannitol glycolipids bearing either (10*S*)- or (10*R*)-mannosyloxy fatty acids at all three acyl positions. Diester analogs were also prepared to assess the impact of the dimannosyloxy fatty acid at the 1-position of the mannitol core. Access to the diastereomeric pair of "44-2" analogs (*S*-44-2 and *R*-44-2) was achieved via chiral resolution of enantiomeric 10-hydroxystearate. The enantiomers of racemic 10-hydroxystearate were obtained by chiral resolution using the characteristic acetal-forming chiral resolving agent (*S*)-5-allyl-2-oxabicyclo[3.3.0]oct-8-ene (ALBO) [39]. Fortunately, the glycosylation using a 4,6-*O*-benzylidene-protected mannosyl donor afforded mono- or di-mannosyloxy ester with high diastereoselectivity. Subsequently, the obtained fatty acids were appropriately condensed to the L-mannitol derivative, and the protecting groups were removed under mild conditions to afford the desired β-mannosyloxymannitol glycolipids (Scheme 1). Our biological evaluations using Mincle reporter assays revealed that the synthetic β-mannosyloxymannitol glycolipids exhibited strong Mincle-mediated signaling activity in both human and murine systems, with only slightly different activities depending on the stereochemistry of the mannosylated fatty acid. In contrast, diester derivatives, lacking the dimannosyloxy fatty acid moiety, showed weak activity, indicating that this motif is critical for potent Mincle activation.

Furthermore, a structure–activity relationship (SAR) study was conducted based on the structure of the mannosyloxy fatty acid moiety of "44-2" [40]. A series of mono- and diglycosylated derivatives of 12-hydroxystearic acid, referred to as glycosyloxystearates, was synthesized and evaluated for Mincle-mediated signaling.

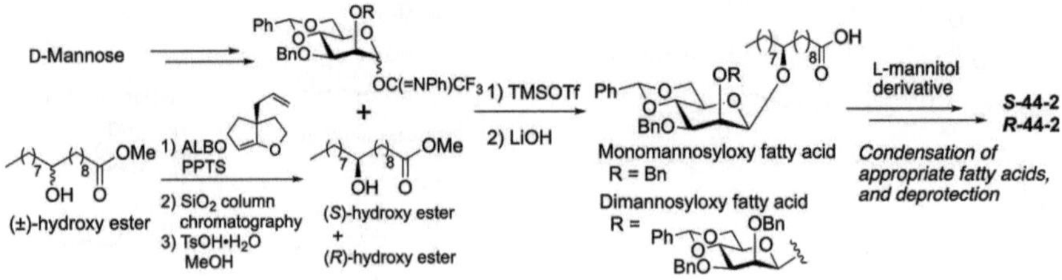


**Scheme 1** The synthesis of β-mannosyloxymannitol glycolipids (*S*-44-2 and *R*-44-2)

Several diglycosylated analogs, including 12-*O*-(2-*O*-α-D-mannopyranosyl)-α-D-mannopyranosyl-oxystearate (12-dimannosyloxy ester described in Fig. 4), exhibited weak signaling activity, whereas monoglycosylated forms showed little or none.

### 2.2.4 GlcCer and Its Derivatives

β-Glucosylceramide (β-GlcCer) has been identified as an endogenous glycolipid ligand for Mincle through a ligand screening approach using culture supernatants from damaged cells, aiming to elucidate how endogenous metabolites generated upon cell damage influence innate immunity (Fig. 5a) [41]. Although β-GlcCer normally resides in the Golgi and endoplasmic reticulum, it functions as an immunostimulatory signal when released extracellularly upon cell disruption. Reporter assays using Mincle-expressing cells demonstrated that synthetic β-GlcCer (C24:1) activates Mincle, and the secretion of inflammatory mediators such as TNF and MIP-2 was evaluated in BMDCs. In mice deficient in the β-GlcCer-degrading enzyme GBA1, β-GlcCer accumulated in tissues and enhanced sterile inflammation, which was significantly attenuated in a Mincle-deficient background. Furthermore, in GBA1-deficient mice, elevated β-GlcCer was shown to promote antigen-specific T-cell responses in a Mincle-dependent manner, acting as a damage-associated self-ligand that activates both innate and adaptive immunity. In addition, Mincle promotes neutrophil extracellular traps (NET) formation in response to β-GlcCer, but not to other ligands like TDM, suggesting a unique role in neutrophil activation and potential relevance in inflammatory disease [42, 43].

The structure–immunoactivity relationships of endogenous major β-GlcCer and its α-anomer (which is usually present at very low levels in vivo) had not been thoroughly investigated, particularly with respect to variations in the structures of their fatty acyl chains. In our group, both the α- and β-anomers of GlcCer were chemically synthesized and highly purified for the evaluation of immunological activities, including Mincle activation and CD1d–NKT cell stimulation. Each glycolipid was designed to contain one of four representative fatty acyl chains (C18:0, C18:1, C24:0, or C24:1), which are abundantly present in the central nervous system (Fig. 5a) [44]. These compounds were efficiently prepared using a synthetic strategy that employed allyl protecting groups and ruthenium-catalyzed deprotection under mild conditions

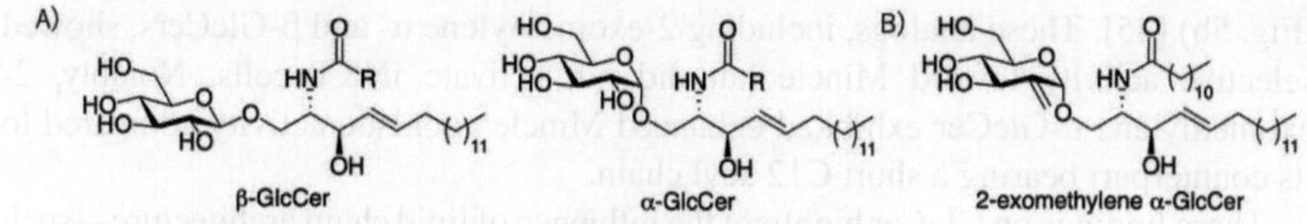


| RC(=O) | β-GlcCer CD1d-NKT cell activation (INF-γ) | β-GlcCer Mincle signal activation | α-GlcCer CD1d-NKT cell activation (INF-γ) | α-GlcCer Mincle signal activation |
|---|---|---|---|---|
| 18:0 | - | ± | - | ++ |
| 18:1 | - | +++ | - | ++ |
| 24:0 | - | ++ | + | ++ |
| 24:1 | - | ± | + | ± |

**Fig. 5** **a** The structures of β/α-GlcCers and the immunomodulatory activities. **b** 2-Exomethylene α-GlcCer. Mincle signaling activities are shown in relative activities against TDM

(Scheme 2). Glycosylation with an allyl-protected glycosyl donor afforded both α- and β-anomers, after which α-Glc-psychosine and β-Glc-psychosine were individually subjected to amine-selective condensation conditions to yield the corresponding α- and β-GlcCer. This method enabled us to obtain the target molecules in high yield without compromising the integrity of the unsaturated lipid moieties.

We investigated the immunological activities of the synthetic GlcCers through the innate immune receptor Mincle. Both the α- and β-anomers of GlcCer exhibited similar trends in Mincle-mediated signaling activity, except for the C18:0 derivative, with C18:1 and C24:0 showing relatively higher activity. In parallel with the evaluation of Mincle agonistic activity, the binding affinities of these glycolipids to CD1d molecules and their ability to induce cytokine production via the CD1d–NKT cell axis were assessed. In contrast to Mincle, CD1d–NKT cell activation is undetectable for the GlcCer β-anomer and for α-anomers with short acyl chains, whereas an α-anomer with a long acyl chain (e.g., C24) elicits a low yet quantifiable response. These findings collectively highlight the functional importance of lipid chain structure in endogenous glycosphingolipids and their capacity to modulate both innate and adaptive immunity.

As in the development of GlcCer derivatives, artificial glycoconjugates with a 2-exomethylene group replacing the sugar's C2 hydroxyl have also been reported

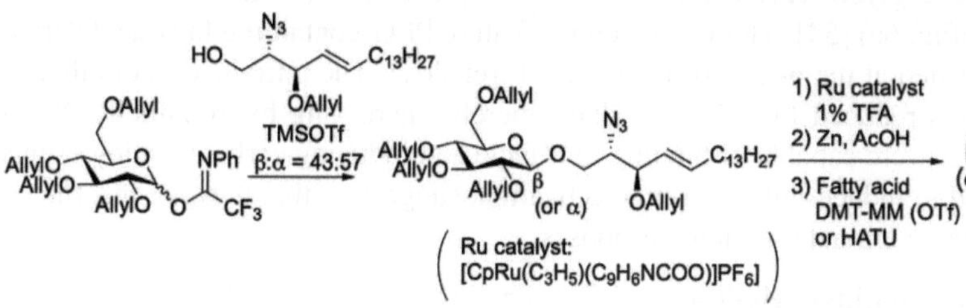


**Scheme 2** The synthesis of α- and β-GlcCer

(Fig. 5b) [45]. These analogs, including 2-exomethylene α- and β-GlcCers, showed selective activity toward Mincle but did not activate iNKT cells. Notably, 2-exomethylene α-GlcCer exhibited enhanced Mincle agonistic activity compared to its counterpart bearing a short C12 acyl chain.

These findings on GlcCer highlight the influence of lipid chain architecture—such as length and saturation—on the immunological functions of endogenous glycosphingolipids. Structural variations were shown to affect the selective activation of innate immune receptors like Mincle, as well as interactions with the CD1d–NKT cell axis. These insights contribute to a better understanding of how glycolipid structure governs immune responses and may inform the future design of immunoregulatory molecules.

### 2.2.5 Other Ligands

Apart from the examples described above, a wide array of Mincle ligands exhibiting structural diversity has been reported [9–12, 15]. This section highlights, in particular, those identified in recent studies.

**Brartemicin and Related Structures**: Brartemicin is a natural trehalose-based compound reported to bind bovine Mincle with high affinity (Fig. 6a) [46]. To enhance its immunomodulatory properties, lipidated derivatives were developed [47], and several compounds, including derivative 1 (C18-lipidated) showed strong Mincle-dependent activity. Further structure–activity studies identified potent agonists, such as derivatives 2–4 [48, 49]. In addition, aryl trehalose derivatives with aryl groups at the 6,6'-positions, such as derivative 5 and amide-linked derivative 6, also showed potent Mincle-dependent activity [50–52].

Bis-aryl triazole trehalolipids have been synthesized as lipid-modified analogs of brartemicin, featuring aromatic rings conjugated to the trehalose core via triazole linkers through copper-catalyzed azide–alkyne cycloaddition (CuAAC) chemistry (Fig. 6b) [53]. Molecular docking studies suggested that the triazole moiety is positioned near arginine residues in Mincle, highlighting its potential as a bioisostere of ester functionalities for tuning stability and hydrophobicity in ligand design.

**Trehalose Glycolipid Bearing PEG Moieties (PEG-TGLs)**: As an alternative strategy for developing water-soluble Mincle ligands, trehalose diesters bearing polyethylene glycol (PEG) moieties on the lipid chains (PEG-TGLs) have been reported (Fig. 6c) [54]. These compounds feature PEG-containing fatty acid derivatives esterified at the 6- and 6'-positions of trehalose. The introduction of ethylene glycol units reduced the CLogP values, thereby increasing hydrophilicity. While PEG-TGLs were capable of binding to Mincle and inducing early signaling events, they failed to induce inflammasome activation, suggesting that they are insufficient to fully activate innate immune responses.

**6,6′-Diphosphatidyltrehalose**: 6-Phosphatidyltrehalose (PT) and 6,6′-diphosphatidyltrehalose (diPT) have been discovered in *Salmonella enterica*

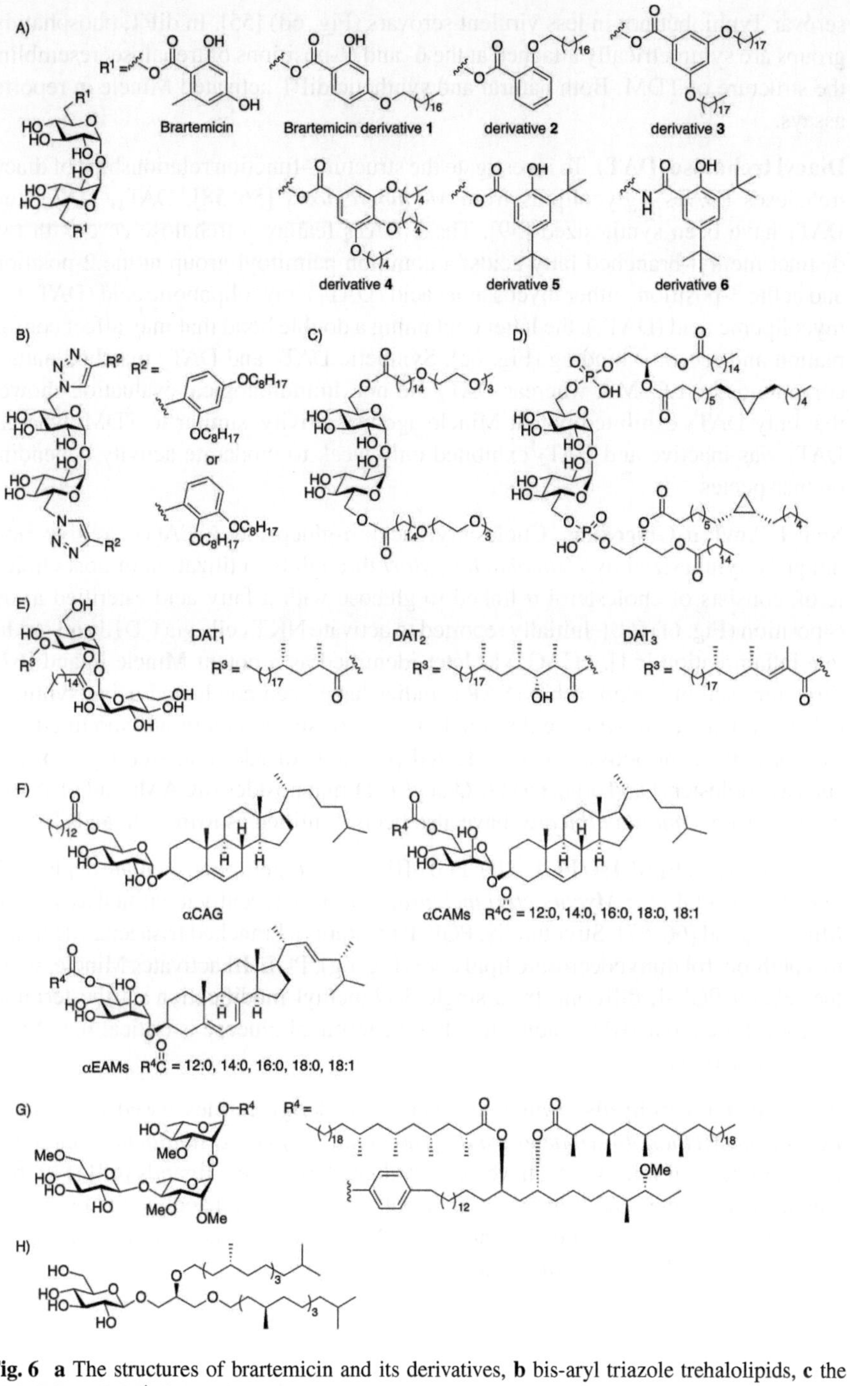


**Fig. 6** **a** The structures of brartemicin and its derivatives, **b** bis-aryl triazole trehalolipids, **c** the PEG-TGLs, **d** 6,6′-diphosphatidyltrehalose, **e** diacyl trehaloses, $DAT_1$, $DAT_2$, and $DAT_3$, **f** αCAG (from *H. pylori*) and synthetic αCAMs and αEAMs, **g** PGL-III, and H) β-glucosyl dialkylglyceride

serovar Typhi, but not in less virulent serovars (Fig. 6d) [55]. In diPT, phosphatidyl groups are symmetrically attached at the 6- and 6′-positions of trehalose, resembling the structure of TDM. Both natural and synthetic diPT activated Mincle in reporter assays.

**Diacyl trehalose (DAT)**: To investigate the structure–function relationships of diacyl trehaloses (DATs), glycolipids from *M. tuberculosis* [56–58], $DAT_1$, $DAT_2$, and $DAT_3$ have been synthesized [59]. These DATs feature a trehalose core with two distinct methyl-branched fatty acids: a common palmitoyl group at the 2-position, and at the 3-position, either mycosanoic acid ($DAT_1$), mycolipanolic acid ($DAT_2$), or mycolipenic acid ($DAT_3$), the latter containing a double bond that may affect conformation and receptor binding (Fig. 6e). Synthetic $DAT_1$ and $DAT_3$ matched natural compounds by LC–MS, whereas $DAT_2$ did not. Immunological evaluation showed that only $DAT_3$ exhibited potent Mincle agonist activity, similar to TDM, whereas $DAT_1$ was inactive and $DAT_2$ exhibited only weak to moderate activity depending on the species.

**Steryl Acyl α-Glycoside**: Cholesteryl acyl α-glucoside (αCAG), a glycolipid uniquely synthesized by *Helicobacter pylori* through the utilization of host cholesterol, consists of cholesterol α-linked to glucose with a fatty acid esterified at the 6-position (Fig. 6f) [60]. Initially reported to activate NKT cells via CD1d and modulate inflammation [61], αCAG was later identified as a potent Mincle ligand [62]. Structure–activity relationship (SAR) studies have been conducted using synthetic αCAG derivatives in which either the fatty acid or sterol moiety was modified, and Mincle-dependent activity was evaluated [63, 64]. In addition, steryl glycosides, such as cholesteryl and ergosteryl 6-*O*-acyl-α-D-mannosides (αCAMs and αEAMs), derived from *Candida albicans*, have also been identified as Mincle ligands [65].

**Phenolic Glycolipid-III (PGL-III)**: PGL-III, a biosynthetic intermediate of phenolic glycolipids (PGLs) in *Mycobacterium leprae*, has been recently identified as a potent Mincle ligand [66, 67]. Structurally, PGL-III features a branched trisaccharide linked to a phthiocerol dimycocerosate lipid core (Fig. 6g). PGL-III activates Mincle, while the related PGL-I, differing by a single 3-*O*-methyl modification on the terminal glucose, failed to activate, indicating that the terminal glucose is critical for Mincle agonistic activity.

**Archaeal Glycerolipids**: Synthetic glycosyl dialkylglycerides based on archaeal lipids from *Methanobrevibacter smithii* and *Methanosphaera stadtmanae*, members of the human gut microbiota, have been evaluated as Mincle ligands [68] (Fig. 6h). Chemically synthesized β-glucosyl archaeol, containing archaeol as the lipid core, bound and activated both human and mouse Mincle. Mincle activation depended on features such as lipid branching and glycosyl headgroup.

## 2.3 Molecular Probes Based on the Structure of Mincle Ligand

Molecular probes are invaluable tools for visualizing and analyzing biological phenomena such as molecular interactions, metabolic pathways, and intracellular localization. In particular, the use of well-designed molecular probes is essential for studying glycolipids, which have highly diverse structures and exhibit biological activity that is highly sensitive to subtle structural variations. This section introduces molecular probes designed for the analysis of Mincle and its ligands. For example, fluorescence-labeled probes, turn-on type probe, and activity-based fluorescent photoaffinity probe (AfBPP) have been reported as useful tools. AfBPP is a chemoproteomic method for covalent labeling and identification of protein targets of bioactive small molecules in living systems [69]. Although no Mincle-specific AfBPP probe has been reported, a trehalose glycolipid-type probe showed nitric oxide production in macrophages [70]. Future applications in Mincle-targeted AfBPP are anticipated.

### 2.3.1 Fluorescence-Labeled Trehalose Esters, GlcCer, LacCer, and GM3

The synthesis of fluorescent trehalose glycolipids (F-TGLs) bearing either a dansyl or a BODIPY fluorophore was reported in 2015 [71]. These probes share a common architecture based on a trehalose monoester core with a single C26 acyl chain (Fig. 7). The fluorophores were introduced at the 6-position of the non-acylated glucose residue through a C11 ester terminal amide, or sulfonylamide linker. Confocal microscopy using a BODIPY-labeled F-TGL, along with flow cytometric analysis, revealed that these fluorescent trehalose esters were internalized by macrophages even in the absence of Mincle (Mincle$^{-}$/$^{-}$BMMs). These results suggest that the endocytic uptake of BODIPY-labeled trehalose glycolipids occurs independently of Mincle, although Mincle is essential for their immunological activation in BMMs.

In 2022, it was revealed that Mincle directly binds to endogenous glycosphingolipids (GSLs), thereby mediating a novel endocytic pathway termed Mincle-mediated endocytosis (MiME) [72]. The study specifically focused on GSLs such as glucosylceramide (GlcCer), lactosylceramide (LacCer), and gangliosides GM3, GD3, and GM1. To evaluate these glycosphingolipids as potential ligands of Mincle, surface plasmon resonance (SPR) analysis was conducted using recombinant human Mincle immobilized on a sensor chip. The results demonstrated direct binding of GlcCer, LacCer, GM3, GD3, and GM1 to Mincle. Among these, lysoglucosylceramide was identified as the minimal structural motif necessary for Mincle recognition. However, lysoglucosylceramide dissociated rapidly from Mincle, whereas diacylated glycolipids showed significantly slower dissociation kinetics, indicating that the acyl chain composition strongly contributes to binding stability. Fluorescence-labeled GSLs were then employed to examine whether these

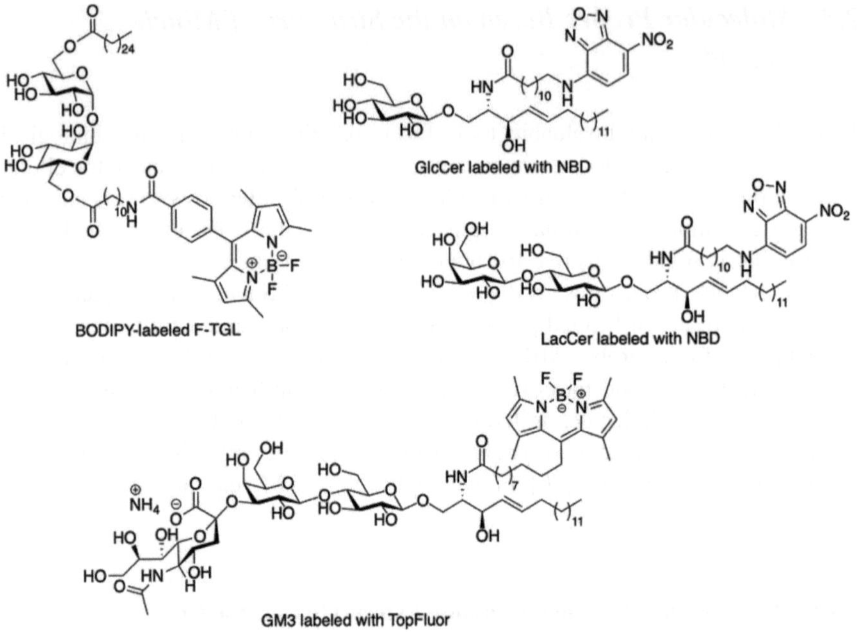


**Fig. 7** The structures of BODIPY-labeled F-TGL and fluorescence-labeled GlcCer, LacCer, and GM3

glycolipids colocalize with Mincle in cells. Specifically, GlcCer and LacCer labeled with 4-nitrobenzo-2-oxa-1,3-diazole (NBD), and GM3 labeled with TopFluor were used (Fig. 7). Upon incubation of these glycolipids with human umbilical vein endothelial cells (HUVECs), followed by labeling with anti-Mincle antibodies, structured illumination superresolution microscopy (SIM) imaging demonstrated that Mincle was found to localize together with fluorescent glucosylceramide, lactosylceramide, and GM3. Superresolution imaging showed that fluorescent GSLs were mainly localized in Mincle-positive vesicles within 1 h, suggesting lipid transfer via these vesicles. To further validate the functional role of Mincle in lipid uptake, uptake experiments were performed using brain-derived endothelial cells from wild-type and Mincle-deficient (Mincle KO) mice. TopFluor-labeled GM3 was added to the cells, and after incubation, the cells were fixed and analyzed by confocal microscopy. Quantification of intracellular fluorescence revealed that GM3 uptake was significantly reduced, with up to a sixfold decrease observed in Mincle KO cells. These results confirmed that Mincle is essential for the endocytosis of GM3.

Recently, our group developed two types of fluorescent molecular probes for the Mincle system. One probe, *N*-NBD-labeled trehalose diester (TDE) was used to track ligand internalization, while the other, *O*-NBD-labeled TDE, covalently labels the Mincle receptor [73]. The *O*-NBD probe was designed based on a turn-on fluorescence strategy [74], in which nucleophilic lysine residues on the protein surface attack the *O*-NBD moiety to form a covalent *N*-NBD linkage, along with the incorporation of a ω-hydroxy fatty acid prepared using a $C(sp^3)$–$C(sp^3)$ cross-coupling reaction [75, 76]. Using these molecular probes, we demonstrated that trehalose diester-type ligands are selectively internalized into cells in a Mincle-dependent manner and that intracellular Mincle can be specifically labeled.

# 3 The Ligands for Dendritic Cell Immunoactivating Receptor (DCAR)

This section provides an overview of the glycolipid ligands of C-type lectin receptor DCAR. Synthetic approaches have enabled the identification of the minimal and most potent ligand structures, facilitating detailed analysis of their binding affinities and immunological functions.

## *3.1 DCAR*

Dendritic cell immunoactivating receptor (DCAR) is a member of the CLR family that functions as a novel activating immunoreceptor expressed on mouse dendritic cells (DCs). Initially reported by Kanazawa et al. in 2003, DCAR is a type II transmembrane protein, and its association with the FcRγ is mediated in a non-covalent manner via a positively charged arginine residue located within its transmembrane domain [77]. This association enables FcRγ to mediate intracellular signaling. Co-immunoprecipitation and flow cytometry experiments demonstrated that DCAR is poorly expressed on the cell surface when expressed alone, but is robustly expressed in the presence of FcRγ. Functional studies using FcRγ–DCAR chimeras showed that receptor cross-linking induces intracellular calcium mobilization and tyrosine phosphorylation in an FcRγ-dependent manner, and that these activating signals are completely abrogated when the ITAM of FcRγ is mutated. DCAR is structurally related to the inhibitory CLR DCIR (dendritic cell immunoreceptor), sharing a high degree of sequence homology, with 91% amino acid identity within the carbohydrate recognition domain (CRD) [77, 78]. DCIR contains an immunoreceptor tyrosine-based inhibitory motif (ITIM) in its cytoplasmic tail and transduces inhibitory signals via SHP-1/2 phosphatases [79]. Given their structural similarity but opposing functions, DCAR and DCIR are proposed to constitute a paired receptor system that regulates immune activation and homeostasis [78].

To date, phosphatidylinositol mannosides (PIMs) and *H. pylori*-derived cholesteryl phosphatidyl-α-glucoside (αCPG) [62] have been reported as ligands. In the following section, we focus on PIMs.

### 3.2 Phosphatidylinositol Mannosides (PIMs)

Although the physiological ligands for DCAR (gene symbol Clec4b1) had remained unidentified for many years, phosphatidylinositol mannosides (PIMs) were identified in 2016 as direct ligands for DCAR [80]. In this study, mycobacterial metabolites were investigated as candidate ligands, and the ligand recognition capacity of DCAR was evaluated using reporter cells co-expressing DCAR and FcRγ. These reporter cells showed no response to stimulation with known mycobacterial CLR ligands, such as TDM or lipoarabinomannan (LAM), but responded specifically to intact mycobacteria and their lipid extracts, suggesting that DCAR recognizes distinct lipid species. Subsequent fractionation of mycobacterial lipid extracts by high-performance thin-layer chromatography (HPTLC), followed by analysis of fractions containing both carbohydrate and phosphate moieties, identified acylated $PIM_2$ ($AcPIM_2$) (Fig. 8) and diacylated $PIM_2$ ($Ac_2PIM_2$) as DCAR ligands. Structurally, $AcPIM_2$ consists of a phosphatidyl-myo-inositol backbone bearing two mannose residues ($PIM_2$) and a diacylglycerol moiety, and is further modified by a single fatty acyl chain. In contrast, $Ac_2PIM_2$ carries two fatty acyl chains. Upon stimulation with these PIMs, resident peritoneal exudate cells (PECs), including small peritoneal macrophages (SPMs) produced inflammatory chemokines such as MCP-1 and MIP-2, whereas this response was markedly diminished in Clec4b1-deficient mice.

Several synthetic methods for PIMs have been reported [81]. In our group, we have successfully synthesized $Ac_1PIM_1$, a potential biosynthetic intermediate of complex PIMs derived from *Mycobacterium tuberculosis*, and demonstrated that $Ac_1PIM_1$ functions as a potent ligand for DCAR (Fig. 8) [82]. $Ac_1PIM_1$ consists of a phosphatidyl-*myo*-inositol core linked to a single mannose residue, bearing a diacylglycerol moiety and an additional acyl group attached to the 6-position of the mannose. A defining structural difference is that $Ac_2PIM_2$, which is based on the $Ac_1PIM_1$ scaffold, contains an additional acyl chain and an extra mannose residue

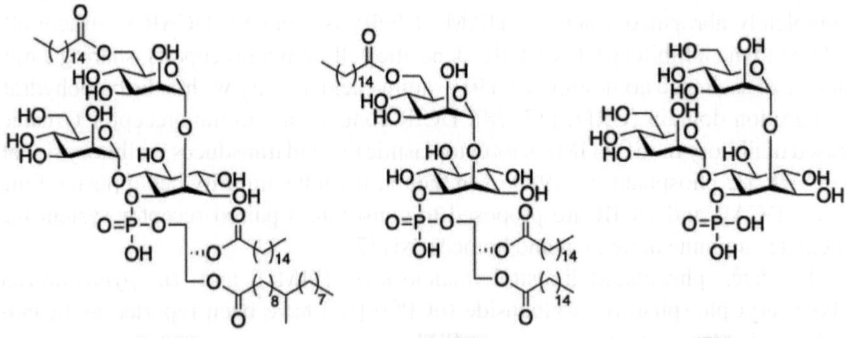

$Ac_1PIM_2$ $Ac_1PIM_1$ $IPM_2$

**Fig. 8** The structures of $Ac_1PIM_2$, $Ac_1PIM_1$, and $IPM_2$

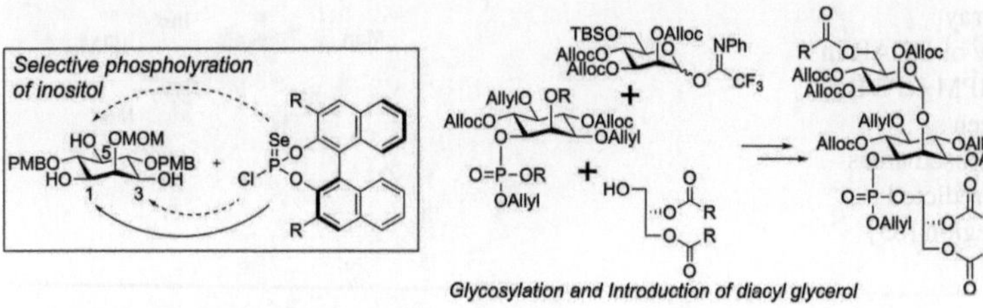


**Scheme 3** The synthesis of $Ac_1PIM_1$

attached to the inositol moiety. Thus, $Ac_1PIM_1$ is structurally simpler than $Ac_2PIM_2$ in both lipid and glycan components.

$Ac_1PIM_1$ was successfully synthesized from *myo*-inositol using a synthetic strategy involving regio- and stereo-selective phosphorylation with a chiral auxiliary-containing phosphorylating reagent (Scheme 3) [75, 83, 84]. Subsequent glycosylation using a mannose donor and construction of the diacylglycerol moiety completed the synthesis of $Ac_1PIM_1$. Importantly, our results demonstrated that $Ac_1PIM_1$ exhibited more potent agonistic activity toward DCAR than $Ac_1PIM_2$. These results suggested that $Ac_1PIM_1$ represents the minimal essential structural unit for DCAR-mediated signaling activity. These insights provided a key example of structure–activity relationships (SAR) in the context of PIM recognition by innate immune receptors.

On the other hand, following the identification of PIMs as physiological ligands for DCAR, structural and mutational studies elucidated the molecular basis underlying this interaction [85]. The co-crystal structure with a water-soluble analog ($IPM_2$), which mimics the polar headgroup of AcPIMs, has revealed $Ca^{2+}$-dependent coordination with the 3- and 4-hydroxyl groups of mannose (Fig. 9). Comparison of the ligand-free and $IPM_2$-bound DCAR structures revealed no significant conformational changes upon ligand binding. Notably, a hydrophobic region adjacent to the carbohydrate recognition site was identified in the DCAR structure, positioned to potentially accommodate the lipid chains of AcPIMs, thereby contributing to the overall stability and specificity of ligand recognition. These findings suggested that the glycan, phosphate, and lipid moieties are critical for DCAR recognition.

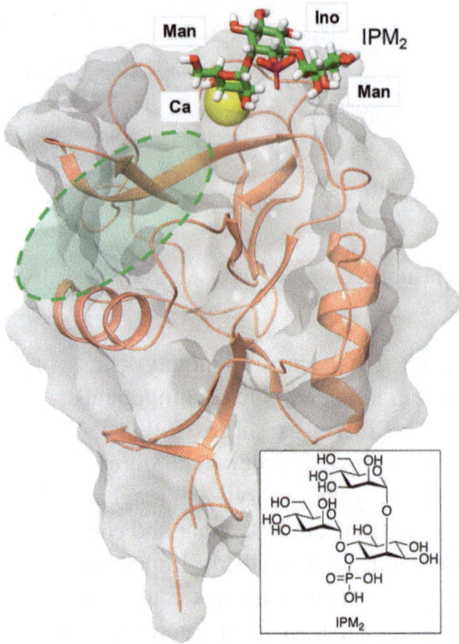


**Fig. 9** The X-ray crystallography of DCAR in complex with $IPM_2$ (PDB: 6LFJ). The green dotted-line-enclosed areas indicated the predicted hydrophobic region [85]

## 4 Conclusion

This review highlights representative and recently developed synthetic ligands and molecular probes for the glycolipid-recognizing C-type lectin receptors Mincle and DCAR. These chemical tools have profoundly advanced our understanding of receptor–ligand interactions and intracellular dynamics. In particular, the development of structurally diverse ligands has enabled detailed structure–activity relationship (SAR) studies, revealing how specific molecular features govern receptor specificity and signaling potency. Additionally, fluorescence-labeled molecular probes have proven indispensable for live-cell imaging and detailed analysis of receptor trafficking. The continued expansion and refinement of this chemical toolbox are expected to deepen mechanistic insights into innate immune recognition and to drive the development of innovative immunomodulators and functional probes, ultimately leading to the discovery of previously unrecognized aspects of immune regulation.

## References

1. Zelensky AN, Gready JE (2005) The C-type lectin-like domain superfamily. FEBS J 272(24):6179–6217. https://doi.org/10.1111/j.1742-4658.2005.05031.x
2. Osorio F, Reis e Sousa C (2011) Myeloid C-type lectin receptors in pathogen recognition and host defense. Immunity 34(5):651–64. https://doi.org/10.1016/j.immuni.2011.05.001

3. Sancho D, Reis e Sousa C (2012) Signaling by myeloid C-type lectin receptors in immunity and homeostasis. Annu Rev Immunol 30:491–529. https://doi.org/10.1146/annurev-immunol-031210-101352
4. Geijtenbeek TB, Gringhuis SI (2009) Signalling through C-type lectin receptors: shaping immune responses. Nat Rev Immunol 9(7):465–479. https://doi.org/10.1038/nri2569
5. Drummond RA, Saijo S, Iwakura Y, Brown GD (2011) The role of Syk/CARD9 coupled C-type lectins in antifungal immunity. Eur J Immunol 41(2):276–281. https://doi.org/10.1002/eji.201041252
6. Fischer S, Stegmann F, Gnanapragassam VS, Lepenies B (2022) From structure to function—ligand recognition by myeloid C-type lectin receptors. Comput Struct Biotechnol J 20:5790–5812. https://doi.org/10.1016/j.csbj.2022.10.019
7. Reis ESC, Yamasaki S, Brown GD (2024) Myeloid C-type lectin receptors in innate immune recognition. Immunity 57(4):700–717. https://doi.org/10.1016/j.immuni.2024.03.005
8. Ishikawa E, Ishikawa T, Morita YS, Toyonaga K, Yamada H, Takeuchi O et al (2009) Direct recognition of the mycobacterial glycolipid, trehalose dimycolate, by C-type lectin Mincle. J Exp Med 206(13):2879–2888. https://doi.org/10.1084/jem.20091750
9. Williams SJ (2017) Sensing lipids with mincle: structure and function. Front Immunol 8:1662. https://doi.org/10.3389/fimmu.2017.01662
10. Braganza CD, Teunissen T, Timmer MSM, Stocker BL (2017) Identification and biological activity of synthetic macrophage inducible C-type lectin ligands. Front Immunol 8:1940. https://doi.org/10.3389/fimmu.2017.01940
11. Lu X, Nagata M, Yamasaki S (2018) Mincle: 20 years of a versatile sensor of insults. Int Immunol 30(6):233–239. https://doi.org/10.1093/intimm/dxy028
12. Weth AF, Dangerfield EM, Timmer MSM, Stocker BL (2024) Recent advances in the development of mincle-targeting vaccine adjuvants. Vaccines 12(12). https://doi.org/10.3390/vaccines12121320
13. Yamasaki S, Ishikawa E, Sakuma M, Hara H, Ogata K, Saito T (2008) Mincle is an ITAM-coupled activating receptor that senses damaged cells. Nat Immunol 9(10):1179–1188. https://doi.org/10.1038/ni.1651
14. Schoenen H, Bodendorfer B, Hitchens K, Manzanero S, Werninghaus K, Nimmerjahn F et al (2010) Cutting edge: mincle is essential for recognition and adjuvanticity of the mycobacterial cord factor and its synthetic analog trehalose-dibehenate. J Immunol 184(6):2756–2760. https://doi.org/10.4049/jimmunol.0904013
15. Matsumaru T (2022) Lipid conjugates as ligands for the C-type lectin receptor mincle. Trends Glycosci Glycotechnol 34(199):E55–E59. https://doi.org/10.4052/tigg.2029.1E
16. Furukawa A, Kamishikiryo J, Mori D, Toyonaga K, Okabe Y, Toji A et al (2013) Structural analysis for glycolipid recognition by the C-type lectins Mincle and MCL. Proc Natl Acad Sci U S A 110(43):17438–17443. https://doi.org/10.1073/pnas.1312649110
17. Feinberg H, Jegouzo SA, Rowntree TJ, Guan Y, Brash MA, Taylor ME et al (2013) Mechanism for recognition of an unusual mycobacterial glycolipid by the macrophage receptor mincle. J Biol Chem 288(40):28457–28465. https://doi.org/10.1074/jbc.M113.497149
18. Jegouzo SA, Harding EC, Acton O, Rex MJ, Fadden AJ, Taylor ME et al (2014) Defining the conformation of human mincle that interacts with mycobacterial trehalose dimycolate. Glycobiology 24(12):1291–1300. https://doi.org/10.1093/glycob/cwu072
19. Feinberg H, Rambaruth ND, Jegouzo SA, Jacobsen KM, Djurhuus R, Poulsen TB et al (2016) Binding sites for acylated trehalose analogs of glycolipid ligands on an extended carbohydrate recognition domain of the macrophage receptor Mincle. J Biol Chem 291(40):21222–21233. https://doi.org/10.1074/jbc.M116.749515
20. Furukawa A, Shuchi Y, Wang J, Guillen-Poza PA, Ishizuka S, Kagoshima M, et al (2023) Structural basis for plastic glycolipid recognition of the C-type lectin Mincle. Structure 31(9):1077–85 e5. https://doi.org/10.1016/j.str.2023.05.018
21. Furukawa A, Kumeta H, Saitoh T, Maenaka K (2024) Protocol to identify the ligand binding site of Mincle using NMR spectroscopy. STAR Protoc 5(2):102996. https://doi.org/10.1016/j.xpro.2024.102996

22. Kato M, Tamura T, Silve G, Asselineau J (1978) Chemical structure and biochemical activity of cord factor analogs. A comparative study of esters of methyl glucoside and non-hydroxylated fatty acids. Eur J Biochem. 87(3):497–504. https://doi.org/10.1111/j.1432-1033.1978.tb12400.x
23. Ryll R, Kumazawa Y, Yano I (2001) Immunological properties of trehalose dimycolate (cord factor) and other mycolic acid-containing glycolipids—a review. Microbiol Immunol 45(12):801–811. https://doi.org/10.1111/j.1348-0421.2001.tb01319.x
24. Asselineau J, Lederer E (1950) Structure of the mycolic acids of Mycobacteria. Nature 166(4227):782–783. https://doi.org/10.1038/166782a0
25. Barry CE 3rd, Lee RE, Mdluli K, Sampson AE, Schroeder BG, Slayden RA et al (1998) Mycolic acids: structure, biosynthesis and physiological functions. Prog Lipid Res 37(2–3):143–179. https://doi.org/10.1016/s0163-7827(98)00008-3
26. Al Dulayymi JR, Baird MS, Maza-Iglesias M, Vander Beken S, Grooten J (2009) The first unique synthetic mycobacterial cord factors. Tetrahedron Lett 50(26):3702–3705. https://doi.org/10.1016/j.tetlet.2009.03.213
27. Taher SG, Muzael M, Al Dulayymi JR, Baird MS (2015) Synthetic trehalose esters of cis-alkene and diene alpha'-mycolic acids of Mycobacteria. Chem Phys Lipids 189:28–38. https://doi.org/10.1016/j.chemphyslip.2015.05.004
28. Al Dulayymi JaR, Baird MS, Maza-Iglesias M, Hameed RaT, Baols KS, Muzael M et al (2014) Synthetic trehalose di- and mono-esters of α-, methoxy- and keto-mycolic acids. Tetrahedron 70(52):9836–9852. https://doi.org/10.1016/j.tet.2014.10.072
29. van der Peet PL, Gunawan C, Torigoe S, Yamasaki S, Williams SJ (2015) Corynomycolic acid-containing glycolipids signal through the pattern recognition receptor Mincle. Chem Commun (Camb) 51(24):5100–5103. https://doi.org/10.1039/c5cc00085h
30. Bird JH, Khan AA, Nishimura N, Yamasaki S, Timmer MSM, Stocker BL (2018) Synthesis of branched trehalose glycolipids and their Mincle agonist activity. J Org Chem 83(15):7593–7605. https://doi.org/10.1021/acs.joc.7b03269
31. Smith AJ, Miller SM, Buhl C, Child R, Whitacre M, Schoener R et al (2019) Species-specific structural requirements of alpha-branched trehalose diester Mincle agonists. Front Immunol 10:338. https://doi.org/10.3389/fimmu.2019.00338
32. Khan A, Kodar K, Timmer MSM, Stocker BL (2018) Lipid length and ISO-branching of trehalose diesters influences Mincle agonist activity. Tetrahedron 74(12):1269–1277. https://doi.org/10.1016/j.tet.2017.11.076
33. Stocker BL, Khan AA, Chee SH, Kamena F, Timmer MS (2014) On one leg: trehalose monoesters activate macrophages in a Mincle-dependant manner. ChemBioChem 15(3):382–388. https://doi.org/10.1002/cbic.201300674
34. Huber A, Kallerup RS, Korsholm KS, Franzyk H, Lepenies B, Christensen D et al (2016) Trehalose diester glycolipids are superior to the monoesters in binding to Mincle, activation of macrophages in vitro and adjuvant activity in vivo. Innate Immun 22(6):405–418. https://doi.org/10.1177/1753425916651132
35. Matsumaru T, Sueyoshi K, Okubo K, Fujii S, Sakuratani K, Saito R et al (2022) Trehalose diesters containing a polar functional group-modified lipid moiety: Synthesis and evaluation of Mincle-mediated signaling activity. Bioorg Med Chem 75:117045. https://doi.org/10.1016/j.bmc.2022.117045
36. Ishikawa T, Itoh F, Yoshida S, Saijo S, Matsuzawa T, Gonoi T et al (2013) Identification of distinct ligands for the C-type lectin receptors Mincle and Dectin-2 in the pathogenic fungus Malassezia. Cell Host Microbe 13(4):477–488. https://doi.org/10.1016/j.chom.2013.03.008
37. van der Peet PL, Gunawan C, Watanabe M, Yamasaki S, Williams SJ (2019) Synthetic beta-1,2-mannosyloxymannitol glycolipid from the fungus Malassezia pachydermatis signals through human Mincle. J Org Chem 84(11):6788–6797. https://doi.org/10.1021/acs.joc.9b00544
38. Matsumaru T, Sakuratani K, Yanaka S, Kato K, Yamasaki S, Fujimoto Y (2022) Fungal β-mannosyloxymannitol glycolipids and their analogues: synthesis and Mincle-mediated signaling activity. Eur J Org Chem 2022(20). https://doi.org/10.1002/ejoc.202200109

39. Nemoto H, Tsutsumi H, Yuzawa S, Peng X, Zhong W, Xie J et al (2004) Highly efficient chiral resolution and determination of absolute configuration of 2-alkanols by using a cyclopenta[b]furan derivative. Tetrahedron Lett 45(8):1667–1670. https://doi.org/10.1016/j.tetlet.2003.12.104
40. Van Huy L, Tanaka C, Imai T, Yamasaki S, Miyamoto T (2019) Synthesis of 12-O-Mono- and Diglycosyl-oxystearates, a new class of agonists for the C-type lectin receptor Mincle. ACS Med Chem Lett 10(1):44–49. https://doi.org/10.1021/acsmedchemlett.8b00413
41. Nagata M, Izumi Y, Ishikawa E, Kiyotake R, Doi R, Iwai S et al (2017) Intracellular metabolite beta-glucosylceramide is an endogenous Mincle ligand possessing immunostimulatory activity. Proc Natl Acad Sci U S A. 114(16):E3285-E94. https://doi.org/10.1073/pnas.1618133114
42. Sharma A, Steichen AL, Jondle CN, Mishra BB, Sharma J (2014) Protective role of Mincle in bacterial pneumonia by regulation of neutrophil mediated phagocytosis and extracellular trap formation. J Infect Dis 209(11):1837–1846. https://doi.org/10.1093/infdis/jit820
43. Sharma A, Chauhan A, Chauhan P, Evans DL, Szlabick RE, Aaland MO et al (2022) Glycolipid metabolite beta-glucosylceramide is a neutrophil extracellular trap-inducing ligand of Mincle released during bacterial infection and inflammation. J Immunol 209(2):391–400. https://doi.org/10.4049/jimmunol.2100855
44. Ueki K, Nozawa R, Matsumaru T, Yamasaki S, Fujimoto Y (2025) Lipid structure-dependent CD1d functional stabilization and immunomodulation of endogenous glucosylceramides. Chem Commun (Camb) 61(27):5154–5157. https://doi.org/10.1039/d5cc00003c
45. Ikazaki T, Ishikawa E, Tamashima H, Akiyama H, Kimuro Y, Yoritate M et al (2023) Ligand-controlled stereoselective synthesis and biological activity of 2-exomethylene pseudo-glycoconjugates: discovery of Mincle-selective ligands. Angew Chem Int Ed Engl 62(22):e202302569. https://doi.org/10.1002/anie.202302569
46. Jacobsen KM, Keiding UB, Clement LL, Schaffert ES, Rambaruth ND, Johannsen M et al (2015) The natural product brartemicin is a high affinity ligand for the carbohydrate-recognition domain of the macrophage receptor mincle. Medchemcomm. 6(4):647–652. https://doi.org/10.1039/c4md00512k
47. Foster AJ, Nagata M, Lu X, Lynch AT, Omahdi Z, Ishikawa E et al (2018) Lipidated Brartemicin analogues are potent Th1-stimulating vaccine adjuvants. J Med Chem 61(3):1045–1060. https://doi.org/10.1021/acs.jmedchem.7b01468
48. Foster AJ, Kodar K, Timmer MSM, Stocker BL (2020) Ortho-substituted lipidated Brartemicin derivative shows promising Mincle-mediated adjuvant activity. Org Biomol Chem 18(6):1095–1103. https://doi.org/10.1039/c9ob02397f
49. Riel AMS, Rungelrath V, Elwaie TA, Rasheed OK, Hicks L, Ettenger G et al (2024) Systematic evaluation of regiochemistry and lipidation of Aryl Trehalose Mincle Agonists. Int J Mol Sci 25(18). https://doi.org/10.3390/ijms251810031
50. Ryter KT, Ettenger G, Rasheed OK, Buhl C, Child R, Miller SM et al (2020) Aryl trehalose derivatives as vaccine adjuvants for mycobacterium tuberculosis. J Med Chem 63(1):309–320. https://doi.org/10.1021/acs.jmedchem.9b01598
51. Rasheed OK, Ettenger G, Buhl C, Child R, Miller SM, Evans JT et al (2020) 6,6’-Aryl trehalose analogs as potential Mincle ligands. Bioorg Med Chem 28(14):115564. https://doi.org/10.1016/j.bmc.2020.115564
52. Rasheed OK, Buhl C, Evans JT, Ryter KT (2021) Design of trehalose-based amide/sulfonamide C-type lectin receptor signaling compounds. ChemMedChem 16(8):1246–1251. https://doi.org/10.1002/cmdc.202000775
53. Foster MJ, Dangerfield EM, Timmer MSM, Stocker BL, Wilkinson BL (2024) Probing isosteric replacement for immunoadjuvant design: Bis-Aryl triazole trehalolipids are Mincle agonists. ACS Med Chem Lett 15(6):899–905. https://doi.org/10.1021/acsmedchemlett.4c00100
54. Manthrirathna M, Dangerfield EM, Ishizuka S, Woods A, Luong BS, Yamasaki S et al (2022) Water-soluble trehalose glycolipids show superior Mincle binding and signaling but impaired phagocytosis and IL-1beta production. Front Mol Biosci 9:1015210. https://doi.org/10.3389/fmolb.2022.1015210

55. Reinink P, Buter J, Mishra VK, Ishikawa E, Cheng TY, Willemsen PTJ et al (2019) Discovery of Salmonella trehalose phospholipids reveals functional convergence with mycobacteria. J Exp Med 216(4):757–771. https://doi.org/10.1084/jem.20181812
56. Minnikin DE, Dobson G, Sesardic D, Ridell M (1985) Mycolipenates and mycolipanolates of trehalose from mycobacterium tuberculosis. J Gen Microbiol 131(6):1369–1374. https://doi.org/10.1099/00221287-131-6-1369
57. Daffe M, Lacave C, Laneelle MA, Gillois M, Laneelle G (1988) Polyphthienoyl trehalose, glycolipids specific for virulent strains of the tubercle bacillus. Eur J Biochem 172(3):579–584. https://doi.org/10.1111/j.1432-1033.1988.tb13928.x
58. Besra GS, Bolton RC, McNeil MR, Ridell M, Simpson KE, Glushka J et al (1992) Structural elucidation of a novel family of acyltrehaloses from *Mycobacterium tuberculosis*. Biochemistry 31(40):9832–9837. https://doi.org/10.1021/bi00155a040
59. Holzheimer M, Reijneveld JF, Ramnarine AK, Misiakos G, Young DC, Ishikawa E et al (2020) Asymmetric total synthesis of mycobacterial Diacyl Trehaloses demonstrates a role for lipid structure in immunogenicity. ACS Chem Biol 15(7):1835–1841. https://doi.org/10.1021/acschembio.0c00030
60. Hirai Y, Haque M, Yoshida T, Yokota K, Yasuda T, Oguma K (1995) Unique cholesteryl glucosides in Helicobacter pylori: composition and structural analysis. J Bacteriol 177(18):5327–5333. https://doi.org/10.1128/jb.177.18.5327-5333.1995
61. Chang YJ, Kim HY, Albacker LA, Lee HH, Baumgarth N, Akira S et al (2011) Influenza infection in suckling mice expands an NKT cell subset that protects against airway hyperreactivity. J Clin Invest 121(1):57–69. https://doi.org/10.1172/JCI44845
62. Nagata M, Toyonaga K, Ishikawa E, Haji S, Okahashi N, Takahashi M et al (2021) Helicobacter pylori metabolites exacerbate gastritis through C-type lectin receptors. J Exp Med 218(1). https://doi.org/10.1084/jem.20200815
63. Smith DGM, Ito E, Yamasaki S, Williams SJ (2020) Cholesteryl 6-O-acyl-alpha-glucosides from diverse Helicobacter spp. signal through the C-type lectin receptor Mincle. Org Biomol Chem 18(39):7907–7915. https://doi.org/10.1039/d0ob01776k
64. Timmer MSM, Teunissen TJ, Kodar K, Foster AJ, Yamasaki S, Stocker BL (2021) Cholesteryl glucosides signal through the carbohydrate recognition domain of the macrophage inducible C-type lectin (mincle). Org Biomol Chem 19(10):2198–2202. https://doi.org/10.1039/d0ob02342f
65. Nguyen T, Hosono Y, Shimizu T, Yamasaki S, Williams SJ (2020) Candida albicans steryl 6-O-acyl-alpha-D-mannosides agonize signalling through Mincle. Chem Commun (Camb) 56(95):15060–15063. https://doi.org/10.1039/d0cc06263d
66. Hunter SW, Brennan PJ (1981) A novel phenolic glycolipid from *Mycobacterium leprae* possibly involved in immunogenicity and pathogenicity. J Bacteriol 147(3):728–735. https://doi.org/10.1128/jb.147.3.728-735.1981
67. Ishizuka S, van Dijk JHM, Kawakita T, Miyamoto Y, Maeda Y, Goto M et al (2023) PGL-III, a rare intermediate of mycobacterium leprae phenolic glycolipid biosynthesis, is a potent mincle ligand. ACS Cent Sci 9(7):1388–1399. https://doi.org/10.1021/acscentsci.3c00040
68. Oka S, Watanabe M, Ito E, Takeyama A, Matsuoka T, Takahashi M et al (2023) Archaeal glycerolipids are recognized by C-type lectin receptor mincle. J Am Chem Soc 145(33):18538–18548. https://doi.org/10.1021/jacs.3c05473
69. Hagenstein MC, Mussgnug JH, Lotte K, Plessow R, Brockhinke A, Kruse O et al (2003) Affinity-based tagging of protein families with reversible inhibitors: a concept for functional proteomics. Angew Chem Int Ed Engl 42(45):5635–5638. https://doi.org/10.1002/anie.200352084
70. Khan AA, Kamena F, Timmer MS, Stocker BL (2013) Development of a benzophenone and alkyne functionalised trehalose probe to study trehalose dimycolate binding proteins. Org Biomol Chem 11(6):881–885. https://doi.org/10.1039/c2ob27257a
71. Kodar K, Eising S, Khan AA, Steiger S, Harper JL, Timmer MS et al (2015) The uptake of trehalose glycolipids by macrophages is independent of Mincle. ChemBioChem 16(4):683–693. https://doi.org/10.1002/cbic.201402506

72. Kostarnoy AV, Gancheva PG, Kireev II, Soloviev AI, Lepenies B, Kulibin AY et al (2022) A mechanism of self-lipid endocytosis mediated by the receptor Mincle. Proc Natl Acad Sci U S A 119(30):e2120489119. https://doi.org/10.1073/pnas.2120489119
73. Matsumaru T, Okubo K, Takayama J, Hirano Y, Huo YR, Lin CC, Yamasaki S, Fujimoto Y (2025) Live-cell imaging analysis of the intracellular behavior of C-type lectin receptor Mincle and its ligand using newly developed fluorescence-labeled molecular probes based on trehalose diester. ChemRxiv. https://doi.org/10.26434/chemrxiv-2025-w7lbt
74. Yamaguchi T, Asanuma M, Nakanishi S, Saito Y, Okazaki M, Dodo K et al (2014) Turn-ON fluorescent affinity labeling using a small bifunctional O-nitrobenzoxadiazole unit. Chem Sci 5(3):1021–1029. https://doi.org/10.1039/c3sc52704b
75. Aiba T, Sato M, Umegaki D, Iwasaki T, Kambe N, Fukase K et al (2016) Regioselective phosphorylation of myo-inositol with BINOL-derived phosphoramidites and its application for protozoan lysophosphatidylinositol. Org Biomol Chem 14(28):6672–6675. https://doi.org/10.1039/c6ob01062h
76. Iwasaki T, Higashikawa K, Reddy VP, Ho WW, Fujimoto Y, Fukase K et al (2013) Nickel-butadiene catalytic system for the cross-coupling of bromoalkanoic acids with alkyl Grignard reagents: a practical and versatile method for preparing fatty acids. Chem Eur J 19(9):2956–2960. https://doi.org/10.1002/chem.201204222
77. Kanazawa N, Tashiro K, Inaba K, Miyachi Y (2003) Dendritic cell immunoactivating receptor, a novel C-type lectin immunoreceptor, acts as an activating receptor through association with Fc receptor gamma chain. J Biol Chem 278(35):32645–32652. https://doi.org/10.1074/jbc.M304226200
78. Kanazawa N (2007) Dendritic cell immunoreceptors: C-type lectin receptors for pattern-recognition and signaling on antigen-presenting cells. J Dermatol Sci 45(2):77–86. https://doi.org/10.1016/j.jdermsci.2006.09.001
79. Bates EEM, Fournier N, Garcia E, Valladeau J, Durand I, Pin J-J et al (1999) APCs express DCIR, a novel C-type lectin surface receptor containing an immunoreceptor tyrosine-based inhibitory motif. J Immunol 163(4):1973–1983. https://doi.org/10.4049/jimmunol.163.4.1973
80. Toyonaga K, Torigoe S, Motomura Y, Kamichi T, Hayashi JM, Morita YS et al (2016) C-type lectin receptor DCAR recognizes mycobacterial phosphatidyl-inositol mannosides to promote a Th1 response during infection. Immunity 45(6):1245–1257. https://doi.org/10.1016/j.immuni.2016.10.012
81. Holzheimer M, Buter J, Minnaard AJ (2021) Chemical synthesis of cell wall constituents of mycobacterium tuberculosis. Chem Rev 121(15):9554–9643. https://doi.org/10.1021/acs.chemrev.1c00043
82. Arai Y, Torigoe S, Matsumaru T, Yamasaki S, Fujimoto Y (2020) The key entity of a DCAR agonist, phosphatidylinositol mannoside Ac(1)PIM(1): its synthesis and immunomodulatory function. Org Biomol Chem 18(19):3659–3663. https://doi.org/10.1039/c9ob02724f
83. Aiba T, Suehara S, Choy SL, Maekawa Y, Lotter H, Murai T et al (2017) Employing BINOL-phosphoroselenoyl chloride for selective inositol phosphorylation and synthesis of glycosyl inositol phospholipid from entamoeba histolytica. Chem Eur J 23(34):8304–8308. https://doi.org/10.1002/chem.201701298
84. Choy SL, Bernin H, Aiba T, Bifeld E, Lender SC, Muhlenpfordt M et al (2017) Synthetic analogs of an Entamoeba histolytica glycolipid designed to combat intracellular Leishmania infection. Sci Rep 7(1):9472. https://doi.org/10.1038/s41598-017-09894-8
85. Omahdi Z, Horikawa Y, Nagae M, Toyonaga K, Imamura A, Takato K et al (2020) Structural insight into the recognition of pathogen-derived phosphoglycolipids by C-type lectin receptor DCAR. J Biol Chem 295(17):5807–5817. https://doi.org/10.1074/jbc.RA120.012491

# Multivalent Probes for Exploring Carbohydrate Binding Proteins

**Shione Kamoshita and Kaori Sakurai**

**Abstract** Carbohydrate-protein interactions play vital roles in a diverse array of biological and pathological processes. However, many carbohydrate-binding proteins remain elusive due to inherent difficulties in their identification. Designing appropriate probes with necessary functional groups for effectively capturing and enriching proteins for target analysis remains challenging. To address these problems, we have developed novel multivalent photoaffinity probes, which employ gold nanoparticles as a probe scaffold. These probes multivalently display both carbohydrate ligands and photoreactive groups. More recently, we expanded the utility of the gold-nanoparticle probes by employing electrophilic groups as alternative labeling groups to photoreactive groups. In this chapter, recent progress in the development of multivalent affinity labeling probes for exploring targets of carbohydrate ligands is highlighted, and their scopes and characteristics are discussed.

**Keywords** Multivalency · Carbohydrate ligands · Affinity labeling · Target proteins · Gold nanoparticle

## 1 Introduction

Carbohydrate-binding proteins, which include lectins, glycosidases, and glycosyltransferases are key mediators in vital biological processes, from immune response and cell signaling to host–pathogen interactions and cancer metastasis [1–5]. Characterization of new carbohydrate-binding proteins is important to better understand their physiological functions and develop new anticancer drugs, or biomarkers and diagnostic tools [6, 7]. These carbohydrate-binding proteins can recognize carbohydrate motifs with exquisite specificity, yet the interactions are often weak and

S. Kamoshita · K. Sakurai (✉)
Department of Biotechnology and Life Science, Faculty of Engineering, Tokyo University of Agriculture and Technology, Naka-cho, Koganei-shi, Tokyo, Japan
e-mail: sakuraik@cc.tuat.ac.jp

S. Kamoshita
e-mail: s256563x@st.go.tuat.ac.jp

 
K. Toshima and Y. Fujimoto (eds.), *Chemical Biology of Carbohydrates*,
https://doi.org/10.1007/978-981-95-5335-8_10

transient [8], complicating efforts to catalog and characterize them in a complex proteomic context. Traditional methods such as affinity chromatography have enhanced our understanding of carbohydrate-binding protein specificities and functions [9]. However, they face significant limitations: low binding affinities result in poor capture of weak or low abundance carbohydrate-binding proteins, and monovalent probes fail to recapitulate the multivalent glycan presentations that nature employs for high avidity interactions [10]. As a result, many protein–glycan associations are thought to remain elusive. Therefore, there is a growing demand for novel chemical tools that deliver both enhanced affinity and selectivity in complex biological environments.

In order to address these challenges, we have taken two approaches to designing new probes that have the capabilities in enhancing affinity as well as stabilizing labile protein-glycan interactions. The first approach involved the use of photoaffinity labeling, which allows for chemical stabilization of sugar–protein complexes [11–13]. Photoaffinity probes are composed of a ligand functionalized with a photoreactive group, which can be activated by UV irradiation to react with proteins at a judicious timing to form a covalently crosslinked complex in a ligand-dependent manner. Once crosslinked, the complexes are chemically stable, making the downstream biochemical analyses tractable. Aryl azides, benzophenones, and diazirines represent the major photoreactive groups used in photoaffinity labeling (Scheme 1). They generate nitrenes, ketyl diradicals, and carbenes as reactive intermediates. A choice of photoreactive groups is critical for the probe design, however, there had been no guideline for which group to be used for a given application. To compare the effects of three major photoreactive groups on the efficiency and selectivity of crosslinking a binding protein with low affinity, we designed and synthesized photoaffinity probes bearing a carbohydrate ligand and different photoreactive groups (**1**–**3**, Fig. 1) [14–16]. We found that the diazirine probe **3** displayed high ligand-dependent reactivity consistent with the ideal mechanism of photoaffinity labeling, and that it enabled highly selective crosslinking of a binding protein in cell lysate. However, the labeling yields for the diazrine probe were too low (~ 1 to 2%) to be used for target identification, which represented a general problem associated with photoreactive groups [17].

One solution to this problem is to increase the protein-binding affinity of the probes by multivalent presentation of ligands [10, 18, 19]. Therefore, as a second approach, we introduced multivalent ligand presentation using gold nanoparticles to promote affinity enhancement [20]. While conventional molecular probes are composed of only one ligand molecule and one photoreactive group, on the surface of gold nanoparticles, tens to thousands of organic molecules can be attached in high density and in a well-defined orientation. Multivalent carbohydrate affinity labeling probes fulfill the need to mimic the clustered presentation of natural glycan, driving cooperatively amplified binding via the avidity effect [10, 19]. This multivalency improves carbohydrate-binding protein capture even when individual ligand protein interactions are weak. A central consideration in designing multivalent probes lies in scaffold chemistry and linker architecture (**4**–**8**, Fig. 2) [21–25], both of which critically influence the probe's binding efficacy, reactivity, and biocompatibility.

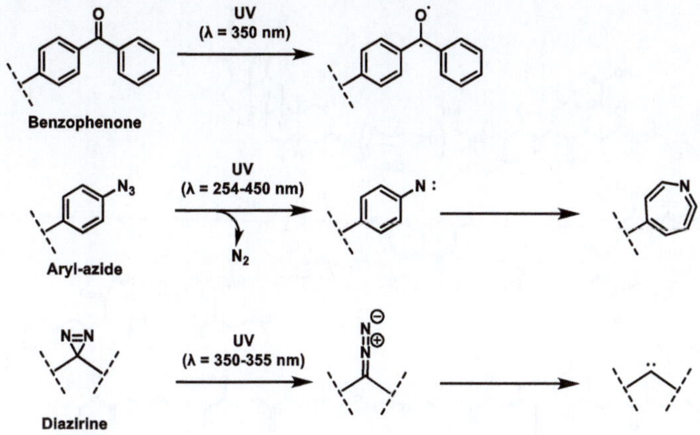


**Scheme 1** Most commonly used photoaffinity probes and the corresponding reactive intermediates

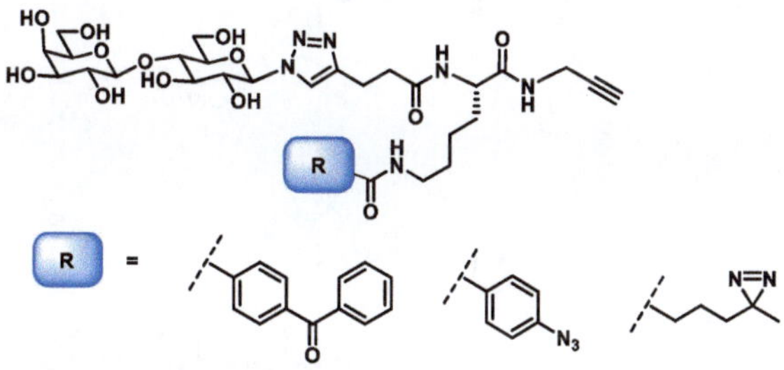


**Fig. 1** Lactose-based photoaffinity probes (**1–3**) with three different types of photoreactive groups

In this chapter, we introducenovel gold-nanoparticle based affinity labeling probes that multivalently display both carbohydrate ligands and protein reactive groups are presented as promising tools for a streamlined analysis of target proteins and discuss their scopes and characteristics.

## 2 Development of Multivalent Affinity Labeling Probes Based on Gold Nanoparticle as a Probe Scaffold [20]

Despite the resurgence of the utility of photoaffinity labeling in target identification, one major limiting aspect of photoaffinity labeling is its low labeling efficiency in general. Consequently, low-affinity or low-abundance proteins have been difficult to pursue by photoaffinity labeling-based target analysis methods. As described above, to address this outstanding problem, strategies have been devised to increase the protein binding affinity of the probes by multivalent presentation of ligands [10, 18, 19]. Y. C. Li et al. first reported a multivalent photoaffinity probe, which displays multiple carbohydrate ligands (**4**, Fig. 2) [21]. Monomeric sugar ligands are known

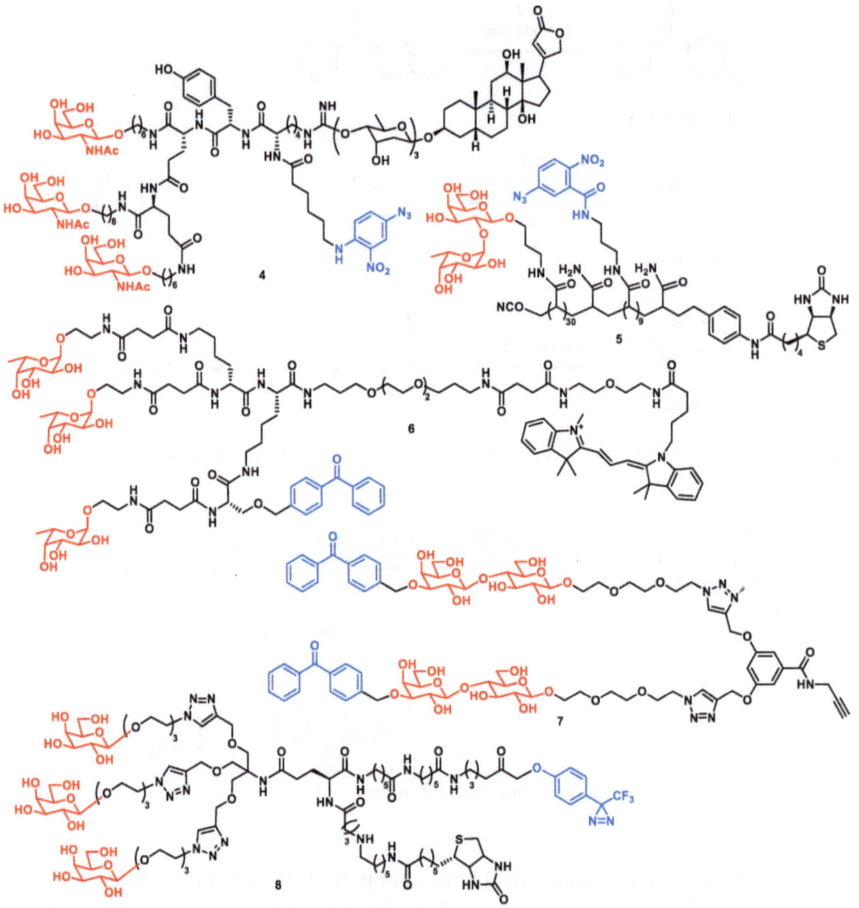

**Fig. 2** Design of various multivalent photoaffinity probes (**4**–**8**) with different scaffold structures for probing carbohydrate-protein interaction analysis

to bind their receptor proteins with typically weak affinity in the range of high μm to mm. They were able to combine the benefit of the multivalent effect for affinity enhancement and the covalent crosslinking for facile protein purification. We developed novel gold nanoparticle-based photoaffinity probes that display ligands and photoreactive groups in a multivalent fashion (Fig. 3) [20]. The gold nanoparticle was chosen as a molecular scaffold to promote high local concentrations of both the ligands and photoreactive groups, and reactions over a large surface area. We hypothesized that these features enabled our novel photoaffinity probes to exert enhanced protein binding affinity and ligand-dependent reactivity.

While conventional molecular probes are composed of only one ligand molecule and one photoreactive group, on the surface of a gold nanoparticle, tens to thousands

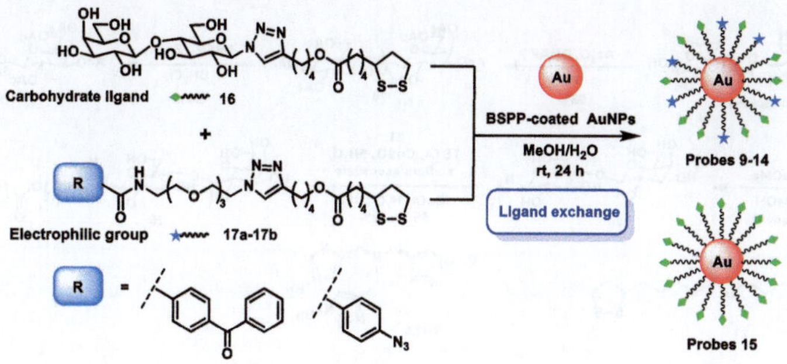


**Fig. 3** Modular assembly of gold-nanoparticle photoaffinity probes

of organic molecules can be attached in high density and in a well-defined orientation. This gives rise to the multivalent effect due to the increased local effective concentration of the ligand and the labeling group. Additional benefits of employing a gold-nanoparticle scaffold are that the functionalities can be efficiently assembled in a modular fashion, which makes it easy to optimize the probe design. Any organic molecule with a terminal thiol can be functionalized on the gold surface by a covalent Au–S bond. In addition, the functionalized gold nanoparticles are dispersible in water just like proteins. At the same time, they can be easily recovered from solution by centrifugation, making them useful as an affinity matrix (Fig. 4).

Initially, we selected lactose as a model carbohydrate ligand and two types of photoreactive groups, namely aryl azide **17a** and benzophenone **17b**, and prepared

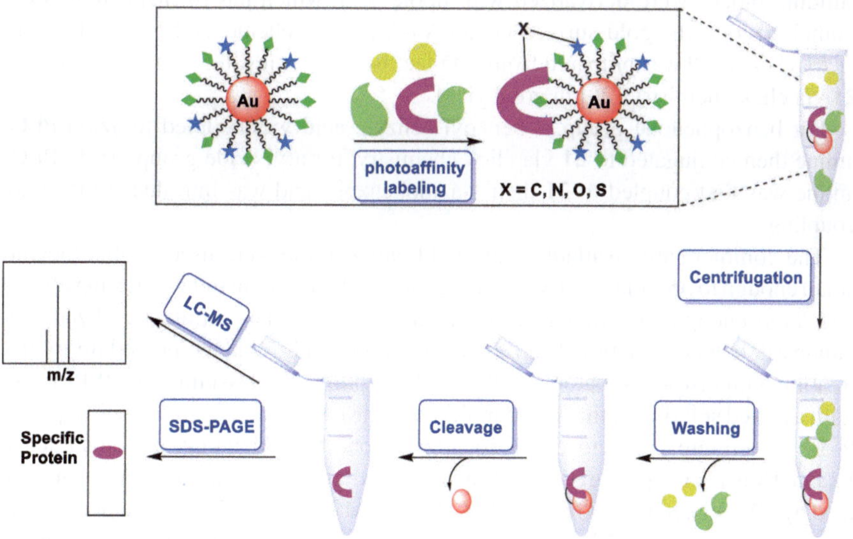


**Fig. 4** Streamlined scheme for detection and identification of carbohydrate-binding proteins using gold nanoparticle-based photoaffinity probes

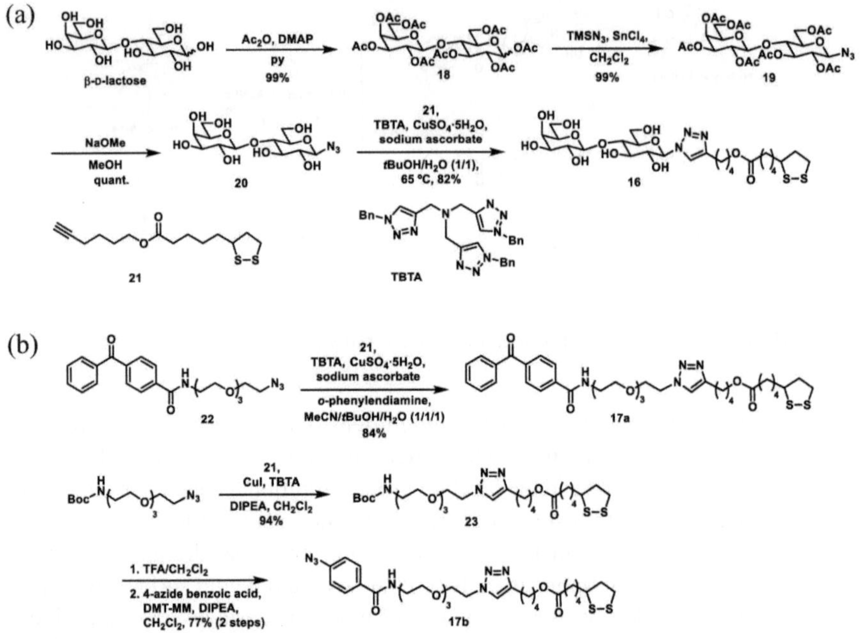

**Scheme 2** Synthesis of **a** ligand moiety **16** and **b** photoreactive moiety (**17a, 17b**) as lipoic acid derivatives

six types of probes (**9–16**) by varying the combination and their ratio. The two building blocks were derivatized with lipoic acid, which has divalent thiols to be immobilized to the gold surface via an Au–S bond (Scheme 2). Specifically, β-D-lactosyl azide **19** was prepared from β-D-lactose and conjugated to hexynoyl lipoate **21** via click chemistry [26] in good yield.

For benzophenone group, 4-benzoyl benzoic acid was coupled to azido-PEG3 amine then conjugated to **21** via click chemistry, for aryl azide group, azido-PEG3 amine was first coupled to **21** then 4-azidobenzoic acid was introduced via amide coupling.

The commercially available 5 nm gold nanoparticle were used so that they are small enough to provide good water-dispersion with minimum light scattering effects, and large enough for efficient centrifugal separation. The citrate stabilized gold nanoparticle were functionalized by a two-step ligand exchange procedure by first treating with bis(*p*-sulfonatophenyl)phenylphosphine dipotassium salt (BSPP), then with a mixed solution of the ligand moiety and the photoreactive group to give probes **9–14**. As a reference, lactose functionalized probe **15**, which only presents Lactose ligand **16** was also prepared. The functionalized gold nanoparticle were characterized by UV–Vis, SDS-PAGE and MALDI-TOF MS analysis. The ligand density was determined for probe **15** by cleaving the ligands from the gold-nanoparticle surface by thiolate exchange reaction, and then quantitated by LC–MS analysis. On average, the probe **15** contained 152 molecules of lactose moiety **9** and 70 molecules of aryl azide moiety **17b** per gold nanoparticle.

The protein binding activity of probes **9–15** were evaluated by an affinity pull-down assay. Due to their photoreactive nature, these probes (**9–14**) were not amenable to standard binding interaction analysis methods involving spectrophotometric monitoring. All the probes efficiently bound a lactose-binding lectin PNA with nanomolar affinity (Table 1), demonstrating that more than 1000-fold affinity enhancement is achieved compared to a monomeric lactose ligand [27].

Next, we evaluated the reactivity of probes **9–14** toward PNA. In a typical photoaffinity labeling reaction, the probe was incubated with a protein to bind at 4 °C for 2 h then was photoirradiated with a UV lamp ($\lambda = 365$ nm) at 0 °C for 1 h (Fig. 4). The resultant probe-labeled protein was purified by repeated cycles of centrifugation and incubation with 3 M guanidinium hydrochloride (Gdn-HCl) buffer, a strong protein denaturing agent. Upon cleavage from the gold nanoparticle by 2-mercaptoethanol, the eluted protein was analyzed by SDS-PAGE and quantified by in-gel fluorescence scanning. Among the six probes tested, probes **10** and **13** with a 2: 1 ratio of lactose moiety and a photoreactive moiety provided the optimal labeling efficiency (29% for **10**, 24% for **13**, Table 1). These data therefore showed that the ligand density, rather than the types of photoreactive groups (benzophenone or aryl azide), affected the outcomes of photoaffinity labeling. Given that the apparent binding affinity of the probes **1–6** did not vary significantly (52–67 nM, Table 1), it is suggested that the local concentration of the photoreactive group on a gold nanoparticle is important for the photoaffinity labeling efficiency.

The generality of the multivalent carbohydrate photoaffinity probes to photolabel other lactose binding proteins were investigated by using probes **10** and **13** against ECA and RCA [28, 29]. While both probes reacted with ECA with similar yields (7%), benzophenone probe **10** displayed higher labeling efficiency toward RCA (**10**: 71%, **13**: 37%, Fig. 5). Given the chemoselectivity known for benzophenone group, [30, 31] one possible reason for this observation may be that target amino acid residues of benzophenone group such as methionine and cysteine happened to be positioned closely at the probe-protein interface. Moreover, both probes more effectively reacted with RCA than with PNA, which may be explained by the higher affinity

**Table 1** PNA binding activity and photoreactivity of probe **9–14**

| Probes | Ligand design | $K_d$ (nM) | Labeling yield (%) |
|---|---|---|---|
| **9** | Lactose/benzophenone = 1/1 | 62 | 19 |
| **10** | Lactose/benzophenone = 2/1 | 63 | 29 |
| **11** | Lactose/benzophenone = 4/1 | 67 | 1 |
| **12** | Lactose/aryl azide = 1/1 | 66 | 9 |
| **13** | Lactose/aryl azide = 2/1 | 52 | 24 |
| **14** | Lactose/aryl azide = 4/1 | 64 | 7 |
| **15** | Lactose only | 71 | – |

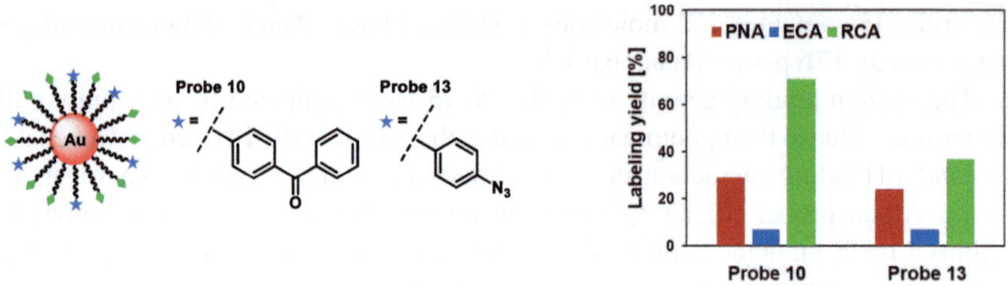


**Fig. 5** Photoaffinity labeling reactivity of probes **10** and **13** against three different lactose-binding proteins

of RCA toward D-lactose [28]. Despite the higher lactose-binding affinity than that of PNA, the labeling yields of ECA by probe **10** and **13** were lower, suggesting that the geometry of carbohydrate-binding sites on the proteins may also influence the labeling efficiency. PNA and RCA form tetramers with two carbohydrate binding sites facing the same direction [28, 32], enabling two lactose ligands on a probe to bind at the same time with possible chelate effect [10, 20], thereby promoting enhanced reactivity. ECA as a dimer have two carbohydrate binding sites facing opposite directions [29], which provides no chelate effect enhancing photolabeling.

Finally, the utility of the gold-nanoparticle photoaffinity probes was evaluated by reacting probe **10** or **13** and PNA mixed in HeLa cell lysates at a w/w ratio of 1:100 (Fig. 6). The probes were used at 50 nM, in a concentration range corresponding to their $EC_{50}$ values, to facilitate ligand-dependent reaction while minimizing nonspecific protein labeling. PNA was effectively detected by SDS-PAGE gel or by MALDI-MS only under the reaction condition involving UV irradiation, demonstrating that these probes at a nanomolar concentration can selectively react with a target protein in a complex protein mixture. Taken together, we successfully showed that probes **10** and **13** enabled highly selective crosslinking and efficient isolation of low affinity carbohydrate-binding proteins.

In summary, the development of gold nanoparticle-based multivalent photaffinity probes was motivated by a need to enhance capture sensitivity while minimizing background interference. The modularity of gold nanoparticle functionalization allows facile tuning of ligand density and photoreactive functionality as a key advantage for optimizing probe performance. The dual functionality of these probes,

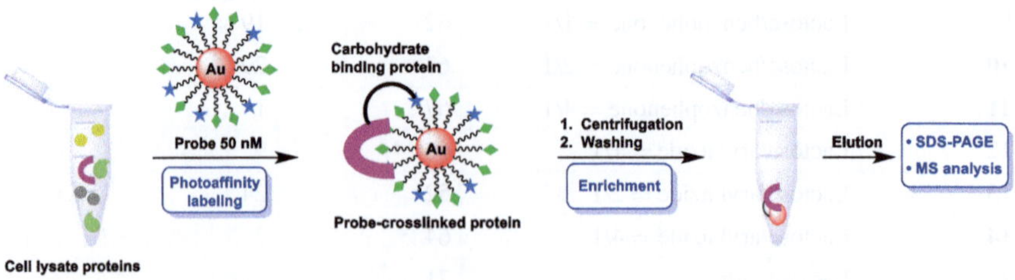


**Fig. 6** Selective photoaffinity labeling of carbohydrate-binding proteins in cell lysate

acting both as solution-phase photoaffinity labeling agents and solid-phase affinity matrix, streamlines the workflow and enables the selective, efficient isolation of low-affinity carbohydrate-binding proteins from complex proteomes. We therefore demonstrated that our novel gold nanoparticle-based probes serve dual functions by facilitating photoaffinity labeling and by directly enriching the crosslinked proteins, which allowed selective photoaffinity labeling and isolation of a low affinity carbohydrate-binding protein in cell lysate.

## 3 Development of Electrophilic Multivalent Affinity Labeling Probes [33–35]

Designing optimal probes is critical to successful affinity labeling of target proteins. Probes should consist of a high affinity ligand, a suitable protein-reactive group, and an affinity tag [18, 36]. For target identification studies, a protein-reactive group of choice is ideally capable of non-selectively reacting with any nearby amino acid residues in a ligand-dependent manner. Photoaffinity labeling has been recognized as a powerful approach, with photoreactive groups in part fulfilling these requirements. However, the low labeling yields and selectivity hampers its routine application and the type of photoreactive groups are still limited. In addition, UV light is damaging to proteins and cells, which is a concern when we conduct live cell studies. In contrast, electrophilic groups come in a wide variety of structures and reactivity, they provide an attractive set of tools for covalent drugs [37–39], affinity-based [40, 41], or reactivity-based protein profiling probes [42, 43], and ligand-directed chemistries [44]. Despite a wide ranging electrophilic groups being used in the field of synthetic organic chemistry, many of them have not been evaluated for the application to affinity labeling of unknown target proteins with no structural information or prediction. It is presumed that their spontaneous reactivity of electrophilic groups with free nucleophiles and chemoselectivity toward certain nucleophilic amino acid residues are not compatible with the required nonselective, ligand-dependent reactivity for affinity labeling. Other complicating factors determining the reactivity and selectivity of electrophile-based affinity labeling include the presence of the "preferred" nucleophilic amino acid residues, which are often dependent on the local microenvironment [45, 46]. Therefore, the generally useful electrophilic labels for target identification should overcome these issues. With the goal of expanding the scope of our gold-nanoparticle probes for target identification, we hypothesized that the multivalent presentation of a suitable electrophilic group should address these problems in combination while harnessing their favorable reactivity. First, ligand-independent, nonselective bimolecular reaction due to the inherent reactivity of electrophilic groups can be suppressed by performing affinity labeling reactions at nanomolar concentration. Secondly, through the avidity effect, broad and simultaneous screening for labeling sites should be possible in the proximal area of the ligand binding site, circumventing the chemoselectivity issue.

To test the hypothesis in a pilot study, we designed a small set of affinity labeling gold-nanoparticle probes bearing four types of electrophilic groups (Fig. 7), namely, benzylchloride (BC, **24**), squaryl (SQ, **25**), *p*-phenylsulfonyl fluoride (*p*-SF, **26**), acrylamide groups (AA, **27**) [33]. These electrophilic groups were chosen to represent distinct reaction mechanisms with different target amino acid residues (Scheme 3). Including the aryl azide (AZ) probe as a control [20], five types of multivalent probes (**23–27**) were synthesized by cofunctionalizing 13 nm gold nanoparticles with a lactose ligand and one of five different groups at a 2:1 ratio via our previously established two-step ligand exchange method [20]. To verify if the multivalent presentation promotes affinity enhancement, the $K_d$ values of lactose-functionalized gold nanoparticle **28** were measured with PNA, ECA, and RCA by the pull-down assay [20], which provided 26 nM for PNA, 8.7 nM for ECA, and 18 nM for RCA. These results clearly showed that the multivalent probe achieved a large affinity enhancement of approximately 15,000-fold compared to monomeric lactose ($K_d = 400$ mM for PNA, 323 mM for ECA, and 37 mM for RCA) [27–29].

To investigate the utility of these electrophilic gold-nanoparticle probes, we applied them to affinity labeling reactions with model target proteins, PNA, ECA and RCA, and BSA as a model nonselective protein. Probes **23–27** at 50 nM were incubated with a protein (0.4 μM as a monomer) at 4 °C for 2 h and the labeled and enriched proteins were analyzed by SDS-PAGE and fluorescence imaging. The results showed that *p*-SF probe **26** reacted with all three lectins with the highest yields among other probes (25–42%, Fig. 8a). In contrast, the yields for photoaffinity labeling by AZ probe **23** were about 16% with three lectins. Despite their known examples of squaryl group and acroyl group in chemical modification of proteins, SQ probe **25** and AA probe **27** were capable of labeling only RCA modestly at around 15% yield, while they were barely reactive toward PNA and ECA. These observations may be explained by the fact that 13 Cys are present in RCA, two of

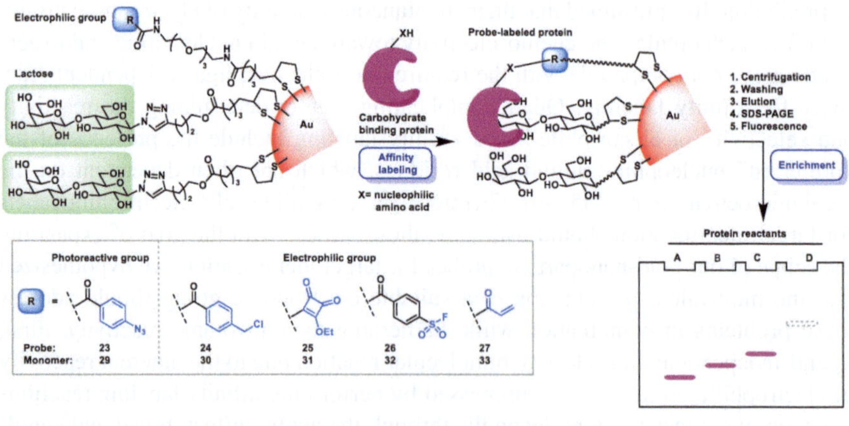


**Fig. 7** Gold-nanoparticle affinity labeling probes (**24–27**) assembled with monomeric lactose moiety (**16**) and different electrophilic moieties (**29–32**) in comparison with aryl azide photoaffinity probe (**23**) using **16** and **29**

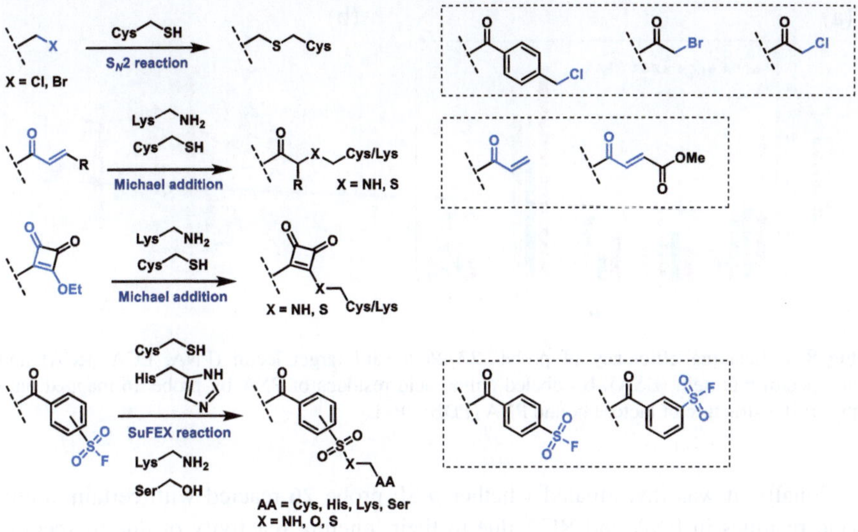


**Scheme 3** Distinct mechanisms of various electrophilic groups toward proteogenic amino acid residues

which are found around the lactose binding sites, whereas PNA and ECA have no Cys [29, 47]. SQ and AA groups are Michael acceptors and therefore, Cys may serve as their target residue to facilitate labeling of RCA by SQ probe **25** and AA probe **27**. Finally, negligible labeling was observed with BSA for all probes, suggesting that nonselective labeling by electrophilic groups, even if multivalently displayed, can be suppressed at nanomolar concentrations. Further comparative reactivity analysis found *p*-SF probe **26** to display high ligand-dependent reactivity toward PNA spiked in cell lysate. It was also found that only SF probe was able to quantitatively label PNA by extending the reaction time up to 16 h, which was a distinct feature in contrast to other probes with different labeling groups.

Using *p*-SF probe **26**, the amino acid residues labeled by the multivalent affinity labeling probes were determined. It was anticipated that due to the multivalent presentation of electrophilic groups around the ligand moiety on a gold-nanoparticle, multiple sites in the proximity of the ligand binding site could be labeled. We chose PNA and RCA and reacted with *p*-SF probe 4 °C for 16 h, and the labeled proteins were enriched and resolved by SDS-PAGE then the gel bands were excised for in-gel trypsin digestion and liquid chromatography–tandem mass spectrometry (LC–MS/MS) analysis (Fig. 8b). Several amino acid residues around the ligand binding site in PNA and RCA were detected, among which, Tyr was found as a major labeling site. This observation was consistent with previous studies employing *p*-SF groups in covalent drugs or ABPP probes [48, 48]. Lys, Asp, and Asn were also found to be labeled, indicating that a variety of amino acid residues could be targeted by *p*-SF probe **26**.

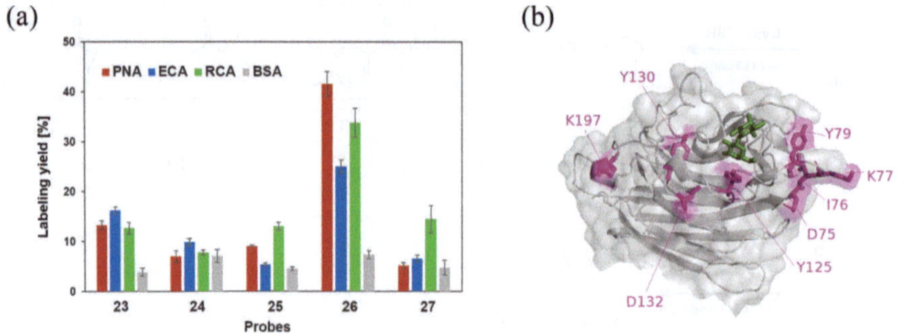


**Fig. 8** **a** Labeling efficiency of probes **23**–**26** toward target lectin (PNA, ECA, RCA) and nonspecitific protein (BSA). **b** Labeled amino acid residues of PNA by probe **26** mapped onto the crystal structure of lactose bound PNA (PDB: 2PEL)

Finally, it was investigated whether *p*-SF probe **26** reacted with certain amino acid residues in PNA and RCA due to their inherent reactivity or due to specific molecular settings at the protein-gold nanoparticle interface, we reacted probe **26** at 50 nM with *N*-acetylated monomeric amino acids at 100 mM in hydroxyethylpiperazine ethanesulfonic acid (HEPES) buffer at pH 7.3 and analyzed the time course of the formation of the crosslinked products by LC/MS. Notably, only a sulfonic acid derivative of SF moiety instead of amino acid adducts were detected, indicating that *p*-SF groups on probe **26** hydrolyzed only slowly over 48 h. It should be noted that no adducts were observed between *p*-SF group and the lactose moiety, which are expected to be the nearest neighbor molecules to each other. Therefore, these results supported our hypothesis that multivalent presentation of both the ligand and the electrophilic group effectively promote affinity labeling due to the avidity effect, providing a large reactive surface area against a target protein.

These findings prompted us to search more widely for other electrophilic groups that might be suitable for affinity labeling of carbohydrate-binding proteins [34]. In addition to SQ (**25**) and *p*-SF groups (**26**), *m*-SF, bromoacetyl (BA), chloroacetyl (CA), and monomethyl fumarate (MF) groups were functionalized onto the gold-nanoparticle probes and reacted with PNA and RCA (Fig. 9). The *m*-SF group was employed as a more selective protein labeling than the *p*-isomer (*p*-SF), having attenuated reactivity, whereas BA, CA, and MF were known warheads in covalent drugs targeting the thiol moiety of Cys at enzyme active sites. It was found that *p*- and *m*-SF are the most reactive groups toward these lectins at a nanomolar probe concentration. Between the SF isomers, *m*-SF exhibited a more selective affinity labeling while being less reactive, as evaluated by a competitive reaction condition in the presence of excess BSA as a nonspecific protein. Furthermore, evaluation of the reactivity under different reaction temperature conditions showed that the labeling efficiency of SF probes can be improved at an elevated temperature condition at 25 °C instead of 4 °C, providing a distinct advantage over photoreactive groups. These results presented *p*- and *m*-isomers of SF groups as more efficient alternative affinity labels for target identification studies groups to photoreactive groups.

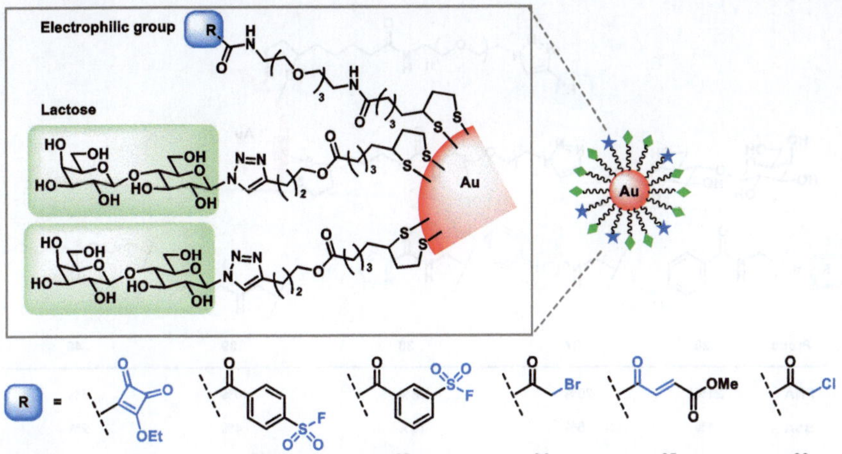


| Probe: | 25 | 26 | 33 | 34 | 35 | 36 |
|---|---|---|---|---|---|---|
| PNA | 9% | 42% | 43% | 20% | 23% | 19% |
| RCA | 8% | 47% | 31% | N.D. | N. D. | N. D. |
| BSA | 18% | 15% | 9% | N.D. | N. D. | N. D. |

**Fig. 9** Extended search for suitable electrophilic groups for gold-nanoparticle affinity labeling. Labeling yields are shown for each probe bearing a distinct electrophilic group for reactions with model target proteins (PNA and RCA) and nonspecific proteins (BSA)

We extended our exploration of electrophilic groups by presenting biphenylphosphinamide (BP), dichlorotriazine (DCT), expoxide (EP), and Arylisoxazole(AI) (Fig. 10) [35]. Known target amino acid residues for BP are mainly Lys and Ser, Thr, for DCT, Lys and Tyr, and for EP, Cys, Lys and histidine (His). Although AI is not a canonical electrophilic group, it can form a highly electrophilic intermediate acylazirine, which reacts with the carboxylic acid side chain of Asp and Glu. In the previous studies, it was observed that some of the more hydrophobic electrophilic groups enhanced the tendency of the functionalized gold nanoparticle to flocculate. Thus, to confer better dispensability in aqueous buffer conditions, a longer PEG linker was employed for the new panel of electrophilic gold nanoparticle probes. It was found that in addition to p-SF probe, DCT probe gave good labeling yield and selectivity toward PNA and PNA/BSA mixture, respectively. Labeled site analysis by MS revealed that that DCT probe reacted with several amino acid residues, not only those that are close to the ligand-binding pocket, but also distant residues. It was suggested that the broad amino acid selectivity of DCT group can be harnessed as a favorable affinity label in the context of multivalent probes, which are used at nanomolar probe conditions. Therefore, this study found DCT as an effective electrophilic group suitable for affinity labeling of a low affinity carbohydrate-binding protein.

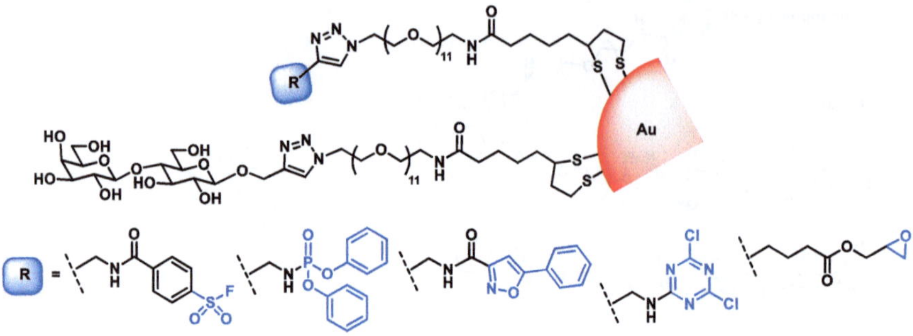


| Probe: | 26 | 37 | 38 | 39 | 40 |
|---|---|---|---|---|---|
| PNA | 21% | 26% | 28% | 40% | 6% |
| BSA | 1% | 5% | 10% | 4% | 2% |

**Fig. 10** Dichlorotriazine group was identified as an additional promising electrophilic label for gold-nanoparticle affinity labeling with PEG11 linker. Labeling yields are shown for each probe bearing a distinct electrophilic group for reactions in a mixed solution of a model target proteins (PNA) and excess nonspecific proteins (BSA)

# 4 Conclusions

To explore elusive carbohydrate-binding proteins in complex cellular contexts, we have developed novel multivalent photoaffinity probes that employ gold nanoparticles as a probe scaffold. It remains a significant challenge to address the issues associated with typically weak carbohydrate-protein interactions. Our gold-nanoparticle probes multivalently display both carbohydrate ligands and photoreactive groups. We demonstrated that the gold nanoparticle serves as a useful probe scaffold, allowing for streamlining of probe synthesis and target identification. While optimization of probe design is often a challenging initial phase in target identification studies, use of gold nanoparticle facilitates not only a facile and modular assembly of various combination of a ligand and a photoreactive group in different ratios, providing rapid access to structure–function relationship analysis. Moreover, presentation of a ligand and a photoreactive groups in high copy numbers and density on a gold-nanoparticle surface promoted enhancement of protein-binding affinity more than three orders of magnitude, as well as labeling efficiency by ten-fold due to multivalent effect. Furthermore, due to its high density, gold-nanoparticle probes can be directly subjected to enrichment of captured target proteins by simple centrifugation, providing a one-pot scheme for photoaffinity labeling and enrichment of target proteins. Subsequently, we expanded the utility of the gold-nanoparticle probes to protein-reactive electrophilic groups as alternative labeling groups to photoreactive groups. We found that *p*- and *m*-sulfonylfluoride isomers, as well as dichlorotriazine group, represent promising labels suitable for affinity labeling of low affinity carbohydrate binding proteins. The application of well-designed multivalent gold-nanoparticle probes should aid the investigation of uncharacterized cellular interactions mediated by carbohydrate ligands under physiologically relevant conditions, toward the understanding of their biological functions.

## References


1. Gagneux P, Hennet T, Varki A (2022) Biological functions of glycans. In: Varki A, Cummings RD, Esko JD et al (eds) Essentials of glycobiology [Internet]. 4th edn. Cold Spring Harbor Laboratory Press, Cold Spring Harbor (NY), Chap. 7. Available from: https://www.ncbi.nlm.nih.gov/books/NBK579984/ https://doi.org/10.1101/glycobiology.4e.7
2. Lee YC, Lee RT (1995) Carbohydrate-protein interactions: basis of glycobiology. Acc Chem Res 28(8):321–327. https://doi.org/10.1021/ar00056a001
3. Pinho SS, Reis CA (2015) Glycosylation in cancer: mechanisms and clinical implications. Nat Rev Cancer 15(9):540–555. https://doi.org/10.1038/nrc3982
4. Dugan AE, Peiffer AL, Kiessling LL (2022) Advances in glycoscience to understand viral infection and colonization. Nat Methods 19(4):384–387. https://doi.org/10.1038/s41592-022-01451-0
5. Wisnovsky S, Bertozzi CR (2022) Reading the glyco-code: new approaches to studying protein–carbohydrate interactions. Curr Opin Struct Biol 75:102395. https://doi.org/10.1016/j.sbi.2022.102395
6. Shivatare SS, Shivatare VS, Wong CH (2022) Glycoconjugates: synthesis, functional studies, and therapeutic developments. Chem Rev 122(20):15603–15671. https://doi.org/10.1021/acs.chemrev.1c01032
7. Smith BAH, Bertozzi CR (2021) The clinical impact of glycobiology: targeting selectins, Siglecs and mammalian glycans. Nat Rev Drug Discov 20:217–243. https://doi.org/10.1038/s41573-020-00093-1
8. Collins BE, Paulson JC (2004) Cell surface biology mediated by low affinity multivalent protein-glycan interactions. Curr Opin Chem Biol 8:617–625. https://doi.org/10.1016/j.cbpa.2004.10.004
9. Griffin ME, Hsieh-Wilson LC (2022) Tools for mammalian glycoscience research. Cell 185:2657–2677. https://doi.org/10.1016/j.cell.2022.06.016
10. Lee RT, Lee YC (2000) Affinity enhancement by multivalent lectin–carbohydrate interaction. Glycoconj J 17:543–551. https://doi.org/10.1023/A:1011070425430
11. Sakurai K (2015) Photoaffinity probes for identification of carbohydrate-binding proteins. Asian J Org Chem 4:116–126. https://doi.org/10.1002/ajoc.201402209
12. Wu H, Kohler J (2019) Photocrosslinking probes for capture of carbohydrate interactions. Curr Opin Chem Biol 53:173–182. https://doi.org/10.1016/j.cbpa.2019.09.002.
13. Babulic JL, De León González FV, Capicciotti CJ (2024) Recent advances in photoaffinity labeling strategies to capture glycan-protein interactions. Curr Opin Chem Biol 80:102456. https://doi.org/10.1016/j.cbpa.2024.102456.
14. Sakurai K, Ozawa S, Yamada R, Yasui T, Mizuno S (2014) Comparison of the reactivity of carbohydrate photoaffinity probes with different photoreactive groups. ChemBioChem 15(10):1399–1403. https://doi.org/10.1002/cbic.201402051
15. Sakurai K, Ozawa S, Yamaguchi T (2015) Photoaffinity labeling studies of the carbohydrate-binding proteins with different affinities. Bioorg Med Chem 23:5319–5325. https://doi.org/10.1016/j.bmc.2015.07.065
16. Sakurai K, Yasui T, Mizuno S (2015) Comparative analysis of the reactivity of diazirine-based photoaffinity probes toward a carbohydrate-binding protein. Asian J Org Chem 4:724–728. https://doi.org/10.1002/ajoc.201500116

17. Kotzyba-Hibert F, Kapfer I, Goeldner M (1995) Recent trends in photoaffinity labeling. Angew Chem Int Ed Engl Ed 34:1296–1312. https://doi.org/10.1002/anie.199512961
18. Nagao, M. Matsumoto H, Miura Y (2023) Design of glycopolymers for controlling the interactions with lectins. Chem Asian J 18:e202300643. https://doi.org/10.1002/asia.202300643
19. Kiessling LL, Grim JC (2013) Glycopolymer probes of signal transduction. Chem Soc Rev 42:4476–4491. https://doi.org/10.1039/C3CS60097A
20. Sakurai K, Hatai Y, Okada A (2016) Gold nanoparticle-based multivalent carbohydrate probes: selective photoaffinity labeling of carbohydrate-binding proteins. Chem Sci 7:702–706. https://doi.org/10.1039/C5SC03275J
21. Lauc G, Lee RT, Dumiæ J, Lee YC (2000) Photoaffinity glycoprobes—a new tool for the identification of lectins. Glycobiology 10:357–364. https://doi.org/10.1093/glycob/10.4.357
22. Wibowo A, Peters EC, Hsieh-Wilson LC (2014) Photoactivatable glycopolymers for the proteome-wide identification of fucose-α(1–2)-galactose binding proteins. J Am Chem Soc 136:9528–31. https://doi.org/ https://doi.org/10.1021/ja502482a
23. Lee MR, Jung DW, Williams D, Shin I (2005) Efficient solid-phase synthesis of trifunctional probes and their application to the detection of carbohydrate-binding proteins. Org Lett 7:5477–80. https://doi.org/10.1021/ol0523188
24. Ballell L, Alink KJ, Slijper M, Versluis C, Liskamp RM, Pieters RJ (2005) A new chemical probe for proteomics of carbohydrate-binding proteins. Chembiochem 6:291–295. https://doi.org/10.1002/cbic.200400209
25. Chang TC, Lai CH, Chien CW, Liang CF, Adak AK, Chuang YJ, Chen YJ, Lin CC (2013) Synthesis and evaluation of a photoactive probe with a multivalent carbohydrate for capturing carbohydrate-lectin interactions. Bioconjug Chem 24:1895–1906. https://doi.org/10.1021/bc400306g
26. Kolb HC, Finn MG, Sharpless KB (2001) Click chemistry: diverse chemical function from a few good reactions. Angew Chem Int Ed Engl 40:2004–2021. https://doi.org/10.1002/1521-3773
27. Cagnoni AJ, Varela O, Uhrig ML, Kovensky J (2013) Efficient synthesis of thiolactoside glycoclusters by ruthenium-catalyzed cycloaddition reaction of disubstituted alkynes on carbohydrate scaffolds. Eur J Org Chem 2013:972–983. https://doi.org/10.1002/ejoc.201201412
28. Sharma S, Bharadwaj S, Surolia A, Podder SK (1998) Evaluation of the stoichiometry and energetics of carbohydrate binding to Ricinus communis agglutinin: a calorimetric study. Biochem J 333:539–542. https://doi.org/10.1042/bj3330539
29. Svensson C, Teneberg S, Nilsson CL, Kjellberg A, Schwarz FP, Sharon N, Krengel U (2002) High-resolution crystal structures of Erythrina cristagalli lectin in complex with lactose and 2'-alpha-L-fucosyllactose and correlation with thermodynamic binding data. J Mol Biol 321:69–83. https://doi.org/10.1016/s0022-2836(02)00554-5
30. Deseke E, Nakatani Y, Ourisson G (1998) Intrinsic reactivities of amino acids towards photoalkylation with benzophenone—a study preliminary to photolabelling of the transmembrane protein glycophorin A. Eur J Org Chem 1998:243–251. https://doi.org/10.1002/(SICI)1099-0690(199802)1998:2%3c243::AID-EJOC243%3e3.0.CO;2-I
31. Wiegand M, Lindhorst TK (2006) Synthesis of photoactive α-mannosides and mannosyl peptides and their evaluation for lectin labeling. Eur J Org Chem 2006:4841–4851. https://doi.org/10.1002/ejoc.200600449
32. Salunke DM, Swamy MJ, Khan MI, Mande SC, Surolia A, Vijayan M (1985) Arrangement of subunits in peanut lectin. Rotation function and chemical cross-linking studies. J Biol Chem 260:13576–13579. https://doi.org/10.1016/S0021-9258(17)38762-8
33. Suto N, Kamoshita S, Hosoya S, Sakurai K (2022) Exploration of the reactivity of multivalent electrophiles for affinity labeling: sulfonyl fluoride as a highly efficient and selective label. Angew Chem Int Ed Engl 60(31):17080–17087. https://doi.org/10.1002/anie.202104347
34. Kamoshita S, Matsui S, Suto N, Sakurai K (2022) Reactivity analysis of new multivalent electrophilic probes for affinity labeling of carbohydrate binding proteins. ChemBioChem 23(4):e202100388. https://doi.org/10.1002/cbic.202100388

35. Tsuruno A, Kamoshita S, Hosoya S, Sakurai K (2024) Dichlorotriazine-based multivalent probe for selective affinity labeling of carbohydrate-binding proteins. Org Biomol Chem 22:7659–7663. https://doi.org/10.1039/d4ob01285b
36. Ziegler S, Pries V, Hedberg C, Waldmann H (2013) Target identification for small bioactive molecules: finding the needle in the haystack. Angew Chem Int Ed Engl 52(10):2744–2792. https://doi.org/10.1002/anie.201208749
37. Cohen MS, Zhang C, Shokat KM, Taunton J (2005) Structural bioinformatics-based design of selective, irreversible kinase inhibitors. Science 308(5726):1318–1321. https://doi.org/10.1126/science1108367
38. Shindo N, Ojida A (2021) Recent progress in covalent warheads for in vivo targeting of endogenous proteins. Bioorg Med Chem 47:116386. https://doi.org/10.1016/j.bmc.2021.116386
39. Boike L, Henning NJ, Nomura DK (2022) Advances in covalent drug discovery. Nat Rev Drug Discov 21(12):881–898. https://doi.org/10.1038/s41573-022-00542-z
40. Fang H, Peng B, Ong SY, Wu Q, Li L, Yao SQ (2021) Recent advances in activity-based probes (ABPs) and affinity-based probes (AfBPs) for profiling of enzymes. Chem Sci 12:8288–8310. https://doi.org/10.1039/d1sc01359a
41. Cravatt BF, Wright AT, Kozarich JW (2008) Activity-based protein profiling: from enzyme chemistry to proteomic chemistry. Annu Rev Biochem 77:383–414. https://doi.org/10.1146/annurev.biochem.77.072407.135656
42. Backus KM, Correia BE, Lum KM, Forli S, Horning BD, González-Páez GE, Chatterjee S, Lanning BR, Teijaro JR, Olson AJ, Wolan DW, Cravatt BF (2016) Proteome-wide covalent ligand discovery in native biological systems. Nature 534:570–574. https://doi.org/10.1038/nature18002
43. Hacker SM, Backus KM, Lazear MR, Forli S, Correia BE, Cravatt BF (2017) Global profiling of lysine reactivity and ligandability in the human proteome. Nat Chem 9(12):1181–1190. https://doi.org/10.1038/nchem.2826
44. Tsukiji S, Miyagawa M, Takaoka Y, Tamura T, Hamachi I (2009) Ligand-directed tosyl chemistry for protein labeling in vivo. Nat Chem Biol 5:341–343. https://doi.org/10.1038/nchembio.157
45. Shiraiwa K, Cheng R, Nonaka H, Tamura T, Hamachi I (2020) Chemical tools for endogenous protein labeling and profiling. Cell Chem Biol 27(8):970–985. https://doi.org/10.1016/j.chembiol.2020.06.016
46. Haedke U, Küttler EV, Vosyka O, Yang Y, Verhelst SH (2013) Tuning probe selectivity for chemical proteomics applications. Curr Opin Chem Biol 17:102–109. https://doi.org/10.1016/j.cbpa.2012.11.024
47. Banerjee R, Das K, Ravishankar R, Suguna K, Surolia A, Vijayan M (1996) Conformation, protein-carbohydrate interactions and a novel subunit association in the refined structure of peanut lectin-lactose complex. J Mol Biol 259:281–296. https://doi.org/10.1006/jmbi.1996.0319
48. Jones LH (2025) Advances in sulfonyl exchange chemical biology: expanding druggable target space. Chem Sci 16(23):10119–10140. https://doi.org/10.1039/d5sc02647d

# Chemical Probes for Analyzing Endoglycosidases

**Nozomi Ishii and Ichiro Matsuo**

**Abstract** Endo-β-*N*-acetylglucosaminidases (ENGases) are used in glycoprotein analysis and glycan remodeling. However, existing ENGase activity assays are complex and unsuitable for high-throughput analysis. This study describes the development of FRET-based glycan probes for simple and real-time detection of ENGase activity. We designed novel probes, containing a fluorophore and a quencher flanking the chitobiose moiety, utilizing Förster resonance energy transfer (FRET) quenching. We synthesized di-, tri-, and pentasaccharide probes and evaluated their quenching efficiencies and suitability for activity detection. The pentasaccharide probe, **MM3D**, was efficiently cleaved by Endo-M, resulting in increased fluorescence. We then tested the **MM3D** probe with various ENGases (Endo-H, Endo-F3, Endo-Om, Endo-CC, Endo-S, and Endo-D). While Endo-M, Endo-CC, and Endo-Om activities were detected, Endo-H activity was not, suggesting that ENGase substrate specificity varies depending on the enzyme's origin. These results demonstrate that the developed FRET-based glycan probes enable simple and real-time detection of ENGase activity. Further, we synthesized a library of probes with diverse glycan structures. Using this library, we evaluated the activities of six commercially available ENGases and observed their distinct substrate specificities. This probe library offers a valuable tool for the simple detection of ENGase activity and will contribute significantly to advances in glycobiology research.

**Keywords** Endoglycosidase · Chemical probe · FRET · Enzyme assay

## 1 Introduction

Endo-β-*N*-acetylglucosaminidases (ENGases) are endoglycosidases that cleave the β1-4 glycosidic bond within the *N*,*N'*-diacetylchitobiose core of asparagine-linked glycans (*N*-glycans). Given their ability to excise *N*-glycans from glycoproteins, ENGases have been utilized as valuable tools for analyzing protein glycan structures.

---

N. Ishii · I. Matsuo (✉)
Graduate School of Science and Technology, Gunma University, Kiryu, Gunma, Japan
e-mail: matsuo@gunma-u.ac.jp



K. Toshima and Y. Fujimoto (eds.), *Chemical Biology of Carbohydrates*,
https://doi.org/10.1007/978-981-95-5335-8_11

Furthermore, certain ENGases exhibit not only hydrolysis activity but also transglycosylation activity, enabling the transfer of *N*-glycans to *N*-acetylglucosamine (GlcNAc) residues on proteins. This bifunctional property has led to their application in glycoprotein glycan remodeling studies [1–3]. Specifically, by removing glycans from proteins using an ENGase and subsequently introducing glycans with the desired structure, it has become possible to synthesize glycoproteins with homogeneous glycan structures. Notably, ENGase-mediated remodeling of glycans on antibodies has been reported to enhance antibody-dependent cell-mediated cytotoxicity (ADCC) and facilitate the preparation of antibody–drug conjugates (ADCs) via glycan conjugation. This highlights the potential for glycan-mediated control of protein functions. Thus, in addition to their utility in elucidating the functions of protein glycans, ENGases are valuable enzymes in the pharmaceutical industry [4–8].

ENGases are widely distributed in nature and are found in organisms such as humans, filamentous fungi, and yeast. Based on sequence similarities among carbohydrate-hydrolyzing enzymes, ENGases are classified into Glycoside Hydrolase (GH) families 18 and 85 [9, 10]. Within GH18 ENGases, notable examples include Endo-H from *Streptomyces plicatus* [11], Endo-F1, Endo-F2, and Endo-F3 from *Flavobacterium meningosepticum* [12], Endo-S and Endo-S2 from *Streptococcus pyogenes* [13, 14], and Endo-CoM from *Cordyceps militaris* [15]. In contrast, GH85 ENGases include Endo-D from *Streptococcus pneumoniae* [16, 17], Endo-A from *Arthrobacter protophormiae* [18, 19], Endo-M from *Mucor hiemalis* [20, 21], Endo-Om from *Ogataea minuta* [22], and Endo-CC from *Coprinopsis cinerea* [23]. Figure 1 illustrates the structures of *N*-linked glycans and the substrate specificities of previously reported ENGases toward these *N*-glycans [1]. As mentioned previously, ENGases are industrially valuable enzymes, prompting ongoing vigorous efforts in their discovery. Recent reports have highlighted the identification of ENGases from sources such as *Oryza sativa* and the human gut microbiome [24–33]. Although the biological functions of these ENGases remain to be fully elucidated, human ENGase (*h*ENGase) has been implicated in the degradation of free *N*-glycans released from misfolded glycoproteins in the cytosol, playing a crucial role in glycan clearance. Additionally, *h*ENGase has been suggested to be involved in NGLY1 disease [34–36]. Furthermore, Endo-S from *Streptococcus pyogenes* exhibits remarkably high recognition affinity for immunoglobulin G (IgG) and is thought to facilitate infection by cleaving glycans on host antibodies. Despite their significant utility of ENGases in glycobiology and industry, detecting their enzymatic activity has historically been challenging. Previous methods primarily involved tracking reactions via HPLC using fluorescently labeled glycans or glycopeptides as substrates, or analyzing changes in glycoprotein molecular weight through SDS-PAGE after enzymatic reactions. Both approaches are cumbersome and labor-intensive. Moreover, when glycoproteins are used as substrates, the inherent heterogeneity of their glycan structures can introduce ambiguity in substrate specificity analyses. Another fundamental limitation is the inability to monitor the enzymatic reactions in real time. Therefore, there has been a strong demand for the development of simple, highly sensitive, and high-throughput

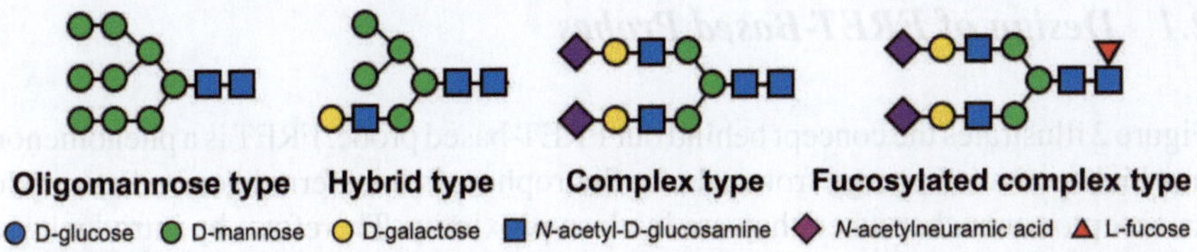


| GH family | ENGase | Oligomannose type | Hybrid type | Complex type | Fucosylated Complex type |
|---|---|---|---|---|---|
| GH18 | Endo-H | ○ | ○ | × | △ |
| | Endo-S | - | - | ○ | ○ |
| | Endo-F1 | ○ | ○ | × | △ |
| | Endo-F2 | △ | - | ○ | ○ |
| | Endo-F3 | × | - | ○ | ○ |
| GH85 | Endo-D | △ | × | △ | ○ |
| | Endo-A | ○ | ○ | × | × |
| | Endo-M | ○ | ○ | ○ | × |
| | Endo-Om | ○ | ○ | ○ | × |
| | Endo-CC | ○ | - | ○ | - |

**Fig. 1** Summary of the well-studied substrate specificities of ENGases. ○: hydrolyzed, △: slightly hydrolyzed, ×: not hydrolyzed, -: not investigated

methods for ENGase activity detection to facilitate extensive research into novel ENGases, their substrate specificities, and the discovery of ENGase inhibitors.

## 2 Development of a Simple, High-Throughput, Real-Time Assay for ENGase Activity

Although many substrates for detecting glycosidase activity are commercially available as kits, assessing exoglycosidase activity is relatively straightforward. This is because the absorbance or fluorescence of the products generated upon enzymatic hydrolysis can be measured, facilitating sensitive, simple, and real-time tracking of the reaction using instruments, such as microplate readers, to detect the target wavelength [37]. However, obtaining suitable substrates for ENGases has been challenging. Furthermore, no probes that generate a fluorescent signal directly upon the cleavage of the substrate glycan are available. Given these challenges, we developed glycan molecular probes that operate on the principle of Förster resonance energy-transfer (FRET) quenching. This probe enables the simple, quantitative, and real-time analysis of ENGase activity.

## 2.1 Design of FRET-Based Probes

Figure 2 illustrates the concept behind our FRET-based probe. FRET is a phenomenon in which excitation energy from a donor fluorophore is transferred non-radiatively to an acceptor quencher when they are in close proximity. Therefore, by introducing a fluorophore (donor) and a quencher (acceptor) close to each other within the glycan substrate, the excitation energy from the fluorophore is transferred to the quencher, resulting in no detectable fluorescence. However, once the glycosidic bond is cleaved by an enzyme, this intramolecular FRET quenching no longer occurs, leading to an increase in fluorescence intensity as the enzymatic reaction progresses. This allows quantification of enzyme activity by detecting the fluorescence increase in real time, enabling the detection of ENGase activity.

We selected small FRET dye molecules to account for the steric hindrance. Specifically, we chose the *N*-methylanthraniloyl (MANT) group as the fluorophore, and the 2,4-dinitrophenyl (DNP) group as the quencher (Fig. 3). The fluorophore and quencher were positioned to flank the chitobiose unit, which serves as the cleavage site for ENGase. The fluorophore was introduced at the non-reducing end, and the quencher was introduced at the reducing end via a propyl linker. For the glycan portion of the probe, we synthesized three types of molecular probes: a disaccharide probe with a chitobiose structure, a trisaccharide probe containing a β-mannose residue, and a common core pentasaccharide structure of *N*-glycans [38]. We then evaluated their quenching efficiency and ability to detect enzyme activity. The structures of these three glycan molecular probes are shown in Fig. 3.

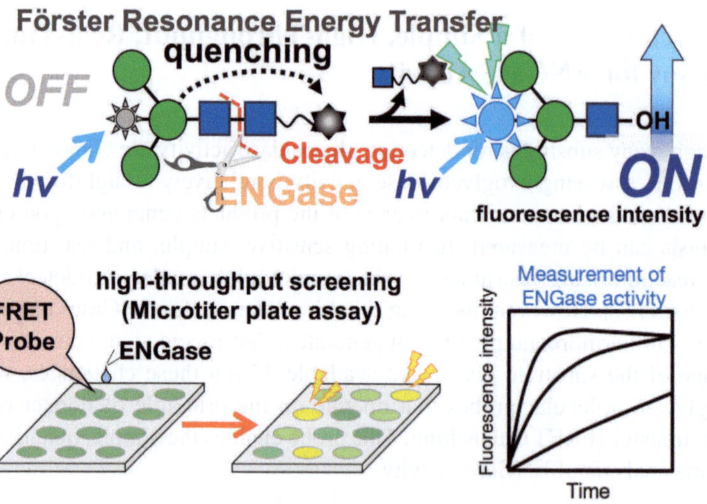


**Fig. 2** Concept of a fluorescence-quenching-based assay system for determining ENGase activity

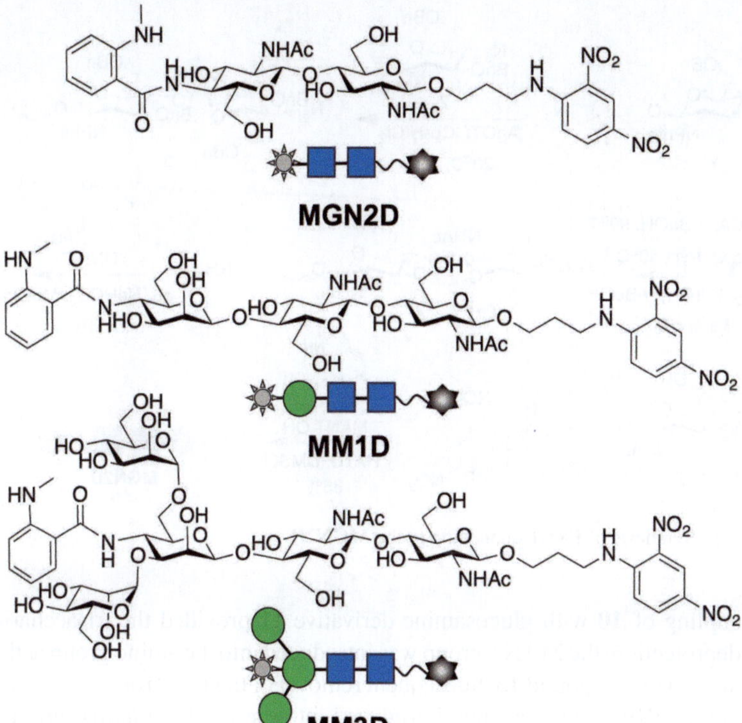


**Fig. 3** Structures of FRET quenching probes

## 2.2 *Synthesis of FRET-Type Glycan Molecular Probes*

The synthesis of the FRET-type glycan probe **MGN2D**, featuring a chitobiose structure, is outlined in Scheme 1. The chitobiose core was constructed by coupling glucosamine acceptor **1** with glucosamine donor **2**, which bears an azide group at the C-4 position. Subsequently, the phthalimide group was converted to an acetamido group [39], and simultaneous deprotection and reduction of the azide group yielded diamine **4**. Taking advantage of the differential reactivity of the amino groups, a DNP group was then introduced at the reducing end to give compound **5**. Finally, the MANT group was introduced at the non-reducing end, leading to the disaccharide probe **MGN2D**.

The synthesis of the trisaccharide probe is depicted in Scheme 2. Starting from the lactosamine derivative, **6**, pivaloyl groups were selectively introduced at the 3- and 6-positions of the galactose residue. The free hydroxy group of compound **7** was then converted to a trifluoromethanesulfonyl (Tf) group to yield compound **8**. Treatment with tetrabutylammonium azide ($TBAN_3$) selectively introduced an azide group at the 4-position of the galactose residue, providing the monotriflate **9**. Further treatment with CsOAc/18-crown-6 under sonication yielded the disaccharide donor

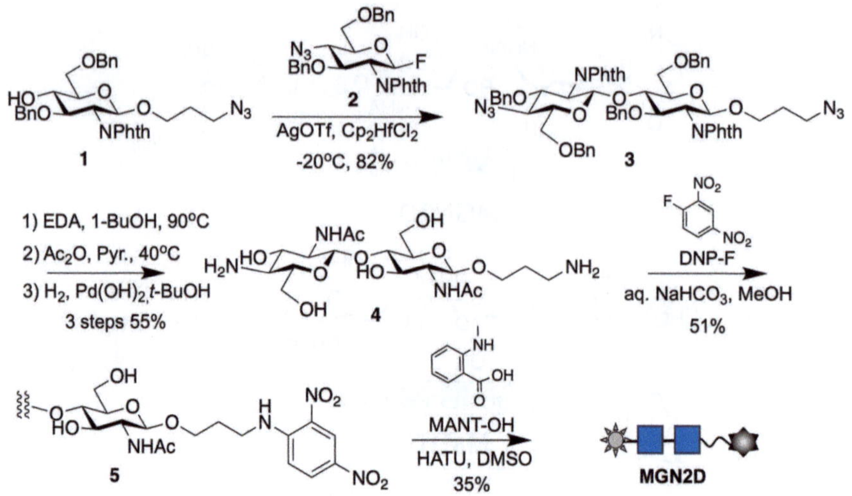

**Scheme 1** Synthesis of FRET quenching probe **MGN2D**

**10**. Coupling of **10** with glucosamine derivative **11** provided the trisaccharide **12**. After deprotection, the MANT group was introduced into the amino group at the non-reducing end of compound **13**. Subsequent removal of the Boc group gave compound **14**, and the DNP group was then introduced into the resulting amino group at the reducing end, affording the trisaccharide probe **MM1D**.

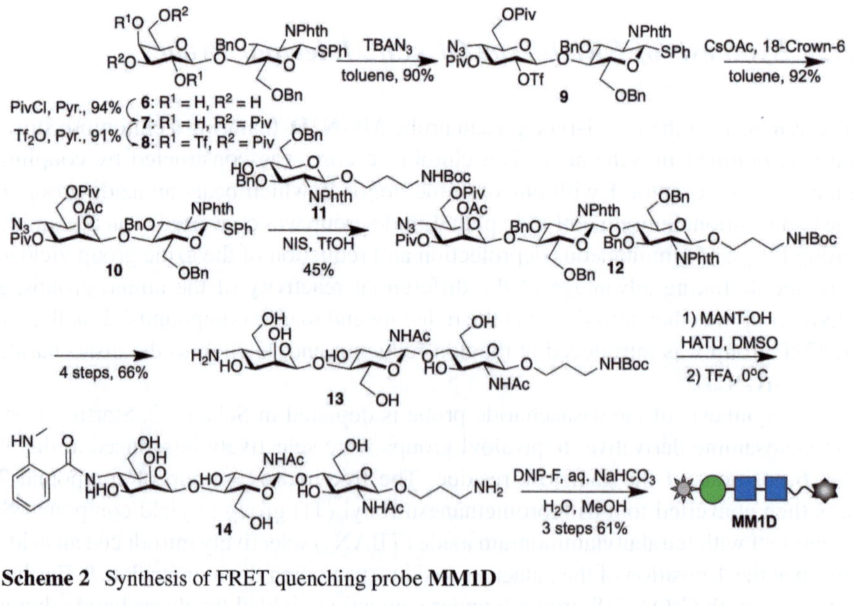

**Scheme 2** Synthesis of FRET quenching probe **MM1D**

The synthesis of the pentasaccharide probe is shown in Scheme 3. The pentasaccharide skeleton was constructed by taking advantage of the different reactivities of the hydroxy groups on the galactose residue of the galactosylchitobiose derivative **15** [40]. Specifically, reacting two equivalents of mannose donor **16** with glycan acceptor **15** yielded the desired pentasaccharide derivative **17**, with mannose residues introduced via α-linkages at the 3- and 6-positions of the galactose residue. Detailed analysis of this reaction revealed minor formation of tetrasaccharide **18**, where the mannose residue was linked at the 3-position. However, no tetrasaccharide derivatives with mannose residue introduced at the 6-position were detected. This observation suggests that the reactivity of the hydroxy groups on the galactose residue is the highest at the 3-position, followed by the 6-position. This differential reactivity proved valuable information for subsequent glycan construction.

Regarding stereoselectivity, we believe that only the α-anomer was obtained due to remote neighboring group participation by the Piv group at the C-3 position of mannose donor **16**. In contrast, we confirmed that a mixture of α and β stereoisomers was obtained when a glycosyl donor with a Bn group at the 3-position was used for the glycosylation [41]. The free hydroxy groups at the 2- and 4-positions of pentasaccharide derivative **17** were triflated. Subsequent treatment with $TBAN_3$ selectively introduced an azide group at the 4-position of the branched moiety, yielding monotriflate **20**. The resulting compound **20** was then treated under CsOAc/18-crown-6 conditions with sonication, which simultaneously introduced an acetoxy group via stereoinversion at the 2-position, leading to the desired pentasaccharide derivative **21**

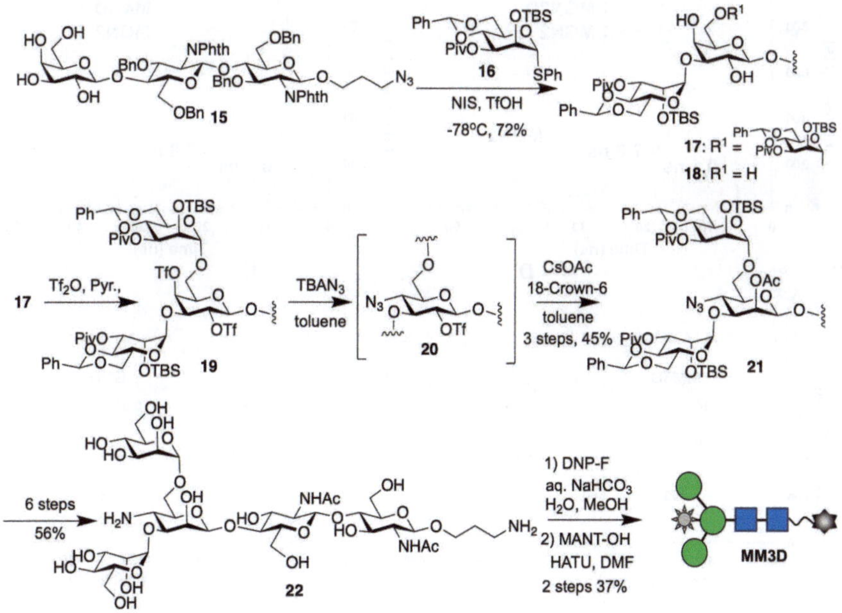


**Scheme 3** Synthesis of FRET quenching probe **MM3D**

with the β-mannoside structure. After deprotection of the obtained pentasaccharide **21** to form diamine **22**, the DNP group was introduced at the reducing end, and the MANT group was sequentially introduced into the amino group at the non-reducing end, yielding the pentasaccharide probe **MM3D**.

## 2.3 *Photophysical Analysis of FRET-Quenched Glycan Molecular Probes*

To confirm that the fluorescence of the MANT group incorporated into the glycans was quenched by the DNP group, we measured the fluorescence lifetimes of the MANT group. The results for the disaccharide probe **MGN2D** and the pentasaccharide probe **MM3D** are shown in Fig. 4. Our measurements showed that while the fluorescence lifetime of the chitobiose derivative **MGN2**, which contains only the MANT group, was 7.8 ns, the fluorescence lifetime of the MANT group on **MGN2D** was 0.1 ns. This clearly confirms that the fluorescence was completely quenched by the DNP group at the reducing end (Fig. 4a). This result indicates that quenching by DNP occurs effectively even when separated by two sugar residues.

In contrast, the fluorescence lifetime of the pentasaccharide probe **MM3D** was 0.3 ns (Fig. 4b). This reduction in quenching efficiency was likely due to the MANT

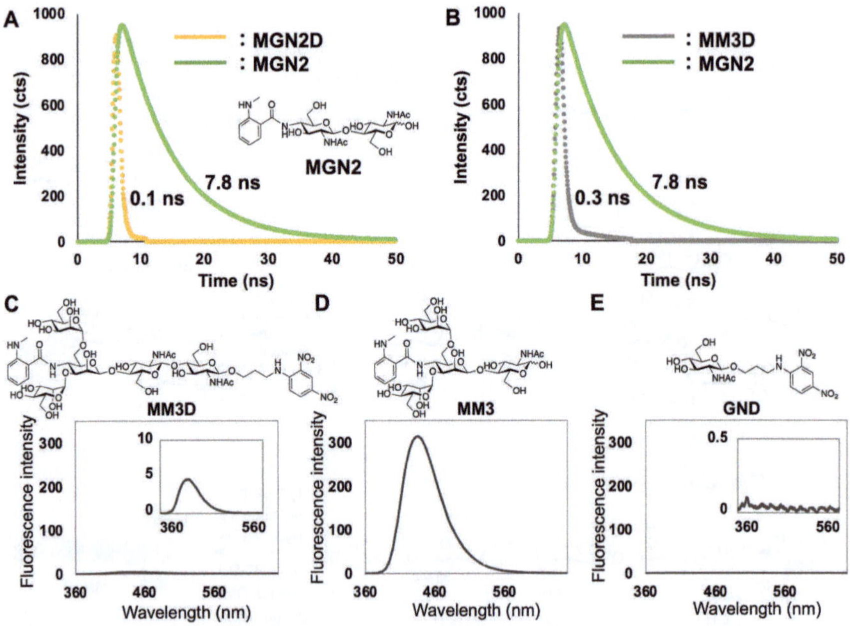


**Fig. 4** Photochemical properties of FRET quenching probes

and DNP groups being separated by three sugar residues. Although MANT fluorescence was not completely quenched, we determined that it was sufficient for tracking enzyme reactions. This is because the expected fluorescence intensity of tetrasaccharide **MM3**, produced upon complete cleavage of the probe by ENGase, was approximately 50 times higher than that of **MM3D** (Fig. 4c, d). Additionally, we confirmed that the DNP group itself does not exhibit any fluorescence (Fig. 4e).

## *2.4 Analysis of ENGases Using FRET-Type Pentasaccharide Probes*

After obtaining three FRET-type probes, we evaluated their suitability as substrates for Endo-M. Initially, we found no change in the fluorescence intensity when the disaccharide probe **MGN2D** was treated with Endo-M. Similarly, the trisaccharide probe **MM1D** exhibited no change in fluorescence. In contrast, the pentasaccharide probe **MM3D** exhibited increased fluorescence intensity immediately after the reaction began.

Given the observed increase in fluorescence intensity of **MM3D** upon Endo-M treatment, we conducted a more detailed analysis. HPLC analysis of the reaction mixture after treating **MM3D** with Endo-M revealed a decrease in the **MM3D** peak over time, accompanied by the appearance of two new peaks at approximately 4 min and 13 min (Fig. 5). MS analysis of these new peaks confirmed that the peak at 4 min corresponded to the tetrasaccharide **MM3**, whereas the peak at 13 min corresponded to the monosaccharide **GND**. This unequivocally confirmed that ENGase cleaved the glycosidic bond within the chitobiose part of **MM3D**. Furthermore, by varying the amount of Endo-M and using **MM3D** as the substrate while tracking the enzymatic reaction using a microplate reader (excitation λex: 340 nm, emission λem: 440 nm), we observed a time-dependent increase in fluorescence intensity (Fig. 6a). This result strongly supported the HPLC analysis, leading us to conclude that the **MM3D** probe enables simple and real-time evaluation of ENGase activity.

Encouraged by the successful detection of Endo-M activity, we then investigated whether ENGases from different origins could also be detected using **MM3D**. Since **MM3D** possesses the common core structure of *N*-glycans that ENGases cleave, we anticipated that it would detect the hydrolysis activity of all ENGases. Therefore, we attempted to detect the enzymatic activities of Endo-H (GH18 family [11]), Endo-F3 (GH18 family [12]), Endo-Om (GH85 family [22]), and Endo-CC (GH85 family [23]). The results are presented in Fig. 6b, c. While the ENGase activities of Endo-M, Endo-CC, and Endo-Om can be detected using **MM3D** as a substrate, the activity of Endo-H could not [38]. This finding suggests that the glycan-recognition mechanisms of ENGases vary depending on their origin. Some ENGases strongly recognize the core pentasaccharide structure, whereas for others, recognition of only the reducing end structure is insufficient, and they may also utilize the glycan structure at the non-reducing end for efficient cleavage of the chitobiose portion.

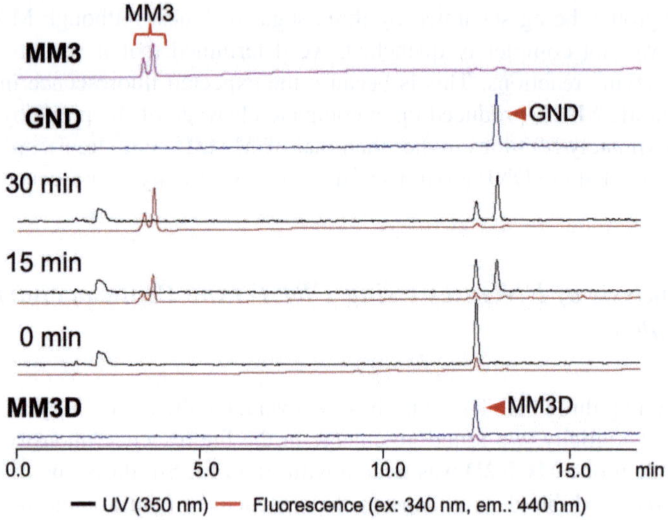


**Fig. 5** HPLC analysis of hydrolysis reaction of synthetic probe **MM3D** catalyzed by Endo-M. Assay conditions: 5 μM **MM3D**, 25 ng Endo-M, 40 mM phosphate buffer (pH 6), 10% DMSO, total vol. 250 μL, 37 °C. Analytical condition: TOSOH TSK-gel ODS-100 V 5 μm, 4.6 mm × 15 cm, 3–40% MeCN/$H_2O$ containing 0.1% TFA, flow rate: 1 mL/min, 15 min

Having established a simple ENGase activity detection method using a plate reader, we investigated whether **MM3D** could be used to evaluate ENGase inhibitors. Specifically, we assessed the reactions of Endo-M in the presence of Rabeprazole [42] and thiazoline derivative [43], both known to exhibit inhibitory activity against ENGases. Our results confirmed that the progression of the enzymatic reaction was delayed in a concentration-dependent manner in the presence of these inhibitors (Fig. 6d, e). This demonstrated the utility of the **MM3D**-based ENGase assay system for the discovery of novel inhibitors. While the **MM3D** probe we synthesized proved sufficient for detecting ENGase activity, our findings indicate that to detect the activity of a wider range of ENGases, it will be necessary to systematically prepare probes with diverse glycan structures, including variations in the non-reducing end glycan structures and the presence or absence of core fucose residues.

## 3 Construction of a Glycan Molecular Probe Library for Detecting ENGase Activity

Although ENGases are enzymes conserved across various species, their recognize glycan structures differs depending on their origin [1, 2]. Despite the pentasaccharide probe **MM3D** possessing the common core structure of *N*-glycans cleaved by ENGases, it has not been proven to be a universal probe capable of detecting

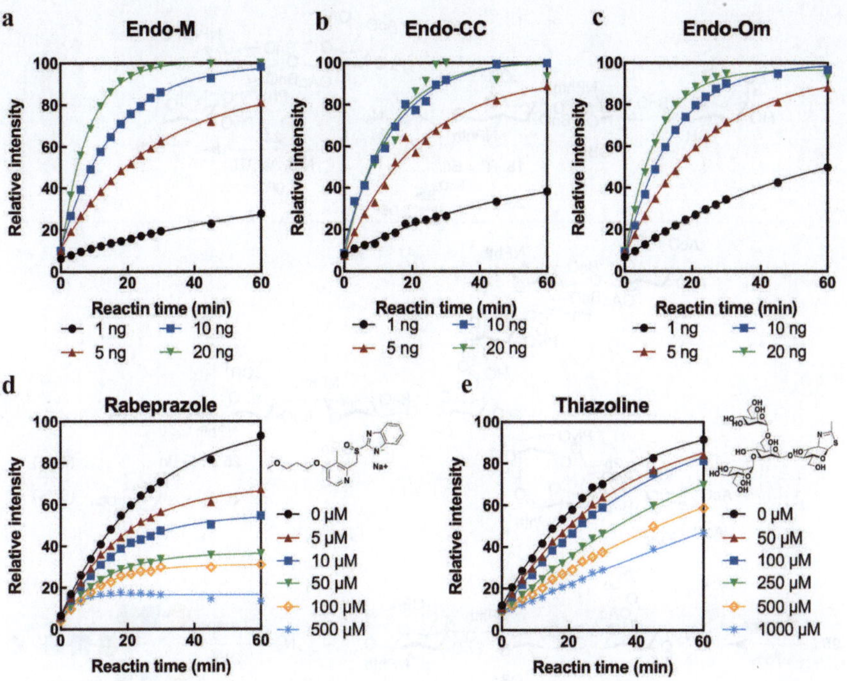


**Fig. 6** Assay of the hydrolysis activities of various ENGases (Endo-M; A, Endo-CC; B, Endo-Om; C), assay conditions: 5 μM **MM3D**, ENGase, 40 mM sodium phosphate buffer (pH 6), 10% DMSO, total vol. 50 μL, 37 °C. Evaluation of the ENGase inhibitor (Rabeprazole; D, thiazoline; E). Inhibition assay conditions: 2.5 μM **MM3D**, inhibitor, 5 ng Endo-M, 40 mM sodium phosphate buffer (pH 6), 10% DMSO, total vol. 100 μL, 37 °C. Fluorescence intensities were measured with a microplate reader. λex = 340 nm, λem = 440 nm

the hydrolysis activity of all ENGases. Therefore, to detect the activity of diverse ENGases, we recognized the systematic need for probes with various glycan structures. Consequently, we decided to synthesize glycan probes representing complex, oligomannose, hybrid, and core-fucosylated *N*-glycan structures.

## *3.1 Synthesis of Molecular Probes with Complex-Type Glycan Structures*

The synthesis of probes featuring complex-type glycan structures is illustrated in Scheme 4. The glycan skeleton was constructed by exploiting the different reactivities of the hydroxy groups on the galactose residue, similar to the synthesis of the pentasaccharide probe. Additionally, the introduction of a Piv group at the 3-position of the mannose residue in the trisaccharide donor was expected to ensure α-selectivity.

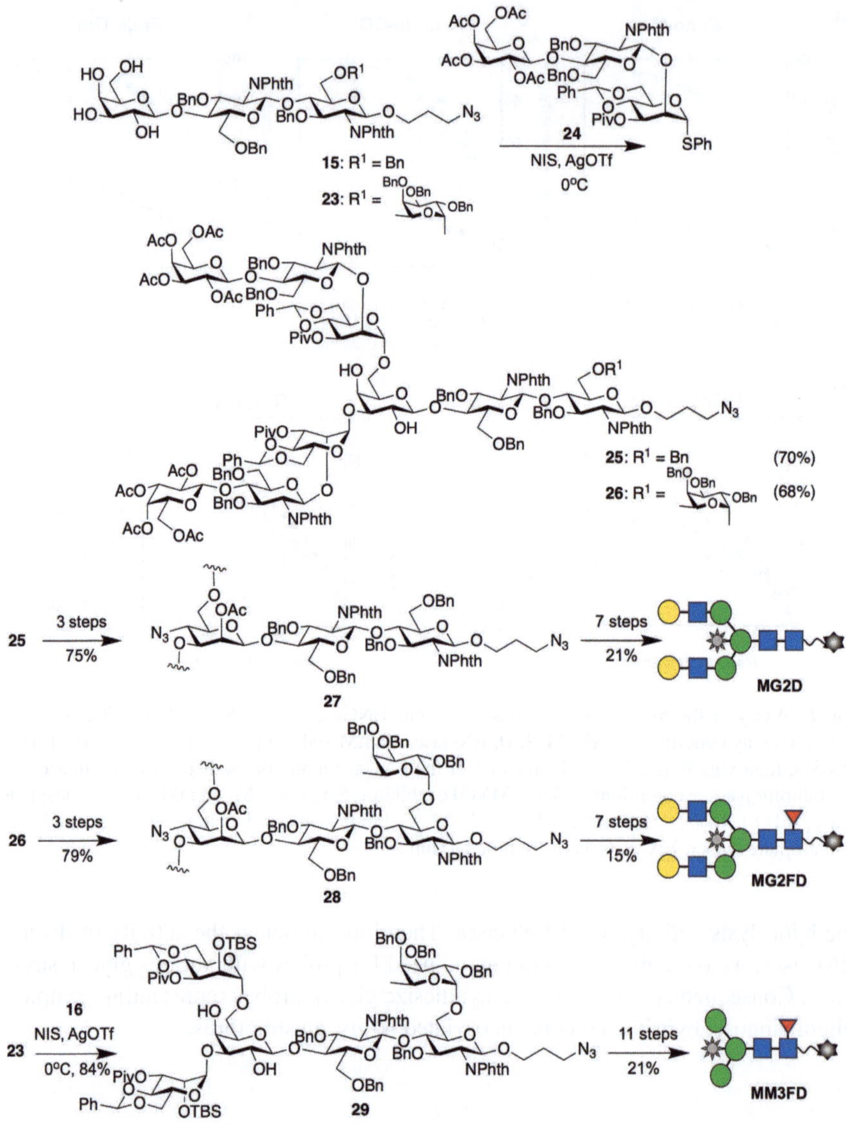


**Scheme 4** Synthesis of complex type FRET quenching probes

When two equivalents of trisaccharide donor **24** were reacted with trisaccharide acceptor **15** under NIS/TfOH conditions, the desired nonasaccharide **25** was obtained in only 9% yield. The major product was a hexasaccharide, in which the galactose residue was α-glycosylated at the 3-position. As observed during the construction of the pentasaccharide skeleton, no hexasaccharides with glycosylation at the 6-position were detected. Consequently, we switched the promoter to NIS/AgOTf.

This modification significantly accelerated the reaction, resulting in the selective formation of the nonasaccharide derivative **25** in 70% yield, where two trisaccharide donors **24** were stereoselectively introduced at both the 3- and 6-positions of the galactose residue. Similarly, by employing the tetrasaccharide acceptor **23**, which has a fucose linked to the glucosamine residue at the reducing end, we successfully constructed the core-fucosylated decasaccharide derivative **26** in 68% yield.

The obtained nonasaccharide **25** and decasaccharide derivative **26** were subsequently converted into glycan probes. Specifically, the free hydroxy groups at the 2- and 4-positions of the galactose residue of the branched moiety were triflated. The resulting triflated product was then treated with $TBAN_3$ to selectively introduce an azide group at the 4-position. Subsequently, stereoinversion of the 2-hydroxy group to the β-Man configuration was achieved by treatment with CsOAc/18-crown-6 ether under sonication. This was followed by conversion of the phthalimido group to an acetamido group and global deprotection to yield a diamine. Leveraging the differential reactivity of the amino groups, a DNP group was selectively introduced to the amino group at the reducing end, and a MANT group was introduced to the amino group at the glycan branch point, leading to the nonasaccharide probe **MG2D**. The decasaccharide probe **MG2FD** was synthesized from derivative **26** via a similar pathway [44].

For the synthesis of the fucosylated core hexasaccharide probe **MM3FD**, mannose donor **16** was reacted with sugar acceptor **23** to yield the hexasaccharide **29**. This was then followed by the introduction of an azide group onto the galactose residue, stereoinversion, DNP introduction, and the introduction of the MANT group to furnish the probe **MM3FD** [45].

## *3.2 Synthesis of Oligomannose-Type and Hybrid-Type Glycan Probes*

The probes synthesized thus far possessed identical glycan structures at both the 3- and 6-positions of the β-mannose residue. However, to synthesize oligomannose-type and hybrid-type glycans, it was necessary to differentiate the 3- and 6-hydroxy groups during the glycosylation of the galactose residue on the trisaccharide acceptor **15**. We previously observed that glycosylation under NIS/TfOH conditions predominantly proceeded at the 3-position of the galactose residue during the synthesis of **MM3D** and **MG2D**. Therefore, to construct glycans with varying non-reducing end structures, we investigated a synthetic route involving position-selective glycosylation at the 3-hydroxy group of trisaccharide acceptor **15**, followed by the introduction of a structurally different glycosyl donor at the 6-hydroxy group (Scheme 5).

When trisaccharide acceptor **15** was reacted with 1 equiv. of mannose donor **16** under NIS/TfOH conditions, tetrasaccharide **18** was obtained as the main product in 16% yield. Subsequently, by reacting the obtained tetrasaccharide **18** with the

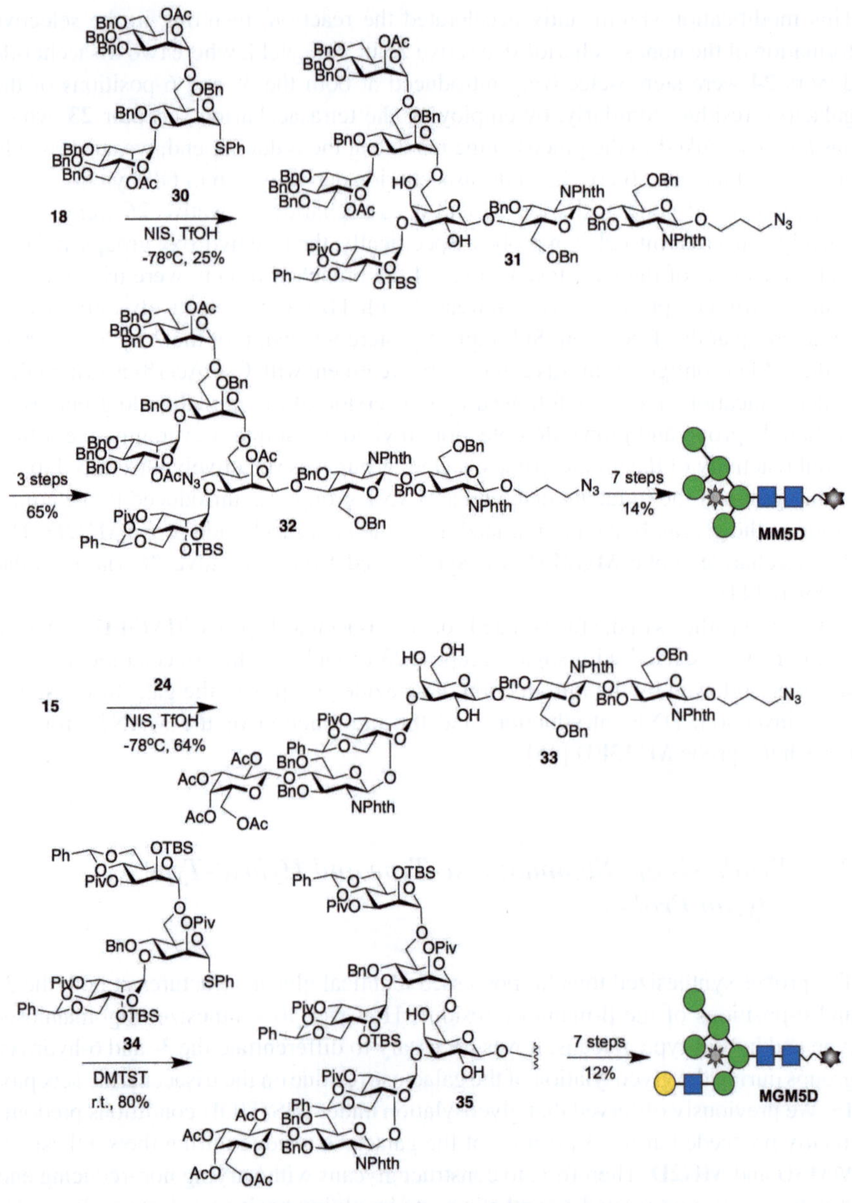


**Scheme 5** Synthesis of oligomannose and hybrid type FRET quenching probes

known branched trisaccharide derivative **30** under NIS/TfOH conditions, we successfully constructed oligomannose-type heptasaccharide structure **31** in two steps. The resulting heptasaccharide derivative **31** was then converted into the heptasaccharide probe **MM5D** [46] using the same probe synthesis method.

For the construction of the hybrid-type glycan structure, reacting trisaccharide acceptor **15** with 1.1 equivalents of trisaccharide donor **24** under NIS/TfOH conditions resulted in selective glycosylation at the 3-position of the galactose residue, yielding hexasaccharide **33** in 64% yield. We designed a new branched mannose donor, trisaccharide donor **34**, featuring a pivaloyl group at the 2-position. The reaction of hexasaccharide acceptor **33** with trisaccharide donor **34** under DMTST conditions successfully afforded nonasaccharide derivative **35** in 80% yield. Notably, the construction of the oligomannose-type glycan also significantly improved the yield of the heptasaccharide backbone to 80% by changing the sugar donor to **34**. The obtained hybrid nonasaccharide **35** was subsequently converted into the hybrid probe **MGM5D** [47] using an established method.

## 4 Detection of ENGase Activity Using a Glycan Molecular Probe Library

We evaluated the activity of various ENGases from different origins using our newly constructed library of molecular probes: the biantennary complex-type nonasaccharide (**MG2D**), the fucosylated decasaccharide probe (**MG2FD**), the oligomannose-type heptasaccharide (**MM5D**), the hybrid-type nonasaccharide probe (**MGM5D**), the core-fucosylated hexasaccharide (**MM3FD**), and the core pentasaccharide (**MM3D**) (Fig. 6). While it is possible to normalize substrate concentrations in ENGase enzymatic reactions from different sources, it is challenging to precisely standardize enzyme amounts or activities. Therefore, a strict comparison of reactivity between different enzymes toward the glycan probes is not feasible. However, for a single enzyme, maintaining a constant enzyme amount and substrate concentration allows for relative comparison of reactivity toward probes with different glycan structures. In essence, differences in the cleavage rates of the glycan probes can be interpreted as relative substrate specificities for each glycan structure.

Under this premise, we performed enzymatic reactions using commercially available ENGases from the GH85 family (Endo-M, Endo-CC, and Endo-D) and the GH18 family (Endo-H, Endo-F3, and Endo-S). The relative cleavage activity, with the best-cleaved substrate set to 100%, is presented in Fig. 7. Endo-M efficiently cleaved **MM3D** and **MM5D**; however, its reactivity toward **MGM5D**, **MG2D**, and other probes with complex-type glycan structures was low. Furthermore, it showed no hydrolysis activity toward the fucosylated probes **MM3FD** and **MG2FD**. These results are consistent with previously reported substrate specificities for Endo-M [48]. Endo-CC mirrored the behavior of Endo-M, acting on non-core-fucosylated glycans and showed a preference for oligomannose-type glycans over complex-type glycans. In contrast, Endo-D did not hydrolyze any of the glycan probes. This finding aligns with the known substrate specificity of Endo-D for linear substrates.

Next, we investigated Endo-H, Endo-F3, and Endo-S from GH18 cells. We found that Endo-H used oligomannose-type and hybrid-type glycans as substrates but showed no activity toward complex-type glycans, core pentasaccharides, or fucosylated glycan probes. Endo-F3, on the contrary, acted on the core-fucosylated glycan

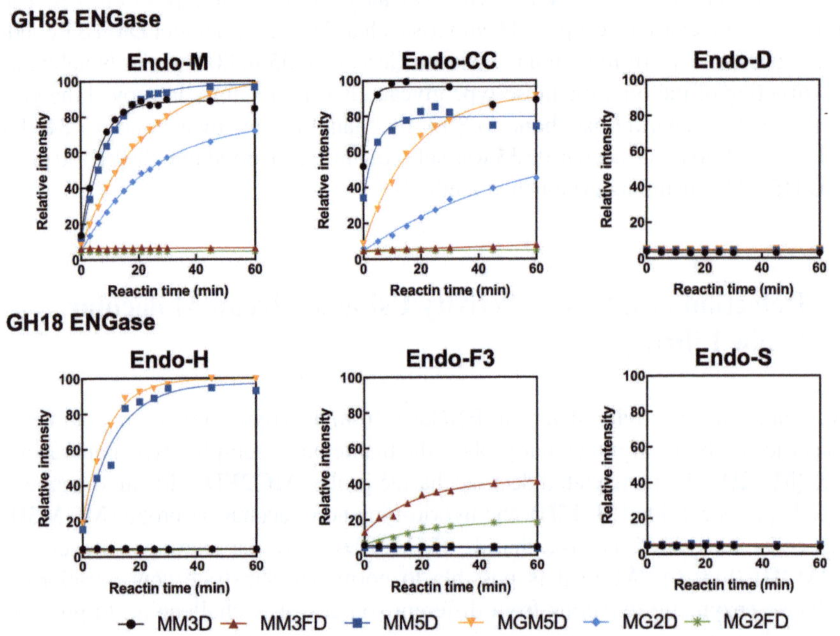


**b**

| ENGase \ probes | 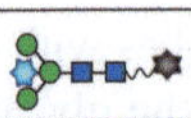 | 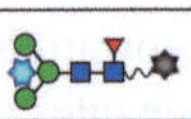 | 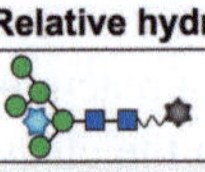 | 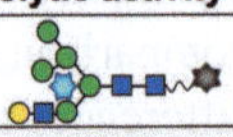 | 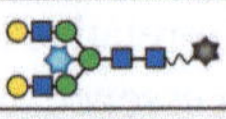 | 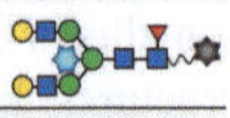 |
|---|---|---|---|---|---|---|
| | Relative hydrolytic activity | | | | | |
| Endo-M | 100 | n.d. | 97 | 63 | 42 | n.d. |
| Endo-CC | 100 | 1 | 77 | 57 | 14 | n.d. |
| Endo-D | n.d. | n.d. | n.d. | n.d. | n.d. | n.d. |
| Endo-H | n.d. | n.d. | 94 | 100 | n.d. | n.d. |
| Endo-F3 | n.d. | 100 | n.d. | n.d. | n.d. | 32 |
| Endo-S | n.d. | n.d. | n.d. | n.d. | n.d. | n.d. |

**Fig. 7** Assay of various ENGases using FRET-quenching probes (**a**). A summary of substrate specificity studies of different ENGases evaluated by FRET-quenching probes (**b**). Assay conditions: 5 μM probe, ENGase (Endo-M 10 μU, Endo-CC 75 μU, Endo-D 5 U, Endo-H 5 units, Endo-F3 0.5 U or Endo-S 100 U), total vol. 50 μL. Fluorescence intensities were measured with a microplate reader. λex = 340 nm, λem = 440 nm

probes **MG2FD** and **MM3FD** but not on the other probes, indicating a strong recognition of core fucose. Moreover, its stronger recognition of the core-fucosylated hexasaccharide **MM3FD** compared to the fucosylated complex-type glycan **MG2FD** suggests a higher affinity for glycan structures at the reducing end.

We also attempted to detect the activity of Endo-S, an industrially relevant enzyme known to act on glycans in antibodies. However, no activity was detected with any of the probes, as none served as substrates. This can be explained by the crystal structure of Endo-S [14]. X-ray crystallographic analysis of Endo-S revealed a loop structure near the β-Man residue at the glycan recognition site. We believe that the steric hindrance caused by the fluorescent substituent introduced on the β-Man residue for FRET probes prevented substrate recognition. This hypothesis is supported by our current unpublished data, which show that Endo-S activity can be detected using probes with a modified fluorescent substituent.

In summary, we demonstrated that ENGase activity can be detected in real time by appropriately utilizing the glycan molecular probe library constructed in this study. Furthermore, given that ENGases exhibit differential reactivity toward probes with varying glycan structures, our synthesized probes are expected to serve as a valuable library for easily evaluating the substrate specificity of ENGases.

## 5 Conclusion

We successfully developed FRET-type glycan molecular probes that enable simple and real-time detection of ENGase activity. Efficient block synthesis was achieved by combining region- and stereoselective glycosylation reactions with a stereoinversion reaction from β-galactoside to β-mannoside. Our developed glycan molecular probes allow for high-throughput and straightforward detection of ENGase activity using a microplate reader. Furthermore, by systematically utilizing probes with different glycan structures, it is possible to analyze trends in ENGase substrate specificity. Consequently, these probes are applicable to novel ENGase discovery research and the analysis of substrate specificities of ENGase mutants generated by genetic engineering.

We also confirmed that for enzymes such as Endo-S and Endo-D, whose activities could not be detected initially, careful re-evaluation of the fluorescent group's introduction site and glycan structure allowed for enzyme activity detection. Moving forward, we plan to expand our probe development efforts to include increasing sensitivity by modifying FRET pairs and developing probes for intracellular ENGase activity detection. Through such chemically synthesized glycan molecular probes, we aim to contribute to the elucidation of ENGase functions and their discovery, which are indispensable for glycobiology research.

## References


1. Fairbanks AJ (2017) The ENGases: versatile biocatalysts for the production of homogeneous N-linked glycopeptides and glycoproteins. Chem Soc Rev 46:5128–5146. https://doi.org/10.1039/C6CS00897F
2. Kurogochi M (2018) Glycan remodeling of glycoproteins using ENGases. Trends Glycosci Glycotechnol 30:E209–E219. https://doi.org/10.4052/tigg.1764.1E
3. Li C, Wang L-X (2018) Chemoenzymatic methods for the synthesis of glycoproteins. Chem Rev 118:8359–8413. https://doi.org/10.1021/acs.chemrev.8b00238
4. Wang L-X (2019) Glycoengineering of antibodies for modulating functions. Annu Rev Biochem 88:433–459. https://doi.org/10.1146/annurev-biochem-062917-012911
5. Yang Q, Liu Y (2025) Technical, preclinical, and clinical developments of Fc-glycan-specific antibody–drug conjugates. RSC Med Chem 16:50–62. https://doi.org/10.1039/d4md00637b
6. Hsieh Y-C et al (2024) Structure-based high-efficiency homogeneous antibody platform by endoglycosidase Sz provides insights into its transglycosylation mechanism. JACS Au 4:2130–2150. https://doi.org/10.1021/jacsau.4c00004
7. Manabe S et al (2024) Systematic preparation of a 66-IgG library with symmetric and asymmetric homogeneous glycans and their functional evaluation. J Am Chem Soc 146:23426–23436. https://doi.org/10.1021/jacs.4c06558
8. Ou C et al (2021) One-pot conversion of free sialoglycans to functionalized glycan oxazolines and efficient synthesis of homogeneous antibody–drug conjugates through site-specific chemoenzymatic glycan remodeling. Bioconjug Chem 32:1888–1897. https://doi.org/10.1021/acs.bioconjchem.1c00314
9. Henrissat B (1991) A classification of glycosyl hydrolases based on amino acid sequence similarities. Biochem J 280:309–316. https://doi.org/10.1042/bj2800309
10. Lombard V et al (2014) The carbohydrate-active enzymes database (CAZy) in 2013. Nucleic Acids Res 42:D490–D495. https://doi.org/10.1093/nar/gkt1178
11. Tarentino AL, Maley F (1974) Purification and properties of an endo-β-N-acetylglucosaminidase from *Streptomyces griseus*. J Biol Chem 249:811–817. https://doi.org/10.1016/S0021-9258(19)43001-9
12. Trimble RB, Tarentino AL (1991) Distinct endoglycosidase activities in *Flavobacterium meningosepticum*. J Biol Chem 266:1646–1651. https://doi.org/10.1016/S0021-9258(18)52343-7
13. Collin M, Olsén A (2001) EndoS from *Streptococcus pyogenes*. EMBO J 20:3046–3055. https://doi.org/10.1093/emboj/20.12.3046
14. Trastoy B et al (2018) Recognition of complex-type N-glycans by endoglycosidase S. Nat Commun 9:1874. https://doi.org/10.1038/s41467-018-04300-x
15. Seki H et al (2019) Structural basis for cleavage of core-fucosylated N-glycans. J Biol Chem 294:17143–17154. https://doi.org/10.1074/jbc.RA119.010842
16. Muramatsu H (1971) Demonstration of an Endo-glycosidase Acting on a Glycoprotein. J Biol Chem 246(17):5535–5537. https://doi.org/10.1016/S0021-9258(18)61940-4
17. Muramatsu H et al (2001) Molecular cloning and expression of endo-β-N-acetylglucosaminidase D. J Biochem 129:923–928. https://doi.org/10.1093/oxfordjournals.jbchem.a002938
18. Takegawa K et al (1989) Induction and purification of endo-β-N-acetylglucosaminidase from *Arthrobacter protophormiae* grown in ovalbumin. Appl Environ Microbiol 55:3107–3112. https://doi.org/10.1128/aem.55.12.3107-3112.1989
19. Takegawa K et al (1997) Cloning, sequencing, and expression of *Arthrobacter protophormiae* endo-β-N-acetylglucosaminidase in *Escherichia coli*. Arch Biochem Biophys 338:22–28. https://doi.org/10.1006/abbi.1996.9803
20. Kadowaki S et al (1990) Novel fungal endo-β-N-acetylglucosaminidase. Agric Biol Chem 54:97–106. https://doi.org/10.1271/bbb1961.54.97
21. Fujita K et al (2004) Molecular cloning of *Mucor hiemalis* endo-β-N-acetylglucosaminidase. Arch Biochem Biophys 432:41–49. https://doi.org/10.1016/j.abb.2004.09.013

22. Murakami S et al (2013) Endo-β-N-acetylglucosaminidase from *Ogataea minuta.* Glycobiology 23:736–744. https://doi.org/10.1093/glycob/cwt012
23. Eshima Y et al (2015) Transglycosylation activity of glycosynthase mutants. PLoS ONE 10:e0132859. https://doi.org/10.1371/journal.pone.0132859
24. Takashima S et al (2020) Novel endo-β-N-acetylglucosaminidases from *Tannerella* species. Glycobiology 30:923.e934. https://doi.org/10.1093/glycob/cwaa037
25. Stals I et al (2010) Identification of a gene coding for a deglycosylating enzyme in *Hypocrea jecorina.* FEMS Microbiol Lett 303:9–17. https://doi.org/10.1111/j.1574-6968.2009.01849.x
26. Huang Y et al (2018) Characterization of novel endo-β-N-acetylglucosaminidases from various species. Sci Rep 8:246. https://doi.org/10.1038/s41598-017-17467-y
27. Bienes KM et al (2022) Characterization of novel endo-β-N-acetylglucosaminidase from *Bacteroides nordii.* J Biosci Bioeng 134:7–13. https://doi.org/10.1016/j.jbiosc.2022.03.011
28. Nishizawa H et al (2022) Identification and characterization of a novel thermo-stable endo-β-N-acetylglucosaminidase from *Rhizomucor pusillus.* J Biosci Bioeng 134(4):295–300. https://doi.org/10.1016/j.jbiosc.2022.06.013
29. Maeda M et al (2021) Purification, characterization, and gene expression of rice endo-β-N-Acetylglucosaminidase, Endo-Os. Front Plant Sci 12:647684. https://doi.org/10.3389/fpls.2021.647684
30. Trastoy B et al (2021) GH18 endo-β-N-acetylglucosaminidases and hybrid-type N-glycans. J Biol Chem 297:101011. https://doi.org/10.1016/j.jbc.2021.101011
31. Doi K et al (2024) Characterization of novel endo-β-N-acetylglucosaminidases from *Barnesiella intestinihominis.* J Biosci Bioeng 137:101–107. https://doi.org/10.1016/j.jbiosc.2023.12.004
32. Sastre DE et al (2024) Human gut microbes express functionally distinct endoglycosidases to metabolize the same N-Glycan substrate. Nat Commun 15:5123. https://doi.org/10.1038/s41467-024-48802-3
33. Bolino M et al (2025) Novel endo-β-N-Acetylglucosaminidases derived from human fecal samples selectively release N-Glycans from model glycoproteins. Foods 14:1288. https://doi.org/10.3390/foods14081288
34. Suzuki T et al (2002) Endo-β-N-acetylglucosaminidase in processing of free oligosaccharides in the cytosol. Proc Natl Acad Sci USA 99:9691–9696. https://doi.org/10.1073/pnas.152333599
35. Huang C et al (2015) Endo-β-N-acetylglucosaminidase forms N-GlcNAc protein aggregates during ER-associated degradation in Ngly1-defective cells. Proc Natl Acad Sci USA 112:1398–1403. https://doi.org/10.1073/pnas.1414593112
36. Fujihira H et al (2017) Lethality of mice bearing a knockout of the Ngly1-gene is partially rescued by the additional deletion of the Engase gene. PLoS Genet 13:e1006696. https://doi.org/10.1371/journal.pgen.1006696
37. Burke HM (2015) Recent advances in the development of synthetic chemical probes for glycosidase enzymes. Chem Commun 51:10576–10588. https://doi.org/10.1039/C5CC02793D
38. Ishii N et al (2018) A new fluorogenic probe for ENGase detection. ChemBioChem 19:660–663. https://doi.org/10.1002/cbic.201700662
39. Kanie O et al (1993) Acceptor-substrate recognition by N-acetylglucosaminyltransferase-V. Carbohydr Res 243:139–164. https://doi.org/10.1016/0008-6215(93)84087-M
40. Matsuo I et al (1999) Synthesis of an asparagine-linked core pentasaccharide. J Carbohydr Chem 18:841–850. https://doi.org/10.1080/07328309908544039
41. Sano K et al (2019) Efficient synthesis of α(1,2)-linked oligomannoside derivatives through one-pot glycosylation. Carbohydr Res 494:108072. https://doi.org/10.1016/j.carres.2020.108072
42. Bi Y et al (2017) Repurposing of Proton Pump Inhibitors as first identified small molecule inhibitors of endo-β-N-acetylglucosaminidase (ENGase) for the treatment of NGLY1 deficiency, a rare genetic disease. Bioorg Med Chem Lett 27(13):2962–2966. https://doi.org/10.1016/j.bmcl.2017.05.010
43. Li B et al (2008) Synthesis and inhibitory activity of oligosaccharide thiazolines as a class of mechanism-based inhibitors for endo-β-N-acetylglucosaminidases. Bioorg Med Chem 16(8):4670–4675. https://doi.org/10.1016/j.bmc.2008.02.032

44. Ishii N et al (2025) Fluorogenic probes for assessing endo-β-N-acetylglucosaminidase activity on complex-type N-glycans. J Carbohydr Chem 44:275–295. https://doi.org/10.1080/07328303.2025.2558652
45. Ishii N et al (2023) A fluorogenic probe for core-fucosylated glycan-preferred ENGase. Carbohydr Res 523:108724. https://doi.org/10.1016/j.carres.2022.108724
46. Ishii N et al (2019) Fluorogenic probe for ENGase H activity. Bioorg Med Chem Lett 29:1643–1646. https://doi.org/10.1016/j.bmcl.2019.04.039
47. Ishii N et al (2024) Fluorescent probe for hybrid-type N-glycan recognizing ENGases. Bioorg Med Chem 100:117612. https://doi.org/10.1016/j.bmc.2024.117612
48. Yamamoto K et al (1994) Novel Specificities of *Mucor hiemalis* Endo-β-N-acetylglucosaminidase acting complex asparagine-linked oligosaccharides. Biosci Biotech Biochem 58(1):72–77. https://doi.org/10.1271/bbb.58.72

# Modification Glycoside on IgG: Preparation of IgG with Homogeneous Glycan and Expansion to Antibody–Drug Conjugate

**Shino Manabe**

**Abstract** The Fc region of immunoglobulin G (IgG) contains a conserved N-linked glycan at Asn297; its structural variations significantly influence interactions with Fc gamma receptors (FcγRs). Recently, glycoengineering technologies have progressed substantially, allowing precise control over glycan structures and enabling the development of antibodies with enhanced effector functions. This review outlines the biological significance of Fc glycosylation, recent advances in glycoengineering strategies, and their applications, including our own findings. Although the pair of N-glycans present in the Fc region of antibodies is structurally heterogeneous in most cases, previous methods for antibody glycan remodeling have been limited to the generation of symmetric structures. We discovered that IgG can be separated according to the number of Fc glycans through FcγRIIIa affinity column chromatography. By adding second glycans to the isolated hemi-glycosylated IgG, we developed a method to produce IgG molecules that possess structurally distinct paired glycans while maintaining glycan homogeneity at each site. Antibody–drug conjugates (ADCs) have demonstrated remarkable antitumor activity in clinical settings and numerous efforts are underway to develop improved formats. To broaden the therapeutic window, site-specific conjugation strategies are attracting increasing attention as alternatives to conventional random conjugation methods. By expanding antibody glycoengineering approaches, we established a site-specific conjugation platform that enables the generation of structurally homogeneous ADCs exhibiting the anticipated cytotoxic effects to antigen expressed cells.

**Keywords** Antibody-dependent cellular cytotoxicity · Antibody–drug conjugate (ADC) · Endo-β-*N*-acetylglucosaminidase · FcγIIIa affinity column chromatography · IgG glcan-remodeling

S. Manabe (✉)
Laboratory of Synthetic Biomolecular Chemistry, School of Pharmacy and Pharmaceutical Sciences, Hoshi University, Shinagawa-ku, Tokyo, Japan
e-mail: s-manabe@hoshi.ac.jp; shino.manabe.e1@tohoku.ac.jp

Research Center for Pharmaceutical Development, Graduate School of Pharmaceutical Sciences and Faculty of Pharmaceutical Sciences, Tohoku University, Sendai, Miyagi, Japan

 
K. Toshima and Y. Fujimoto (eds.), *Chemical Biology of Carbohydrates*,
https://doi.org/10.1007/978-981-95-5335-8_12

# 1 Introduction

More than half of all proteins undergo glycosylation, a unique and pervasive post- and co-translational modification that critically influences their biological activity, conformational stability, and biodistribution [1]. A prototypical example is erythropoietin (EPO), which bears three N-linked and one O-linked glycans [2]. Notably, the terminal sialic acid residues on these glycans prevent recognition by hepatic clearance receptors such as the asialoglycoprotein receptor, thereby mitigating rapid hepatic clearance and prolonging systemic circulation. Conversely, desialylation leads to the accelerated hepatic uptake of EPO, resulting in a diminished *in vivo* half-life and biological efficacy.

The intricate, multi-enzymatic biosynthetic pathways governing glycan assembly inherently produce microheterogeneity in glycan structures. This structural heterogeneity is also prominent in therapeutic antibodies, where batch-to-batch variability in glycoform distribution is well-documented [3]. Conventional glycan structural analyses rely predominantly on the enzymatic or chemical release of glycans from glycoproteins, a process that disrupts critical information regarding glycosylation site occupancy and glycan co-occurrence patterns. The absence of such topological and combinatorial glycan information poses significant challenges for the rigorous structural characterization of glycoproteins and impedes the elucidation of definitive correlations between glycan microheterogeneity and antibody effector functions.

Currently, immunoglobulin G (IgG) molecules constitute a substantial fraction of therapeutics [4]. Each IgG molecule carries a pair of biantennary complex-type N-glycans attached to Asn297 within the Fc region (Fig. 1). Common glycoform variations include core-fucosylation, galactosylation, and sialylation. As a minor component, high-mannose structures are identified. These glycan modifications modulate IgG interactions with Fc gamma receptors (FcγRs), complement component C1q, and the neonatal Fc receptor, thereby affecting downstream effector mechanisms, such as antibody-dependent cellular cytotoxicity (ADCC), complement-dependent cytotoxicity, and pharmacokinetics, including serum half-life. For instance, afucosylated IgG exhibits substantially increased ADCC due to its heightened affinity for FcγRIIIa. In addition, non-human type glycans of antibody-producing cell origin, such as *N*-glycolylneuraminic acid and α1,3-Gal at the reducing terminus, are also known to possess immunogenicity.

Recent advancements in host cell line engineering, particularly with Chinese hamster ovary (CHO) cells, have facilitated the precise manipulation of glycosylation pathways. For example, α-1,6-fucosyltransferase (FUT)8-knockout CHO cell lines can produce afucosylated antibodies with significantly enhanced ADCC, and several afucosylated IgGs have been administrated in clinical settings [5]. Recent studies employing NMR and molecular dynamics simulations have reported that the presence or absence of fucose and galactose regulates binding to FcγRIIIa [6]. Thus, a comprehensive understanding and deliberate modulation of glycan structures are pivotal for the rational design and development of next-generation therapeutic antibodies.

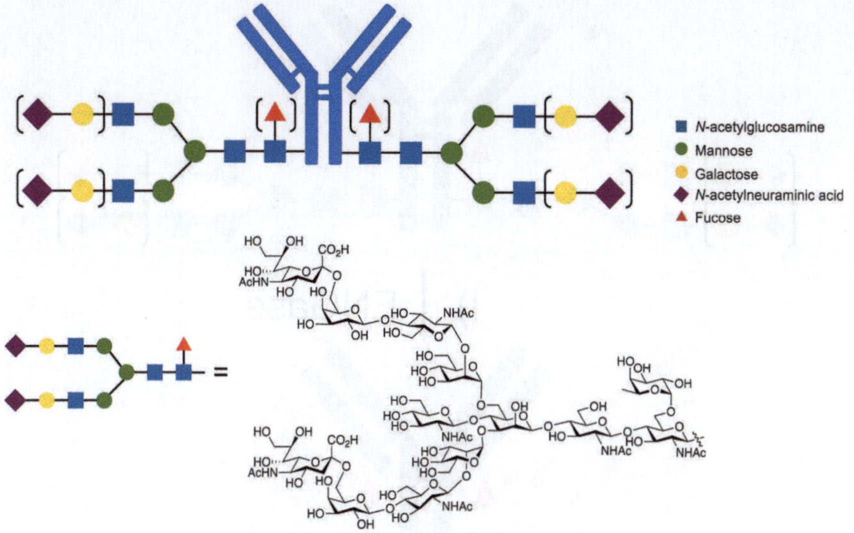


**Fig. 1** A representative biantennary complex-type N-glycan attached to the Fc region of IgG. Parentheses denote structural heterogeneity from biosynthetic pathway variations

# 2 Preparation of IgG with Homogeneous Glycans

## 2.1 *Conventional IgG Glycan-Remodeling Strategy: Symmetric Pattern*

To elucidate the effect that glycan structures have on IgG function, methodologies have been established to generate IgG bearing homogeneous glycan structures (Fig. 2). This technology employs endo-β-*N*-acetylglucosaminidase (ENGase) mutants and glycan oxazolines as glycan donors [7]. ENGases are glycosidic hydrolases that cleave the β-1,4-glycosidic linkage within the *N,N′*-diacetylchitobiose core of *N*-glycans, leaving a single *N*-acetylglucosamine residue attached to the asparagine residue of the Fc region. The displacement of amino acids in the active site generates ENGase mutants with suppressed glycan hydrolysis activity that retain their glycan transfer activity. Thus, ENGase mutants transfer glycans *en block* to glycan-truncated antibodies, resulting in the production of antibodies with homogeneous glycan structures. Various ENGases have been isolated from different sources [8]. Among the known ENGases, EndoS [9] and Endo S2 mutants [10] have especially been employed for antibody glycan remodeling because of their high affinity for IgG. Glycan cleavage effectively removes heterogeneous non-reducing terminal residues, as well as potentially immunogenic moieties. Subsequently, a structurally defined glycan can be attached to the glycan-truncated IgG, resulting in an IgG molecule with a homogeneous glycan structure. For this purpose, glycans are converted into their corresponding oxazoline derivatives, which serve as substrates for ENGase mutants.

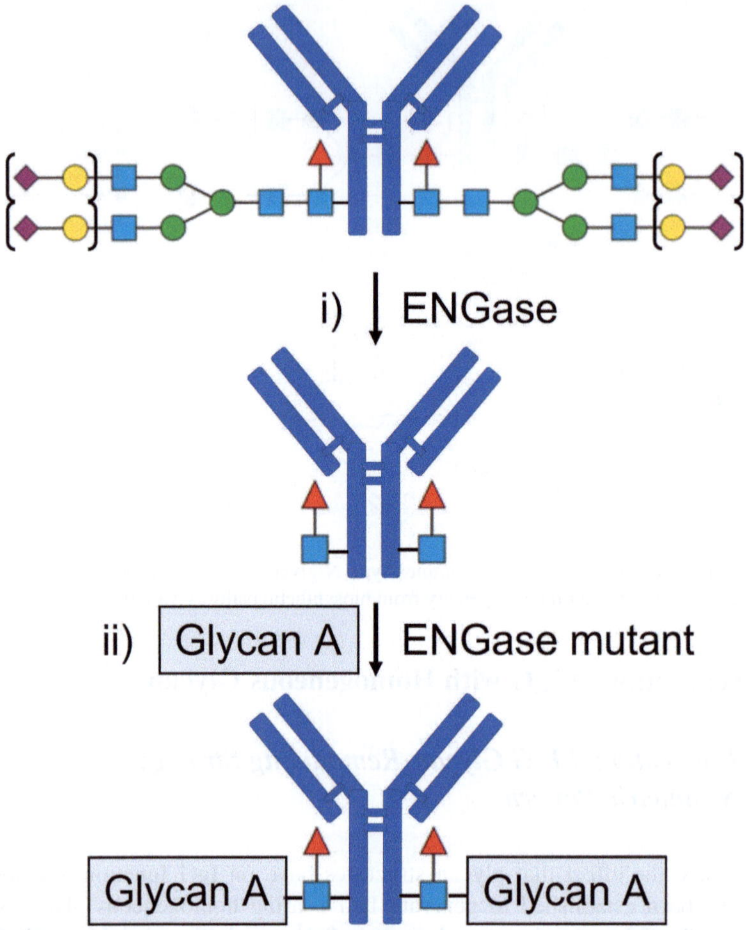


**Fig. 2** Conventional IgG glycan-remodeling. (i) Removal of the N-glycan at the core chitobiose linkage by endo-β-*N*-acetylglucosaminidase (ENGase); (ii) Subsequent transfer of a structurally defined N-glycan onto the GlcNAc acceptor site by ENGase-catalyzed transglycosylation. Parentheses denote structural heterogeneity

Glycan oxazolines are analogs of the oxazolinium ion intermediates produced during enzymatic hydrolysis (Fig. 3). This high-energy oxazoline species represents a transient state generated during hydrolysis, so ENGase mutants can utilize the oxazolines as glycosyl donors. Several IgG variants carrying homogeneous glycans have been successfully synthesized using a combination of ENGase mutants and glycan oxazolines.

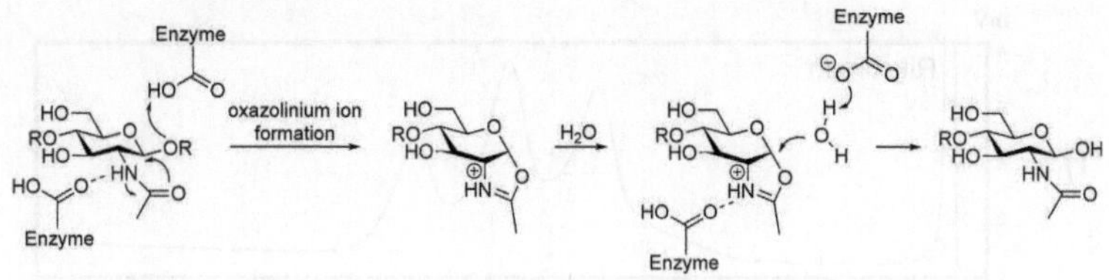


**Fig. 3** Mechanism of ENGase-catalyzed glycan remodeling and oxazolinium ion intermediates. Cleavage of the N-glycan at the core chitobiose bond by ENGase proceeds through the formation of an oxazolinium ion intermediate. Engineered ENGase mutants can exploit this reactive oxazoline as a glycosyl donor, thereby catalyzing transglycosylation to transfer a structurally defined glycan onto the acceptor GlcNAc residue of the antibody Fc domain

## 2.2 FcγRIIIa Affinity Column Chromatography

FcγRIIIa affinity column chromatography was developed for measuring ADCC [11]. ADCC is an immune mechanism through which effector cells recognize and kill target cells that have been opsonized with antibodies. Specifically, antibodies bind to antigens expressed on the surface of target cells via their Fab regions. The Fc regions of these bound antibodies are then recognized by FcγRs expressed on the surface of effector cells, such as natural killer cells, macrophages, and neutrophils. Upon FcγR engagement, effector cells become activated and release cytotoxic molecules, including perforin and granzymes, which induce apoptosis in the target cell. This process contributes to the elimination of infected or malignant cells and is an important mechanism of action of several therapeutic antibodies.

Typical therapeutic IgGs such as rituximab, trastuzumab, and mogamulizimab, exhibit three peaks on FcγRIIIa affinity chromatograms due to glycan heterogeneity (Fig. 4). Mogamulizimab, afucosylated IgG to enhance ADCC showed longer retention time. In contrast, IgGs bearing homogeneous glycans generated through IgG glyco-remodeling technology produce a single peak on FcγRIIIa affinity chromatograms [12]. A similar pattern was observed for the Fc domain alone. Fc domains obtained via papain digestion of IgGs with heterogeneous glycans yielded three peaks [13], whereas those derived from IgGs with homogeneous glycans displayed a single peak. In both cases, the retention times of Fc fragments were longer than those of the intact IgG.

## 2.3 IgG Glycan-Remodeling Strategy: Asymmetric Pattern

Conventional IgG glycan remodeling is limited to the generation of IgGs with identical glycan pairs (symmetric type). In natural IgGs, most antibodies have different glycans in each pair (asymmetric type). Nevertheless, developing a method for producing asymmetric glycan-homogeneous IgG remains unexplored.

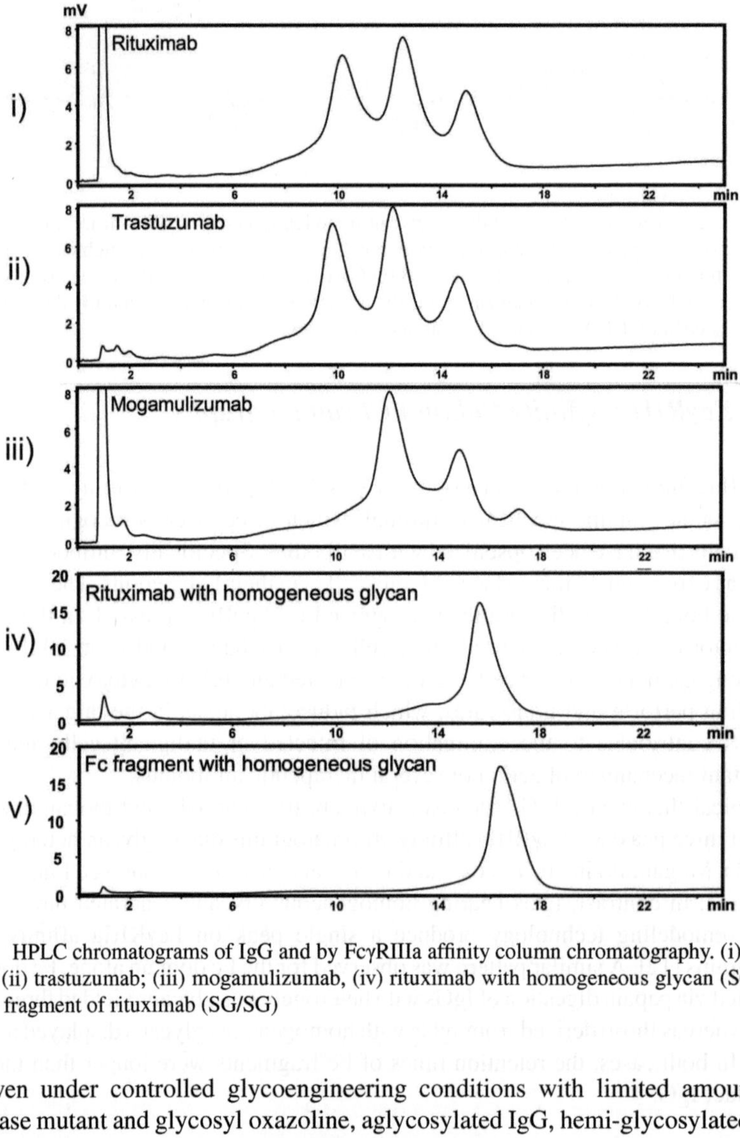


**Fig. 4** HPLC chromatograms of IgG and by FcγRIIIa affinity column chromatography. (i) Rituximab; (ii) trastuzumab; (iii) mogamulizumab, (iv) rituximab with homogeneous glycan (SG/SG); (v) Fc fragment of rituximab (SG/SG)

Even under controlled glycoengineering conditions with limited amounts of ENGase mutant and glycosyl oxazoline, aglycosylated IgG, hemi-glycosylated IgG, and IgG containing two glycans could be generated.

We originally discovered that the retention time of IgG in FcγRIIIa affinity chromatography correlates with the number of attached glycans (Fig. 5). Leveraging this phenomenon, hemi-glycosylated IgG was successfully isolated. Subsequently, a second glycan was enzymatically attached to the hemi-glycosylated IgG using ENGase and glycan oxazoline donors, yielding homogeneous IgG molecules with asymmetric glycan structures (Fig. 6). Using purification via FcγRIIIa affinity column chromatography, we synthesized a homogeneous glycan library from 66-trastuzumab (Table 1) [14].

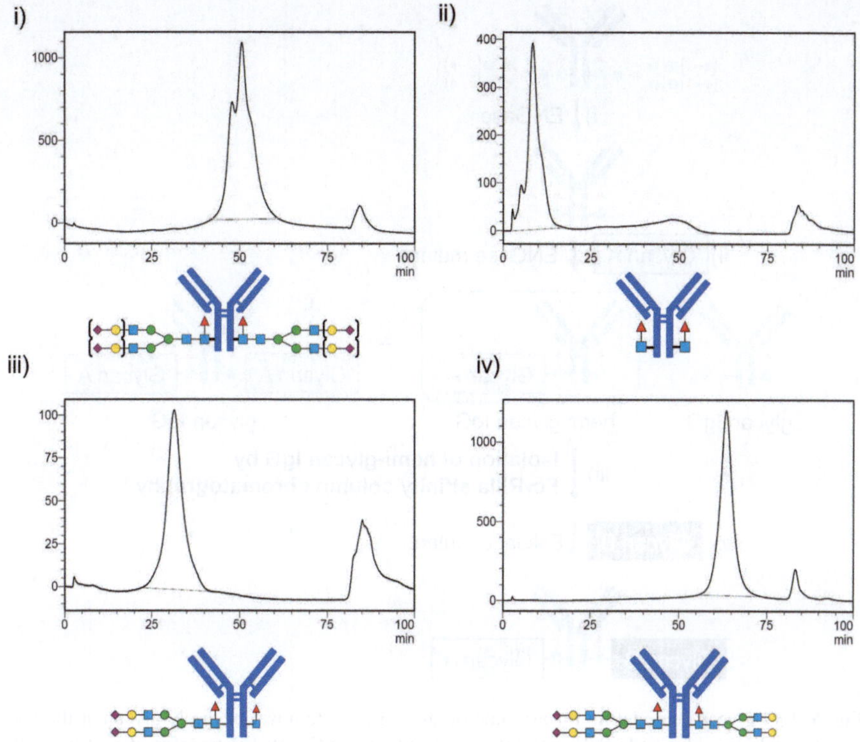


**Fig. 5** HPLC spectra of FcγRIIIa affinity column chromatography depending on the number of N-glycans on trastuzumab: (i) Trastuzumab with heterogeneous glycan structures, retention time 50.64 min; (ii) Trastuzumab after glycan removal by ENGase, retention time 8.37 min; (iii) Trastuzumab with a single N-glycan, retention tine 31.48 min; (iv) Trastuzumab with a homogeneous N-glycan, retention time 63.56 min

## 2.4 *ADCC of IgG with Homogeneous Glycan*

Trastuzumab is a therapeutic IgG widely used for breast and gastric cancer. The prepared library showed that ADCC tended to increase as the glycan became larger. In glycan-truncated IgG, ADCC was markedly reduced, and in single-chain glycan IgG, ADCC tended to decrease. Among the trastuzumab molecules with an asymmetric glycan type, several with a high ADCC were found (Table 1). As previously reported, the G2 series had a higher ADCC than the G1 and G0 series did. In addition, even when the number of galactose residues was the same, the A1a-F and G1a-F (Manα1–6 arm) series exhibited a higher ADCC than the A1b-F and G1b-F (Manα1–3 arm) series did. A similar tendency was observed when we compared the A1a-F and A1b-F series. In general, the influence of *N*-acetylneuraminic acid residues on ADCC remains unclear [15–17]. Our library revealed that the range of ADCC values was reduced when the number of *N*-acetylneuraminic acid residues was greater than two. The A1a-F series tended to exhibit a higher ADCC than the A1b-F series did.

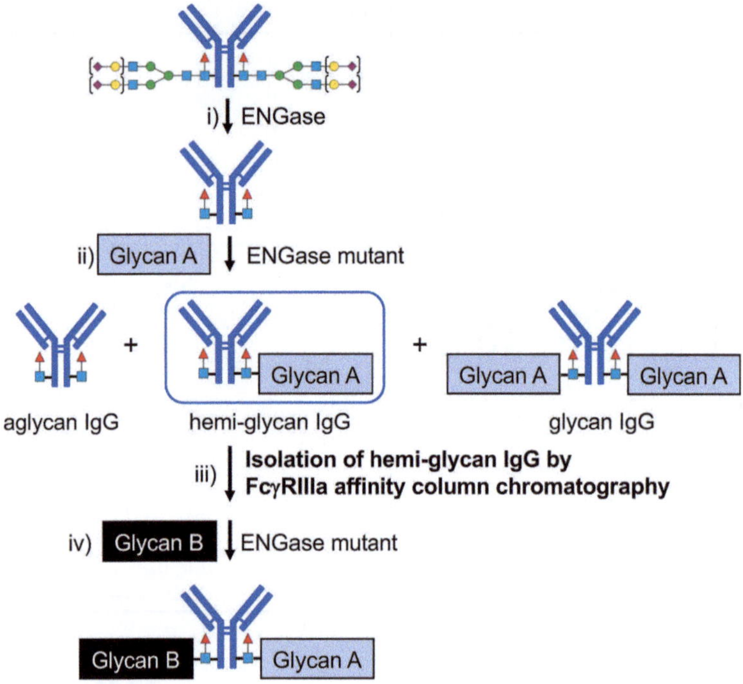


**Fig. 6** IgG glycan-remodeling for asymmetric glycans. (i) Removal of the N-glycan at the core chitobiose linkage by endo-β-*N*-acetylglucosaminidase (ENGase); (ii) transfer of a structurally defined N-glycan onto the GlcNAc acceptor site by ENGase-catalyzed transglycosylation; (iii) isolation of hemi-glycan IgG by FcgIIIa affinity column chromatography; (iv) transfer of second N-glycan onto the GlcNAc acceptor site by ENGase-catalyzed transglycosylation. Parentheses denote structural heterogeneity

## 2.5 *Thermal Stability of IgG with Homogeneous Glycan*

Glycans are known to influence the thermal stability of IgG. To evaluate the correlation between glycosylation and thermal stability, we measured the thermal unfolding temperature (Tm) and calorimetric unfolding enthalpy (ΔHcal) using differential scanning calorimetry. The Tm of glycan-truncated trastuzumab was substantially reduced to 64.9 °C. Although the thermal stability of IgG carrying only one glycan chain markedly decreased, in the case of IgG with two glycans, the stability generally improved with increasing glycan length. The addition of a single sialic acid residue increased Tm values and decreased the ΔHcal; however, when the number of sialic acid residues exceeded two, the Tm values tended to decrease. The influence of sialic acid might be attributed to charge of the carboxylic acid and/or high dynamic properties. The reason is not clear yet.

A strong direct linear correlation was observed between the Tm and ΔHcal values of trastuzumab and the A1b-F, A1b-Gal-F, and M3-F glycan series. The reason for this phenomenon still remains unclear, requiring further elucidation using a structural

**Table 1** ADCC of trastuzumabs with homogeneous glycans

| | SG-F | A1a-F | A1b-F | A1a-Gal-F | A1b-Gal-F | G2-F | G1a-F | G1b-F | G0-F | M3-F | GlcNAc |
|---|---|---|---|---|---|---|---|---|---|---|---|
| SG-F | 3133 | | | | | | | | | | |
| A1a-F | 27643 | 39280 | | | | | | | | | |
| A1b-F | 41017 | 38193 | 39547 | | | | | | | | |
| A1a-Gal-F | 33170 | 33777 | 33373 | 38490 | | | | | | | |
| A1b-Gal-F | 29577 | 35690 | 36510 | 29983 | 36930 | | | | | | |
| G2-F | 38107 | 44187 | 40923 | 37900 | 37417 | 42130 | | | | | |
| G1a-F | 39320 | 42807 | 38637 | 28817 | 25830 | 41340 | 35523 | | | | |
| G1b-F | 34920 | 41317 | 36523 | 37573 | 40503 | 37167 | 39343 | 17920 | | | |
| G0-F | 30937 | 41557 | 25007 | 29507 | 33773 | 42967 | 32157 | 29753 | 38633 | | |
| M3-F | 36403 | 34967 | 13400 | 35940 | 36080 | 40213 | 33367 | 23283 | 28880 | 17303 | |
| GlcNAc | 11510 | 23990 | 11173 | 15983 | 13450 | 28227 | 18807 | 8930 | 6193 | 5647 | 11450 |

biology approach. Our approach allowed for the analysis of glycan structures and IgG functions without the need for glycan cleavage. By avoiding glycan removal from IgG, we preserved information about the paired glycans that would otherwise have been lost.

# 3 Application of Glycoengineering to Site-Specific Antibody–Drug Conjugate (ADC)

## *3.1 Introduction of ADC*

Recently, antibody–drug conjugates (ADCs) have emerged as a major therapeutic strategy for cancer treatment. Currently, more than 13 ADCs have been approved for clinical use, and over 100 are in various stages of development [18].

An ADC comprises three key components: an IgG antibody, cytotoxic payload, and linker. The IgG molecule serves to deliver the potent cytotoxic agent specifically to the tumor site. The therapeutic mechanism of ADCs begins with their penetration into tumor tissue via tumor-associated blood vessels, facilitated by the enhanced permeability and retention effect [19]. Once the ADC extravasates from the tumor vasculature, it binds to specific antigens on the cell surface. Subsequently, the antigen–ADC complex is internalized by the cell. Within the lysosome, metabolic enzymes release the cytotoxic payload, ultimately leading to cell death.

### 3.2 Importance of Site-Specific Conjugation in ADC Design

The conjugation strategy is a critical aspect of ADC design. Currently, lysine or cysteine residues on antibodies are commonly used as attachment sites for the cytotoxic payload. However, this conventional approach lacks precise control over the drug-to-antibody ratio (DAR) and specific conjugation sites, resulting in structural heterogeneity.

To enhance the therapeutic window of ADC and respond to regulatory science, site-specific conjugation was investigated [20]. Several strategies have been proposed for generating site-specific conjugated ADCs, including the incorporation of noncanonical amino acids for selective modification [21, 22] and enzymatic conjugation for specific peptide sequence recognition [23]. Moreover, homogeneous ADCs created using Fc-affinity peptide-based reagents have markedly improved therapeutic indices [24].

### 3.3 N-Glycans for Site-Specific Conjugation in ADC Design

Glycans are suitable targets for potential conjugation sites. As glycans are located at the Fc region, they does not affect the binding affinity between Fab and an antigen. Strategies for glycan-conjugated ADC preparation can be categorized into three principal approaches. The first involves conjugation of the cytotoxic payload to a carbonyl group generated through oxidative cleavage of the *cis*-diol present in fucose or *N*-acetylneuraminic acid. The second approach utilizes monosaccharyl transferases, such as galactosyl and sialyl transferases, with azide group-containing monosaccharide donors, leveraging the tolerance of these transferases for modified glycosyl donor structures. The third strategy involves the expansion of IgG glycan remodeling. When glycans with a conjugation site are attached to IgG, the cytotoxic payload can be attached to the glycan site in a site-specific manner.

### 3.4 Diol Oxidation Strategy

As fucose and sialic acid have *cis*-diol molecules, formyl group generation via the oxidative cleavage of fucose and subsequent conjugation between formyl group and hydrazide was reported [25]. Overoxidation of one formyl group to carboxylic acid was also observed. The modified antibody showed comparable dissociation to antigen that of the unmodified antibody based on a surface plasmon resonance (SPR) experiment. The stability of hydrazone linkers depends on their structural characteristics, and the dolastatin analog was released slowly. ADC preparation through formyl group generation via sialic acid oxidation was also possible [26]. The acyclic diol moiety of sialic acid is flexible and unhindered compared to those found in other monosaccharides. Sialic acid can be oxidized under relatively mild conditions. However, normal IgGs normally contain low levels (< 10–14%) of sialic acid. To increase sialic acid levels, the oligosaccharide was elongated using commercially available galactosyltransferase and α2,6-sialyltransferase using uridine diphosphate-Gal and cytidine monophosphate (CMP)-sialic acid as glycosyl donors. IgG contains oxidation-prone residues such as Met and Trp, and exposure to oxidative conditions raises concerns about potential antibody damage.

### *3.5 Monosaccharyl Transferase Strategy*

As galactosyltransferase tolerates a modified galactosyl donor structure, galactose with a handle for functionalization can be added by galactosyl transferase (Fig. 7). C2-keto-galactose was incorporated by the mutant galactosyl transferase mutant, β(1,4)Gal-T1-Y289L [27]. In addition to C2-keto-galactose, azide-GalNAc was added by the enzyme after the removal of native glycan by ENGase. Currently, at least three glycan-conjugated ADCs, including ADCT-601 [28] and XMT-1592 [29], are undergoing Phase I clinical trials. These ADCs incorporate pyrrolobenzodiazepine derivatives, known for their potent cytotoxicity. Sialyltransferases are alternative enzyme for modified monomer addition. Sialyltransferases tolerate modifications at the C-5 and C-9 positions of sialic acids. 9-azide-functionalized CMP-sialic acid was introduced for a handle for conjugation and ADC was prepared via biorthogonal reaction [30].

Metabolic N-glycan modifications are possible; when fucose analogs are added outside the cells, fucosyltransferase adds fucose analogs to the core GlcNAc molecule. Fucose analogs are typically added in their peracetylated form as this modification is known to increase cellular uptake efficiency. Acetate is subsequently hydrolyzed in the cytosol and converted to GDP-sugar via the fucose salvage pathway. Extensive studies have examined applications of the metabolic modification of *N*-glycan to site-specific anticancer conjugation techniques [31]. Thiols carrying a fucose derivative were incorporated instead of fucose, showing 70% efficacy. Anti-CD70 (h1F6) and anti-CD30 (cAc10) antibodies were used to generate ADCs via the incorporation of thiofucose into CHO cells.

### *3.6 IgG Glycan-Remodeling Strategy*

For ADC preparation using ENGase mutant, the carboxylic acid in *N*-acetylneuraminic acid can act as a conjugation site. Through carboxylic acid, amido bond formation with the azide group for drug conjugation is possible. The azide group is a representative bio-orthogonal functional group that reacts with strained alkyne groups.

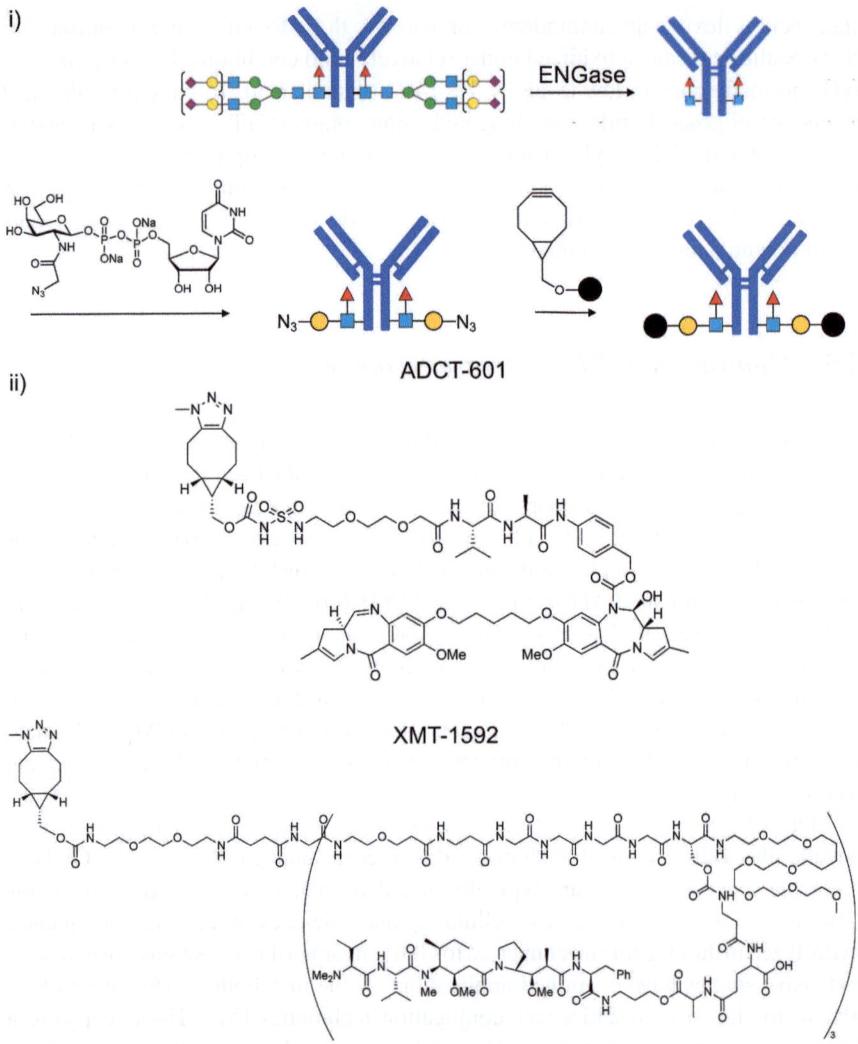


**Fig. 7** (i) Scheme of site-specific conjugation using a galactosyltransferase mutant; (ii) Structures of the ADCT-601 and XMT-1592 payloads. Parentheses denote structural heterogeneity

We added an azide group by using bifunctional amino-PEG-azide to the sialyl glycan (Scheme 1) [32]. The glycan was prepared from sialylglycopeptide isolated from egg yolk by action of ENGase. Condensation between the carboxylic acid in *N*-acetylneuraminic acid and amino group with an azide was carried out using 1*H*-benzotriazol-1-yloxytripyrrolidinophosphonium hexafluorophosphate. Subsequently, the oxazoline was prepared using 2-chloro-1,3-dimethyl-1*H*-benzimidazol-3-ium chloride [33]. The obtained glycan oxazoline was added by endoS D233Q to glycan-truncated trastuzumab to generate IgG with an azide group at the glycan moiety. The cytotoxic payload with the strained alkyne was then conjugated to the azide group. This process enabled the preparation of site-specific conjugated homogeneous ADCs with a DAR of 4 (Fig. 8). For the cytotoxic payload, we chose monomethyl auristatin E (MMAE), a common payload used for therapeutic ADC, which also include brentuximab vedotin, polatuzumab vedotin, and enfortumab vedotin. MMAE is a highly potent synthetic antimitotic agent derived from the natural product dolastatin 10. It functions as a tubulin polymerization inhibitor, binding to tubulin and pre MMAE is a highly potent synthetic antimitotic agent derived from the natural product, dolastatin 10. It functions as a tubulin polymerization inhibitor that binds to tubulin, preventing microtubule assembly and causing subsequent apoptotic cell death. Owing to its extreme cytotoxicity, MMAE cannot be administered as a free drug but is instead utilized as a payload in ADCs, where its activity is selectively delivered to tumor cells. MMAE is typically linked to antibodies via a protease-cleavable dipeptide linker (commonly a valine–citrulline [Val-Cit] linker) [34]. This linker is stable in systemic circulation but is efficiently cleaved by cathepsin B and related lysosomal proteases within target cells following internalization of the ADC. Upon cleavage, free MMAE is released into the cytosol where it exerts its microtubule-disrupting activity. We confirmed the homogeneity of ADC using hydrophilic interaction chromatography and electrospray ionization mass spectrometry.

The low HER2-expressing breast cancer cell line, MCF–7, and gastric cancer cell line, MKN–45, were insensitive to antibody or ADC exposure. In contrast, exposure

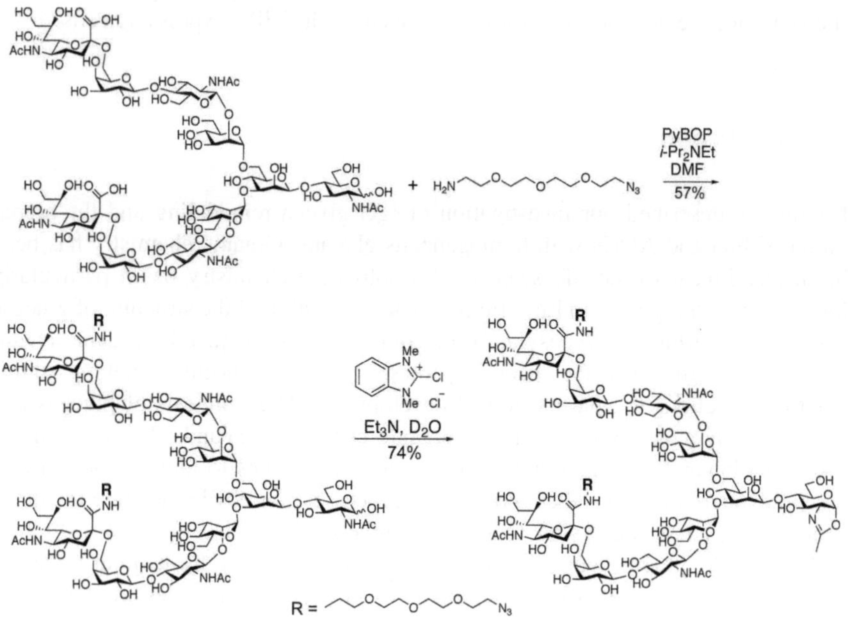


**Scheme 1** Preparation of azide-carrying glycan oxazoline

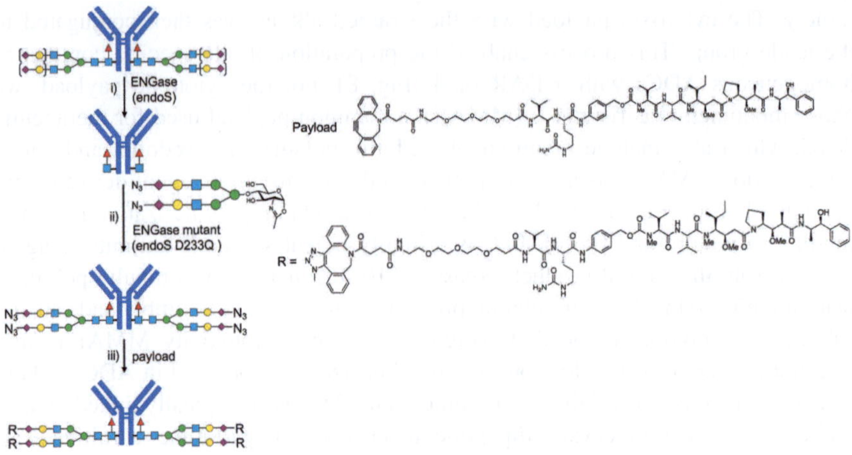


**Fig. 8** Scheme for glycan specific conjugated homogeneous ADC preparation. (i) Removal of the N-glycan at the core chitobiose linkage by endo-β-*N*-acetylglucosaminidase (ENGase); (ii) subsequent transfer of a structurally defined N-glycan with azide group onto the GlcNAc acceptor site by ENGase-catalyzed transglycosylation; (iii) addition of payload via biorthogonal reaction. Parentheses denote structural heterogeneity

of high HER2-expressing cells to ADCs caused a dose-dependent decrease in cell proliferation, and the $IC_{50}$ values were calculated as 0.3, 0.9, and 1.8 nM (MMAE equivalent) against SK–BR–3, OE–19, and N87 cells, respectively. These results showed selective delivery of cytotoxic agents to high HER2-expressing cells.

# 4 Conclusion

Herein, we described our investigation of IgG glycan remodeling and the preparation of IgG and ADCs with homogeneous glycans. Organic chemistry has been intimately linked to the life sciences. Carbohydrate chemistry has a particularly long history, dating back to Emil Fischer's determination of the structure of glucose and his pioneering on glycosylation reactions [35]. Subsequently, organic chemistry has contributed to the precise synthesis of small molecules, making significant contributions to pharmaceutical development. The synthesis of glycans and ADCs in a homogenous form using organic chemistry expands the scope of the discipline beyond its traditional focus on small molecules to include biologically essential macromolecules that underlie life processes, thereby broadening the range of molecular targets accessible to organic synthesis.

Meanwhile, pharmaceutical modalities have been shifting from small-molecule compounds to biologics, such as antibodies. Carbohydrate moieties are ubiquitous and found not only as glycosides in small-molecule antibiotics but also as essential components of proteins, exemplified by antibodies. Carbohydrate chemistry is anticipated to play an important role in cutting-edge fields, such as regenerative medicine and cell therapy. Therefore, an increasing demand exists for organic chemistry that can enable the precise and versatile synthesis of complex biomacromolecules.

## References

1. Toustou C, Walet-Balieu ML, Kiefer-Meyer MC, Houdou M, Lerouge P, Foulquier F, Bardor M (2022) Towards understanding the extensive diversity of protein N-glycan structures in eukaryotes. Biol Rev Camb Philos Soc 97:732–748. https://doi.org/10.1111/brv.12820
2. Takeuchi M, Kobata A (1991) Structures and functional roles of the sugar chains of human erythropoietins. Glycobiology 1:337–346. https://doi.org/10.1093/glycob/1.4.337
3. Kim S, Song J, Park S, Ham S, Paek K, Kang M, Chae Y, Seo H, Kim HC, Flores M (2017) Drifts in ADCC-related quality attributes of Herceptin®: impact on development of a trastuzumab biosimilar. MAbs 9:704–714. https://doi.org/10.1080/19420862.2017.1305530
4. Chan AC, Martyn GD, Carter PJ (2025) Fifty years of monoclonals: the past, present and future of antibody therapeutics. Nat Rev Immunol. https://doi.org/10.1038/s41577-025-01207-9
5. Golay J, Andrea AE, Cattaneo I (2022) Role of Fc core fucosylation in the effector function of IgG1 antibodies. Front Immunol 13:929895. https://doi.org/10.3389/fimmu.2022.929895
6. Yanaka S, Sakae Y, Miyanoiri Y, Yamaguchi T, Isono Y, Kondo S, Iwasaki M, Onitsuka M, Yagi H, Kato K (2025) Exploring glycoform-dependent dynamic modulations in human immunoglobulin G via computational and experimental approaches. Proc Natl Acad Sci USA 122:e2505473122. https://doi.org/10.1073/pnas.2505473122
7. Fairbanks AJ (2018) Synthetic and semi-synthetic approaches to unprotected N-glycan oxazolines. Beilstein J Org Chem 14:416–429. https://doi.org/10.3762/bjoc.14.30
8. Fairbanks, AJ (2017) The ENGases: versatile biocatalysts for the production of homogeneous N-linked glycopeptides and glycoproteins. Chem Soc Rev 46:5128–5146. https://doi.org/10.1039/C6CS00897F
9. Huang W, Giddens J, Fan S-Q, Toonstra C, Wang L-X (2012) Chemoenzymatic Glycoengineering of Intact IgG antibodies for gain of functions. J Am Chem Soc 134:12308–12318. https://doi.org/10.1021/ja3051266
10. Li T, Tong X, Yang Q, Giddens JP, Wang LX (2016) Glycosynthase mtants of endoglycosidase S2 show potent transglycosylation activity and remarkably relaxed substrate specificity for antibody glycosylation remodeling. J Biol Chem 291:16508–18. https://doi.org/10.1074/jbc.M116.738765
11. Kiyoshi M, Caaveiro JMM, Tada M, Tamura H, Tanaka T, Terao Y, Morante K, Harazono A, Hashii N, Shibata H, Kuroda D, Nagatoishi S, Oe S, Ide T, Tsumoto K, Ishii-Watabe A (2018) Assessing the heterogeneity of the Fc-glycan of a therapeutic antibody using an engineered Fcγreceptor IIIa-immobilized column. Sci Rep 8:3955. https://doi.org/10.1038/s41598-018-22199-8
12. Wada R, Matsui M, Kawasaki N (2018) Influence of N-glycosylation on effector functions and thermal stability of glycoengineered IgG1 monoclonal antibody with homogeneous glycoforms. mAbs 11:350–372. https://doi.org/10.1080/19420862.2018.1551044
13. Hiranyakorn M, Iwamoto S, Hoshinoo A, Tsumura R, Takashima H, Yasunaga M, Manabe S (2023) Chromatographic analysis of the N-glycan profile on therapeutic antibodies using FcγRIIIa affinity column chromatography. ACS Omega 8:16513–16518. https://doi.org/10.1021/acsomega.3c02374
14. Manabe S, Iwamoto S, Nagatoishi S, Hoshinoo A, Mitani A, Sumiyoshi W, Kinoshita T, Yamaguchi Y, Tsumoto K (2024) Systematic preparation of a 66-IgG library with symmetric and asymmetric homogeneous glycans and their functional evaluation. J Am Chem Soc 146:23426–23436. https://doi.org/10.1021/jacs.4c06558

15. Lin CW, Tsai M, Li ST, Tsai TI, Chu KC, Liu YC, Lai MY, Wu CY, Tseng YC, Shivatare SS, Wang CH, Chao P, Wang SY, Shih HW, Zeng YF, You TH, Liao JY, Tu YC, Lin YS, Chuang HY, Chen CL, Tsai CS, Huang CC, Lin NH, Ma C, Wu CY, Wong CH (2015) A common glycan structure on immunoglobulin G for enhancement of effector functions. Proc Natl Acad Sci USA 112:10611–10616. https://doi.org/10.1073/pnas.1513456112
16. Li T, DiLillo DJ, Bournazos S, Giddens JP, Ravetch JV, Wang LX (2017) Modulating IgG effector function by Fc glycan engineering. Proc Natl Acad Sci USA 114:3485–3490. https://doi.org/10.1073/pnas.1702173114
17. Scallon BJ, Tam SH, McCarthy SG, Cai AN, Raju TS (2007) Higher levels of sialylated Fc glycans in immunoglobulin G molecules can adversely impact functionality. Mol Immunol 44:1524–1534. https://doi.org/10.1016/j.molimm.2006.09.005
18. Liu K, Li M, Li Y, Li Y, Chen Z, Tang Y, Yang M, Deng G, Liu H (2024) A review of the clinical efficacy of FDA-approved antibody-drug conjugates in human cancers. Mol Cancer 23:62. https://doi.org/10.1186/s12943-024-01963-7
19. Matsumura Y, Maeda H (1986) A new concept for macromolecular therapeutics in cancer chemotherapy: mechanism of tumoritropic accumulation of proteins and the antitumor agent smancs. Cancer Res 46:6387–6392
20. Tsuchikama K, Anami Y, Ha SYY, Yamazaki CM (2024) Exploring the next generation of antibody-drug conjugates. Nat Rev Clin Oncol 21:203–223. https://doi.org/10.1038/s41571-023-00850-2
21. Axup JY, Bajjuri KM, Ritland M, Hutchins BM, Kim CH, Kazane SA, Halder R, Forsyth JS, Santidrian AF, Stafin K, Lu Y, Tran H, Seller AJ, Biroc SL, Szydlik A, Pinkstaff JK, Tian F, Sinha SC, Felding-Habermann B, Smider VV, Schultz PG (2012) Synthesis of site-specific antibody-drug conjugates using unnatural amino acids. Proc Natl Acad Sci USA 109:16101–16106. https://doi.org/10.1073/pnas.1211023109
22. Junutula JR, Raab H, Clark S, Bhakta S, Leipold DD, Weir S, Chen Y, Simpson M, Tsai SP, Dennis MS, Lu Y, MengYG NC, Yang J, Lee CC, Duenas E, Gorrell J, Katta V, Kim A, McDorman K, Flagella K, Venook R, Ross S, Spencer SD, Mallet W (2008) Site-specific conjugation of a cytotoxic drug to an antibody improves the therapeutic index. Nat Biotech 26:925–932. https://doi.org/10.1038/nbt.1480
23. Anami EY, Xiong W, Gui X, Deng M, Zhang CC, Zhang N, Ana Z, Tsuchikama K (2017) Enzymatic conjugation using branched linkers for constructing homogeneous antibody–drug conjugates with high potency. Org Biomol Chem 15:5635–5642. https://doi.org/10.1039/C7OB01027C
24. Fujii T, Matsuda Y, Seki T, Shikida N, Iwai Y, Ooba Y, Takahashi K, Isokawa M, Kawaguchi S, Hatada N, Watanabe T, Takasugi R, Nakayama A, Shimbo K, Mendelsohn BA, Okuzumi T, Yamada K (2023) AJICAP second generation: Improved chemical site-specific conjugation technology for antibody-drug conjugate production. Bioconjug Chem 34:728–738. https://doi.org/10.1021/acs.bioconjchem.3c00040
25. Zuberbühler K, Casi G, Bernardesa GJL, Neri D (2012) Fucose-specific conjugation of hydrazide derivatives to a vascular-targeting monoclonal antibody in IgG format. Chem Commun 48:7100–7102. https://doi.org/10.1039/C2CC32412A
26. Zhou Q, Stefano JE, Manning C, Kyazike J, Chen B, Gianolio DA, Park A, Busch M, Bird J, Zheng X, Simonds-Mannes H, Kim J, Gregory RC, Miller RJ, Brondyk WH, K. Dhal PK (2014) Site-specific antibody-drug conjugation though glycoengineering. Bioconjug Chem 25:510–520. https://doi.org/10.1021/bc400505q
27. Boeggeman E, Ramakrishnan B, Pasek M et al (2009) Site specific conjugation of fluoroprobes to the remodeled Fc *N*-glycans of monoclonal antibodies using mutant glycosyltransferases: application for cell surface antigen detection. Bioconjug Chem 20:1228–1236. https://doi.org/10.1021/bc900103p
28. Zammarchi F, Havenith KE, Chivers S, Hogg P, Bertelli F, Tyrer P, Janghra N, Reinert HW, Hartley JA, van Berkel PH (2022) Preclinical development of ADCT-601, a novel pyrrolobenzodiazepine dimer-based antibody–drug conjugate targeting AXL-expressing cancers. Mol Cancer Ther 21:582–593. https://doi.org/10.1158/1535-7163.MCT-21-0715

29. Toader D, Fessler SP, Collins SD, Conlon PR, Bollu R, Catcott KC, Chin C-N, Dirksen A, Du B, Duvall JR, Higgins S, Kozytska MV, Bellovoda K, Faircloth C, Lee D, Li F, Qin L, Routhier C, Shaw P, Stevenson CA, Wang J, Wongthida P, Ter-Ovanesyan E, Ditty E, Bradley SP, Xu L, Yin M, Yurkovetskiy AV, Mosher R, Damelin M, Lowinger TB (2023) Discovery and preclinical characterization of XMT-1660, an optimized B7-H4-targeted antibody–drug conjugate for the treatment of cancer. Mol Cancer Ther 22:999–1012. https://doi.org/10.1158/1535-7163.MCT-21-0715.
30. Li X, Fang T, Boons GJ (2014) Preparation of well-defined antibody-drug conjugates through glycan remodeling and strain-promoted azide-alkyne cycloadditions. Angew Chem Int Ed 53:7179–7182. https://doi.org/10.1002/anie.201402606
31. Manabe S, Yamaguchi Y, Matsumoto K, Fuchigami H, Kawase T, Hirose K, Mitani A, Sumiyoshi W, Kinoshita T, Abe J, Yasunaga M, Matsumura Y, Ito Y (2019) Characterization of antibody products obtained through enzymatic and nonenzymatic glycosylation reactions with a glycan oxazoline and preparation of a homogeneous antibody-drug conjugate via Fc N-glycan. Bioconjugate Chem 30:1343–1355. https://doi.org/10.1021/acs.bioconjchem.9b00132
32. Manabe S, Ito Y (2019) 1,2-*cis*-selective formation of a unique amino-containing amino glycoside by endocyclic cleavage strategy. Heterocycles 99:1304–1314. https://doi.org/10.3987/COM-18-S(F)101
33. Noguchi M, Fujieda T, Huang WC, Ishihara M, Kobayashi A, Shoda SI (2012) A practical one-step synthesis of 1,2-oxazoline derivatives from unprotected sugars and its application to chemoenzymatic β-*N*-Acetylglucosaminidation of disialo-oligosaccharide. 95:1928–1936. https://doi.org/10.1002/hlca.201200414
34. Doronina SO, Toki BE, Torgov MY, Mendelsohn BA, Cerveny CG, Chace DF, DeBlanc RL, Gearing RP, Bovee TD, Siegall CB, Francisco JA, Wahl AF, Meyer DL, Senter PD (2023) Development of potent monoclonal antibody auristatin conjugates for cancer therapy. Nat Biotechnol 21:778–84. https://doi.org/10.1038/nbt832
35. Fischer E (1983) Über die Glucoside der Alkohole. Ber Dtsch Chem Ges 26:2400–2412. https://doi.org/10.1002/cber.18930260327

# Regulation of the Dynamics of Antibodies and Cell Membrane Molecules Through Synthetic Glycan Modification

**Kazuya Kabayama and Yoshiyuki Manabe**

**Abstract** Glycans play crucial roles in regulating the behavior of cell surface and membrane-associated proteins. Recent developments in synthetic glycan modification techniques have enabled precise manipulation of glycan structures on specific proteins, providing new insights into the roles of glycans in controlling protein dynamics. This chapter summarizes recent progress in this field, focusing on two major areas: (1) the regulation of membrane protein dynamics using HaloTag-mediated glycan display systems, and (2) the enhancement of antibody therapeutic functions via glycan–lectin interactions. By integrating chemical biology, molecular imaging, and glycobiology, researchers have begun to dissect the structure–function relationships of glycans with unprecedented resolution. Particular attention is paid to the functional roles of galectin-3-mediated lattice formation, membrane diffusion. These technologies not only improve our understanding of glycan biology but also open new avenues for therapeutic engineering.

**Keywords** Synthetic glycans · Glycan–lectin interaction · Antibody dynamics · HaloTag · Galectin lattice · Endocytosis · Glycoengineering

## 1 Introduction

Glycans, or carbohydrates, are complex macromolecules composed of monosaccharide units connected by glycosidic bonds. Unlike nucleic acids and proteins, glycans are not directly templated by genetic information. Instead, their biosynthesis involves a network of glycosyltransferases and glycosidases that function in a spatially and temporally regulated manner. This results in an extraordinary diversity and heterogeneity of glycan structures on the cell surface and secreted proteins. Cell-surface

K. Kabayama (✉)
Institute for Radiation Sciences, The University of Osaka, Toyonaka, Japan
e-mail: kaba@irs.osaka-u.ac.jp

K. Kabayama · Y. Manabe
Graduate School of Science, The University of Osaka, Toyonaka, Japan

 
K. Toshima and Y. Fujimoto (eds.), *Chemical Biology of Carbohydrates*,
https://doi.org/10.1007/978-981-95-5335-8_13

glycans play pivotal roles in intercellular recognition, adhesion, signal transduction, immune responses, and pathogen interaction [1].

The pericellular layer rich in glycoproteins, glycolipids, and proteoglycans is referred to as the glycocalyx. This dense and dynamic network of glycans is recognized by a broad array of glycan-binding proteins, or lectins, which include selectins, siglecs, and galectins. Among them, galectins, especially galectin-3 (Gal-3), are known to form multivalent cross-linked lattices with β-galactoside-containing glycans, thereby regulating membrane organization and signaling [2]. The interplay between glycan structures and lectins exerts a significant influence on the dynamics of membrane proteins, affecting their clustering, lateral diffusion, internalization, and retention.

Antibodies are another class of glycoproteins whose function is closely related to their glycosylation pattern. The N-glycans attached to the Fc region of IgG antibodies are essential for their effector functions, including antibody-dependent cellular cytotoxicity (ADCC) and complement-dependent cytotoxicity (CDC). Modifying these glycan structures can dramatically alter antibody efficacy and pharmacokinetics. However, the influence of glycan-mediated interactions, particularly with lectins such as Gal-3, on antibody membrane dynamics has remained largely unexplored until recently.

To address these challenges, recent chemical biology approaches have enabled the installation of defined glycan structures on specific proteins. These methods include enzymatic remodeling, metabolic labeling, and the direct chemical conjugation of synthetic glycans. Notably, the HaloTag-based system allows for the selective modification of a target protein on the plasma membrane with a structurally defined glycan, without disturbing the endogenous glycan environment. This chapter will explore these innovative technologies and their application to studying and controlling the behavior of antibodies and cell membrane proteins.

## 2 De Novo Glycan Display on Cell Surfaces Using HaloTag

Cell-surface glycans form a dense layer known as the glycocalyx, which plays vital roles in diverse biological processes. However, dissecting their functions remains difficult due to the complexity and heterogeneity of glycan structures and their interactions with various biomolecules. To address this, we developed a HaloTag [3]-based strategy that allows for the site-specific introduction of chemically defined glycans onto a protein of interest (POI), along with fluorescent labeling. This system enables precise functional analysis of specific glycan structures in their native cellular context. Using this platform, we demonstrated that interactions between Galectin-3 (Gal-3) and various N-glycans regulate membrane protein mobility in a structure-dependent manner, likely through the formation of Gal-3-mediated glycan lattices. Additionally, we found that N-glycan modification influences the incorporation of membrane proteins into extracellular vesicles. Importantly, our method preserves the native glycan environment, offering a complementary approach to conventional glycan-editing techniques and a powerful tool for unraveling glycan functions on the cell surface [4].

## *2.1 Cell-Surface-Glycan Editing Methods*

Figure 1 illustrates the concept and workflow of the HaloTag-based glycan display system developed in this study. In this approach, a protein of interest (POI) fused to HaloTag is expressed on the cell surface, allowing for the site-specific conjugation of chemically synthesized N-glycans via HaloTag ligands that also carry fluorescent tags. This enables both the controlled presentation of structurally defined glycans and the visualization of glycosylated POIs. The introduced glycans retain their ability to interact with glycan-binding proteins such as galectin-3 (Gal-3), facilitating the investigation of how these interactions influence membrane protein behavior—particularly lateral diffusion and extracellular vesicle loading—in a native glycan environment.

## *2.2 Preparation of HaloTag Ligands 1–7 and Their Introduction on Living Cell Surface*

We synthesized a series of HaloTag ligands (1–7), each bearing a distinct N-glycan structure and a fluorescent group (TMR) linked via a PEG-based spacer to an alkyl halide for efficient conjugation. These glycans—ranging from core and bisecting GlcNAc, core fucose, and various sialylated forms to terminal galactose and polylactosamine—were chosen for their known roles in modulating protein function through interactions with lectins and other biomolecules. The PEG linker was introduced to improve glycan accessibility, and the ligands were prepared by functionalizing the reducing ends of the glycans through a modular synthetic strategy (Fig. 2).

## *2.3 Glycan Display on the Cell Surface Using HaloTag, and Recognition of the Displayed Glycans by the Corresponding Lectins*

We successfully displayed HaloTag ligands 1–7 on the surface of living HeLa cells expressing a GPI-anchored HaloTag protein (GPI(Thy)-Halo) designed to minimize homodimerization. Following a 15-min incubation at 37 °C with each ligand, uniform fluorescence was observed across the cell surface, confirming efficient glycan display. Although ligand 7, bearing a bulky glycan, showed slightly reduced efficiency, Western blotting further validated the incorporation of all ligands. These results highlight the versatility and simplicity of the HaloTag-based system for site-specific,

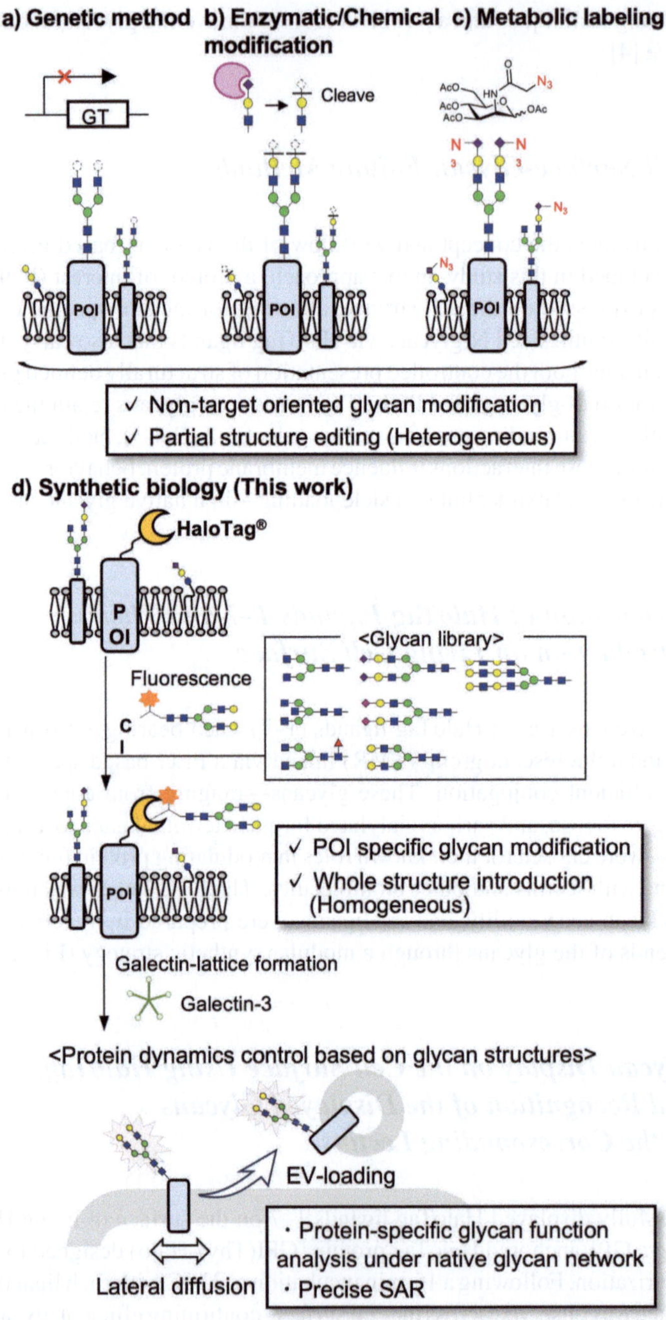
a) Genetic method
b) Enzymatic/Chemical modification
c) Metabolic labeling
GT
Cleave
POI
✓ Non-target oriented glycan modification
✓ Partial structure editing (Heterogeneous)
d) Synthetic biology (This work)
HaloTag®
Fluorescence
<Glycan library>
✓ POI specific glycan modification
✓ Whole structure introduction (Homogeneous)
Galectin-lattice formation
Galectin-3
<Protein dynamics control based on glycan structures>
EV-loading
Lateral diffusion
· Protein-specific glycan function analysis under native glycan network
· Precise SAR

◂**Fig. 1** Cell-surface-glycan editing methods. **a** Gene manipulation of enzymes involved in glycan biosynthesis. **b** Glycan modification using enzymatic/chemical methods. **c** Metabolic labeling using azide-functionalized sugars. **d** POI-specific glycan modification using HaloTag and its application for analyzing protein dynamics controlled by glycan modification (this work). Expression of the HaloTag-fused protein followed by treatment with HaloTag ligands bearing glycans enables the display of homogeneous glycans on the cell surface in a POI-specific manner

cell-surface glycan engineering. And we evaluated the specificity of glycan–lectin interactions using our HaloTag-based glycan-display system. In N-glycan-deficient HeLa and HepG2 cells (via Mgat1 or FUT8 knockout), bisecting GlcNAc-containing glycan 2 and core fucose-containing glycan 3 were selectively recognized by their respective lectins, PHA-E4 and PhoSL, while control glycans lacking these features showed no binding. These interactions were also confirmed with other HaloTag-fusion proteins, such as TfR-Halo. Furthermore, α2,3- and α2,6-sialylated glycans (glycans 5 and 4) were specifically recognized by MAM and SSA in CHO-Lec8 cells. These results demonstrate the high fidelity and functional accessibility of the displayed glycans, supporting the utility of this platform for probing glycan–lectin interactions in a controlled cellular context [4] (Fig. 3).

### 2.4 *Analysis of Lateral Diffusion by Fluorescence Recovery After Photobleaching*

Using the established HaloTag-based glycan-display system, we investigated how specific N-glycan structures affect membrane protein dynamics via interactions with Galectin-3 (Gal-3). FRAP analysis revealed that Gal-3 significantly slowed the lateral diffusion of GPI(Thy)-Halo proteins displaying N-glycans, with the effect correlating to glycan affinity for Gal-3 ($7 > 6 > 4 > 1$). The strongest suppression was observed with ligand 7, bearing polylactosamine. This reduced mobility was reversed by excess lactose, confirming Gal-3-specific binding. Similar diffusion suppression was also seen with SSA, a multivalent lectin for sialylated glycans, indicating that multivalent lectin–glycan interactions broadly modulate membrane protein dynamics through network formation like the galectin lattice (Fig. 4).

### 2.5 *Analysis of Lateral Diffusion by Single-Fluorescent-Molecule Imaging*

To further explore how N-glycans influence protein dynamics, we performed single-fluorescent-molecule imaging (SFMI) using GPI(LFA)-Halo, a monomeric membrane protein optimized for diffusion tracking. In the presence of Gal-3, the lateral diffusion of glycosylated GPI(LFA)-Halo was suppressed in a glycan-dependent manner, consistent with FRAP results. High-affinity ligands (especially

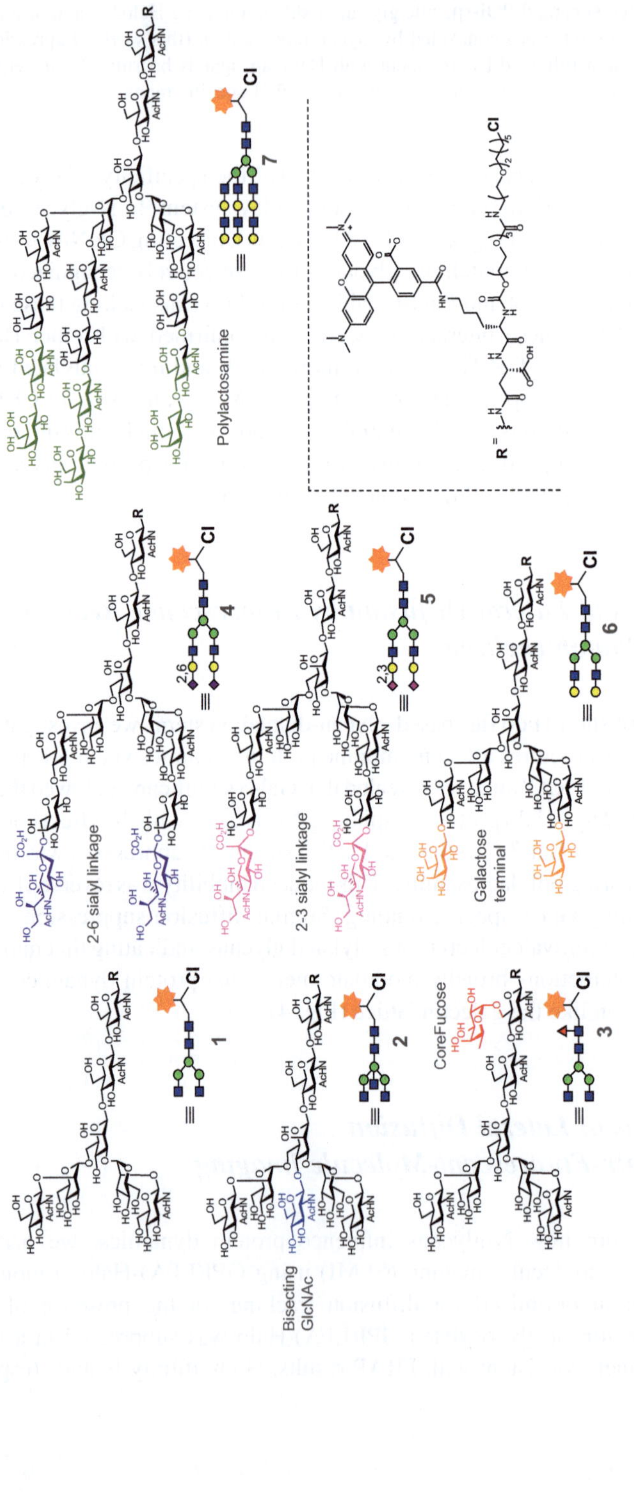


**Fig. 2** Structures of HaloTag ligands with N-glycans (1–7)

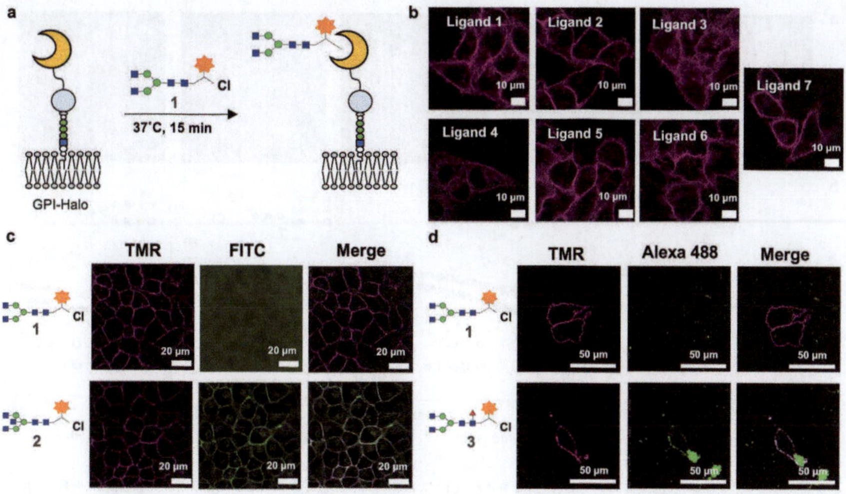


**Fig. 3** Glycan display on the cell surface using HaloTag, and recognition of the displayed glycans by the corresponding lectins. **a** Scheme of glycan display using HaloTag. **b** Fluorescence imaging of glycan-displaying cells. HeLa cells that express GPI(Thy)-Halo were treated with HaloTag ligands 1–7 at 37 °C for 15 min. **c** Fluorescence imaging of HaloTag ligands 1/2 and FITC-labeled PHA-E4 lectin. Mgat1-KO HeLa cells that express GPI(Thy)-Halo were treated with 100 nM HaloTag ligands 1/2 at 37 °C for 15 min and subsequently with 20 μg/mL FITC-labeled PHA-E4 at 37 °C for 30 min. From left to right: fluorescence (TMR), fluorescence (FITC), and merged image of TMR and FITC. **d** Fluorescence imaging of HaloTag ligands 1/3 and Alexa-488-labeled PhoSL lectin. Fut8-KO HepG2 cells were treated with 500 nM HaloTag ligand at 37 °C for 15 min and subsequently with 1 μg/mL Alexa-488-labeled PhoSL at 37 °C for 30 min. From left to right: fluorescence (TMR), fluorescence (Alexa 488), and merged image of TMR and Alexa 488

ligand 7 with polylactosamine) caused a significant shift toward slower or immobilized populations, as shown by reduced diffusion coefficients and increased immobile fractions. These findings reinforce the role of Gal-3–glycan interactions in modulating membrane protein mobility through lattice-like structures. And, trajectory analysis based on mean square displacement revealed that GPI(LFA)-Halo molecules primarily exhibited free diffusion (directed or Brownian) in the absence of Gal-3, regardless of the glycan structure. However, upon Gal-3 addition, a significant shift toward confined diffusion was observed—especially with ligands 7 and 4—indicating that Gal-3 binding restricts protein mobility through glycan-specific interactions (Fig. 5).

## 2.6 *Interim Summary*

We developed a HaloTag-based glycan-display system to investigate cell surface glycan functions with molecular precision [4]. Seven distinct N-glycans were displayed on living cells and shown to interact with specific lectins. Using this

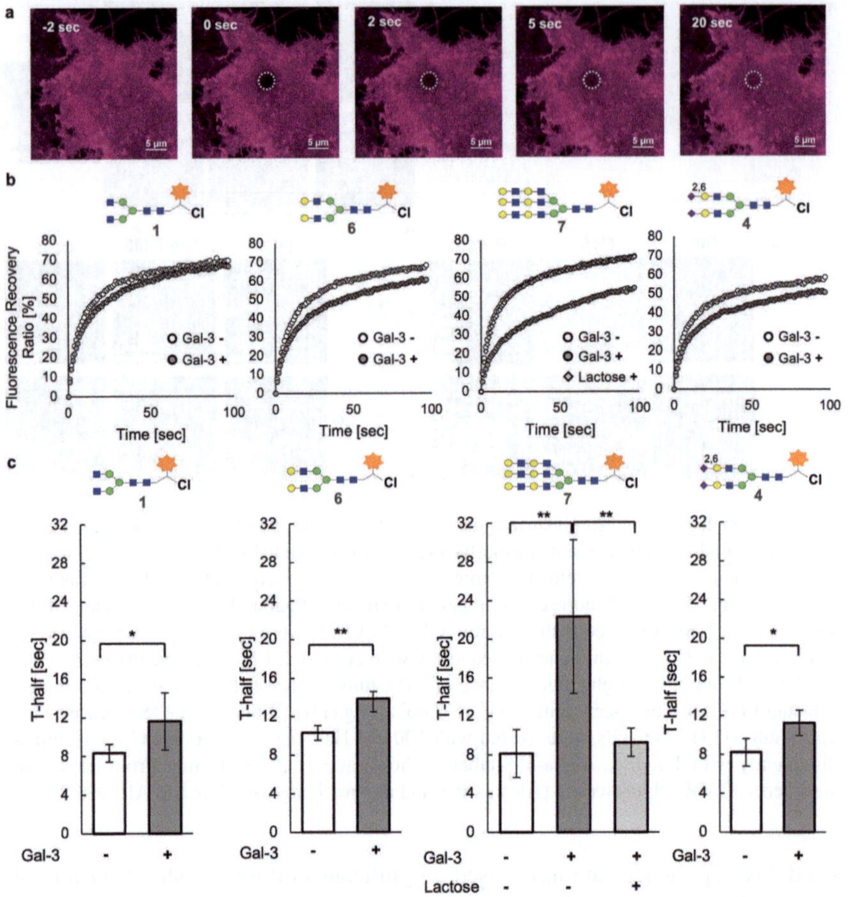


**Fig. 4** Lateral-diffusion analysis of GPI(Thy)-Halo using FRAP. **a** Images of photobleaching and fluorescence recovery at the region of interest. HaloTag ligand 7 was introduced to GPI(Thy)-Halo on HeLa cells, and photobleaching was carried out at 0 s. **b** Fluorescence recovery curves in the presence/absence of Gal-3 (gray/white circles) or the presence of Gal-3 and lactose (diamonds). HeLa cells expressing GPI(Thy)-Halo were treated with 500 nM of HaloTag ligands 1, 6, 7, or 4 at 37 °C for 15 min, the cells were washed twice, and imaging analysis was performed. 100 µg/mL Gal-3 and 50 mM lactose were used. The graphs were generated by measuring the fluorescence intensity for 100 s. **c** Half-time to recovery ($T1/2$) calculated by fitting with an exponential function. Data represent the results from 11 ($n = 11$, ligands 1 and 7) or 6 ($n = 6$, ligands 6 and 4) biologically independent replicates. The error bars represent the standard deviation of the mean value. $*p < 0.01$, $**p < 0.001$

system, we found that Gal-3 reduces membrane protein mobility through galectin lattice formation, regardless of glycan type, and that high-affinity glycans induce a biphasic diffusion pattern reflecting Gal-3 binding strength. Additionally, glycan

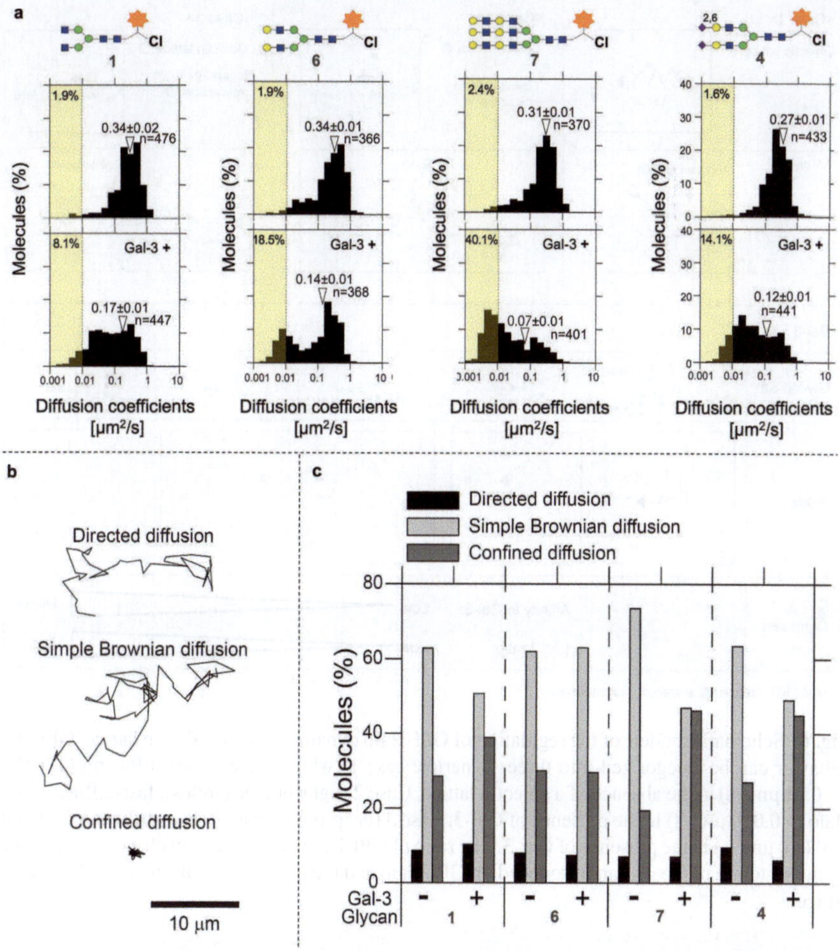


**Fig. 5** Lateral diffusion analysis of N-glycan introduced GPI(LFA)-Halo using SFMI. The distributions of the diffusion coefficient of GPI(LFA)-Halo molecules in the absence/presence of Gal-3 are shown. T24 cells that express GPI(LFA)-Halo were treated with 200 nM of HaloTag ligands 1, 6, 7, or 4 at 37 °C for 15 min, the cells were washed twice, and single-molecule observation of GPI(LFA)-Halo was performed with/without Gal-3 (250 μg/mL). **a** The histograms show the distribution of the diffusion coefficients of individual GPI(LFA)-Halo molecules in the time window of 200 ms (D200 ms). Immobile fractions (D200 ms < 0.010 μm$^2$/s) were highlighted in yellow, and their proportion was written at the upper left. **b** Representative trajectories demonstrating directed diffusion, simple Brownian diffusion, and confined diffusion are presented. **c** The histograms show the proportion of GPI(LFA)-Halo trajectories categorized into directed diffusion, simple Brownian diffusion, and confined diffusion

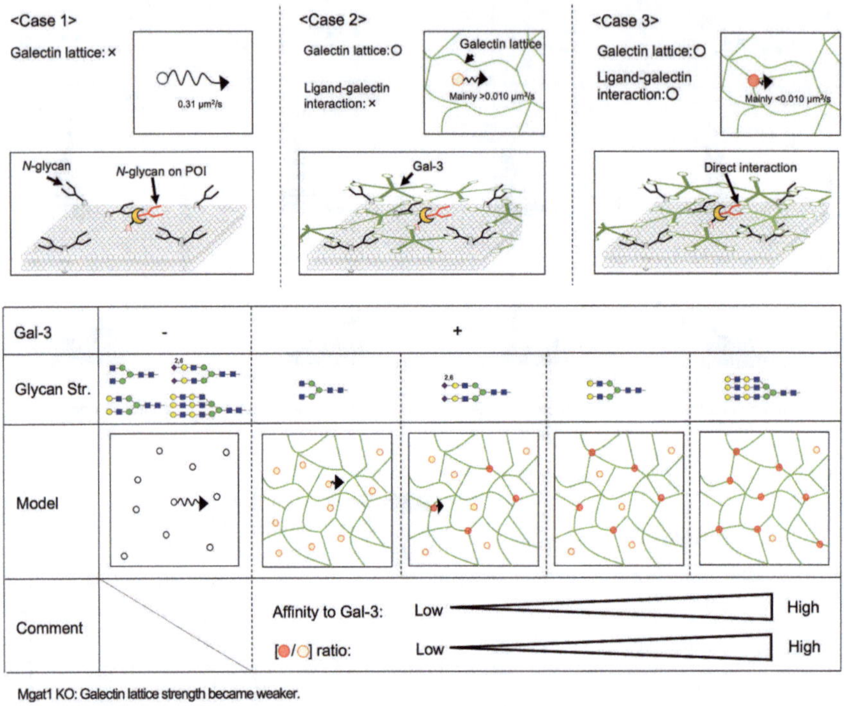


**Fig. 6** Schematic models of the regulation of GPI-Halo dynamics by the galectin lattice. GPI-Halo behavior can be categorized into three scenarios: case 1 (white circles): fast-diffusing GPI-Halo (~ 0.31 μm²/s) in the absence of a galectin lattice. Case 2 (light-orange circles): fast-diffusing GPI-Halo (> 0.010 μm²/s) in the presence of Gal-3. Case 3 (deep-pink circles): slow-diffusing GPI-Halo (< 0.010 μm²/s) in the presence of Gal-3. The ratio of GPI-Halo belonging to each category depends on the structure of the glycan introduced on GPI-Halo and the presence (or absence) of the galectin lattice

structure was shown to influence membrane protein loading into extracellular vesicles. This platform offers a powerful tool for precise, POI-specific glycan analysis while preserving native glycan networks (Fig. 6).

## 3 Improvement of Antibody Activity by Controlling Its Dynamics Using the Glycan–Lectin Interaction

The glycocalyx, composed of glycoconjugates like glycolipids and glycoproteins, regulates key membrane protein behaviors such as endocytosis, diffusion, and clustering through interactions with glycan-binding proteins like galectins. Galectin-3, in particular, forms lattices that influence protein dynamics. In this study, we modulated antibody function by conjugating a galactose-containing N-glycan to the antibody, enabling interaction with galectin-3 to suppress internalization and enhance CDC activity. Using an anti-HER2 antibody as a model, we demonstrated that glycan-based control of antibody dynamics offers a novel strategy to improve therapeutic efficacy, especially in tumors where galectin-3 is overexpressed [5]. This approach highlights the potential of glycan–lectin interactions as a means to regulate antibody activity beyond traditional effector binding (Fig. 7).

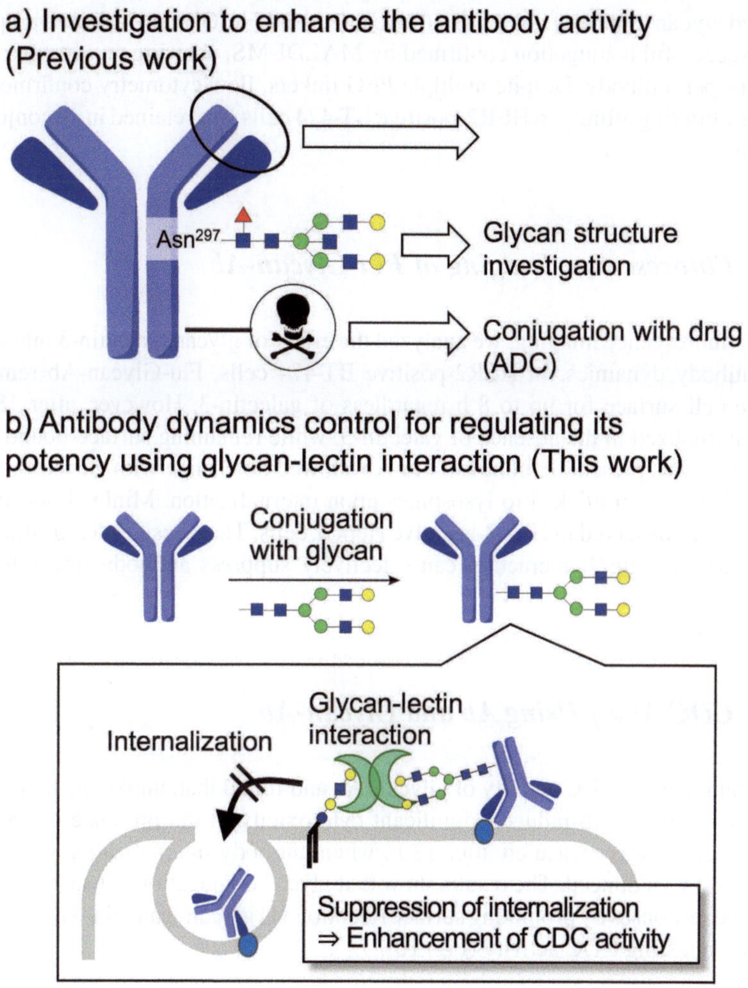


**Fig. 7** Approaches for improving antibody activity. **a** Investigations to enhance the antibody activity include epitope investigation, glycan structure investigation, and conjugation with drugs. **b** Preparation of the glycan-antibody conjugate to control the antibody dynamics and activity (strategy of this study)

### 3.1 Preparation of Glycan-Antibody Conjugates (Glycan-Ab and Flu-Glycan-Ab)

Galactose-containing N-glycans were chemically conjugated to an anti-HER2 antibody to control its dynamics via galectin-3 interaction. The N-glycan was functionalized with a thiol group and attached to the antibody using a PEG linker to enhance accessibility and effectiveness over native glycans. Both unlabeled and fluorescently labeled glycan–antibody conjugates (Glycan-Ab and Flu-Glycan-Ab) were prepared, with successful conjugation confirmed by MALDI-MS, showing an average of five glycans per antibody. Despite multiple PEG linkers, flow cytometry confirmed that antigen-binding affinity to HER2-positive BT-474 cells was retained in all conjugates (Fig. 8).

### 3.2 Fluorescence Imaging of Flu-Glycan-Ab

Using fluorescence imaging, we analyzed the effect of glycan–galectin-3 interaction on antibody dynamics. In HER2-positive BT-474 cells, Flu-Glycan-Ab remained on the cell surface for up to 8 h regardless of galectin-3. However, after 18 h, it was internalized in the absence of galectin-3, while remaining surface-bound when galectin-3 was present. In contrast, Flu-Ab showed no change with galectin-3. Both antibodies were trafficked to lysosomes upon internalization. Minimal nonspecific binding was observed in HER2-negative HepG2 cells. These results demonstrate that N-glycan–galectin-3 interaction can effectively suppress antibody internalization (Fig. 9).

### 3.3 CDC Assay Using Ab and Glycan-Ab

We evaluated the CDC activity of Glycan-Ab and found that, unlike the unmodified antibody, Glycan-Ab induced significant cytotoxicity in the presence of galectin-3. This effect was measured after 18 h, when antibody internalization differences were most pronounced. The results show that glycan conjugation enhances antibody immune responses by promoting surface retention via glycan–galectin-3 interactions, thereby boosting CDC activity (Fig. 10).

### 3.4 Interim Summary

This study introduces a novel strategy to enhance antibody potency by controlling antibody dynamics through glycan conjugation. By attaching a galactose-containing

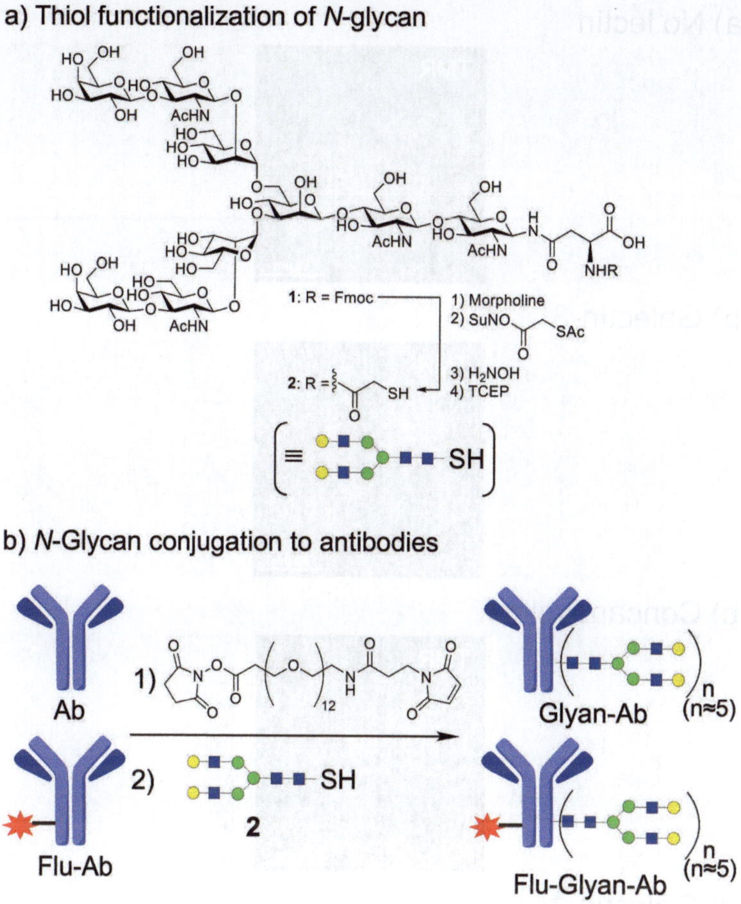


**Fig. 8** Preparation of glycan-antibody conjugates (Glycan-Ab and Flu-Glycan-Ab). **a** Thiol functionalization of galactose-containing N-glycan. **b** Conjugation between antibodies with N-glycan. The average loading ratio of the N-glycan was estimated to be five in both Glycan-Ab and Flu-Glycan-Ab

N-glycan to an anti-HER2 antibody, the antibody was retained on the cell surface via interaction with galectin-3, resulting in enhanced CDC activity through endocytosis inhibition. Unlike conventional modifications of Asn297 glycans, this approach adds an external glycan, representing the first demonstration of dynamic control via glycan conjugation. The method's efficacy depends on the conjugation strategy, glycan number, and structure, with multivalent and polylactosamine-rich glycans expected to further improve performance. Beyond in vitro effects, glycan conjugation could optimize in vivo stability and pharmacokinetics, offering a safer alternative to PEGylation. Overall, this platform provides a promising approach for re-engineering antibodies with suboptimal activity by leveraging endogenous glycan–lectin interactions [5] (Fig. 11).

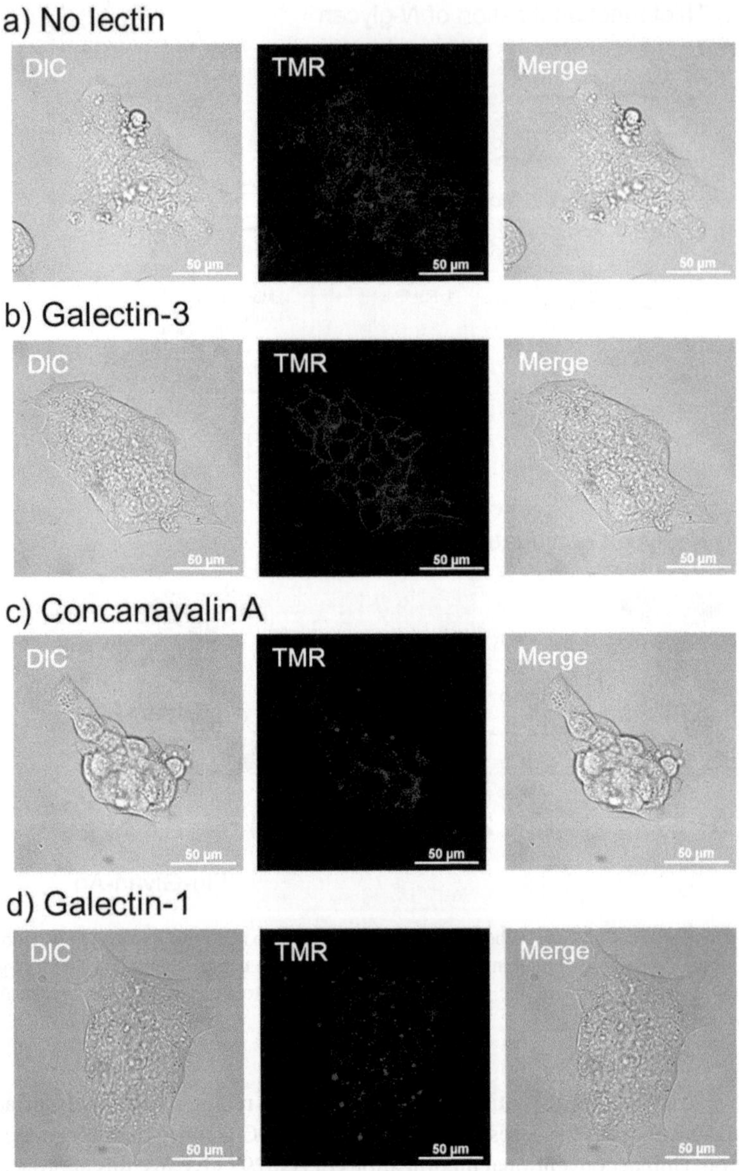


**Fig. 9** Fluorescence imaging of Flu-Glycan-Ab. After incubating the BT-474 cells with or without lectins (20 μg/mL) for 30 min, Flu-Glycan-Ab (2.0 μg/mL) was added. Representative images obtained by fluorescence microscopy after incubation for 18 h in the absence of galectin-3 (**a**) or in the presence of galectin-3 (**b**), concanavalin A (**c**), or galectin-1 (**d**). From left to right: differential interference contrast (DIC), fluorescence (TMR), and the merge of DIC and TMR (Merge)

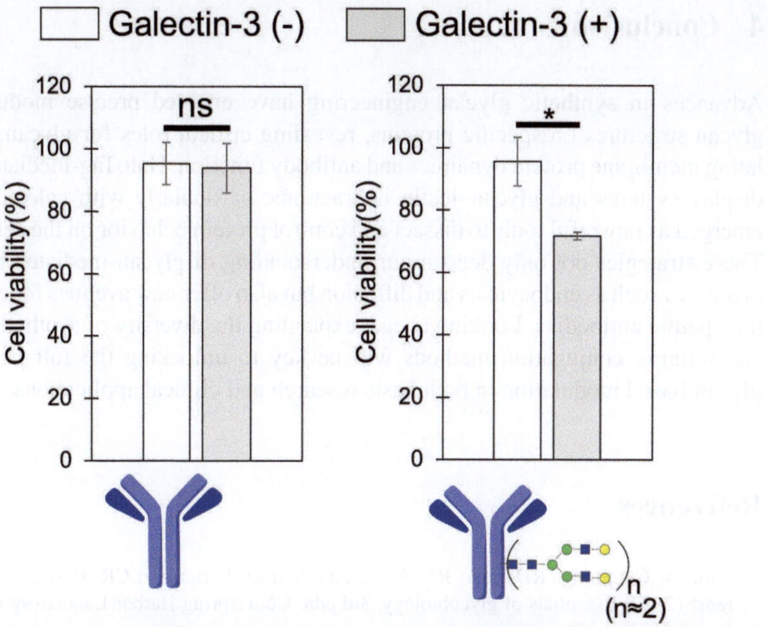


**Fig. 10** CDC assay using Ab and Glycan-Ab. After incubating the BT-474 cells with or without galectin-3 (20 μg/mL) for 30 min, the antibodies (0.5 μg/mL) were added. After incubation for 18 h, the cells were treated with complement. The viability of the cells treated with Ab (left) and Glycan-Ab (right) in the absence (white) and presence (dark) of galectin-3 is shown. All experiments were carried out in triplicate ($n = 3$). Standard deviation (SD) is indicated by the error bars. $^{*}p < 0.05$

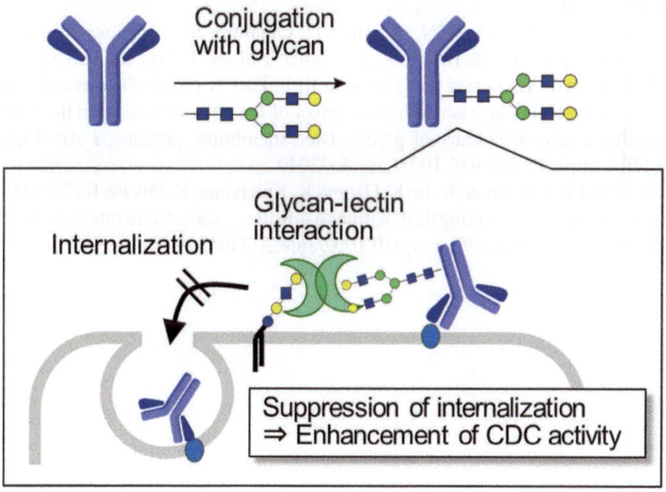


**Fig. 11** A novel strategy to enhance antibody potency by controlling antibody dynamics through glycan conjugation

# 4 Conclusion

Advances in synthetic glycan engineering have enabled precise modulation of glycan structures on specific proteins, revealing critical roles for glycans in regulating membrane protein dynamics and antibody function. HaloTag-mediated glycan display systems and glycan–lectin interactions, particularly with galectin-3, have emerged as powerful tools to dissect and control protein behavior on the cell surface. These strategies not only deepen our understanding of glycan-mediated biological processes such as endocytosis and diffusion but also offer new avenues for enhancing therapeutic antibodies. Looking ahead, expanding the diversity of synthetic glycans and refining conjugation methods will be key to unlocking the full potential of glycan-based modulation in both basic research and clinical applications.

## References

1. Varki A, Cummings RD, Esko JD, Freeze HH, Stanley P, Bertozzi CR, Hart GW, Etzler ME (eds) (2017) Essentials of glycobiology, 3rd edn. Cold Spring Harbor Laboratory Press, Cold Spring Harbor, NY
2. Nabi IR, Shankar J, Dennis JW (2015) The galectin lattice at a glance. J Cell Sci 128: 2213–2219. https://doi.org/10.1242/jcs.151159
3. Los GV, Encell LP, McDougall MG, Hartzell DD, Karassina N, Zimprich C, Wood MG, Learish R, Friedman-Ohana RF, Urh M, Simpson D, Mendez J, Zimmerman K, Otto P, Vidugiris G, Zhu J, Darzins A, Klaubert DH, Bulleit RF, Wood KV (2008) HaloTag: a novel protein labeling technology for cell imaging and protein analysis. ACS Chem Biol 3: 373–382. https://doi.org/10.1021/cb800025k.
4. Miura A, Manabe Y, Suzuki KGN, Shomura H, Okamura S, Shirakawa A, Yano K, Miyake S, Mayusumi K, Lin C-C, Morimoto K, Ishitobi J, Nakase I, Arai K, Kobayashi S, Ishikawa U, Kanoh H, Miyoshi E, Yamaji T, Kabayama K, Fukase K (2024) De novo glycan display on cell surfaces using HaloTag: visualizing the effect of the galectin lattice on the lateral diffusion and extracellular vesicle loading of glycosylated membrane proteins. J Am Chem Soc 146: 22193–22207. https://doi.org/10.1021/jacs.4c02040
5. Manabe Y, Iizuka Y, Yamamoto R, Ito K, Hatano K, Kabayama K, Fukase K (2023) Improvement of antibody activity by controlling its dynamics using the glycan–lectin interaction. Angew Chem Int Ed 62:e202304779. https://doi.org/10.1002/anie.202304779

# Peptide-Based Molecules Specific to Ganglioside Nanoclusters Toward Innovative Therapeutic Strategies

**Teruhiko Matsubara**

**Abstract** Gangliosides form high-density nanoclusters in the neuronal cell membrane, and toxic amyloid β protein (Aβ) aggregates are induced by the interaction of Aβ monomers with ganglioside nanoclusters. The shapes of Aβ aggregates, including oligomers and fibrils, were identified via surface topographic studies using atomic force microscopy (AFM) of the planar lipid bilayer. Ganglioside nanocluster-binding molecules are potential candidate inhibitors against Aβ aggregation. We identified artificial peptides that bind to the sugar chains of gangliosides from peptide libraries by biopanning (affinity selection) using phage display technology. AFM and thioflavin T studies indicated that our peptides could bind to Aβ-sensitive ganglioside nanoclusters (ASIGN) and effectively inhibit ganglioside-induced Aβ assemblies. This inhibition suggested that our peptides reduce formation of the ganglioside-bound Aβ (GAβ) complex, which acts as a seed to promote toxic Aβ fibrils. Our results indicate that ganglioside nanocluster-binding peptides (GCBPs) are prospective effective inhibitors against Aβ aggregation in the nervous system.

**Keywords** Ganglioside nanocluster · Amyloid β protein · Peptide library · Phage display · Planar lipid bilayer · Membrane raft

## 1 Introduction

Gangliosides, sialic acid-containing glycosphingolipids (GSLs), are abundant in the cell membranes of animal nervous systems [1, 2]. Gangliosides and other GSLs are localized in functional lipid structures such as membrane rafts (microdomains), which are composed of sphingolipids, cholesterol, and membrane proteins (Fig. 1) [3–5]. Membrane rafts act as a platform for cell–cell communication [6]; this is because gangliosides and GSLs are clustered at the nanometer scale with high density, and this so-called "cluster effect" is critical for communication [7, 8].

T. Matsubara (✉)
Department of Biosciences and Informatics, Faculty of Science and Technology, Keio University, Yokohama, Kanagawa, Japan
e-mail: matsubara@bio.keio.ac.jp

 
K. Toshima and Y. Fujimoto (eds.), *Chemical Biology of Carbohydrates*,
https://doi.org/10.1007/978-981-95-5335-8_14

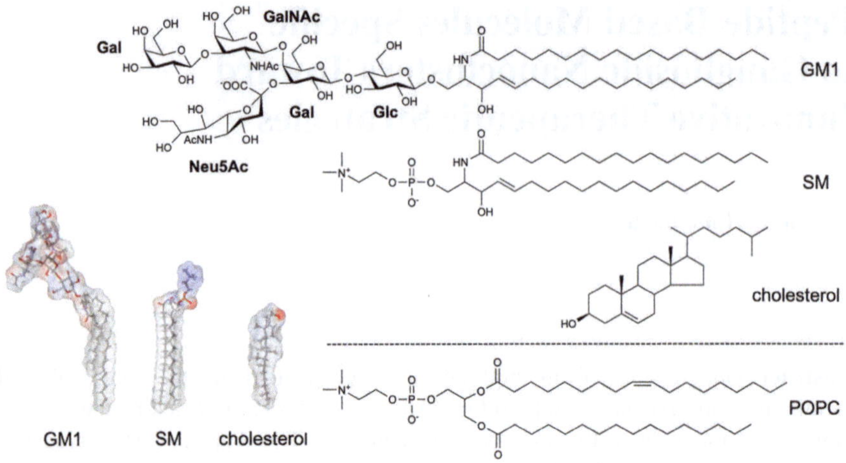


**Fig. 1** Chemical structure (*right*) and molecular modeling (*left*) of the major lipid components of cell membrane. Ganglioside GM1, cholesterol, sphingomyelin (SM), 1-palmitoyl-2-oleoyl-*sn*-glycelo-3-phosphocholine (POPC)

The monosialoganglioside GM1, Galβ1–4GalNAcβ1–3(Neu5Acα2–3)Galβ1–4Glc1–1′Cer, is a major ganglioside recognized by cholera toxin B subunit (CTB). Therefore, GM1 is often used as a marker for membrane rafts (lipid rafts). In 1995, Yanagisawa et al. reported that the ganglioside GM1 could bind to amyloid β protein (Aβ) monomers to form a ganglioside-bound amyloid β (GAβ) complex, and that the complex could act as a seed to induce toxic Aβ aggregates [9]. Aβ is a polypeptide with 39–43 residues and is generated by the secretase digestion of its precursor protein. Multiple Aβ monomers form oligomers, globules, and fibers under various conditions; toxic assemblies induce cell death and are related to the onset of progressive neurodegenerative diseases like Alzheimer's disease (AD). We thus investigated the Aβ aggregation pathway involving gangliosides in nerve cell membrane, and further designed peptide-based inhibitors.

## 2 Planar Lipid Bilayer for Atomic Force Microscopy Observation

The lateral distribution of gangliosides at the nanometer scale (10–200 nm) can be observed using atomic force microscopy (AFM) [10]. In many studies, supported lipid membranes (SLB) prepared from GSL-containing liposomes have been used for AFM studies [11]. Liposomes can be used to prepare SLB easily; however, their lipid compositions are limited. Therefore, we prepared a planar lipid bilayer on mica and analyzed it using AFM (Fig. 2). The lipid bilayer was obtained by accumulating POPC-coated mica with a ganglioside-containing lipid monolayer at

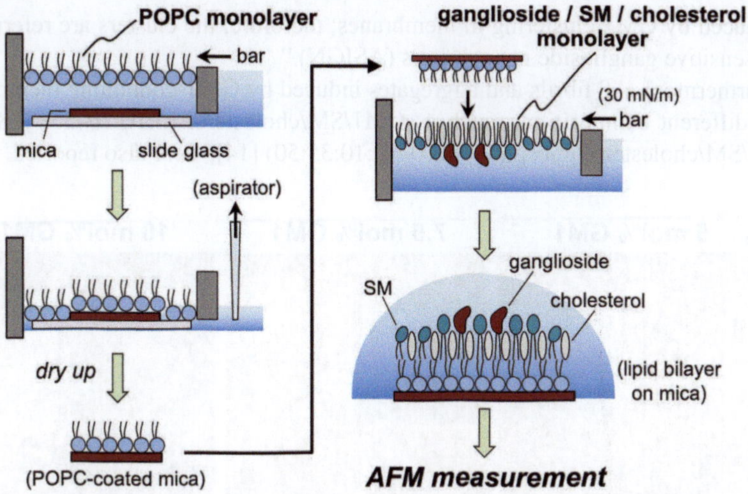


**Fig. 2** Preparation of a planar lipid bilayer containing ganglioside for AFM observation. A monolayer of POPC at the air–water interface in a Langmuir trough was deposited on mica to prepare POPC-coated mica (*left*). A ganglioside-containing lipid monolayer (*e.g.*, ganglioside/SM/cholesterol) was then deposited on the POPC-coated mica to form a lipid bilayer (*right*)

an air–water interface using a Langmuir (or Langmuir–Blodgett) trough. The surface tension of the lipid monolayer was maintained at 30 mN/m, corresponding to the lateral plasma membrane pressure. Using the Langmuir trough, a wide range of lipid compositions, including a mixture of ganglioside/sphingomyelin (SM)/cholesterol and lipid extracts from animal cells, are available for AFM observation [12–14]. The surface topographies of membranes reconstituted using lipids extracted from B16 cells [15], mouse brain [16], and human brain [17] have also been reported.

## 3 Generation of Aβ Assemblies on Ganglioside GM1-Containing Lipid Membranes: Aβ-Sensitive Ganglioside Nanoclusters (ASIGN)

To investigate Aβ assemblies induced by ganglioside-containing membranes, a seed-free monomer Aβ(1–40)($A\beta_{40}$) was incubated with 1–10 mol% GM1 membrane (GM1/SM/cholesterol, 10:45:45) and the surface topography was determined using AFM [12]. The AFM results indicated that Aβ deposition on 10 mol% GM1 membranes was observed after incubation with Aβ for 15 min, and that Aβ began to form protofibrils at 12 h. After 24 h of incubation, no fibrils were found on the 1 and 5 mol% GM1 membranes, but many Aβ fibrils were observed on the 7.5 and 10 mol% GM1 membranes (Fig. 3). Notably, glucosyl ceramide (Glc1–1′Cer) did not induce Aβ fibrils (Fig. 4) [13]. These results indicate that Aβ fibril formation is induced by GM1 clustering in membranes; therefore, the clusters are referred as "Aβ-sensitive ganglioside nanoclusters (ASIGN)."

Furthermore, Aβ fibrils and aggregates induced by GM1-containing membranes with different compositions, such as GM1/SM/cholesterol (5:10:10–35) [18] and GM1/SM/cholesterol/phospholipid (0–10:10:35:50) [14], were also reported.

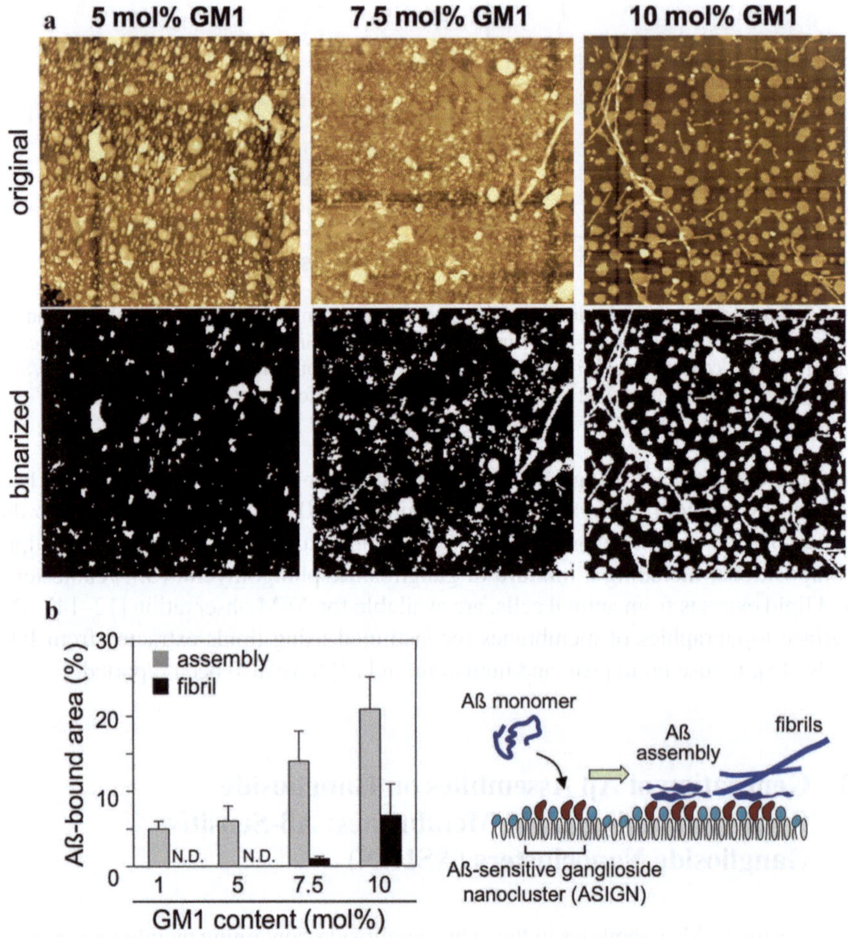


**Fig. 3** Aβ protofibril formation, induced by ganglioside nanoclusters on GM1-containing membranes, observed by AFM. **a** AFM images of 5–10 mol% GM1-containing membranes after incubation with 10 μM Aβ(1–40) at 37 °C for 48 h. Binarized AFM images were obtained by thresholding at 6 nm from the bottom. The molar ratios of GM1/SM/cholesterol were 5:47.5:47.5 (5 mol% GM1) and 10:45:45 (10 mol% GM1). **b** Proportion of Aβ protofibrils to the Aβ-bound area. Aβ assemblies were defined as fibrils with a long/short axis aspect ratio > 3. N.D., not detected. Reprinted with permission from Ref. [12]. Copyright © 2017, American Chemical Society

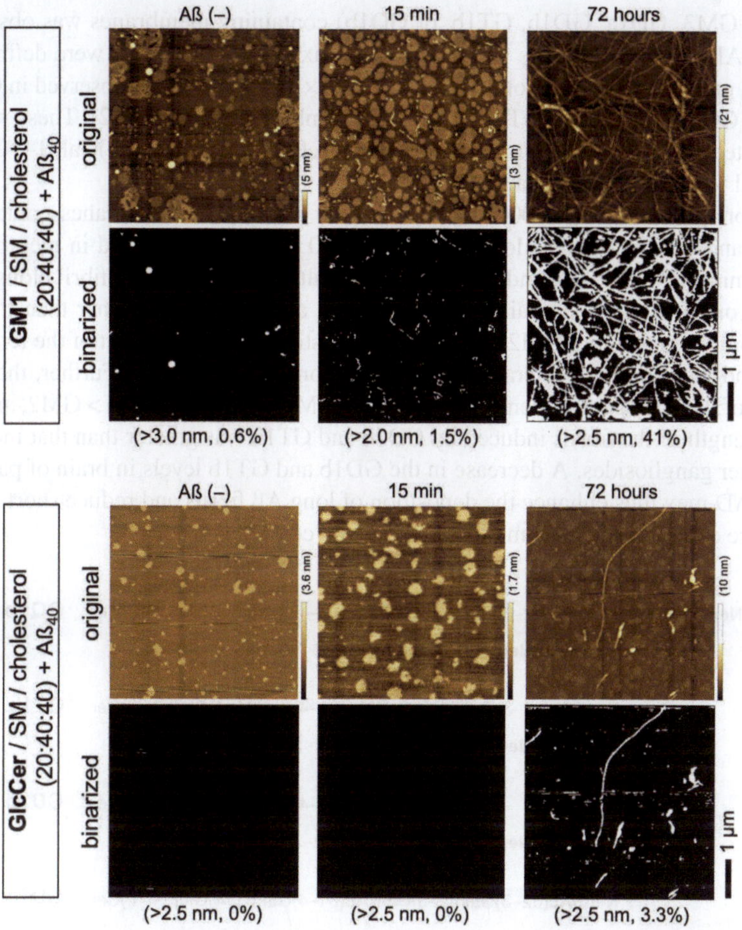


**Fig. 4** Aβ(1–40) fibril formation induced by 20 mol% GM1-(*upper*) and GlcCer-containing membranes (*lower*). After incubation with 10 μM Aβ(1–40) at 37 °C for 0 min (Aβ(–)), 15 min, and 72 h, binarized AFM images were obtained by thresholding at 2.0–3.0 nm from the bottom. Reprinted with permission from Ref. [13]. Copyright © 2018, Matsubara et al.

# 4 Effect of Glycan Structure of Gangliosides on Aβ Fibril Formation

Many kinds of a- and b-series gangliosides are found in animal brains, such as GM1, GD1a/b, and so on (Figs. 5 and 6a). Taki et al. reported significant decreases in b-series gangliosides, such as GD1b and GT1b, in human brain lipid extracts from patients with Alzheimer's disease (AD) [19]. Aβ fibrils are generated in the presence of brain lipids [17, 18], but the roles of each of the gangliosides remain unclear. The size and shape of Aβ fibrils with 10 mol% a- and b-series ganglioside (GM1,

GM2, GM3, GD1a, GD1b, GT1b, or GQ1b)-containing membranes was observed using AFM. Aβ assemblies with a long/short axis aspect ratio > 3 were defined as fibrils, and more than 20 protofibrils in a 2 μm × 2 μm area were observed in GM1, GM2, GD1a, GD1b, and GT1b-containing membranes (Fig. 6b) [12]. These results indicate that the GM2 oligosaccharide, GalNAcβ1–3(Neu5Acα2–3)Galβ1–4Glc, is critical for protofibril formation.

Short protofibrils (< 100 nm) were mainly induced in membranes containing GM2 and GD1b, whereas longer fibrils (> 300 nm) were observed in membranes containing GM1, GD1a, and GT1b. These results indicate that the fibril elongation speed on membranes containing GM1, GD1a, and GT1b was higher than that on membranes containing GM2 and GD1b, suggesting the involvement of the terminal Gal and Neu5Ac in accelerating protofibril elongation (Fig. 6c). Further, the rank order of the average fibril length was GD1a > GM1 > GT1b > GD1b > GM2, and the total length of the fibrils induced by GD1b and GT1b was greater than that induced by other gangliosides. A decrease in the GD1b and GT1b levels in brain of patients with AD may thus enhance the deposition of long Aβ fibrils and reduce short fibrils that are easy to remove from the membrane by cerebrospinal fluid.

Neu5Ac(α2–8)Neu5Ac(α2–3)Gal(β1–3)GalNAc(β1–4)Gal(β1–4)Glcβ(1–1')Cer **GQ1b**
/
Neu5Ac(α2–8)Neu5Ac(α2–3)

Neu5Ac(α2–3)Gal(β1–3)GalNAc(β1–4)Gal(β1–4)Glcβ(1–1')Cer **GT1b**
/
Neu5Ac(α2–8)Neu5Ac(α2–3)

Gal(β1–3)GalNAc(β1–4)Gal(β1–4)Glcβ(1–1')Cer **GD1b**
/
Neu5Ac(α2–8)Neu5Ac(α2–3)

Neu5Ac(α2–3)Gal(β1–3)GalNAc(β1–4)Gal(β1–4)Glcβ(1–1')Cer **GD1a**
/
Neu5Ac(α2–3)

Gal(β1–3)GalNAc(β1–4)Gal(β1–4)Glcβ(1–1')Cer **GM1**
/
Neu5Ac(α2–3)

**Fig. 5** Structure of a- and b-series gangliosides

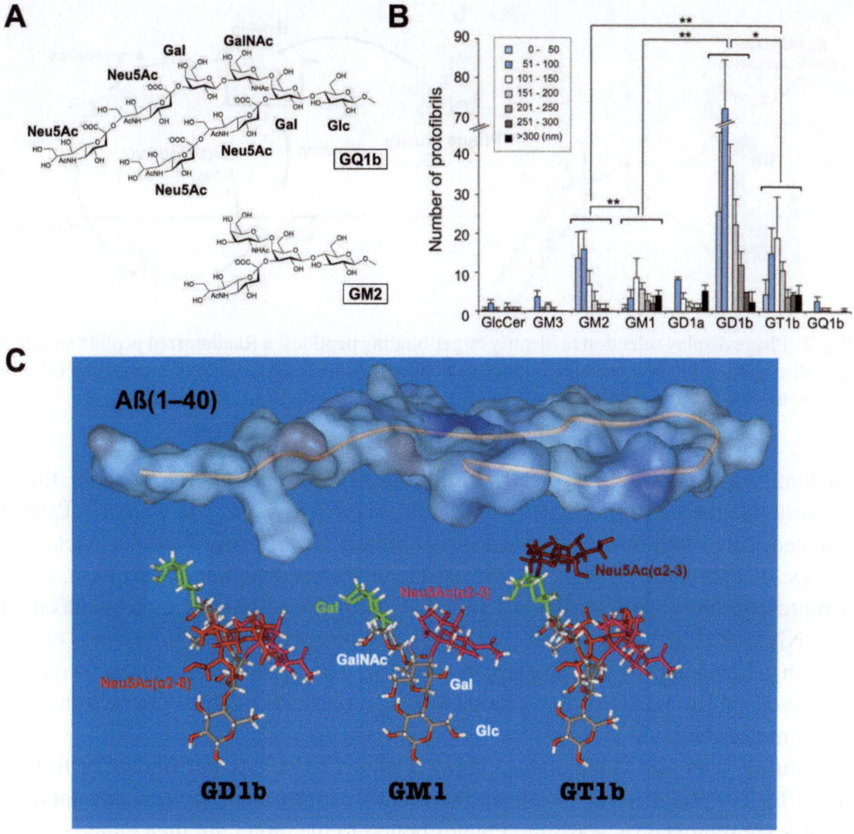


**Fig. 6** Aβ(1–40) fibril formation induced by ganglioside-containing membrane. **a** Glycan structure of GQ1b and GM2. **b** Size distribution of fibrils on 10 mol% ganglioside-containing membrane (ganglioside/SM/cholesterol, 10:45:45) observed using AFM (10 μM Aβ, 48 h; 2 × 2 μm, $n = 3$). **c** Superimposed image of Aβ(1–40) and oligosaccharides of GD1b, GM1, and GT1b. The conformations of Aβ (entry 1FV2) and oligosaccharides (entry 2M4J) were obtained from the Protein Data Bank (PDB). Reprinted with permission from Ref. [12]. Copyright © 2017, American Chemical Society

## 5 Ganglioside-Binding Peptides Identified from a Randomized Library Using Phage-Display Technology

### *5.1 Phage Library*

Phage display is a useful affinity selection technology for identifying amino acid sequences that bind to a target molecule, from sequence pools with a large repertoire [20, 21]. As the size of the pools (called "library") depends on the volume of culture

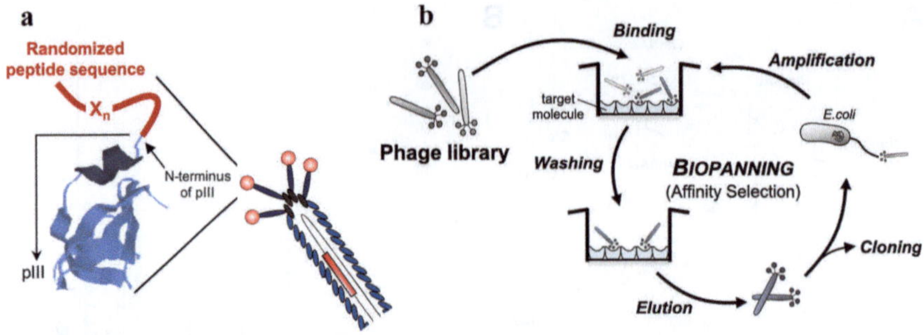


**Fig. 7** Phage display selection to identify target-binding peptides. **a** Randomized peptide sequence ($X_n$) displayed at the N-terminus of phage coat protein III (pIII). **b** Outline of biopanning (affinity selection). See text for details

medium, its maximum is typically in billions ($10^9$) at the laboratory scale; this is because phages are propagated by host cells, like *Escherichia coli* (*E. coli*). Library sequences are often fused to the head (coat protein III) of phage particles, yielding a "phage-display library" (Fig. 7a). We used a randomized peptide library ($X_n$), where $X$ represents any amino acid ($n = 15$) [22]. $X_n$ sequences are generated by the codon type NNK ($N = A, C, G$, or $T$; $K = T$ or $G$), and inserted into the N-terminus of coat protein III encoded by the phage genome [23]. The randomized peptide sequence displayed on the heads of the generated phage particles enables interactions with target molecules.

Affinity selection (biopanning) using a phage library is performed as follows (Fig. 7b). The phage library is allowed to interact with target molecules, and unbound phages are removed by washing. Phages bound to the target are then eluted, amplified using *E. coli*, and subjected to the next cycle. This cycle is typically repeated several times to yield target-binding phage particles. After selection, individual phage clones are isolated and subjected to DNA sequencing of the phage genome. Using recent advances, a mixture of phage genomes may be subjected to next-generation sequencing without cloning.

### 5.2 Affinity Selection of Ganglioside-Binding Peptides

To identify the GM1-binding peptide, a GM1 cast film was fixed on a solid plate and allowed to interact with a phage library ($X_{15}$). Despite several rounds of affinity cycling, the peptide sequences of the collected phages were not enriched (unpublished results). We considered that this was because the interaction between the peptide and ganglioside was not strong. Therefore, we fixed a GM1 monolayer, prepared at an air–water interface, on a gold electrode and monitored the binding of phage particles using a quartz crystal microbalance (QCM) sensor (Fig. 8a) [24]. By orientationally fixing on the hydrophobic surface of the electrode, the sialic acid-containing

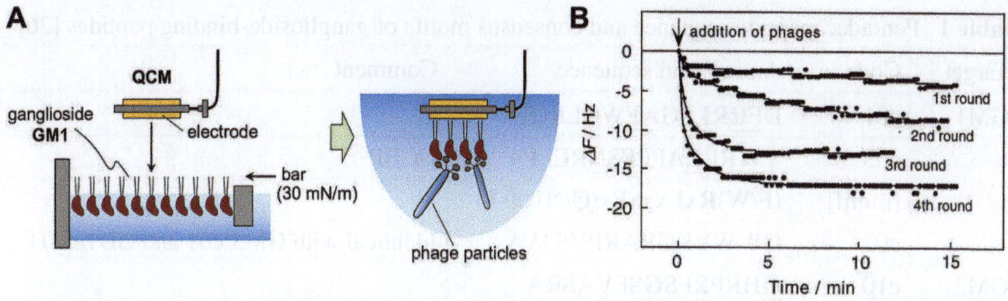


**Fig. 8** Binding of phage particles to the GM1 monolayer during biopanning cycles. The GM1 monolayer was immobilized on the QCM electrode (**a**) and the frequency changes ($\Delta F$, Hz) indicating phage binding were monitored during the binding process (**b**). The arrow indicates the time at which the phage library was injected into the solution. Reprinted with permission from Ref. [24] Copyright © 1999 Federation of European Biochemical Societies. Published by Elsevier B.V. All rights reserved

pentasaccharide of GM1 was clustered and easily interacted with the phage particles. In contrast, the hydrophobic lipid moiety of ceramide was not exposed to the water layer. The QCM signals (frequency changes) indicated that phage particles were bound to the GM1 monolayer, and the amount of bound phages increased with an increase in the number of selection cycles (Fig. 8b). After five cycles, 18 enriched clones were isolated, and three pentadecapeptides (GM1/c01, c02, and c03) were identified (Table 1).

Similar selections against other gangliosides (GM2, GM3, GD1a, and GT1b) were repeatedly performed to identify a series of peptides that bind to gangliosides [25, 26]. The GM1/c02 sequence was again isolated as a GM3- (GM3/c01) and GD1a-binding peptide (GD1a/c03).

Among these peptides, Arg, Pro, and Ala were frequently found; therefore, consensus motifs were predicted using a multiple sequence alignment algorithm (CLUSTALW program) [26]. The GM1- and GM3-binding motifs were (F/W)RxLxxxFx(Q/N)xxxP, Wx(K/H)xxAxPxx(A/S)V, and WRxVxF (Table 1). Arg and hydrophobic amino acids (Ala, Leu, Phe, Pro, Trp, and Val) appeared at high frequencies in the GM1-, GM2-, GM3-, GD1a-, and GT1b-binding peptides.

## 6 Ganglioside Cluster-Binding Peptide (GCBP)

The GM1-binding peptide (GM1/c03), VWRLLAPPFSNRLLP, showed the highest affinity for GM1, and its selectivity was as follows: GM1 > GD1a = GM3 > asialo GM1 > GM2 (Fig. 9) [27]. Because the selectivity corresponded to the sugar structure of gangliosides, the peptide had an affinity for oligosaccharides, but not ceramide.

When the affinity of the peptide for GM1/GlcCer mixed membranes (0–100 mol% GM1) was measured, the amount was found to increase exponentially in proportion with the GM1 content; a large amount of peptide binding was observed at high

**Table 1** Pentadecapeptide sequence and consensus motifs of ganglioside-binding peptides [26]

| Target | Code | Amino acid sequence | Comment |
|---|---|---|---|
| GM1 | c01 | DF**R**R**L**PGA**F**WQLRQ**P** | |
| | c03 | VW**R**L**L**APP**F**SNRLL**P** | GCBP |
| | [motif] | (F/W)**R**x**L**xxx**F**x(Q/N)xxx**P** | |
| | c02 | GWWYKGRARPVSAVA | Identical with GM3/c01 and GD1a/c03 |
| GM2 | c10 | **G**HRPRFS**G**SFVASRA | |
| | c01 | **G**TYYKRF**G**HSIPLVG | |
| | [motif] | **G**xxx(R/K)xx**G**xx(V/I) | |
| | c07 | V**N**RAL**P**A**R**WELWYPR | |
| | c02 | R**N**WNW**P**L**R**ARVLSDA | |
| | [motif] | **N**xxx**P**x**R**xx(L/V) | |
| GM3 | c01 | GW**W**YKGR**A**R**P**VSA**V**A | Identical with GM1/c02 |
| | c03 | RAV**W**RHSV**A**T**P**SHS**V** | |
| | [motif] | **W**x(K/H)xx**A**x**P**xx(A/S)**V** | |
| | c15 | L**WR**P**V**L**F**HSAVRALG | |
| | c30 | **WR**G**V**Y**F**GDRWLGSQP | |
| | [motif] | **WR**x**V**x**F** | |
| GD1a | c03 | G**W**WYKG**R**ARPVSAVA | Identical with GM1/c02 |
| | c18 | R**W**GALL**R**GGAAALFQ | |
| | [motif] | **W**xxxx**R**xxxx(S/A) | |
| GT1b | c03 | **R**IR**A**LFGR**S**P**V**PCCV | |
| | c10 | M**R**RP**A**PWLA**S**R**L**MRP | |
| | [motif] | **R**xx**A**x(F/W)xx**S**x(V/L) | |

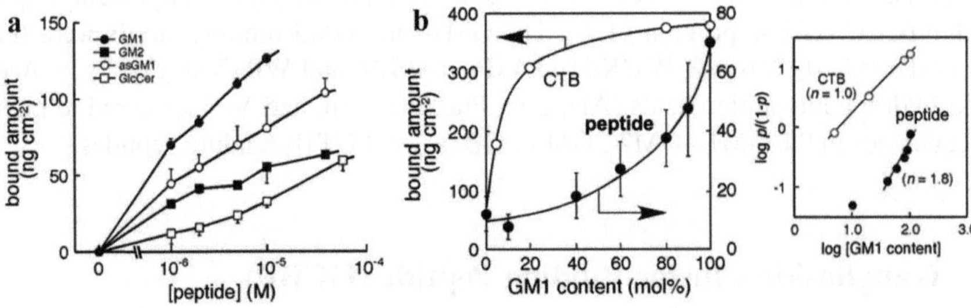


**Fig. 9** Binding of the GM1/c03 peptide to high-density GM1 membranes. **a** Amounts of peptide bound to a ganglioside and GSL monolayer. asGM1, asialo-GM1. **b** Binding of the peptide (1 μM) and cholera toxin B subunit (CTB) (5 nM) to GM1/GlcCer-mixed membranes, where GM1 content was 5–100 mol% (*left*). Hill plots against the GM1 content (*right*). The saturation function ($p$) value was defined from the saturated amounts of the peptide. $n$, Hill coefficient. Reprinted with permission from Ref. [27]. Copyright © 2007, American Chemical Society

GM1 levels (more than 80 mol%). From the Hill plot, the $n$ value of the peptide was determined to be 1.8, indicating the presence of two potential binding sites. As multiple GM1 molecules simultaneously bind to one molecule of the peptide, we named this peptide as ganglioside cluster-binding peptide (GCBP). The binding feature of GCBP is significantly different from that of the cholera toxin B subunit (CTB), which is specific to the GM1 oligosaccharide.

Next, GCBP binding to the GM1 nanoclusters was directly observed using AFM. GM1 nanoclusters (domain $\gamma$, 0.7 nm in height from domain $\beta$) were generated from planar lipid membranes composed of GM1, SM, cholesterol, and unsaturated phospholipid (POPC) (Fig. 10) [14]. Two-dimensional nuclear magnetic resonance (2D NMR) analyses indicated that the size of the GCBP in the presence of GM1 was approximately 1.1 × 1.7 × 1.8 nm [28]. Section analysis indicated that the 1.1 nm increase of domain $\delta$ (1.8 nm) from domain $\beta$ corresponds to GCBP deposition on the GM1 nanocluster. Further, the area of domain $\delta$ (11%) was approximately the same as that of domain $\gamma$ (9%). In contrast, CTB, instead of GCBP, interacts with GM1-containing membranes. CTB is often used as a membrane raft marker. AFM results indicated that CTB binds to the GM1 nanocluster (domain $\gamma$) as well as the membrane raft (domain $\beta$). Comparing the observations of GCBP with those of CTB, we found that GCBP bound to GM1 nanoclusters in GM1-containing membranes. A difference in the binding between GCBP and CTB was also observed on the PC12 cell membrane (Fig. 10d).

## 7 Inhibition of Aβ Fibril Formation by Linear and Cyclic GCBPs

### 7.1 *Inhibition of Aβ Fibril Formation by GCBP*

Aβ monomers are recognized by the GM1 nanocluster and the generated GAβ complex acts as a seed for Aβ fibril formation (Sect. 3). As ganglioside-binding peptide also recognizes GM1 nanoclusters (Sect. 6), we consider that GCBP can inhibit Aβ fibril formation.

When GCBP was incubated with GM1 nanoclusters in Aβ(1–40) solution, the number of Aβ fibrils was significantly decreased (Fig. 11a) [29]. The inhibitory activity of GCBP was further confirmed using a thioflavin T (ThT) assay of GM1-containing liposomes. Furthermore, when the cell damage of toxic Aβ fibrils was evaluated, we found that GCBP can reduce Aβ-induced cytotoxicity (Fig. 11b and c).

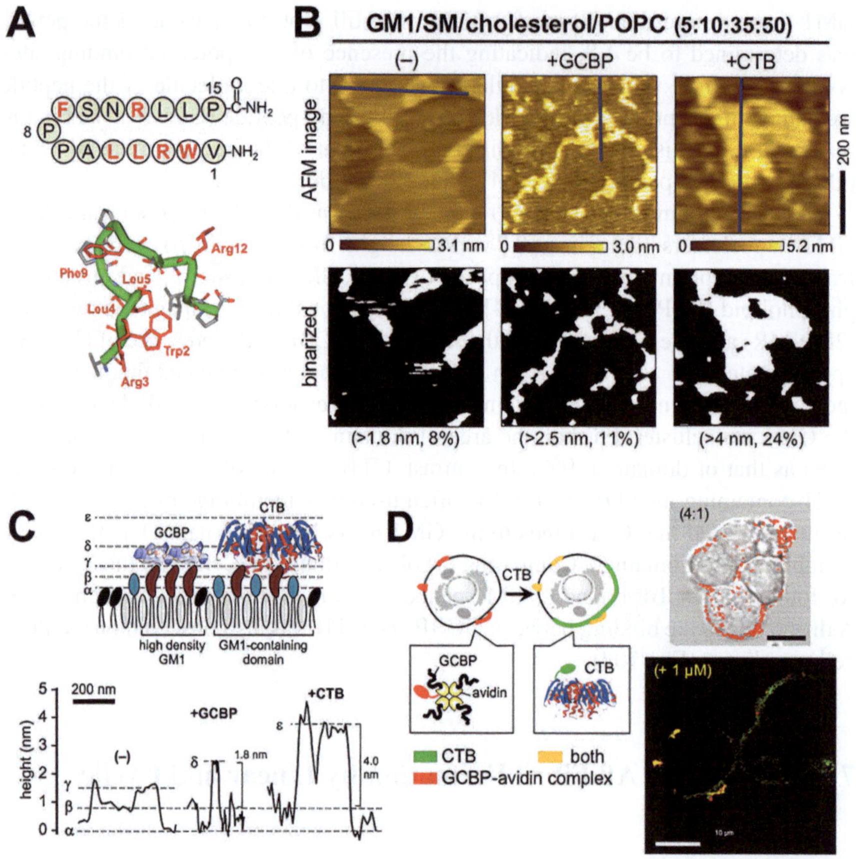


**Fig. 10** Binding of GCBP to the GM1 nanocluster. **a** Amino acid sequence of GCBP (GM1/c03) and molecular modeling. **b** AFM images for the GM1/SM/cholesterol/POPC (5:10:35:50, molar ratio) membranes after incubation with GCBP (1 μM) or CTB (25 nM) for 10 min. **c** Section analysis plots of AFM images for the GM1 – GCBP and GM1 – CTB complexes. **d** Cell surface localization of the GCBP – avidin complex (peptide/avidin, 4:1) and CTB in PC12 cells. PC12 cells were incubated with 1 μM of the GCBP complex (*red*) on ice for 30 min in the presence of CTB (1 μM) (*green*). Yellow areas indicate signals of both red and green. Reprinted with permission from Ref. [14]. Copyright © 2021, American Chemical Society

## 7.2 Cyclic GCBP and Inhibition of Aβ Fibril Formation

To alter the physical and biological characteristics of GCBP, it was cyclized through disulfide bond formation. Conformational restriction by cyclization is often used to increase the bioactivity of molecules; e.g., by avoiding protease digestion [30, 31]. The results of alanine scanning mutagenesis indicated that six residues (W2, R3, L4, L5, F9, and R12) were responsible for GM1 binding and correlated well with the consensus motif of WRLLxxxFxxR (Fig. 12a). One cysteine was replaced

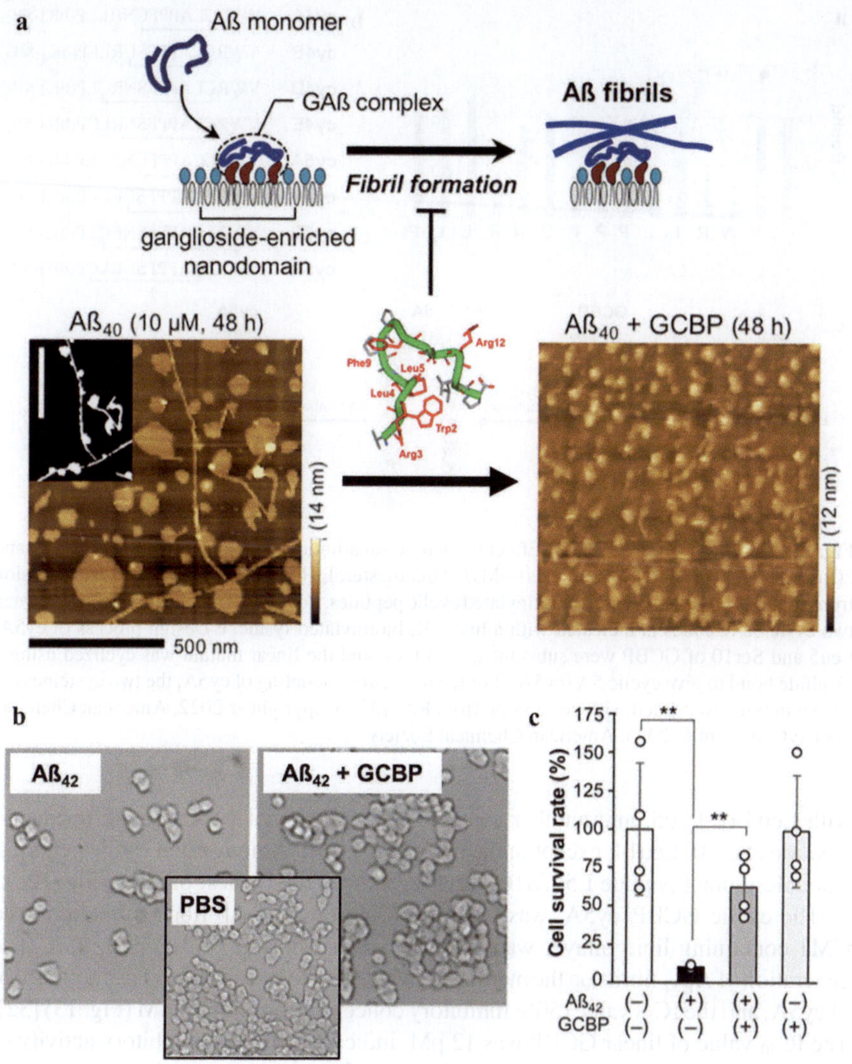


**Fig. 11** Inhibition of the generation of toxic Aβ assemblies on 10 mol% GM1 membrane (GM1/SM/cholesterol, 10:45:45) by GCBP. **a** AFM image of the GM1 membrane after incubation with 10 μM $A\beta_{40}$ and 10 μM GCBP at 37 °C for 48 h. Generation of Aβ fibrils is not observed in the presence of GCBP. **b** Micrographs of PC12 cells on 10 mol% GM1 membranes after 48 h. Prior to seeding PC12 cells, the GM1 membrane was incubated with PBS, 2.5 μM $A\beta_{42}$ solution, or 2.5 μM $A\beta_{42}$ solution containing 50 μM GCBP for 24 h. **c** Prevention of Aβ-induced cell damage by GCBP. Reprinted with permission from Ref. [29]. Copyright © 2022, American Chemical Society

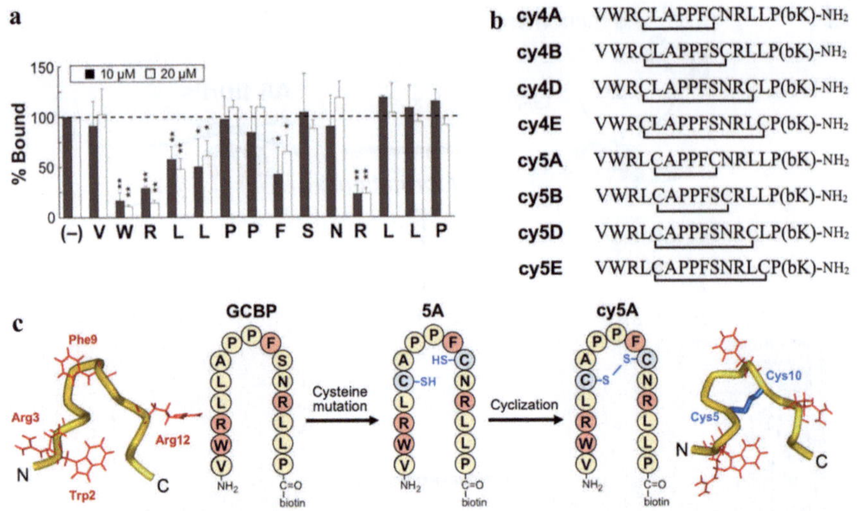


**Fig. 12** Design of cyclic GCBP. **a** Effect of alanine substitution on the binding of GCBP (10 and 20 μM) to 10 mol% GM1 membranes (GM1/SM/cholesterol, 10:45:45). Reprinted with permission from Ref. [29]. **b** Structure of the biotinylated cyclic peptides. Position of the disulfide bond between two cysteine residues is indicated with a line. bK, biotinylated lysine. **c** Design process of cy5A. Leu5 and Ser10 of GCBP were substituted with Cys, and the linear mutant was cyclized using a disulfide bond to give cyclic 5A (cy5A). For the molecular modeling of cy5A, the two cysteines are shown in blue. Reprinted with permission from Ref. [32]. Copyright © 2022, American Chemical SocietyCopyright © 2023, American Chemical Society

with Leu4 or Leu5, and another cysteine was substituted with a single residue at positions Ser10–Leu14, except in the motif (Fig. 12b). Among eight cyclic peptides, a double-mutant peptide L5C/S10C (cy5A) showed the highest binding (Fig. 12c).

The cyclic GCBP, cy5A, was used as an inhibitor of Aβ fibril formation. The GM1-containing lipid bilayer was incubated with Aβ(1–42)($Aβ_{42}$) and cy5A. The generation of $Aβ_{42}$ fibrils on the membrane was found to be suppressed in the presence of cy5A, and the $IC_{50}$ value (50% inhibitory concentration) was 1.2 fM (Fig. 13) [32]. The $IC_{50}$ value of linear GCBP was 12 pM, indicating that the inhibitory activity of GCBP increased 10,000-fold after cyclization. The enhancement by cyclization is attributed to the inhibitory effect of cy5A on GAβ complex formation.

## 7.3 *Linear and Cyclic GCBPs Inhibit Aβ Fibril Formation and Enhance the Release of Fibrils from Human Neuronal Membranes*

The features of Aβ assemblies in patients with AD depend on the brain region; a large amount of Aβ deposition is found in the precuneus (PC), whereas the calcarine

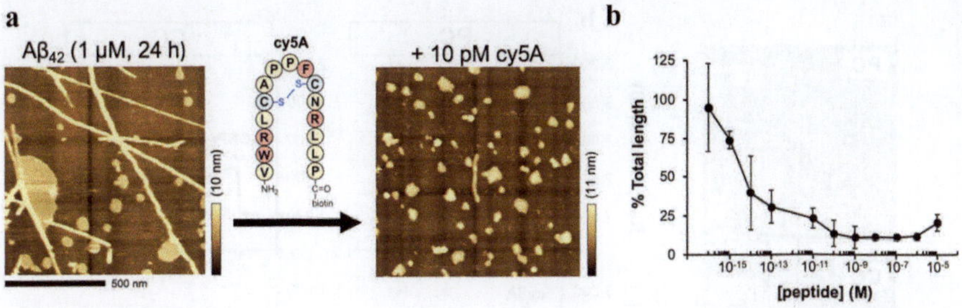


**Fig. 13** Inhibition of Aβ fibrils by cyclic GCBP (cy5A). **a** AFM images of GM1-containing membranes (GM1/SM/cholesterol (10:45:45) incubated with $A\beta_{42}$ (1 μM) in the absence (*left*) and presence (*right*) of cy5A (10 pM) at 37 °C for 24 h. **b** Plot of the % total length of $A\beta_{42}$ fibrils against the concentrations of cy5A. The $IC_{50}$ value of cy5A was 1.2 fM. Reprinted with permission from Ref. [32]. Copyright © 2023, American Chemical Society

cortex (CC) is one of the most resistant regions [33–36]. Aβ accumulation on reconstituted membranes with synaptosomal plasma membrane (SPM) lipids extracted from human autopsied brains was observed using AFM; this accumulation was explained by changes in the fatty-acid chain length of ganglioside GD1b (Fig. 14a) [17]. When Aβ deposition in the early phase of AD brain is reduced in the presence of an effective inhibitor, it is expected to retard the onset of AD.

Aβ assemblies were generated on reconstituted membranes composed of SPM lipids, and the interaction of linear and cyclic GCBPs was then detected using surface plasmon resonance (SPR) (Fig. 14b). When an Aβ(1–42)-accumulated lipid bilayer was interacted with linear (GCBP) and cyclic GCBP (cy5A), an increase in the SPR signal (600–700 resonance unit (RU)), indicating peptide binding to the membrane, was immediately observed at 50 s. The signal then decreased; the decreased signal of Aβ fibrils for PC lipids was greater than that for CC lipids. We thus concluded that the decrease indicates the release of Aβ(1–42) accumulated on the membrane by the peptide interaction. This result indicates the clearance of Aβ assemblies from the neuronal membrane in the presence of linear and cyclic GCBPs (Fig. 14c).

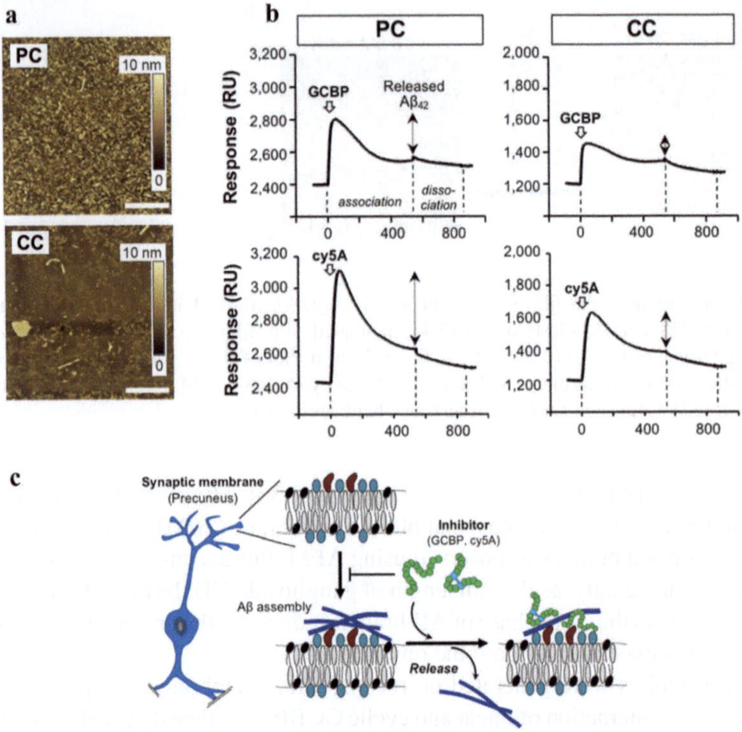


**Fig. 14** Clearance of Aβ assemblies from neuronal lipid membranes by GCBP and cy5A. **a** Aβ assembly on reconstituted lipid membranes prepared using lipid extracts from amyloid-bearing precuneus (PC) and calcarine cortex (CC) regions of human autopsied brains. Reprinted with permission from Ref. [17]. **b** SPR sensorgrams of the $A\beta_{42}$-accumulated lipid bilayer after adding GCBP (*upper*) and cy5A (*lower*). After 540 s for the association phase, the dissociation phase of 300 s is shown. Printed from Ref. [37]. **c** Proposed model for the release of Aβ assemblies on the synaptic membrane by GCBP inhibitors Copyright © 2015, Oikawa et al.

## 8 Conclusion

Because gangliosides are abundant in the nervous system than in other organs, they are potentially involved in the onset and progression of various diseases in these systems. We investigated the clustered structure and function of gangliosides in cell membranes from different perspectives. Many researchers have used liposomes and supported lipid bilayers to investigate ganglioside functions in membranes; however, we attempted to clarify the topological gathering of gangliosides at the nanometer scale using a planar lipid membrane that is physicochemically reconstituted. The topography of the ganglioside nanocluster was imaged using atomic force microscopy (AFM), and the density of gangliosides was distinguished by the difference in binding between CTB and GCBP. The shape and distribution of Aβ aggregates induced by the ganglioside cluster were also identified by AFM imaging in water.

Such information should be useful for investigating the mechanisms underlying AD onset.

Low-molecular-weight compounds such as acetylcholinesterase inhibitors (donepezil) and NMDA inhibitors (memantine) help transmit signals between synapses in nerve cells to suppress the progression of AD symptoms. A recent report indicated that Lecanemab, a humanized IgG monoclonal antibody that binds to Aβ soluble protofibrils, has the ability to reduce amyloid markers in early Alzheimer's disease [38]. In this review, we demonstrate that GCBP and its cyclized peptides can inhibit Aβ aggregation. Inhibitors targeting ganglioside clusters are unique and similar molecules are expected to provide new perspectives for molecular design against ganglioside-related diseases.

## References


1. Lis H, Sharon N (1998) Lectins: carbohydrate-specific proteins that mediate cellular recognition. Chem Rev 98:637–674. https://doi.org/10.1021/cr940413g
2. Kolter T (2012) Ganglioside biochemistry. ISRN Biochem 2012:506160. https://doi.org/10.5402/2012/506160
3. Simons K, Ikonen E (1997) Functional rafts in cell membranes. Nature 387:569–572. https://doi.org/10.1038/42408
4. Brown DA (2006) Lipid rafts, detergent-resistant membranes, and raft targeting signals. Physiol (Bethesda, Md.) 21: 430–439. https://doi.org/10.1152/physiol.00032.2006
5. Pike LJ (2006) Rafts defined: a report on the keystone symposium on lipid rafts and cell function. J Lipid Res 47:1597–1598. https://doi.org/10.1194/jlr.E600002-JLR200
6. Iwabuchi K, Handa K, Hakomori S (1998) Separation of "glycosphingolipid signaling domain" from caveolin-containing membrane fraction in mouse melanoma B16 cells and its role in cell adhesion coupled with signaling. J Biol Chem 273:33766–33773. https://doi.org/10.1074/jbc.273.50.33766
7. Lundquist JJ, Toone EJ (2002) The cluster glycoside effect. Chem Rev 102:555–578. https://doi.org/10.1021/cr000418f
8. London E (2005) How principles of domain formation in model membranes may explain ambiguities concerning lipid raft formation in cells. Biochim Biophys Acta 1746:203–220. https://doi.org/10.1016/j.bbamcr.2005.09.002
9. Yanagisawa K, Odaka A, Suzuki N, Ihara Y (1995) GM1 ganglioside-bound amyloid β-protein (Aβ): a possible form of preamyloid in Alzheimer's disease. Nat Med 1:1062–1066. https://doi.org/10.1038/nm1095-1062
10. Johnston LJ (2007) Nanoscale imaging of domains in supported lipid membranes. Langmuir 23:5886–5895. https://doi.org/10.1021/la070108t
11. Alessandrini A, Facci P (2014) Phase transitions in supported lipid bilayers studied by AFM. Soft Matter 10:7145–7164. https://doi.org/10.1039/c4sm01104j
12. Matsubara T, Nishihara M, Yasumori H, Nakai M, Yanagisawa K, Sato T (2017) Size and shape of amyloid fibrils induced by ganglioside nanoclusters: role of sialyl oligosaccharide in fibril formation. Langmuir 33:13874–13881. https://doi.org/10.1021/acs.langmuir.7b02091
13. Matsubara T, Yasumori H, Ito K, Shimoaka T, Hasegawa T, Sato T (2018) Amyloid-β fibrils assembled on ganglioside-enriched membranes contain both parallel β-sheets and turns. J Biol Chem 293:14146–14154. https://doi.org/10.1074/jbc.RA118.002787
14. Matsubara T, Iijima K, Kojima T, Hirai M, Miyamoto E, Sato T (2021) Heterogeneous ganglioside-enriched nanoclusters with different densities in membrane rafts detected by

a peptidyl molecular probe. Langmuir 37:646–654. https://doi.org/10.1021/acs.langmuir.0c02387
15. Iijima K, Soga N, Matsubara T, Sato T (2009) Observations of the distribution of GM3 in membrane microdomains by atomic force microscopy. J Colloid Interface Sci 337:369–374. https://doi.org/10.1016/j.jcis.2009.05.032
16. Matsubara T, Iijima K, Yamamoto N, Yanagisawa K, Sato T (2013) Density of GM1 in nanoclusters is a critical factor in the formation of a spherical assembly of amyloid beta-protein on synaptic plasma membranes. Langmuir 29:2258–2264. https://doi.org/10.1021/la3038999
17. Oikawa N, Matsubara T, Fukuda R, Yasumori H, Hatsuta H, Murayama S, Sato T, Suzuki A, Yanagisawa K (2015) Imbalance in fatty-acid-chain length of gangliosides triggers Alzheimer amyloid deposition in the precuneus. PLoS ONE 10:e0121356. https://doi.org/10.1371/journal.pone.0121356
18. Matsubara T, Kojima T, Fukuda R, Iijma K, Hirai M, Yamamoto N, Yanagisawa K, Sato T (2018) Responsibility of lipid compositions for the amyloid ß assembly induced by ganglioside nanoclusters in mouse synaptosomal membranes. Polym J 50:745–752. https://doi.org/10.1038/s41428-018-0041-y
19. Valdes-Gonzalez T, Goto-Inoue N, Hirano W, Ishiyama H, Hayasaka T, Setou M, Taki T (2011) New approach for glyco- and lipidomics–molecular scanning of human brain gangliosides by TLC-Blot and MALDI-QIT-TOF MS. J Neurochem 116:678–683. https://doi.org/10.1111/j.1471-4159.2010.07152.x
20. Smith GP, Petrenko VA (1997) Phage display. Chem Rev 97:391–410. https://doi.org/10.1021/cr960065d
21. Scott JK, Smith GP (1990) Searching for peptide ligands with an epitope library. Science 249:386–390. https://doi.org/10.1126/science.1696028
22. Matsubara T (2022) Peptide mimotopes to emulate carbohydrates. Chem Soc Rev 51:8160–8173. https://doi.org/10.1039/d2cs00470d
23. Noren KA, Noren CJ (2001) Construction of high-complexity combinatorial phage display peptide libraries. Methods (San Diego, Calif.) 23:169–178. https://doi.org/10.1006/meth.2000.1118
24. Matsubara T, Ishikawa D, Taki T, Okahata Y, Sato T (1999) Selection of ganglioside GM1-binding peptides by using a phage library. FEBS Lett 456:253–256. https://doi.org/10.1016/S0014-5793(99)00962-X
25. Matsubara T, Sumi M, Kubota H, Taki T, Okahata Y, Sato T (2009) Inhibition of influenza virus infections by sialylgalactose-binding peptides selected from a phage library. J Med Chem 52:4247–4256. https://doi.org/10.1021/jm801570y
26. Matsubara T, Onishi A, Sato T (2012) Carbohydrate recognition by pentadecapeptide ligands for a series of sialylated oligosaccharides. Bioorg Med Chem 20:6452–6458. https://doi.org/10.1016/j.bmc.2012.08.025
27. Matsubara T, Iijima K, Nakamura M, Taki T, Okahata Y, Sato T (2007) Specific binding of GM1-binding peptides to high-density GM1 in lipid membranes. Langmuir 23:708–714. https://doi.org/10.1021/la0619067
28. Fujitani N, Shimizu H, Matsubara T, Ohta T, Komata Y, Miura N, Sato T, Nishimura S (2007) Structural transition of a 15 amino acid residue peptide induced by GM1. Carbohydr Res 342:1895–1903. https://doi.org/10.1016/j.carres.2007.05.026
29. Matsubara T, Nakai M, Nishihara M, Miyamoto E, Sato T (2022) Ganglioside nanocluster-targeting peptidyl inhibitor prevents amyloid β fibril formation on the neuronal membrane. ACS Chem Neurosci 13:1868–1876. https://doi.org/10.1021/acschemneuro.2c00047
30. Ali AM, Atmaj J, Van Oosterwijk N, Groves MR, Dömling A (2019) Stapled peptides inhibitors: a new window for target drug discovery. Comput Struct Biotechnol J 17:263–281. https://doi.org/10.1016/j.csbj.2019.01.012
31. Bhardwaj G, Mulligan VK, Bahl CD, Gilmore JM, Harvey PJ, Cheneval O, Buchko GW, Pulavarti SVSRK, Kaas Q, Eletsky A, Huang P-S, Johnsen WA, Greisen P Jr, Rocklin GJ, Song Y, Linsky TW, Watkins A, Rettie SA, Xu X, Carter LP, Bonneau R, Olson JM, Coutsias E, Correnti CE, Szyperski T, Craik DJ, Baker D (2016) Accurate de novo design of hyperstable constrained peptides. Nature 538:329–335. https://doi.org/10.1038/nature19791

32. Miyamoto E, Sato T, Matsubara T (2023) Cyclization of peptides enhances the inhibitory activity against ganglioside-induced Aβ fibril formation. ACS Chem Neurosci 14:4199–4207. https://doi.org/10.1021/acschemneuro.3c00589
33. Palmqvist S, Schöll M, Strandberg O, Mattsson N, Stomrud E, Zetterberg H, Blennow K, Landau S, Jagust W, Hansson O (2017) Earliest accumulation of β-amyloid occurs within the default-mode network and concurrently affects brain connectivity. Nat Commun 8:1214. https://doi.org/10.1038/s41467-017-01150-x
34. Ikonomovic MD, Klunk WE, Abrahamson EE, Wuu J, Mathis CA, Scheff SW, Mufson EJ, DeKosky ST (2011) Precuneus amyloid burden is associated with reduced cholinergic activity in Alzheimer disease. Neurology 77:39–47. https://doi.org/10.1212/WNL.0b013e3182231419
35. Sperling RA, Dickerson BC, Pihlajamaki M, Vannini P, LaViolette PS, Vitolo OV, Hedden T, Becker JA, Rentz DM, Selkoe DJ, Johnson KA (2010) Functional alterations in memory networks in early Alzheimer's disease. Neuromolecular Med 12:27–43. https://doi.org/10.1007/s12017-009-8109-7
36. Braak H, Braak E (1991) Neuropathological stageing of Alzheimer-related changes. Acta Neuropathol 82:239–259. https://doi.org/10.1007/bf00308809
37. Miyamoto E, Hayashi H, Murayama S, Yanagisawa K, Sato T, Matsubara T (2024) Prevention of amyloid β fibril deposition on the synaptic membrane in the precuneus by ganglioside nanocluster-targeting inhibitors. RSC Chem Biol. https://doi.org/10.1039/D4CB00038B
38. van Dyck CH, Swanson CJ, Aisen P, Bateman RJ, Chen C, Gee M, Kanekiyo M, Li D, Reyderman L, Cohen S, Froelich L, Katayama S, Sabbagh M, Vellas B, Watson D, Dhadda S, Irizarry M, Kramer LD, Iwatsubo T (2023) Lecanemab in early Alzheimer's disease. N Engl J Med 388:9–21. https://doi.org/10.1056/NEJMoa2212948

# Glycomaterials for Biological Applications

**Yoshiko Miura, Masanori Nagao, and Tomonari Tanaka**

**Abstract** Saccharides play important roles in biological phenomena, as bioligands. However, the molecular recognition ability of saccharide is generally weak. It is known that the molecular recognition ability of saccharides is amplified by the cluster glycoside effects. Glycopolymers, which have saccharides on their side chains, are examples of molecules that effectively exhibit the cluster glycoside effect Since glycopolymers are large molecules, they exhibit effective molecular recognition ability. In addition, glycopolymers are polymers, they also function as biomaterials. This section provides an overview of methods for synthesizing glycopolymers, the development of precision synthesis, and their applications.

**Keywords** Glycopolymer · Cluster glycoside effect · Controlled polymerization · Click chemistry · Molecular array

## 1 Introduction

Saccharides are one of the abandoned building blocks of living organisms, both in plants and animals, and are well known as one of the three major biopolymers, along with proteins and nucleic acids. Saccharides have a variety of functions, including serving as carbohydrate food, acting as the skeleton of plants and crustaceans, and providing water solubility to proteins and other useful substances. Oligosaccharides and bioactive polysaccharides also function as biological signals by interacting with

Y. Miura (✉) · M. Nagao
Department of Chemical Engineering, Graduate School of Engineering, Kyushu University, Fukuoka, Japan
e-mail: miuray@chem-eng.kyushu-u.ac.jp

M. Nagao
e-mail: nagaom@chem-eng.kyushu-u.ac.jp

T. Tanaka (✉)
Department of Biobased Materials Science, Graduate School of Science and Technology, Kyoto Institute of Technology, Kyoto, Japan
e-mail: t-tanaka@kit.ac.jp



K. Toshima and Y. Fujimoto (eds.), *Chemical Biology of Carbohydrates*,
https://doi.org/10.1007/978-981-95-5335-8_15

proteins and cells [1]. Saccharides, in particular, as vital signaling molecules, are involved in many important biological phenomena, including cell adhesion, differentiation, pathogen cell infection, and cancer metastasis. Therefore, utilizing sugar-ligands is not only important for the advancement of basic science, including the understanding of biological phenomena, but is also attracting attention from the perspective of biotechnology applications such as drug discovery and biomaterials. Despite the importance of interactions between saccharides and proteins, the interactions are generally very weak when considered in terms of individual saccharides. The cluster glycoside effect was first discovered in the 1980s by Lee et al. [2, 3], who demonstrated the molecular recognition of sugars in the liver. Hepatocytes, which control the quality of red blood cells, have been reported to exhibit extreme, almost exponential, image amplification of their molecular recognition of GalNAc or Gal upon glyco-clusters. This phenomenon is called the cluster glycoside effect. Whiteside et al. subsequently systematically summarized the enhancement of molecular recognition effects achieved by glyco-cluster and discussed the multivalent effect as a general molecular recognition amplification mechanism not only for saccharides but for molecular recognition in general [4]. Unlike nucleic acids and proteins, saccharides are not directly encoded by genes and have therefore been considered to be variable and volatile. This variability and ambiguity in molecular recognition is a characteristic of saccharides, and the cluster glycoside effect is deeply involved in this mechanism.

To artificially utilize sugar signals, glyco-clusters should be synthesized. Various artificial glycocluster molecules have been reported, including liposomes and glycopeptides. One such artificial glycocluster is a glycopolymer. Glycopolymers have a larger glycoclusters than other artificial glycoclusters and therefore exhibit stronger molecular recognition capabilities [5]. Glycopolymers were first reported by Kobayashi et al. in 1985 [6]. Glycopolymers are a general term for polymers with sugars attached to their side chains. The first glycopolymers developed were those with lactose or maltose attached to the side chains of polystyrene. Lactose-attached polystyrene, in particular, easily coated the polystyrene and exhibited unique properties as a culture medium for liver cells [7]. These studies demonstrated numerous practical applications for fundamental analysis of glycobiology and practical applications such as cell culture materials.

Glycopolymers are not only remarkable glycoclusters that amplify the molecular recognition ability of sugars but are also polymeric materials. As polymeric materials, they can also be used as biomaterials. Knowledge of polymer chemistry and materials chemistry also enables the development of precise molecular structures and composite materials. As glycoclusters with excellent molecular recognition capabilities and as polymeric materials with unique structures, glycopolymers provide a variety of biofunctional molecules and materials.

## 2 Synthesis of Glycopolymers

### 2.1 Chemical Structure of Glycopolymers

In this section, the molecular definition of glycopolymer is polymer in which a sugar chain is attached to the side chain of a polymer [5]. Therefore, there are a variety of types of polymers that can become glycopolymers, and various synthetic methods for these polymers. While some articles refer to amylose, mannan, and glycosaminoglycans as glycopolymers, this section will not include these polymers. The structure of glycopolymers is broadly categorized in relation to their synthetic methods. Sugars generally contain hydroxyl groups and functional groups such as amino, sulfate, and phosphate, making them highly nucleophilic. Therefore, polymer synthesis is typically performed with methods that do not conflict with these constraints. Polymerizable compounds with sugar chains attached are prepared, and then polymers are synthesized using these compounds (Fig. 1). The most classic and major method for synthesizing polymers without nucleophilicity is radical polymerization, which utilizes various addition polymerization monomers to obtain glycopolymers (Fig. 1a). Polymer polymerization using metathesis reactions is also being utilized (Fig. 1b).

Glycopolymers can be any polymer that exhibits the cluster glycoside effect, and so they can also be synthesized by attaching saccharide to the existing polymers. For example, polymers have been developed in which glycosylation is attached to the side chains of the polypeptide such as polyglutamic acid [8]. Glycopolymers using chitin and chitosan have also been developed. Glycopolymers using natural polymers are expected to be used in polymer medicines and drug delivery systems, as there is little concern about polymer degradation. For synthetic polymers such as polyesters and polyamides, modification of saccharide is often performed by attaching saccharide to the synthesized polymer, as the polymer synthesis and the functional groups on the saccharide compete with each other. Regarding the controlled polymerization of glycopolymers, which will be described in a later section, a synthesis method in which saccharide is attached after polymerization has also been reported. Because reactions with polymers are less reactive than those with small molecules, quantitatively modifying polymers to form glycopolymers needs highly reactive chemical reactions. In recent years, the use of specific and highly reactive reactions such as click chemistry has attracted attention. Therefore, the usefulness of a method of modifying synthesized polymers with physiologically active sugars using the click reaction has been recognized.

### 2.2 Protecting-Group-Free Synthesis of Glycopolymers

The typical synthetic methods for producing glycopolymers are classified as either the polymerization of glycomonomers and the attachment of saccharide derivatives onto the polymer backbone, called postpolymerization modification. However, both

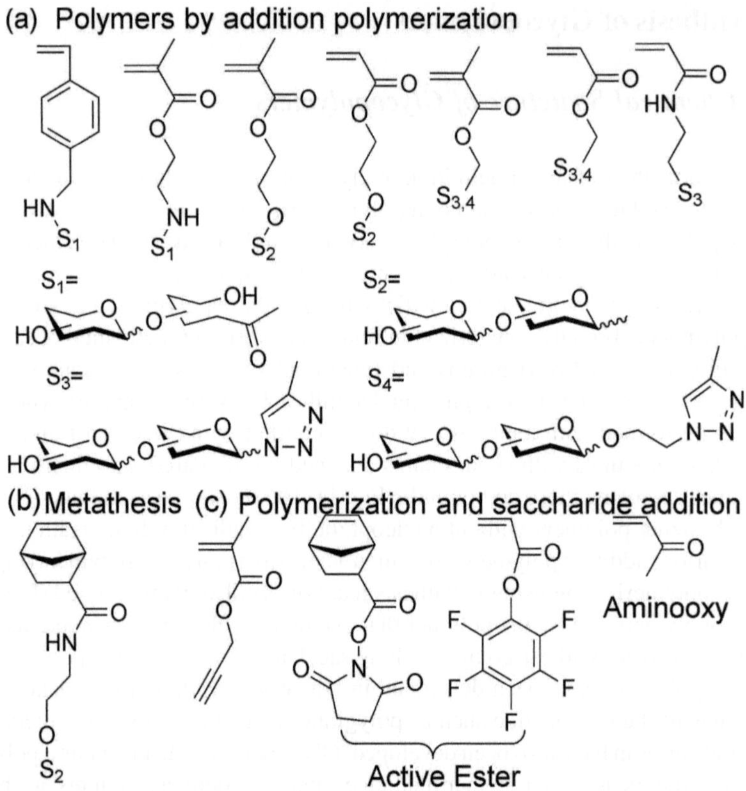


**Fig. 1** Representative monomer structure of glycopolymers

approaches are laborious, involve synthesizing saccharide derivatives from unprotected sugars. In general, unprotected sugars, such as glucose, contain different types of hydroxy groups, including primary, secondary, and hemiacetal alcohols (Fig. 2). In case of the polymerization of glycomonomer method, multistep chemical synthesis is required to obtain glycomonomers in which the hemiacetal hydroxy group is converted to polymerizable functional group at the anomeric position in the saccharide moiety. The first step is the protection of all hydroxy groups. The next step is activation of anomeric position and then introduction of polymerizable functional group. The final step is deprotection. This process enables us to apply it to mono- and disaccharides. However, oligosaccharides with higher molecular weight and complex structure are not easy to apply. Therefore, protecting-group-free and simple synthetic methods for the glycomonomers were demanded.

The direct anomeric activation of unprotected sugars in water was developed by Shoda's group. This method uses water soluble dehydration condensing agents and is called Shoda activation. In 2009, Shoda's group reported the direct synthesis of sugar oxazoline derivatives from unprotected *N*-acetyl-2-amino sugars, such as *N*-acetylglucosamine, using 2-chloro-1,3-dimethylimidazolinium chloride (DMC) in

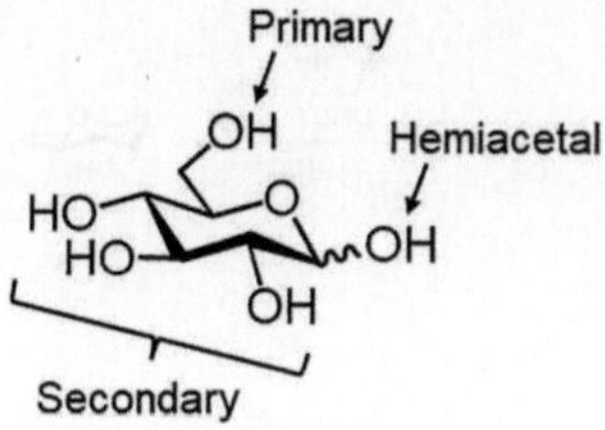


**Fig. 2** Hydroxy groups of glucose

water [9]. This reaction proceeds via an intramolecular dehydration reaction between a hydroxy group at an anomeric position (1-position) and an acetamido group at 2-position under basic conditions. The reaction is selectively proceeded at the anomeric hemiacetal hydroxy group due to the higher acidity than that of other primary and secondary hydroxy groups. At the same time, Shoda's group reported the direct synthesis of various saccharide derivatives, including glycosyl azides, thioglycosides, and 1,6-anhydro sugars using DMC in aqueous media [10–12]. Glycosyl azides were stereoselectively yielded with high yield via direct anomeric activation using DMC with sodium azide in water under basic conditions. The Shoda activation reactions are applicable not only mono- and disaccharides but also oligosaccharides with higher molecular weight and complex structure.

Typical synthetic approaches for the synthesis of glycopolymers are classified into two categories, the polymerization of glycomonomers and attaching saccharide derivatives onto a polymer backbone called post-polymerization modification method. In general, both approaches involve the laborious multistep process for the synthesis of glycosyl derivatives from unprotected saccharides. In the former approach, glycosyl azides which are directly synthesized from unprotected saccharides are useful for the protecting-group-free synthesis of glycomonomers. This approach simplifies the process of glycomonomer synthesis. After the reaction of glycosyl azide synthesis using DMC with sodium azide, copper(I)-catalyzed azide–alkyne cycloaddition reaction called click chemistry was subjected to synthesizing triazole-linked glycomonomers using alkynyl group-containing acryl amide derivatives (Scheme 1) [13–15]. The resulting acrylamide-type glycomonomers were subjected to controlled radical polymerization, such as reversible addition–fragmentation chain transfer (RAFT) polymerization with acrylamide (AAm) as a comonomer using a radical initiator (Ini.) and a chain transfer agent (CTA) to obtain glycopolymers. This protecting-group-free method is applicable to not only mono- and disaccharides but also to higher molecular weight linear and branched oligosaccharides, including sialyllactoses, maltooligosaccharides, and biantennary sialyl *N*-glycans. Similar protecting-group-free synthesis of glycomonomers and glycopolymers via the protecting-group-free synthesis of thioglycosides were reported [16].

The latter approach, post-polymerization modification method, to synthesizing glycopolymers uses attachment reactions, such as click chemistry. When the polymer backbones bearing alkynyl groups in the polymer side chain were synthesized by RAFT polymerization, it was necessary to protect the alkynyl group, such as

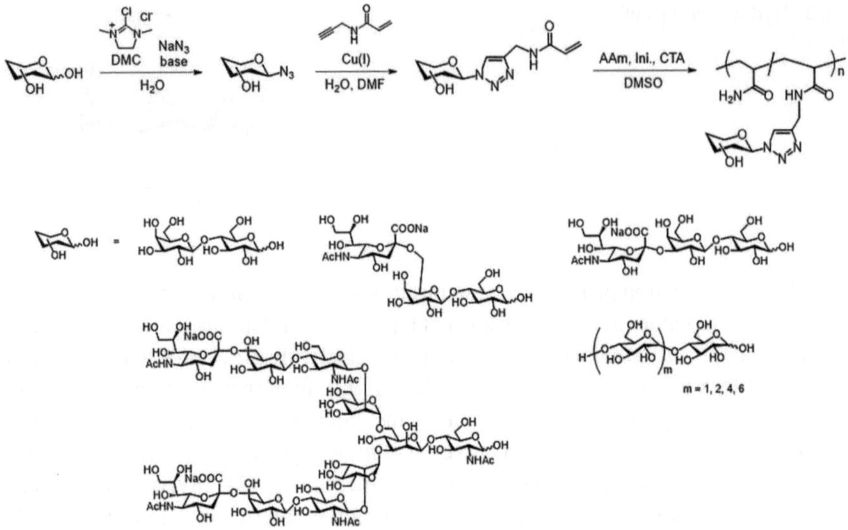

**Scheme 1** Protecting-group-free synthesis of glycomonomers and subsequent polymerization to form glycopolymers via glycosyl azides

by a trimethylsilyl (TMS) group (Scheme 2) [17]. After the deprotection of the TMS groups using tetrabutylammonium fluoride (TBAF), the glycopolymers were obtained the post-polymerization modification via click chemistry using glycosyl azides.

The post-polymerization modification approach often involves the use of activated esters, such as *N*-hydroxy succinimide (NHS) ester and pentafluoro phenyl ester, to synthesize functional polymers including glycopolymers [18–20]. An amide bond is formed by the reaction between an activated ester and an amino group. The reaction proceeds smoothly and is highly selective, giving it a high synthetic chemical

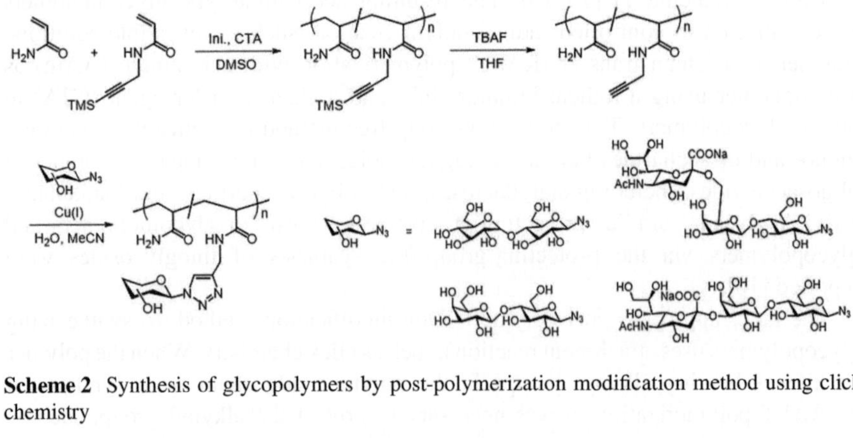

**Scheme 2** Synthesis of glycopolymers by post-polymerization modification method using click chemistry

value. Polymers bearing activated esters are frequently used in a post-polymerization modification mothed to synthesize glycopolymers. Because of the hydrophobicity of these activated esters, the attachment reactions are usually carried out in organic solvents. Although a monomer bearing a water-soluble activated ester, *N*-hydroxy sulfosuccinimide (sulfoNHS) ester, has been reported for post-polymerization modification, the monomer is unstable in water, with a half-life of one hour [21]. The sulfoNHS-bearing monomer could not be isolated due to the hydrolysis. Tanaka's group developed novel and stable monomers bearing water-soluble activated esters. The monomers bearing the water-soluble activated ester, sulfoNHS ester and 4-sulfo-2,3,5,6-tetrafluoro benzyl (sulfoTFB) ester, are stable in water with a half-life of 15 h and 50 h, respectively [22–24]. The alkanediyl linker, especially longer than propanediyl, increases the stability of the monomers and polymers in water. The polymers bearing sulfoNHS and sulfoTFB esters are more stable than their corresponding monomers. The post-polymerization modification using the polymers bearing sulfoNHS and sulfoTFB esters and amino group-containing saccharide derivatives successfully provided glycopolymers in water (Scheme 3). A one-pot and open-to-air synthesis of glycopolymer was achieved via oxygen-tolerant photoinduced controlled radical polymerization, photoinduced electron transfer–reversible addition–fragmentation chain transfer (PET-RAFT) polymerization, using the sulfoTFB-bearing monomer [25]. The subsequent post-polymerization modification in water using amino group-containing saccharide derivatives, including the monosaccharide and sialyl *N*-glycan-bearing peptide, yielded the glycopolymers without isolating the polymer bearing water-soluble activated esters.

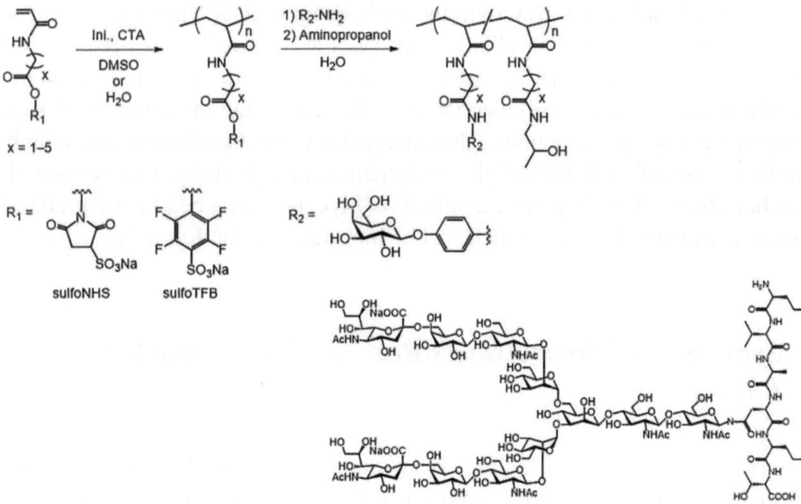


**Scheme 3** Synthesis of glycopolymers by post-polymerization modification using polymer backbones bearing **a** activated esters and **b** water-soluble activated esters

## 2.3 Precision Synthesis of Glycopolymers

In this section, we describe the precise synthesis of polymers bearing saccharide side chains. Polymerization methods are generally categorized into addition and condensation polymerizations, with addition polymerization further classified into radical and ionic polymerizations. Among them, radical polymerization is most widely used due to its broad range of reaction conditions and the variety of applicable monomers. However, polymers prepared by conventional radical polymerization typically exhibit dispersity in their degree of polymerization. Furthermore, considering the stereochemical configuration along the main chain, such polymers are far from uniform. To address this, controlled polymerization techniques (also known as living radical polymerization) have been developed since the late 1990s to synthesize polymers with more uniform structures. In these techniques, dormant species are incorporated at the growing polymer ends to suppress termination reactions via end-to-end coupling, thereby promoting a more uniform polymer growth. These techniques also enable block polymerization, where different types of monomers are introduced in a segmented fashion. By applying such controlled polymerization strategies, it has become possible to design the structure of glycopolymers to a certain extent. In early studies, the hydroxyl groups on sugars were protected with acetyl groups, and polymerization was carried out in organic solvents, followed by deprotection. However, this approach increases the number of synthetic steps, and in the case of side chains containing many hydroxyl groups, such as oligosaccharides, complete protection and deprotection can be difficult. Moreover, oligosaccharides tend to be rare and expensive, making it unfavorable to increase synthetic steps or decrease overall yields. To overcome these challenges, Haddleton and colleagues introduced the concept of "post-click chemistry," [26] originally proposed by Craig Hawker [27], to glycopolymer synthesis. In this method, the polymer backbone is first synthesized via living polymerization, followed by the introduction of sugar side chains through click reactions. This approach enables the efficient and straightforward synthesis of well-defined glycopolymers with high yields. In particular, the authors have focused on the precise synthesis of glycopolymers bearing trisaccharide sialyllactose as the side chain, which will be introduced in the following sections.

## 2.4 Synthesis of Glycopolymers Bearing Sialyllactose Side Chains

As described in the previous section, Tanaka and co-workers developed a method to selectively introduce an azido group at the anomeric position of sugars [10]. This approach enabled the derivatization of structurally complex oligosaccharides into azido-functionalized compounds. The next objective, therefore, was to synthesize polymers bearing alkyne side chains via living radical polymerization (particularly RAFT polymerization). Trisaccharides such as sialyllactose possess numerous hydroxyl groups and are highly hydrophilic. When introducing such saccharides into the side chains via click chemistry, it is necessary to select a solvent in which both the alkyne-functionalized polymer backbone and the sialyllactose are simultaneously soluble. If the polymer backbone is hydrophobic, the choice of compatible solvents is limited. Moreover, the resulting amphiphilic glycopolymers may exhibit unexpected behaviors such as micellization during subsequent functional evaluations. To address this issue, the authors developed an acrylamide-type alkyne polymer, which had been scarcely reported at the time. Initially, the project began with the use of propargyl acrylamide, which had already been described in the literature (Fig. 3) [27]. The authors copolymerized acrylamide with a trimethylsilyl (TMS)-protected propargyl acrylamide (TMS PrAAm) but observed that the rate of radical polymerization decreased as the proportion of TMS PrAAm increased. This trend was also observed in free radical polymerization without a RAFT agent. The authors hypothesized that the low polymerizability of TMS PrAAm was due to conjugation between the vinyl group and the alkyne via the allylic carbon in the radical state. Although this hypothesis remains unconfirmed, a new monomer (TMS BtnAAm) was synthesized in which two carbon atoms were inserted between the amide and alkyne groups to interrupt the conjugation. Remarkably, TMS BtnAAm exhibited a significantly higher polymerization efficiency compared to TMS PrAAm [28]. As a result, RAFT copolymerization of acrylamide and TMS BtnAAm was achieved in arbitrary ratios. The polymers obtained via RAFT polymerization exhibited a dispersity below 1.4, indicating a more controlled molecular weight distribution compared to polymers obtained via free radical polymerization.

After deprotection of the TMS groups on the side chains, either 6′-sialyllactose (6′-SALac) or 3′-sialyllactose (3′-SALac) bearing azido groups at the anomeric position was introduced via copper-catalyzed azide-alkyne cycloaddition (CuAAC)

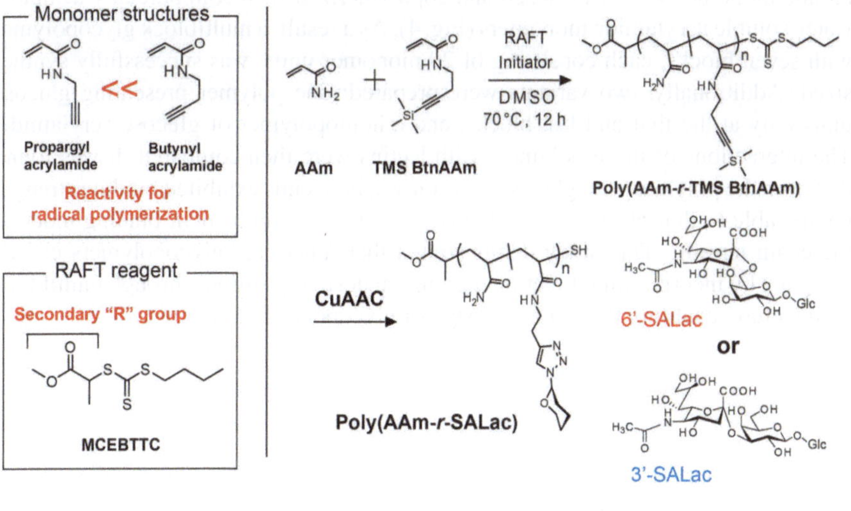


**Fig. 3** Synthetic scheme of glycopolymers carrying sialyllactose by "post-click" chemistry

[29]. Even after the saccharides were introduced into the side chains, the polymer maintained its low dispersity, demonstrating the successful synthesis of well-defined glycopolymers with oligosaccharide side chains. The introduced sialyllactose moieties are natural ligand sugars for influenza viruses, with the 6′- and 3′-linkages corresponding to human-type and avian-type influenza viruses, respectively. The resulting glycopolymers showed selective binding toward each type of virus, suggesting their potential function as virus inhibitors.

## *2.5 Synthesis of Multiblock Glycopolymers*

A characteristic feature of living polymerization is the ability to precisely control the sequence of polymer segments in a block-wise manner. While methacrylate-based polymers have been commonly studied in the past, it has been challenging to increase the number of blocks without broadening the dispersity, especially in RAFT polymerization due to its initiation mechanism. Perrier and co-workers succeeded in synthesizing polymers with 20 blocks by taking advantage of the RAFT mechanism and conducting the polymerization using a very low equivalent of initiator relative to the RAFT agent [30]. Although each block contained an average of 10 repeating units—still short of the monomer-by-monomer control seen in proteins—this work demonstrated a significantly higher degree of sequence control than previously achieved with synthetic polymers. Building on this approach, the authors applied this RAFT polymerization method to glycopolymers, aiming to construct glycopolymers with multiblock structures [31]. To achieve near-quantitative monomer conversion in a short time, it was necessary to use acrylamide monomers with high propagation rate constants and conduct the polymerization in water [32]. Accordingly, a glucose acrylamide monomer was synthesized and copolymerized with commercially available water-soluble acrylamide monomers (Fig. 4). As a result, a multiblock glycopolymer with seven blocks, each consisting of 20 monomer units, was successfully synthesized. Additionally, two variants were prepared: one polymer presenting glucose units only at the first and last blocks, and a homopolymer of glucose acrylamide. The interactions of these polymers with lectins were then compared. It was found that even the polymer with glucose units only at both ends exhibited binding strength comparable to that of the homopolymer, suggesting a multivalent binding mode at the chain termini. This study demonstrated that functional glycopolymers can be designed by incorporating functional groups at desired positions through multiblock architecture, enabling the rational design of glycopolymer functions.

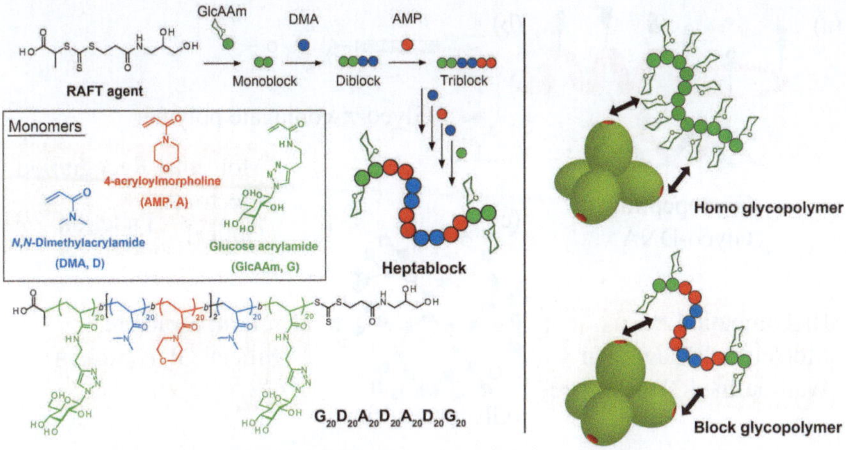


**Fig. 4** Schematic illustration of multiblock glycopolymer and interactions with a target protein

## 2.6 *Glycopolymer with Other Backbones*

When considering the molecular recognition ability of glycopolymers, the requirement is only to form glyco-clusters. Therefore, the backbone of the polymer could be diverse.

### 2.6.1 Glycoconjugate with Peptide and Nucleic Acid

Choosing a polypeptide as the backbone type of glycopolymer allows the fabrication of a glycopolymer that incorporates the properties of peptides, such as biodegradability, potential biocompatibility, and three-dimensional structural formation (Fig. 5).

Polyglutamic acid is a polypeptide with carboxylic acid side chains, and glycopolymers are obtained by modification of saccharides via amide bond formation to the side chains. Totani et al. prepared a polyglutamate-glycopolymer by the modification of various saccharides including sialyloligosaccharides [8]. This glycopolymer demonstrated significant inhibitory effects against the influenza virus. In addition, the glycopeptides are considered as the mucin mimics, and investigations are underway to develop glycovaccines using glycoconjugated peptides as antigens [33].

Another key feature of polypeptides is the properties of the formation of well-defined structures. Since peptides have well-defined structures like a-helixes and b-sheets, 3D structure was precisely designed by de novo designed peptides [34]. This allows for precise control of the spatial presentation of glyco-clusters. For example, glycopeptides, in which are saccharides attached to α-helices, precisely form locally dense structures. Hasegawa et al. synthesized de novo designed peptides in which lactose was periodically attached to α-helical peptides, inducing sugar-sugar

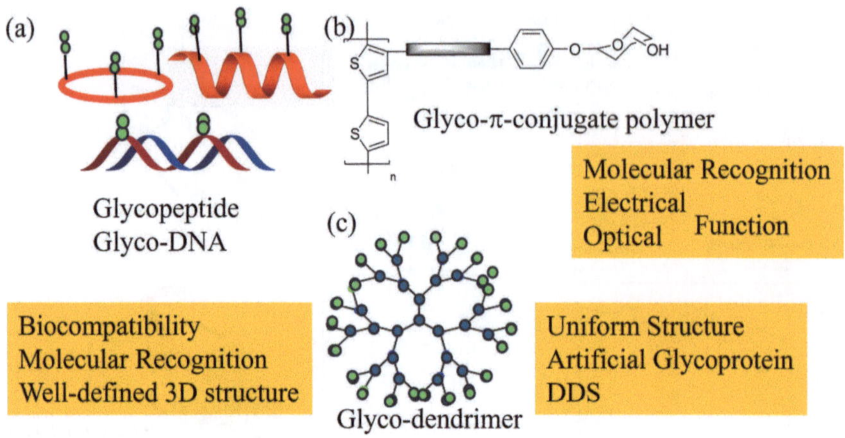


**Fig. 5** Glycomaterials with various backbones: **a** Glycopolymers with peptide and DNA backbone, **b** glyco-π-conjugate polymers, and **c** glycodendrimers

interactions [35]. The Mihara group created a glycopeptide library in which mannose or galactose was regularly attached to α-helical peptides [36]. This glycopeptide library allows precise control of the spatial arrangement of saccharides, and they found that structures in which saccharides are effectively assembled significantly enhance the binding of lectins (mannose and Con A, galactose and PNA) to sugars. In addition to typical structures, polypeptides can also form the unusual specific shapes, such as cyclic peptides. Cyclic peptides have less structural flexibility, allowing for the formation of precise glycoclusters. Nishimura et al. have demonstrated that a cyclic glycopeptide, in which sialyloligosaccharides are linked to a cyclic peptide, has a strong inhibitory effect on influenza viruses [37].

Nucleic acids, particularly DNA, have attracted attention as a backbone whose structure can be designed more clearly than peptides, and this is known as "DNA nanotechnology", allowing the development of glycoclusters that specifically match the structure of lectins. Matsuura et al. linked galactose to single-stranded DNA and used DNA hybridization to design the spatial arrangement of galactose [38].

### 2.6.2 Glycoconjugate with π-Conjugate Polymer

Polymers with π-conjugate polymers, possess characteristics such as fluorescence and electrical conductivity. By using π-conjugate polymers as the backbone of glycopolymers, it is possible to create glycoclusters that combine the optoelectronic properties with the molecular recognition properties of glycans. Such glycopolymers can be used as molecular probes, or, when combined with electrodes, as biosensors.

For example, polythiophenes conjugated with physiologically active glycans such as sialic acid and mannose have been reported to bind to lectins, viruses, and bacteria, resulting in a red-shifted color change that can detect proteins and pathogens [39].

Functionalizing electrodes with glycopolymers and pi-conjugated polymers has also been reported to electrically detect proteins and pathogens. Utilizing the FRET properties of glycopolymer-pi-conjugated polymers, they have been reported to interact with bacterial lectins, bind to bacterial membranes, and detect them.

#### 2.6.3 Glycodendrimers

Dendrimers are one type of special polymer structure. Unlike polymers obtained through addition polymerization, dendrimers have no molecular weight distribution and a dense, compact structure, making them promising candidates for glycopolymers and glycoclusters. Dendrimers do not require a specific chemical structure; they can be any molecule with a regular dendritic structure [40]. However, due to the complexity of their synthesis, most research has focused on commercially available amidoamine-type dendrimers (PAMAM dendrimers).

Roy et al. synthesized glyco-dendrimers, in which the termini of dendrimers were modified with glycans, and have conducted research on them as artificial glycoproteins. Dendrimers modified with various physiologically active glycans at the termini of PAMAM dendrimers have demonstrated high lectin recognition ability. The high molecular recognition ability and compact, uniform shape of glycodendrimers have been reported to inhibit influenza viruses, *E. coli*, and other bacteria. These effects are similar to those of linear glycopolymers, but they form compact, dense glycoclusters. It has also been reported that conjugates of glycondrimes and the anticancer drug doxorubicin effectively exerted their anticancer effects due to the molecular recognition ability of sugars [41].

## 3 Application of Glycopolymers and Glycomaterials

### *3.1 Multivalent Interactions of Glycopolymers*

Sugars interact with a group of proteins called lectins through hydrogen bonding and CH–π interactions. In general, the interaction between a monovalent sugar unit and a single carbohydrate recognition domain (CRD) of a lectin is weak, with an association constant of approximately $10^3$ $M^{-1}$. To compensate for this weakness, nature adopts multivalent binding. Many lectins possess multiple sugar-binding domains, and their interactions with several glycans on the cell surface are significantly enhanced through multivalent interactions, often increasing the binding strength by orders of magnitude. Similarly, pathogens such as viruses display multiple lectins on their surfaces, enabling multivalent interactions with host cell surface glycans. Understanding such multivalent binding effects (also known as the glycoside cluster effect) and incorporating them into material design allows for the development of glycopolymers with enhanced binding affinity to target lectins [4, 5]. Intermolecular interactions are thermodynamically expressed by the Gibbs free energy equation, $\Delta G = \Delta H - T\Delta S$, where the direction of binding is determined by enthalpic contributions ($\Delta H$), such as exothermic interactions, and entropic contributions ($\Delta S$), reflecting the loss of molecular motion or conformational freedom upon binding. Multivalent binding tends to be enthalpically favorable due to the accumulation of multiple interaction points, driving $\Delta G$ more negative (favorable for binding). On the other hand, molecular binding restricts conformational motion, resulting in a negative $\Delta S$ and a positive–$T\Delta S$ term (unfavorable for binding). The authors have reported polymer designs based on this thermodynamic framework for multivalent interactions, and several representative examples are introduced below.

### 3.1.1 Design of Star-Shaped Glycopolymers for Target Lectin (Hemagglutinin)

The spike protein (hemagglutinin) on the influenza virus is a homotrimeric lectin with three glycan-binding domains at its tip [42]. As mentioned above, sialyllactose contains the natural ligand structure for hemagglutinin. These three binding sites are arranged at the vertices of a triangle, and crystallographic studies have shown that the distance between them is approximately 4.5 nm. When designing a molecule that strongly binds to hemagglutinin, a three-point binding interaction that can simultaneously engage with all three binding domains is envisioned. Previous studies have used DNA or peptides to precisely control the molecular length and display three glycan units [43, 44]. However, such molecular design was not achieved using synthetic polymers. To address this, we synthesized star-shaped glycopolymers with a three-arm structure that could strongly bind to hemagglutinin, using living radical polymerization [45, 46] (Fig. 6). The size of the molecule (i.e., the length of each polymer arm) was tuned by controlling the degree of polymerization, aiming for a size that matched the distance between the glycan-binding domains of hemagglutinin. The size of the star-shaped polymers in solution was estimated based on classical equations. Although these equations include constants like the characteristic ratio that depend on the nature of the monomers, a correlation between the number of repeating units and the radius of gyration was obtained. The distance from the center of the triangular arrangement of hemagglutinin binding sites to each vertex was calculated to be 2.5 nm. Therefore, the target degree of polymerization was set at 40, corresponding to a radius of gyration of 2.5 nm. After polymerization, an alkyne group was introduced at each RAFT chain end, followed by modification with sialyllactose. The interaction between influenza virus and the star-shaped glycopolymer was evaluated by hemagglutination inhibition assay. When comparing polymers with degrees of polymerization of 20, 40, and 90, the 40-mer polymer, as predicted, showed the strongest interaction. For shorter polymers, the chain length was insufficient for multivalent binding, while longer polymers likely required chain folding for multivalent interaction, which is entropically unfavorable, leading to weaker interactions. This study demonstrated that the structure of synthetic polymers can be controlled

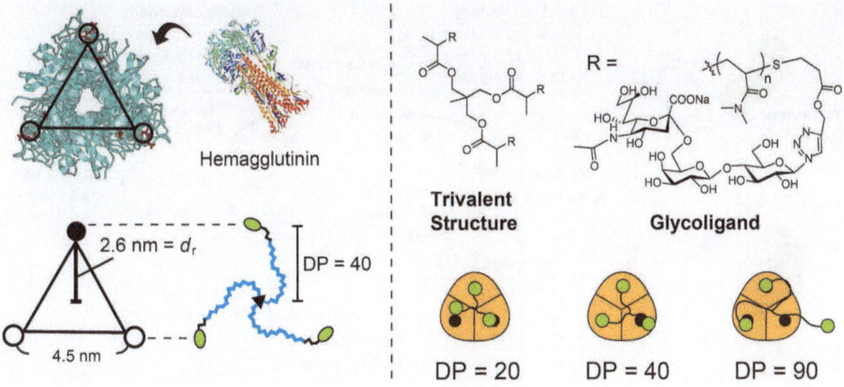


**Fig. 6** Illustration of star glycopolymer and its chemical structure

at the nanometer scale using living radical polymerization, and that such structural control enables regulation of functional interactions.

### 3.1.2 Thermodynamic Evaluation of Multivalent Binding Using Glycooligomers

An increase in the number of binding sites is favorable for molecular interactions from an enthalpic perspective but unfavorable from the viewpoint of molecular entropy. This entropic disadvantage is especially pronounced when glycans are connected by flexible linkers. If the linkers are inherently rigid, the conformational change upon binding is minimal, and thus the entropy change ($\Delta S$) is small. Therefore, to enhance overall interaction via multivalent binding, the enthalpic gain must outweigh the entropic penalty. However, in the case of glycans, as mentioned earlier, the intrinsic binding affinity in a 1:1 interaction is generally weak. To compensate for this, one approach is to present multiple glycan units to a single binding domain. That is, if one glycan detaches, another adjacent glycan can immediately bind to the same site, resulting in an apparent increase in binding strength. Although this mechanism has long been suggested in studies of multivalent glycan ligands, it had not been quantified through thermodynamic parameters. To address this, we synthesized glycooligomers and analyzed their interactions using isothermal titration calorimetry (ITC) [47] (Fig. 7). Mannose acrylamide was polymerized via RAFT polymerization, and the resulting polymers were separated by oligomer length using reverse-phase column chromatography. As a result, mannose oligomers with defined mono- or dimer structures were successfully obtained. These oligomers were then analyzed for their interactions with a model lectin, concanavalin A (ConA), using ITC. While native ConA is a tetramer, it was dissociated into dimers using acetate buffer. Indeed, the mannose dimer showed stronger interaction with ConA than the monomer. Although the enthalpic values were nearly identical, differences in entropy

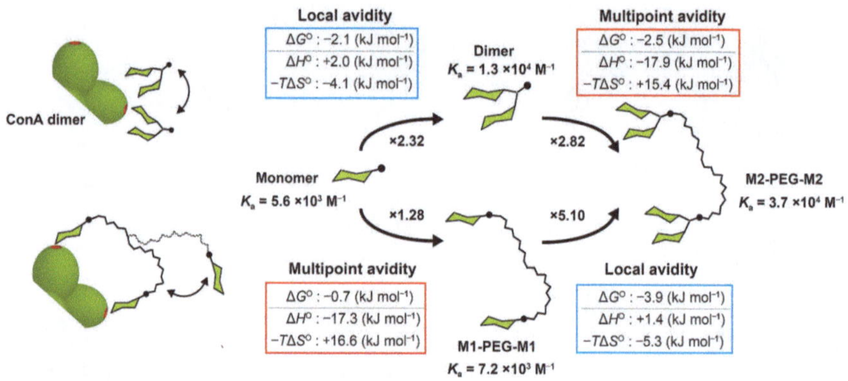


**Fig. 7** Summary of the interactions of glycooligomers with ConA

were observed, leading to differences in overall binding. The entropy term includes contributions from hydration entropy, rotational-translational entropy, and conformational entropy of the molecules. By performing measurements at different temperatures, hydration entropy was calculated, confirming that the conformational entropy differed between the monomer and dimer. In the case of the dimer, the number of molecular configurations capable of binding to ConA increased, thereby reducing the entropy loss.

Next, the RAFT chain ends of these oligomers were converted to thiols and reacted with PEG containing terminal maleimide groups to produce divalent glycan ligands. When the interactions between these ligands and ConA were measured, it was found that the binding constant increased regardless of whether the PEG termini presented monomeric or dimeric mannose. However, the increase in binding strength was more significant in the case where mannose dimers were presented at both ends of PEG. This suggests, as mentioned above, that unless the local binding affinity is sufficiently strong, the entropic penalty from the flexible PEG linker cannot be compensated by the enthalpic gain from multivalent binding. Therefore, as a design principle for multivalent glycan ligands, it is essential to locally cluster glycans if the binding affinity of a single glycan unit is low. Alternatively, making the linker structure more rigid can be effective; however, in this case, the spatial mismatch between the glycan and the target protein must be precisely addressed through molecular design, as rigid linkers reduce the adaptability of the ligand to fit various binding site geometries.

## 3.2 *Pathogen Inhibition*

Glycopolymers can function as inhibitors because they present glycans in a multivalent manner, enabling strong binding to pathogens. In this chapter, we present examples of glycopolymers synthesized by the authors and their evaluation in inhibiting viral infection.

### 3.2.1 Evaluation of Virus Inhibition Using Glycopolymers Bearing Sialyllactose

Using a post-click chemistry strategy, glycopolymers bearing sialyllactose side chains were synthesized [29]. By varying the feed ratio of comonomers, the proportion of glycan side chains within the polymer was controlled. In hemagglutination inhibition (HI) assays with influenza virus, the polymer containing 70 mol% of glycan units exhibited the strongest interaction with the virus. This outcome is attributed to two opposing effects: (i) a higher glycan density increases the number of ligands available for binding hemagglutinin, and (ii) steric hindrance among bulky glycan side chains may interfere with binding. These glycopolymers were further evaluated in viral infection inhibition experiments. Madin-Darby canine kidney (MDCK) epithelial cells were seeded onto culture dishes and grown to confluence. A mixture of influenza virus and the glycopolymer was then applied to the cells. After incubation, the cells were washed and overlaid with an agarose gel to restrict viral diffusion. Two days later, the gel was removed, and the number of plaques formed by infected cells was counted to assess inhibition. Only the glycopolymer with 70 mol% glycan content, which showed the strongest interaction in the HI assay, demonstrated a statistically significant reduction in plaque number. These results indicate that glycopolymers capable of binding hemagglutinin can effectively inhibit viral infection. Moreover, the correlation between hemagglutination inhibition activity and infection inhibition suggests that HI assays may serve as a predictive screening tool. While it may seem that higher molecular weight polymers bearing more sialic acids would be more effective at virus inhibition, such materials may also exhibit undesirable off-target interactions in vivo. Therefore, the development of glycopolymers that selectively and strongly bind to the hemagglutinin of influenza viruses is essential for potential therapeutic applications. The design principles discussed in the previous chapter on multivalent binding effects provide valuable insight toward achieving this goal.

## *3.3 Glycomaterials for Glycomicoarray and Biosensors*

Detecting the molecular recognition of saccharide is important in proteome analysis, and glycan-arrays have been investigated. Glycan-arrays, like those for nucleic acids and peptides, aim to detect molecular recognition. However, because glycan-protein molecular recognition is weak, molecular arrays using glycopolymers that exhibit the cluster glycoside effect have been investigated [5].

Godula and Bertozzi have investigated glycocalyx using glycopolymers. They synthesized glycopolymers by polymerizing poly(methylvinyl ketone) using the RAFT method and then reacting them with glycans bearing oxoamino groups at the anomeric sugar position [48, 49]. They reported that this method allows for quantitative addition of glycans through a polymer reaction, resulting in glycopolymers. They also demonstrated that lipids or biotin can be introduced to the molecular termini during polymer synthesis. For example, glycopolymer arrays bearing biotin at the termini can be used as glycan arrays using streptavidin-immobilized substrates [48]. Glycocalyx arrays can be used to investigate the molecular interactions between lectins, cells, and glycans. Not only are the sugar–protein interactions detected, but also the arrays can also detect differences in recognition due to molecular density, as well as molecular cross-linking by lectins and galectins [50]. Glycan-specific and density-dependent molecular recognition has also been confirmed for pathogens such as influenza viruses. Bertozzi et al. have used glycocalyx to create cell-mimicking environments and analyze the principles of biological activity. For example, they used glycan arrays to analyze the glycan-related immune effects of Siglecs [51]. We have also reported glycopolymer molecular arrays on gold substrates using RAFT-polymerized glycopolymers [52–54]. RAFT-polymerized polymers contain thioesters derived from the RAFT agent at their molecular termini, and decomposition of these thiols yields polymers with terminal thiols. Such glycopolymers are bound to gold substrates to form glycopolymer arrays. The glycopolymer arrays immobilizing a molecular library of glycopolymers on gold substrates can be used for comprehensive analysis using SPR. Thiols can also be immobilized using other methods, such as the thiol-ene reaction. Glycopolymers immobilized using various methods can be used as molecular arrays.

Glycopolymers immobilized on gold substrates have functions of the optical (such as SPR) and electronic properties of the substrate (Fig. 8). SPR allows for comprehensive analysis, making it possible to detect binding with proteins and pathogens. Various glycopolymers are useful not only for the detection and analysis of lectins, but also for the detection of viruses and toxin proteins that recognize sugars. For example, Shiga toxin can be detected using SPR and QCM on gold substrates immobilized with glycopolymers containing globoglycans [55]. Cholera toxin sensing has been reported by combining a galactose-containing glycopolymer-immobilized substrate with an electrochemical sensor [56].

## 3.4 *Glycomaterials as Molecular Device for Cells*

One of the earliest applications of glycopolymers is well-known as a liver cell culture substrate based on a glycopolymer-immobilized substrate [7]. Liver cells have asialoglycoprotein receptors and exhibit strong adhesive properties toward Gal or GalNAc [57]. Several glycoconjugates, such as the strong interaction between mannose and macrophages and the interaction between sialyloligosaccharides and immune cells, have been reported [5] (Fig. 9).

Lactose-containing polystyrene-type glycopolymers can be coated onto polystyrene via hydrophobic interactions to obtain Gal presenting substrates [7]. These polymers specifically adsorbed liver cells of hepatocyte, making them useful as hepatocyte culture materials. Hepatocyte adhesion varied depending on the Galactose density, that is glycopolymer absorbed amount [58]. Recently, hepatocyte

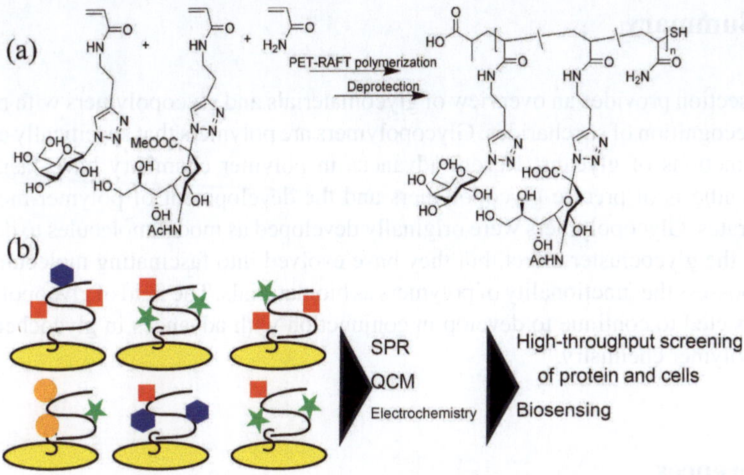


**Fig. 8** **a** Preparation of thiol terminated glycopolymer library by RAFT polymerization, and **b** glycopolymer microarrays for the high-throughput screening of protein and cells and biosensing

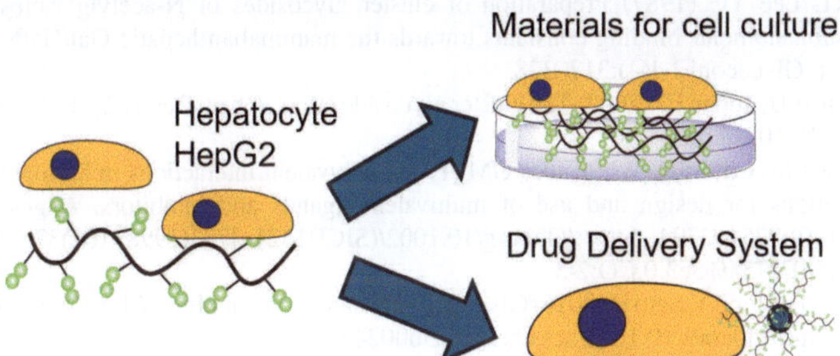


**Fig. 9** Glycomaterials with the cell specific affinities and application of them

culture substrates based on glycopolymers using natural polymers, such as galactose-conjugated gelatin and chitin-chitosan, have been reported [59, 60]. Cell culture substrates based on decellularized matrices with galactose-conjugated substrates have also been reported [61]. In addition, GlcNAc containing polymers were utilized as the artificial extracellular matrix of various mesenchymal and skeletal muscle cells via the interaction with the filament proteins [62].

Drug delivery systems utilizing the cell-specific molecular recognition ability of such glycopolymers have been developed. Specific drug delivery to hepatocytes has been reported for glycopolymers modified with Gal or GalNAc [63]. Micelles based on block copolymers of galactose-containing glycopolymers have also been reported, and efficient drug carriers for hepatocytes and liver cancer cells (HepG2) based on the molecular recognition of glycans have been reported [64].

# 4 Summary

This section provides an overview of glycomaterials and glycopolymers with molecular recognition of saccharides. Glycopolymers are polymers that specifically exhibit the functions of glycans. Recent advances in polymer chemistry have facilitated the synthesis of precise glycopolymers and the development of polymer-modified substrates. Glycopolymers were originally developed as model molecules to demonstrate the glycocluster effect, but they have evolved into fascinating molecules that also possess the functionality of polymers as biomaterials. The field of glycopolymers is expected to continue to develop in conjunction with advances in glycochemistry and polymer chemistry.

## References

1. Varki A (2017) Biological roles of glycans. Glycobiology 27(1):3–49. https://doi.org/10.1093/glycob/cww086
2. Lee RT, Lee YC (1987) Preparation of cluster glycosides of N-acetylgalactosamine that have subnanomolar binding constants towards the mammalian hepatic Gal/GalNAc-specific receptor. Glycoconj J 4(4):317–328
3. Lundquist JJ, Toone EJ (2002) The cluster glycoside effect. Chem Rev 102(2):555–578. https://doi.org/10.1021/cr000418f
4. Mammen M, Choi SK, Whitesides GM (1998) Polyvalent interactions in biological systems: implications for design and use of multivalent ligands and inhibitors. Angew Chem Int Ed 37(20):2754–2794. https://doi.org/10.1002/(SICI)1521-3773(19981102)37:20%3c2754::AID-ANIE2754%3e3.0.CO;2-3
5. Miura Y, Hoshino Y, Seto H (2016) Glycopolymer nanobiotechnology. Chem Rev 116(4):1673–1692. https://doi.org/10.1021/acs.chemrev.5b00247
6. Kobayashi K, Sumitomo H, Ina Y (1985) Synthesis and functions of polystyrene derivatives having pendant oligosaccharides. Polym J 17(4):567–575
7. Ohno K, Tsujii Y, Miyamoto T, Fukuda T, Goto M, Kobayashi K, Akaike T (1998) Synthesis of a well-defined glycopolymer by nitroxide-controlled free radical polymerization. Macromolecules 31(4):1064–1069. https://doi.org/10.1021/ma971329g
8. Totani K, Kubota T, Kuroda T, Murata T, Hidari KIPJ, Suzuki T, Kobayashi K, Usui T (2003) Chemoenzymatic synthesis and application of glycopolymers containing multivalent sialyloligosaccharides with a poly (L-glutamic acid) backbone for inhibition of infection by influenza viruses. Glycobiology 13(5):315–326
9. Noguchi M, Tanaka T, Gyakushi H, Kobayashi A, Shoda S (2009) Efficient synthesis of sugar oxazolines from unprotected *N*-Acetyl-2-amino sugars by using chloroformamidinium reagent in water. J Org Chem 74:2210–2212. https://doi.org/10.1021/jo8024708
10. Tanaka T, Nagai H, Noguchi M, Kobayashi A, Shoda S (2009) One-step conversion of unprotected sugars to β-glycosyl azides using 2-chloroimidazolinium salt in aqueous solution. Chem Commun 23:3378–3379. https://doi.org/10.1039/B905761G
11. Tanaka T, Matsumoto T, Noguchi M, Kobayashi A, Shoda S (2009) Direct transformation of unprotected sugars to aryl 1-thio-β-glycosides in aqueous media using 2-chloro-1,3-dimethylimidazolinium chloride. Chem Lett 38:458–459. https://doi.org/10.1246/cl.2009.458

12. Tanaka T, Huang WC, Noguchi M, Kobayashi A, Shoda S (2009) Direct synthesis of 1,6-anhydro sugars from unprotected glycopyranoses by using 2-chloro-1,3-dimethylimidazolinium chloride. Tetrahedron Lett 50:2154–2157. https://doi.org/10.1016/j.tetlet.2009.02.171
13. Tanaka T, Ishitani H, Miura Y, Oishi K, Takahashi T, Suzuki T, Shoda S, Kimura Y (2014) Protecting-group-free synthesis of glycopolymers bearing sialyloligosaccharide and their high binding with the influenza virus. ACS Macro Lett 3:1074–1078. https://doi.org/10.1021/mz500555x
14. Tanaka T, Fukumoto H, Ishitani H (2017) Synthesis of glycopolymer gold nanoparticles decorated with oligosaccharides via a protecting-group-free process and their specific recognition by lectins. Trans Mater Res Soc Jpn 42:113–118. https://doi.org/10.14723/tmrsj.42.113
15. Tanaka T, Zhou Y, Tamoto C, Kurebayashi Y, Takahashi T, Suzuki T (2017) An α2,3-linked sialylglycopolymer as a multivalent glycoligand against avian and human influenza viruses. J Appl Glycosci 64:43–48. https://doi.org/10.5458/jag.jag.JAG-2017_003
16. Tanaka T, Inoue G, Shoda S, Kimura Y (2014) Protecting-group-free synthesis of glycopolymers bearing thioglycosides via one-pot monomer synthesis from free saccharides. J Polym Sci Part A Polym Chem 52:3513–3520. https://doi.org/10.1002/pola.27417
17. Nagao M, Kurebayashi Y, Seto H, Tanaka T, Takahashi T, Suzuki T, Hoshino Y, Miura Y (2016) Synthesis of well-controlled glycopolymers bearing oligosaccharides and their interactions with influenza viruses. Polym J 48:745–749. https://doi.org/10.1038/pj.2016.14
18. Das A, Theato P (2016) Activated ester containing polymers: opportunities and challenges for the design of functional macromolecules. Chem Rev 116:1434–1495. https://doi.org/10.1021/acs.chemrev.5b00291
19. Tanaka T, Nakashima K, Tsuji S, Han X, Zhao J, Honda Y, Sakakibara K, Kurebayashi Y, Takahashi T, Suzuki T (2019) Controlled synthesis of glycopolymers with pendant complex-type sialylglycopeptides and their binding affinity with a lectin and an influenza virus. Polym Chem 10:5124−5130. https://doi.org/10.1039/c9py00745h
20. Tanaka T (2024) Recent advances in polymers bearing activated esters for the synthesis of glycopolymers by postpolymerization modification. Polymers 16:1100. https://doi.org/10.3390/polym16081100
21. Niu J, Page ZA, Dolinski ND, Anastasaki A, Hsueh AT, Soh HT, Hawker CJ (2017) Rapid visible light-mediated controlled aqueous polymerization with in situ monitoring. ACS Macro Lett 6:1109–1113. https://doi.org/10.1021/acsmacrolett.7b00587
22. Tsuji S, Aso Y, Ohara H, Tanaka T (2019) Polymeric water-soluble activated esters: synthesis of polymer backbones with pendant *N*-hydoxysulfosuccinimide esters for post-polymerization modification in water. Polym J 51:1015–1022. https://doi.org/10.1038/s41428-019-0221-4
23. Tsuji S, Aso Y, Ohara H, Tanaka T (2020) Aqueous synthesis of sialylglycopeptide-grafted glycopolymers with high affinity for the lectin and the influenza virus hemagglutinin. J Polym Sci 58:548–556. https://doi.org/10.1002/pol.20190184
24. Tsuji S, Kobayashi K, Fujii T, Imoto H, Naka K, Aso Y, Ohara H, Tanaka T (2022) Polymers with pendant water-soluble tetrafluorobenzene sulfonic acid activated esters: synthesis, stability, and use for glycopolymers in water. Macromol Chem Phys 223:2200072. https://doi.org/10.1002/macp.202200072
25. Tsuji S, Aso Y, Tanaka T (2023) Aqueous one-pot and oxygen-tolerant synthesis of glycopolymers using polymer-backbone-bearing water-soluble activated esters. Chem Lett 52:67–70. https://doi.org/10.1246/cl.220480
26. Ladmiral V, Mantovani G, Clarkson GJ, Cauet S, Irwin JL, Haddleton DM (2006) Synthesis of neoglycopolymers by a combination of "click chemistry" and living radical polymerization. J Am Chem Soc 128:4823–4830. https://doi.org/10.1021/ja058364k
27. Malkoch M, Thibault RJ, Drockenmuller E, Messerschmidt M, Voit B, Russell TP, Hawker CJ (2005) Orthogonal approaches to the simultaneous and cascade functionalization of macromolecules using click chemistry. J Am Chem Soc 127:14942–14949. https://doi.org/10.1021/ja0549751

28. Nagao M, Kurebayashi Y, Seto H, Takahashi T, Suzuki T, Hoshino Y, Miura Y (2016) Polyacrylamide backbones for polyvalent bioconjugates using "post-click" chemistry. Polym Chem 7:5920–5924. https://doi.org/10.1039/C6PY00904B
29. Nagao M, Fujiwara Y, Matsubara T, Hoshino Y, Sato T, Miura Y (2017) Design of glycopolymers carrying sialyl oligosaccharides for controlling the interaction with the influenza virus. Biomacromol 18:4385–4392. https://doi.org/10.1021/acs.biomac.7b01426
30. Gody G, Maschmeyer T, Zetterlund P B, Perrier S (2013) Rapid and quantitative one-pot synthesis of sequence-controlled polymers by radical polymerization. Nat Commun 4:2505. https://doi.org/10.1038/ncomms3505
31. Nagao M, Hoshino Y, Miura Y (2019) Quantitative preparation of multiblock glycopolymers bearing glycounits at the terminal segments by aqueous reversible addition–fragmentation chain transfer polymerization of acrylamide monomers. J Polym Sci Part A Polym Chem 57:857–861. https://doi.org/10.1002/pola.29344
32. Valdebenito A, Encinas MV (2009) Effect of solvent on the free radical polymerization of *N,N*-dimethylacrylamide. Polm Int 59:1246–1251. https://doi.org/10.1002/pi.2856
33. Galan MC, Dumy P, Renaudet O (2013) Multivalent glyco (cyclo) peptides. Chem Soc Rev 42:4599–4612. https://doi.org/10.1039/C2CS35413F
34. Huang PS, Boyken SE, Baker D (2016) The coming of age of de novo protein design. Nature 537:320–327. https://doi.org/10.1038/nature19946
35. Hasegawa T, Sasaki T (2003) Glyco-helix: parallel lactose-lactose interactions stabilize an α-helical structure of multi-glycosylated peptide. Chem Commun 8:978–979. https://doi.org/10.1039/B212364A
36. Chang IV, Tsutsumi H, Mihara H (2017) Screening for concanavalin A binders from a mannose-modified α-helix peptide phage library. Mol Biosyst 13:2222−2225. https://pubs.rsc.org/en/content/articlelanding/2017/mb/c7mb00495h
37. Ohta T, Miura N, Fujitani N, Nakajima F, Niikura K, Sadamoto R, Guo CT, Suzuki T, Suzuki Y, Monde K, Nishimura SI (2003) Glycotentacles: synthesis of cyclic glycopeptides, toward a tailored blocker of influenza virus hemagglutinin. Angew Chem 115:5344–5347. https://doi.org/10.1002/ange.200351640
38. Matsuura K, Hibino M, Yamada Y, Kobayashi K (2001) Construction of glyco-clusters by self-organization of site-specifically glycosylated oligonucleotides and their cooperative amplification of lectin-recognition. J Am Chem Soc 123:357–358. https://doi.org/10.1021/ja001945j
39. Zeng X, Qu K, Rehman A (2016) Glycosylated conductive polymer: a multimodal biointerface for studying carbohydrate–protein interactions. Acc Chem Res 49:1624–1633. https://doi.org/10.1021/acs.accounts.6b00181
40. Shiao TC, Roy R (2012) Glycodendrimers as functional antigens and antitumor vaccines. New J Chem 36:324−339. https://pubs.rsc.org/en/content/articlelanding/2012/nj/c2nj20873c
41. Mullerova M, Maciel D, Nunes N, Wrobel D, Stofik M, Červenková Šťastná L, Strasak T (2021) Carbosilane glycodendrimers for anticancer drug delivery: synthetic route, characterization, and biological effect of glycodendrimer–doxorubicin complexes. Biomacromolecules 23:276−290. https://doi.org/10.1021/acs.biomac.1c01264
42. Sauter NK, Hanson JE, Glick GD, Brown JH, Crowther RL, Park S, Skehelyo JJ, Wiley DC (1992) Binding of influenza virus hemagglutinin to analogs of its cell-surface receptor, sialic acid: analysis by proton nuclear magnetic resonance spectroscopy and x-ray crystallography. Biochemistry 31:9609–9621. https://doi.org/10.1021/bi00155a013
43. Yamabe M, Kaihatsu K, Ebara Y (2018) Sialyllactose-modified three-way junction DNA as binding inhibitor of influenza virus hemagglutinin. Bioconjugate Chem 29:1490–1494. https://doi.org/10.1021/acs.bioconjchem.8b00045
44. Waldmann M, Jirmann R, Hoelscher K, Wienke M, Niemeyer FC, Rehders D, Meyer B (2014) A nanomolar multivalent ligand as entry inhibitor of the hemagglutinin of avian influenza. J Am Chem Soc 136:783–788. https://doi.org/10.1021/ja410918a
45. Nagao M, Matsubara T, Hoshino Y, Sato T, Miura Y (2019) Topological design of star glycopolymers for controlling the interaction with the influenza virus. Bioconjugate Chem 30:1192–1198. https://doi.org/10.1021/acs.bioconjchem.9b00134

46. Nagao M, Yamaguchi A, Matsubara T, Hoshino Y, Sato T, Miura Y (2022) De novo design of star-shaped glycoligands with synthetic polymer structures toward an influenza hemagglutinin inhibitors. Biomacromolecules 23:1232–1241. https://doi.org/10.1021/acs.biomac.1c01483
47. Nagao M, Hoshino Y, Miura Y (2024) Quantification of thermodynamic effects of carbohydrate multivalency on avidity using synthetic discrete glycooligomers. Chem Commun 60:7021–7024. https://doi.org/10.1039/d4cc02409e
48. Godula K, Bertozzi CR (2012) Density variant glycan microarray for evaluating cross-linking of mucin-like glycoconjugates by lectins. J Am Chem Soc 134(38):15732–15742. https://doi.org/10.1021/ja302193u
49. Woods EC, Yee NA, Shen J, Bertozzi CR (2015) Glycocalyx engineering with a recycling glycopolymer that increases cell survival in vivo. Angew Chem Int Ed 54(52):15782–15788. https://doi.org/10.1002/anie.201508783
50. Honigfort DJ, Altman MO, Gagneux P, Godula K (2021) Glycocalyx crowding with mucin mimetics strengthens binding of soluble and virus-associated lectins to host cell glycan receptors. Proc Natl Acad Sci 118(40):e2107896118. https://doi.org/10.1073/pnas.2107896118
51. Hudak JE, Canham SM, Bertozzi CR (2014) Glycocalyx engineering reveals a siglec-based mechanism for NK cell immunoevasion. Nat Chem Biol 10(1):69–75. https://doi.org/10.1038/nchembio.1388
52. Terada Y, Seto H, Hoshino Y, Murakami T, Shinohara S, Tamada K, Miura Y (2017) SPR study for analysis of a water-soluble glycopolymer interface and molecular recognition properties. Polym J 49(2):255–262. https://doi.org/10.1038/pj.2016.99
53. Terada Y, Hoshino Y, Miura Y (2019) Glycopolymers mimicking GM1 gangliosides: cooperativity of galactose and neuraminic acid for cholera toxin recognition. C Chem –Asian J 14(7):1021–1027. https://doi.org/10.1002/asia.201900053
54. Nagao M, Uemura T, Horiuchi T, Hoshino Y, Miura Y (2021) Screening of a glycopolymer library for GM1 mimetics synthesized by the "carbohydrate module method. Chem Commun 57(83):10871–10874. https://doi.org/10.1039/d1cc04394c
55. Uzawa H, Ito H, Neri P, Mori H, Nishida Y (2007) Glycochips from polyanionic glycopolymers as tools for detecting Shiga toxins. ChemBioChem 8(17):2117–2124. https://doi.org/10.1002/cbic.200700439
56. Leslie KG, Jolliffe KA, Müllner M, New EJ, Turnbull WB, Fascione MA, Mahon CS (2024) A glycopolymer sensor array that differentiates lectins and bacteria. Biomacromol 25(11):7466–7474. https://doi.org/10.1021/acs.biomac.4c01129
57. Khorev O, Stokmaier D, Schwardt O, Cutting B, Ernst B (2008) Trivalent, Gal/GalNAc-containing ligands designed for the asialoglycoprotein receptor. Bioorg Med Chem 16(9):5216–5231. https://doi.org/10.1016/j.bmc.2008.03.017
58. Adachi N, Maruyama A, Ishihara T, Akaike T (1995) Cellular distribution of polymer particles bearing various densities of carbohydrate ligands. J Biomater Sci Polym Ed 6(5):463–479. https://doi.org/10.1163/156856294X00446
59. Hong SR, Lee YM, Akaike T (2003) Evaluation of a galactose-carrying gelatin sponge for hepatocytes culture and transplantation. J Biomed Mater Res Part A: Off J Soc Biomater, Jpn Soc Biomater, Aust Soc Biomater Korean Soc Biomater 67(3):733–741. https://doi.org/10.1002/jbm.a.10138
60. Park IK, Yang J, Jeong HJ, Bom HS, Harada I, Akaike T, Cho CS (2003) Galactosylated chitosan as a synthetic extracellular matrix for hepatocytes attachment. Biomaterials 24(13):2331–2337. https://doi.org/10.1016/S0142-9612(03)00108-X
61. Sasikumar S, Boden A, Chameettachal S, Cipolla L, Cromer B, Kingshott P, Pati F (2022) Galactose tethered decellularized liver matrix: toward a biomimetic and biofunctional matrix for liver tissue engineering. ACS Appl Bio Mater 5(6):3023–3037. https://doi.org/10.1021/acsabm.2c00330
62. Ise H, Kobayashi S, Goto M, Sato T, Kawakubo M, Takahashi M, Akaike T (2010) Vimentin and desmin possess GlcNAc-binding lectin-like properties on cell surfaces. Glycobiology 20(7):843–864. https://doi.org/10.1093/glycob/cwq039

63. Stenzel MH (2022) Glycopolymers for drug delivery: opportunities and challenges. Macromolecules 55(12):4867–4890. https://doi.org/10.1021/acs.macromol.2c00557
64. Li J, Zhang Y, Cai C, Rong X, Shao M, Li J, Yu G (2020) Collaborative assembly of doxorubicin and galactosyl diblock glycopolymers for targeted drug delivery of hepatocellular carcinoma. Biomater Sci 8(1):189–200. https://doi.org/10.1039/C9BM01604J

# Atypical Protein Glycosylation: *C*-Mannosylation and Glucosyl-Galactosyl-Hydroxylation

**Kento Mori and Siro Simizu**

**Abstract** Protein glycosylation plays a pivotal role in regulating protein folding, stability, and secretion. Dysregulation of glycosylation has been implicated in a wide range of diseases. Among the less common types of glycosylation, *C*-mannosylation—defined by the covalent attachment of a mannose residue to the indole C2 carbon of tryptophan via a carbon–carbon bond—remains relatively poorly characterized. In this review, we comprehensively examine the landscape of *C*-mannosylation across diverse proteins, identifying previously unrecognized *C*-mannosylated substrates and elucidating the functional consequences of this modification, including its impact on protein secretion and enzymatic activity. We further characterize DPY19L3, a multi-pass endoplasmic reticulum membrane protein, as a bona fide *C*-mannosyltransferase. Our findings delineate its enzymatic activity, membrane topology, and functional involvement in myogenic differentiation and tumor cell vasculogenic mimicry. In the course of identifying novel *C*-mannosylated proteins, mass spectrometry analyses of CCN1 and FGL1 revealed the presence of another atypical glycosylation—glucosyl-galactosyl-hydroxylation (GGH)—a modification typically observed in collagen-like domains. Notably, neither CCN1 nor FGL1 contains such domains. Functional studies using GLT25D1 and LH3 knockout models confirmed that galactosylation and subsequent glucosylation of hydroxylated lysines in FGL1 are sequentially catalyzed by these enzymes. Collectively, our findings expand the current understanding of atypical glycosylation, establish mechanistic links between specific glycosyltransferases and protein function, and highlight potential therapeutic avenues targeting *C*-mannosylation and GGH in pathological contexts.

**Keywords** Glycobiology · Protein glycosylation · *C*-mannosylation · Glucosyl-galactosyl-hydroxylation

K. Mori
Laboratory of Glycobiology, The Noguchi Institute, Tokyo, Japan
e-mail: kento_mori@noguchi.or.jp

K. Mori · S. Simizu (✉)
Department of Applied Chemistry, Faculty of Science and Technology, Keio University, Yokohama, Kanagawa, Japan
e-mail: simizu@applc.keio.ac.jp



K. Toshima and Y. Fujimoto (eds.), *Chemical Biology of Carbohydrates*,
https://doi.org/10.1007/978-981-95-5335-8_16

# 1 Introduction

Protein glycosylation, the covalent attachment of sugar chains to proteins, plays essential roles in regulating protein secretion, folding, and stability [1, 2]. Aberrant glycosylation is implicated in a variety of diseases, including cancer, and glycosylation pathways and their associated enzymes—particularly glycosyltransferases—are considered potential therapeutic targets [3, 4]. Our research has focused on elucidating the mechanisms and biological implications of several forms of "atypical" glycosylation.

*C*-mannosylation is a unique glycosylation event in which a mannose is covalently linked via a carbon–carbon (C–C) bond to the C2 position of the indole ring of a tryptophan residue [5] (Fig. 1a). This modification typically occurs in proteins that possess an N-terminal signal peptide and the consensus sequence Trp-Xaa-Xaa-Trp/Cys (where Xaa denotes any amino acid), with the N-terminal tryptophan serving as the site of *C*-mannosylation [6]. However, some proteins containing these features do not undergo *C*-mannosylation, suggesting that additional, as-yet-unidentified determinants are required [7, 8]. To date, *C*-mannosylation has been reported in approximately 40 proteins, a relatively small number compared to other glycosylation types. Notably, these modifications are predominantly localized to specific domains, such as thrombospondin type 1 repeats (TSRs) and type I cytokine receptor domains, where the consensus motif is highly conserved [9–13].

In 2013, Buettner et al. first demonstrated that *Caenorhabditis elegans* Dpy19 exhibits *C*-mannosyltransferase activity, based on its structural similarity to Stt3, a key component of the oligosaccharyltransferase complex involved in *N*-glycosylation [14]. In mammals, four homologs of Dpy19 exist (DPY19L1–L4), with DPY19L1 and DPY19L3 confirmed to possess *C*-mannosyltransferase activity [10, 15]. Through our ongoing investigations, we have expanded the repertoire of known *C*-mannosylated proteins and uncovered new insights into the membrane topology and biological functions of mammalian DPY19L3, including its role as an active *C*-mannosyltransferase.

Glucosyl-galactosyl-hydroxylation (GGH) is another specialized glycosylation, typically associated with collagen-like domain-containing proteins such as collagens and adiponectin [16–19] (Fig. 1b). In this pathway, lysine residues are first hydroxylated by lysyl hydroxylase (LH) family enzymes [20], followed by sequential glycosylation: galactose is transferred by procollagen galactosyltransferases (GLT25D1/2) [21], and glucose is subsequently added by lysyl hydroxylase 3 (LH3) [20]. Notably, LH3 exhibits bifunctional enzymatic activity, catalyzing both hydroxylation and glucosylation of hydroxylysine. GGH plays crucial roles in collagen crosslinking and adiponectin multimerization [19, 22], and dysfunction of GGH-related enzymes has been linked to structural abnormalities and connective tissue disorders, including those caused by mutations in LH3 [23].

During our investigation of *C*-mannosylated proteins, we unexpectedly identified two candidate proteins that exhibited GGH-like modifications, despite lacking collagen-like domains. One of these proteins, FGL1, was experimentally confirmed

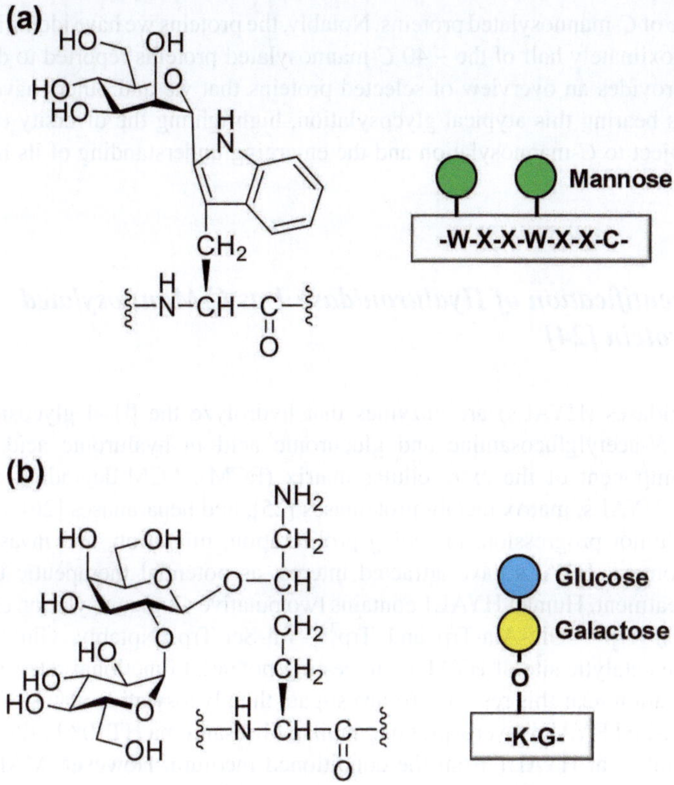


**Fig. 1** **a** The structure and schematic of *C*-mannosylation, **b** The structure and schematic of glucosyl-galactosyl-hydroxylation

to be modified by both GLT25D1 and LH3, representing the first report of GGH occurring in a non-collagen-like protein.

These findings broaden the current understanding of atypical glycosylation mechanisms, reveal previously unrecognized substrates and enzymatic pathways, and highlight the potential physiological significance and therapeutic relevance of both *C*-mannosylation and GGH in non-canonical contexts.

## 2 Identification of *C*-mannosylation in Various Types of Proteins

At the outset of our investigations into *C*-mannosylation, only a limited number of *C*-mannosylated proteins had been reported, and no *C*-mannosyltransferases had yet been identified in mammals. Since then, we have significantly expanded the known repertoire of *C*-mannosylated proteins. Notably, the proteins we have identified represent approximately half of the ~ 40 *C*-mannosylated proteins reported to date. This section provides an overview of selected proteins that we and others have characterized as bearing this atypical glycosylation, highlighting the diversity of protein types subject to *C*-mannosylation and the emerging understanding of its biological relevance.

## 2.1 Identification of Hyaluronidase 1 as C-Mannosylated Protein [24]

Hyaluronidases (HYALs) are enzymes that hydrolyze the β1–4 glycosidic bond between *N*-acetylglucosamine and glucuronic acid in hyaluronic acid (HA), a major component of the extracellular matrix (ECM). ECM-degrading enzymes, including HYALs, matrix metalloproteinases [25], and heparanases [26], are implicated in tumor progression, including proliferation, migration, and invasion [27]. In this context, HYALs have attracted interest as potential therapeutic targets in cancer treatment. Human HYAL1 contains two putative *C*-mannosylation consensus sequences: $Trp^{130}$-Glu-Ala-Trp and $Trp^{321}$-Val-Ser-Trp. Notably, $Glu^{131}$ resides within the catalytic site of HYAL1, suggesting potential functional relevance of *C*-mannosylation near this residue. To investigate the glycosylation status of HYAL1, we established HYAL1-overexpressing human fibrosarcoma HT1080 cells and purified recombinant HYAL1 from the conditioned medium. However, MALDI-TOF mass spectrometry analysis revealed no evidence of *C*-mannosylation in the secreted form of the protein. Interestingly, this analysis identified three *N*-glycosylation sites on HYAL1, which were subsequently shown to regulate its secretion and enzymatic activity [28] (Fig. 2a). We next examined intracellular HYAL1 extracted from cell lysates. MALDI-TOF MS analysis demonstrated that the intracellular form of HYAL1 was *C*-mannosylated exclusively at $Trp^{130}$, but not at $Trp^{321}$ (Fig. 2a). Notably, SDS-PAGE using HA-containing gels confirmed that the secreted HYAL1 retained enzymatic activity. These findings suggest that *C*-mannosylation at $Trp^{130}$ suppresses secretion of HYAL1 and is not essential for its enzymatic function. To further assess the impact of *C*-mannosylation at $Trp^{130}$ on HYAL1 function—particularly its effect on the active site residue $Glu^{131}$ [29]—we performed molecular docking simulations of the HYAL1–HA complex. The simulation indicated that *C*-mannosylation at $Trp^{130}$ alters the conformation of the catalytic pocket, including $Glu^{131}$ (Fig. 2b). These results imply that *C*-mannosylation negatively regulates HYAL1 activity by modulating its active site architecture.

This study represents our initial identification of a novel *C*-mannosylated protein. Unlike the majority of *C*-mannosylated proteins reported to date—which are typically found within thrombospondin type 1 repeat (TSR) domains or type I cytokine receptor families—HYAL1 does not belong to either category. To date, *C*-mannosylation has predominantly been associated with enhanced protein stability

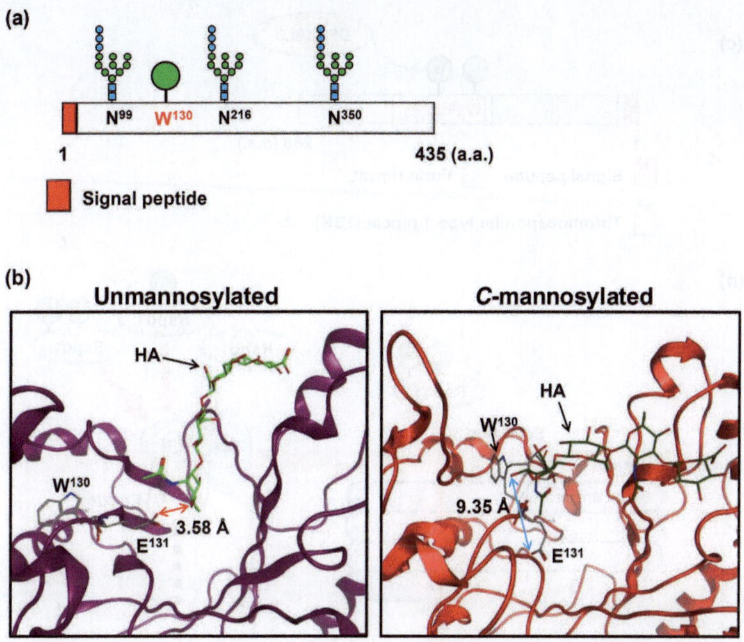


**Fig. 2** **a** Schematic of human HYAL1 and its glycosylation sites, **b** *C*-mannosylation at Trp$^{130}$ alters the conformation of the catalytic pocket of HYAL1, **c** schematic of human Rspo1 and its *C*-mannosylation sites, **d** *C*-mannosylation of Rspo1 regulates its secretion and Wnt/β catenin signaling, **e** Schematic of human LPL and its *C*-mannosylation site, **f** *C*-mannosylation of LPL regulates its secretion and enzymatic activity, **g** AXL and its downstream signaling pathway mediated by GAS6 stimulation, **h** schematic of human AXL and its *C*-mannosylation site, **i** human breast cancer MDA-MB-231 cells under normal culture condition (left) and their in vitro VM formation (right)

and function. In contrast, our findings position HYAL1 as a unique example in which *C*-mannosylation negatively affects both secretion and enzymatic activity.

These results highlight a broader functional diversity of *C*-mannosylation than previously recognized. We are currently expanding our investigation to identify additional "non-canonical" *C*-mannosylated proteins that fall outside of traditional domain families, with the goal of uncovering novel regulatory roles of this atypical glycosylation.

## 2.2 *C-mannosylation of R-spondin 1 is Catalyzed by DPY19L3 [10]*

R-spondins (Rspos) are secreted proteins that activate the Wnt/β-catenin signaling pathway, playing critical roles in embryonic development, cell proliferation, and

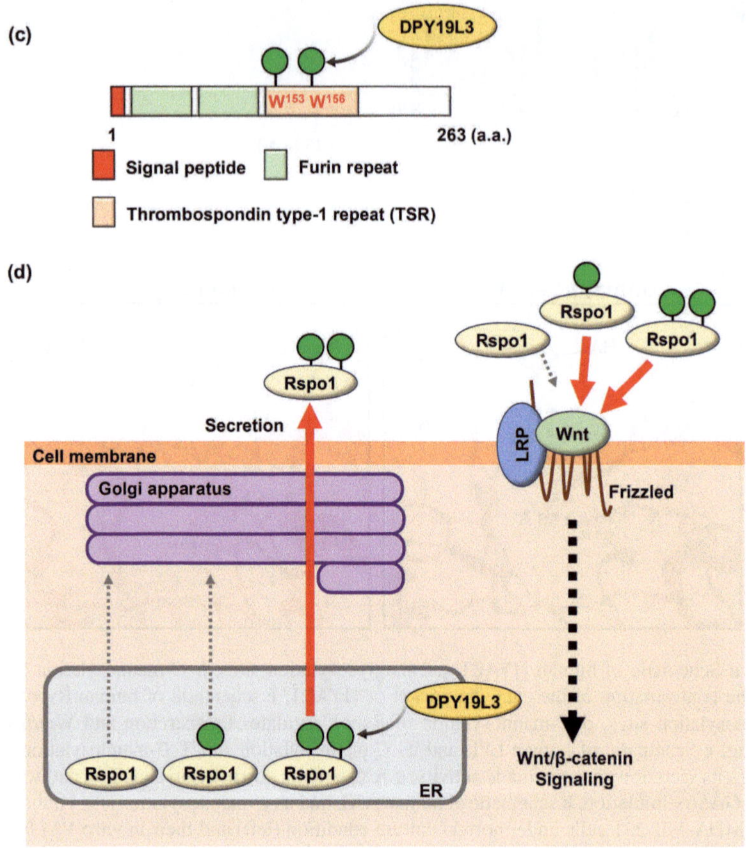


**Fig. 2** (continued)

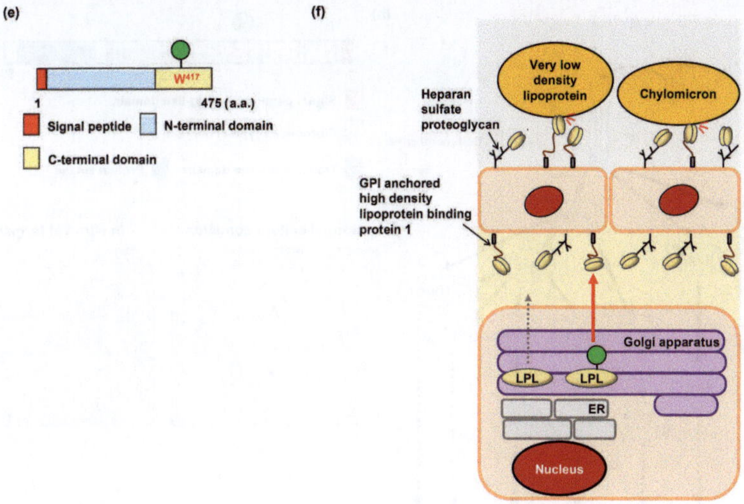


**Fig. 2** (continued)

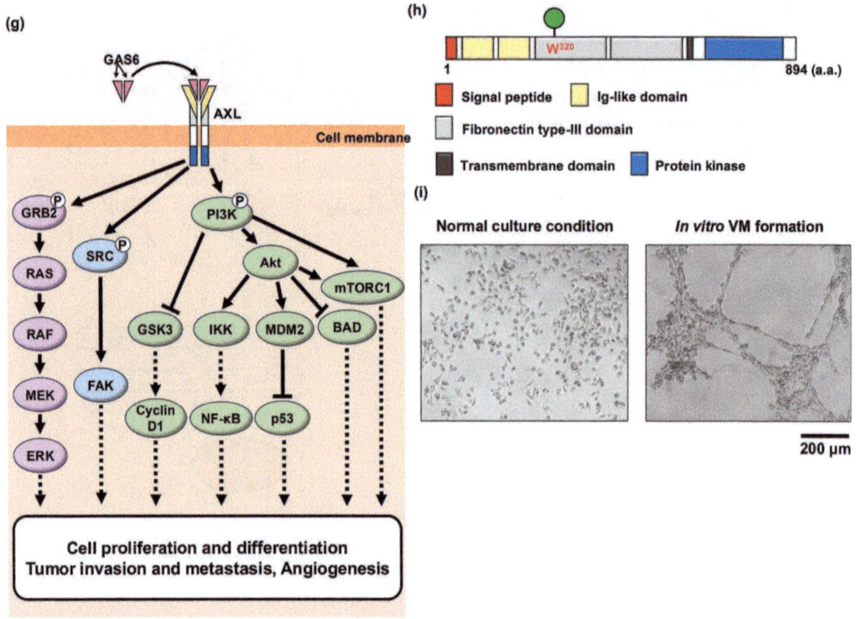


**Fig. 2** (continued)

migration [30]. Aberrant expression of Rspo family members has been implicated in oncogenic transformation and tumor progression [31, 32]. Human R-spondin 1 (Rspo1) contains a thrombospondin type 1 repeat (TSR) domain with two sequential *C*-mannosylation consensus sequences: Trp$^{153}$-Ser-Pro-Trp$^{156}$-Gly-Pro-Cys. To determine whether Rspo1 is *C*-mannosylated at Trp$^{153}$ and Trp$^{156}$, we performed LC–MS/MS analysis on recombinant Rspo1 purified from the conditioned medium of Rspo1-overexpressing HT1080 cells. The results revealed that all secreted Rspo1 molecules were *C*-mannosylated, with the majority modified at both tryptophan residues (Fig. 2c). A subset of mono-mannosylated Rspo1 was modified only at Trp$^{153}$. To identify the specific *C*-mannosyltransferase responsible for Rspo1 modification, we utilized *Drosophila* S2 cells, which lack endogenous *C*-mannosyltransferase activity [14, 33]. S2 cell lines were engineered to stably express one of the four human DPY19 homologs (DPY19L1–L4), and human Rspo1 cDNA was transiently transfected into each line. Recombinant Rspo1 was purified from conditioned medium and analyzed by LC–MS. Mono-mannosylated Rspo1 was detected only in samples derived from DPY19L3-expressing cells. Further LC–MS/MS analysis confirmed that DPY19L3 specifically modified Trp$^{156}$, but not Trp$^{153}$ (Fig. 2c). To verify this activity in human cells, we performed DPY19L1, DPY19L3, and DPY19L4 knockdown (KD) in Rspo1-overexpressing HT1080 cells and purified the secreted Rspo1 proteins. MALDI-TOF MS/MS analysis showed that knockdown of DPY19L3 led to an increased proportion of peptides mannosylated only at Trp$^{153}$, indicating that Trp$^{156}$ is specifically modified by DPY19L3, while other

DPY19 family members do not contribute to Rspo1 *C*-mannosylation. We then investigated the functional consequences of *C*-mannosylation on Rspo1 secretion using a double mutant (Trp153Ala and Trp156Ala; 2WA) and DPY19L3 KD cells. Both the 2WA mutation and DPY19L3 knockdown resulted in reduced secretion of Rspo1, demonstrating that *C*-mannosylation plays a role in promoting its efficient secretion (Fig. 2d). To explore whether *C*-mannosylation affects Rspo1-mediated activation of canonical Wnt signaling, we used the TOPFlash/FOPFlash reporter assay [34]. The Rspo1 2WA mutant exhibited significantly reduced Wnt signaling activity compared to wild-type Rspo1. Interestingly, wild-type Rspo1 purified from DPY19L3 KD cells (mannosylated only at $Trp^{153}$) retained comparable Wnt activity to fully mannosylated Rspo1. Furthermore, wild-type Rspo1 produced in DPY19L3-expressing S2 cells (mannosylated only at $Trp^{156}$) activated Wnt signaling more strongly than non-mannosylated Rspo1 from control S2 cells. These findings suggest that *C*-mannosylation of Rspo1 at either $Trp^{153}$ or $Trp^{156}$ is sufficient to enhance Wnt signaling activity (Fig. 2d).

This work represents the first identification of a mammalian *C*-mannosyltransferase and its specific substrate. Following this discovery, we also identified *C*-mannosylation in Rspo2 [35] and Rspo3 [36] and investigated its impact on secretion and Wnt signaling regulation. Notably, in 2017, Shcherbakova et al. reported that the mouse netrin receptor UNC5A, containing extended Trp-Xaa-Xaa-Trp motifs in two TSR domains, was *C*-mannosylated by both DPY19L1 and DPY19L3, with the former modifying the first two Trp residues and the latter preferring the third [15]. These collective findings have significantly advanced our understanding of the substrate specificity and functional relevance of *C*-mannosyltransferases, paving the way for further exploration into their structural and biological roles.

### 2.3 Regulation of Lipoprotein Lipase Activity by C-mannosylation [37]

Lipoprotein lipase (LPL) is predominantly expressed in adipose tissue, cardiac muscle, and macrophages, where it plays a central role in lipid transport and metabolism [38]. Its primary function is to hydrolyze core triglycerides (TGs) in circulating chylomicrons and very-low-density lipoproteins (VLDLs) into free fatty acids at the luminal surface of capillary endothelial cells, a process that requires apolipoprotein C-II (apoC-II) as a cofactor [39, 40]. Plasma TGs represent a key source of fatty acids for peripheral tissues, and impaired LPL function is closely associated with various metabolic disorders.

We identified a putative *C*-mannosylation consensus motif, $Trp^{417}$-Ser-Asp-Trp, within the C-terminal domain of human LPL; however, no prior evidence of *C*-mannosylation in LPL had been reported. To investigate this possibility, we established LPL-overexpressing HT1080 cells, purified recombinant LPL from conditioned media, and performed MALDI-TOF mass spectrometry. This analysis confirmed that LPL is C-mannosylated at $Trp^{417}$ (Fig. 2e). To elucidate the functional significance of this modification, we generated CHO-K1 cell lines stably expressing either wild-type LPL (wt) or a *C*-mannosylation-deficient mutant in which $Trp^{417}$ was substituted with phenylalanine (WF). Semi-quantitative reverse transcription PCR confirmed that both cell lines expressed comparable levels of exogenous LPL mRNA. However, Western blot analysis revealed that secretion of the WF mutant protein was markedly reduced, accompanied by an accumulation of intracellular LPL, suggesting that *C*-mannosylation at $Trp^{417}$ facilitates efficient secretion of the enzyme. We further assessed the impact of *C*-mannosylation on LPL enzymatic activity by measuring the hydrolysis of triolein, a triglyceride substrate, and quantifying the release of free fatty acids [41]. The WF mutant exhibited significantly lower lipolytic activity compared to wild-type LPL, indicating that *C*-mannosylation not only enhances secretion but also positively regulates enzymatic function (Fig. 2f).

These findings provide novel insights into the post-translational regulation of LPL and highlight the importance of *C*-mannosylation in modulating its stability, secretion, and catalytic activity. Understanding this regulatory mechanism may offer new avenues for therapeutic intervention in lipid metabolism disorders involving LPL dysfunction.

## 2.4 *C-Mannosylation in AXL, a Receptor Tyrosine Kinase [42]*

AXL is a member of the TAM family of receptor tyrosine kinases (RTKs) and is activated by its ligand GAS6 or via ligand-independent mechanisms [43, 44]. Upon activation, AXL modulates several intracellular signaling pathways, including MAPK/ERK and PI3K/AKT [45] (Fig. 2g). Aberrant AXL expression is frequently observed in a variety of tumors and is associated with increased malignancy and poor prognosis [44]. In human breast cancer MDA-MB-231 cells, inhibition of GAS6-induced AXL activation has been shown to suppress tumor cell migration [45, 46]. In non-small cell lung cancers (NSCLCs) harboring EGFR mutations, third-generation EGFR tyrosine kinase inhibitors (EGFR-TKIs) such as osimertinib are commonly used. However, resistance to osimertinib can arise through activation of bypass signaling pathways, notably via AXL [47].

Within the fibronectin type III domain of AXL, $Trp^{320}$ conforms to the consensus sequence for *C*-mannosylation (Trp-Thr-His-Trp), yet *C*-mannosylation of AXL had not been previously investigated. To examine this, we generated MDA-MB-231 cells overexpressing the extracellular domain of AXL (AXL-ECD) and purified recombinant protein from the conditioned medium. LC–MS/MS analysis revealed that AXL-ECD is *C*-mannosylated at $Trp^{320}$ (Fig. 2h). To confirm the generality of this modification, we expressed AXL-ECD in additional cancer cell lines, including NSCLC-derived PC9 and H3122 cells, as well as HT1080 fibrosarcoma cells. In all cases, LC–MS/MS confirmed *C*-mannosylation at $Trp^{320}$, demonstrating that this modification occurs consistently across multiple cell types. To determine whether any human DPY19 family members function as the *C*-mannosyltransferase for AXL, we co-expressed AXL-ECD with individual DPY19L1–L4 isoforms in S2 cells, which lack endogenous *C*-mannosylation activity. However, LC–MS analysis revealed that none of the DPY19 proteins catalyzed AXL *C*-mannosylation, suggesting the existence of an unidentified *C*-mannosyltransferase specific for AXL. To explore the functional relevance of *C*-mannosylation in AXL, we used CRISPR-Cas9 to generate AXL-knockout (KO) MDA-MB-231 cells and reintroduced either wild-type AXL (wt) or a *C*-mannosylation-defective mutant (Trp320Phe; WF). Following GAS6 stimulation using conditioned medium from GAS6-overexpressing CHO-K1 cells, we assessed downstream signaling by measuring ERK and AKT phosphorylation. The results showed no difference in phosphorylation levels between wt and WF cells, indicating that *C*-mannosylation does not influence canonical AXL signaling pathways in response to GAS6.

We next focused on vasculogenic mimicry (VM), a process wherein aggressive cancer cells form vascular-like structures independent of endothelial cells [48, 49] (Fig. 2i). Previous studies have shown that AXL promotes VM formation in MDA-MB-231 cells, although the molecular mechanisms remain unclear [50]. In vitro VM assays revealed that VM formation was significantly reduced in WF-expressing cells compared to those expressing wild-type AXL. Importantly, this reduction was not accompanied by changes in ERK or AKT phosphorylation under basal culture conditions, suggesting that the effect of *C*-mannosylation on VM is independent of classical AXL downstream signaling.

This study represents the first report of *C*-mannosylation in a receptor tyrosine kinase and adds to our growing understanding of non-TSR, non-type I cytokine receptor proteins undergoing this modification, alongside previously characterized proteins [8, 24, 37, 51–53] including HYAL1 and LPL. The identification of AXL as a novel *C*-mannosylated protein opens new avenues for investigating the functional diversity of this modification and may inform the development of therapeutic strategies targeting AXL in breast cancer and other malignancies.

## 3 The Functional Analysis of *C*-Mannosyltransferase DPY19L3

We identified the *C*-mannosyltransferase activity of human DPY19L3 through our investigation of R-spondin 1 (Rspo1). At the time, most functional studies of the mammalian DPY19 family were limited to phenotypic analyses of gene knockout models in mice, with relatively little emphasis on the molecular functions of individual human DPY19 family members [54, 55]. Moreover, few studies had explored specific phenotypes in relation to distinct DPY19 isoforms. In this context, we focused our efforts on the functional characterization of mammalian DPY19L3, conducting a comprehensive analysis to elucidate its enzymatic activity, substrate specificity, and biological significance.

### *3.1 Topological Analysis of DPY19L3 [56]*

Our previous study demonstrated that human DPY19L3 possesses *C*-mannosyltransferase activity, originally identified through investigation of R-spondin 1 (Rspo1). While *C. elegans* Dpy19 was the first *C*-mannosyltransferase characterized—based on its predicted transmembrane (TM) structure similarity to the catalytic subunit Stt3 of the oligosaccharyltransferase (OST) complex involved in *N*-glycosylation [14]—the precise number and orientation of TM domains in human DPY19L3 remained unknown. At the time, most functional studies of mammalian DPY19 family members were based on mouse knockout phenotypes, with limited information available regarding the molecular and topological features of the human proteins. To gain mechanistic insights into DPY19L3 structure and function, we conducted a comprehensive topological analysis using both computational and experimental approaches. Initial in silico predictions using multiple TM domain algorithms [57–66] suggested that DPY19L3 contains between 11 and 14 TM segments. To experimentally validate these predictions, we employed a redox-sensitive *Gaussia* luciferase (Gluc) reporter assay [67, 68]. Since Gluc activity depends on its localization to the oxidative environment of the endoplasmic reticulum (ER) lumen, we constructed a series of DPY19L3-Gluc fusion proteins. Gluc was fused to either the N- or C-terminus of the full-length protein, or to the C-termini of 14 truncated constructs. Following transient transfection into HEK293T cells, Gluc bioluminescence was measured in cell lysates. Seven Gluc fusion constructs ($E^{80}$, $K^{151}$, $Q^{262}$, $N^{307}$, $A^{400}$, $L^{486}$, and the C-terminus) exhibited high luminescence, suggesting that these regions face the ER lumen, while other regions and the N-terminus were cytosolic. These data provided initial evidence for the orientation and compartmentalization of the protein within the ER membrane. DPY19L3 contains four predicted *N*-glycosylation sites ($Asn^{118}$, $Asn^{319}$, $Asn^{439}$, and $Asn^{704}$). Based on the topological data, $Asn^{118}$ and $Asn^{704}$ were hypothesized to be localized within the ER lumen. To test this, we generated site-directed mutants (N118Q, N319Q, N439Q, N704Q, and a double mutant N118Q/N704Q [2NQ]) and expressed them in HEK293T cells. Western blot analysis confirmed that only $Asn^{118}$ and $Asn^{704}$ were glycosylated. These results were further validated using tunicamycin (an *N*-glycosylation inhibitor) and PNGase F (an *N*-glycosidase), confirming that these two sites are the only glycosylated residues. Because the redox-luciferase and *N*-glycosylation results were consistent, we further investigated ambiguous regions, including a potential reentrant loop (RL) that may not be detectable by Gluc assay. By inserting artificial *N*-glycosylation motifs into these regions, we confirmed the presence of RLs. Integrating all data, we concluded that DPY19L3 comprises 11 TM domains and 2 RLs, forming 5 luminal and 7 cytoplasmic loops (Fig. 3).

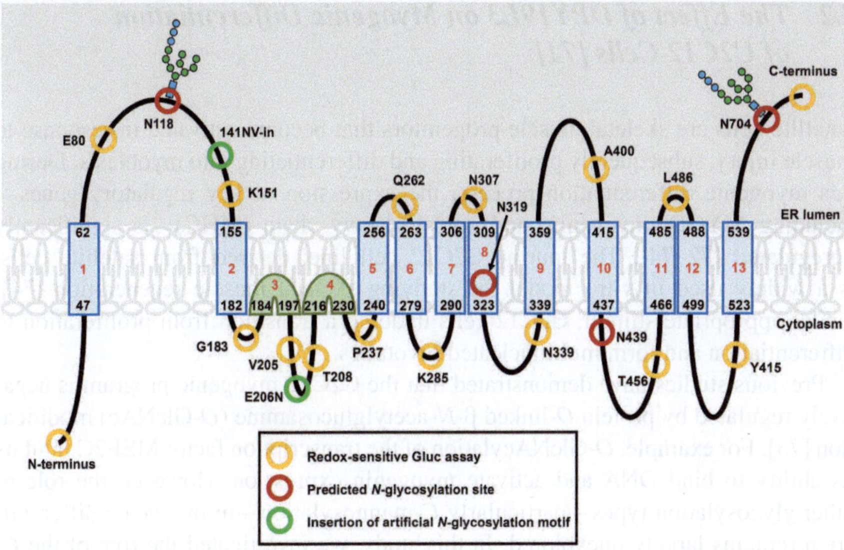


**Fig. 3** Topological model of human DPY19L3 constructed by three types of experimental study

DPY19L3 also has a splicing variant, isoform 2, which lacks exons 16–19 and part of the C-terminal region [69]. Topological predictions using MemBrain [57] suggested that the final TM domain in wild-type DPY19L3 is converted to an RL in isoform 2, thereby altering the localization of its C-terminus from the ER lumen to the cytosol. This was experimentally verified via redox-sensitive Gluc assay, which showed high luminescence when Gluc was fused to the C-terminus of isoform 2, indicating its ER-luminal orientation. To assess whether this structural alteration affected enzymatic activity, we expressed isoform 2 in S2 cells and evaluated *C*-mannosylation of co-expressed Rspo1 using LC–MS/MS. No *C*-mannosylation was detected, demonstrating that the C-terminal region absent in isoform 2 is essential for DPY19L3's *C*-mannosyltransferase activity.

In 2023, Bloch et al. reported the cryogenic electron microscopy (cryo-EM) structure of *C. elegans* Dpy19 [70]. Except for the RLs between TM3 and TM4, the topology described in that study was largely consistent with our experimental model of DPY19L3. Furthermore, their structure–function analysis highlighted that the C-terminal region is critical for enzymatic activity, aligning with our finding that isoform 2 lacks function due to loss of this region.

Our comprehensive topological analysis of human DPY19L3 revealed critical structural elements underlying its function as a *C*-mannosyltransferase. These findings not only validate the structural importance of the C-terminal region but also contribute to the broader understanding of glycosyltransferase topology and function. Our work complements and extends structural insights gained from cryo-EM and provides a foundation for future studies on the mechanistic diversity of DPY19 family members.

### 3.2 The Effect of DPY19L3 on Myogenic Differentiation of C2C12 Cells [71]

Satellite cells are skeletal muscle progenitors that become activated in response to muscle injury, subsequently proliferating and differentiating into myoblasts. During this myogenic differentiation process, the expression of key regulatory genes—including MYOD, myogenin, and myosin heavy chain (MHC)—is significantly upregulated [72–74]. The murine C2C12 cell line, derived from satellite cells, is a widely used in vitro model for studying skeletal muscle regeneration [74]. Under appropriate stimuli, C2C12 cells undergo a transition from proliferation to differentiation and form multinucleated myotubes.

Previous studies have demonstrated that the C2C12 myogenic program is negatively regulated by protein *O*-linked β-*N*-acetylglucosamine (*O*-GlcNAc) modification [75]. For example, *O*-GlcNAcylation of the transcription factor MEF2C inhibits its ability to bind DNA and activate myogenin expression. However, the role of other glycosylation types—particularly *C*-mannosylation—in myogenic differentiation remains largely unexplored. In this study, we investigated the role of the *C*-mannosyltransferase DPY19L3 in the regulation of myogenic differentiation using C2C12 cells. We first assessed the expression of DPY19L3 mRNA in both proliferating and differentiating C2C12 cells. Successful myogenic differentiation was confirmed by morphological changes observed under microscopy and upregulation of known differentiation markers. DPY19L3 was found to be expressed under both growth and differentiation conditions. To evaluate the functional relevance of DPY19L3 in myogenesis, we generated DPY19L3-knockout (KO) C2C12 cells using CRISPR-Cas9 technology. Under proliferative conditions, DPY19L3-KO cells showed no overt growth defects. However, under differentiation-inducing conditions, these cells exhibited significantly impaired myotube formation compared to parental cells, indicating that DPY19L3 is essential for efficient myogenic differentiation. To elucidate the molecular mechanisms underlying this phenotype, we examined phosphorylation levels of signaling molecules known to regulate myogenesis, including MEK, ERK, and p70S6K [76–78]. In differentiating DPY19L3-KO cells, the phosphorylation levels of these kinases were markedly reduced. Furthermore, the expression of key muscle differentiation marker genes was also downregulated in the absence of DPY19L3. These findings suggest that DPY19L3 promotes myogenic differentiation, at least in part, through modulation of MEK/ERK and p70S6K signaling pathways and the appropriate transcriptional activation of myogenic genes.

Our study provides the first evidence that DPY19L3 plays a pivotal role in skeletal muscle differentiation. Given that Rspo1, a known substrate of DPY19L3, has previously been implicated in regulating myogenesis in vivo [79], it is likely that the *C*-mannosyltransferase activity of DPY19L3 is mechanistically involved in this process. Future studies are warranted to identify additional *C*-mannosylated substrates relevant to muscle development and to further define the functional consequences of *C*-mannosylation in myogenic signaling pathways.

### 3.3 DPY19L3 C-Mannosyltransferase Activity Promotes Vasculogenic Mimicry

We have been investigating vasculogenic mimicry (VM) independently of glycosylation pathways for several years and have identified several proteins involved in the regulation of VM [80–85]. However, the molecular mechanisms underlying VM remain incompletely understood. In light of our findings on the enzymatic activity of DPY19L3 as a *C*-mannosyltransferase, we hypothesized that *C*-mannosylation may contribute to VM regulation. To explore this possibility, we examined the relationship between VM and DPY19L3, which had previously been studied in unrelated contexts. To evaluate the functional role of DPY19L3 in VM, we generated DPY19L3-knockout (KO) HT1080 fibrosarcoma cells. Compared to control (mock) cells, DPY19L3-KO cells exhibited markedly reduced VM formation and decreased cell proliferation. To assess whether these effects depended on the enzymatic activity of DPY19L3, we re-expressed either the wild-type (wt) form or the inactive splicing isoform2 of DPY19L3 in KO cells. Re-expression of wt DPY19L3 successfully restored both VM formation and proliferation, while isoform2 did not rescue either phenotype. Notably, expression and phosphorylation levels of AKT were significantly upregulated in wt DPY19L3-reconstituted cells compared to isoform2-expressing and control cells. These findings indicate that the *C*-mannosyltransferase activity of DPY19L3 is essential for both VM formation and cell proliferation in HT1080 cells, likely through activation of the AKT signaling pathway. To investigate whether this mechanism is conserved in other cancer cell types, we knocked down DPY19L3 in MDA-MB-231 breast cancer cells using shRNA. Consistent with our observations in HT1080 cells, DPY19L3 knockdown suppressed both VM formation and proliferation in MDA-MB-231 cells, reinforcing the role of DPY19L3 in these processes. Together, these findings suggest that the enzymatic activity of DPY19L3 is mechanistically linked to VM formation, potentially through the *C*-mannosylation of one or more as-yet-unidentified substrate proteins. While DPY19L3 is expressed in many tissues, no direct correlation has been observed between its expression in tumor versus normal tissues, implying that the functional impact of DPY19L3 in cancer may depend on its specific glycosylation targets.

Building on this insight, we extended our investigation and identified AXL as a novel *C*-mannosylated protein involved in VM regulation, as described in the preceding section. These collective findings highlight the emerging significance of *C*-mannosylation and its associated transferases as potential regulators of cancer-specific processes such as VM. In the future, targeting *C*-mannosylation pathways and their enzymatic mediators may represent a promising strategy for cancer therapy.

# 4 Discovery of Glucosyl-Galactosyl-Hydroxylation in Non-Collagen-Like Proteins

As previously described, we have spent over a decade investigating *C*-mannosylated proteins. During this pursuit, numerous candidate proteins were analyzed using mass spectrometry; however, in many cases, *C*-mannosylation could not be confirmed. Interestingly, some of these mass spectrometry analyses led to the unexpected identification of another form of glycosylation.

This serendipitous discovery prompted us to examine this novel modification further, ultimately identifying it as glucosyl-galactosyl-hydroxylation (GGH)—an atypical glycosylation previously reported primarily in collagen-like proteins. In this section, we describe our work on GGH, a modification uncovered during our search for *C*-mannosylated proteins, and highlight the new insights it has provided into non-canonical glycosylation pathways.

## *4.1 First Identification of Collagen-Like Glycosylation in a Non-Collagen-Like Protein, CCN1 [7]*

Cellular Communication Network Factor 1 (CCN1) is a secreted matricellular protein belonging to the CCN family and is implicated in various biological processes, including cell proliferation, migration, and angiogenesis, through multiple signaling pathways [86, 87]. Given that CCN1 contains a thrombospondin type 1 repeat (TSR) domain with a consensus sequence for *C*-mannosylation, we initially hypothesized that CCN1 might be modified by *C*-mannosylation.

To investigate this, we established a CCN1-overexpressing HT1080 cell line, purified recombinant CCN1 from the conditioned medium, and performed LC–MS/MS analysis. Contrary to our expectations, no *C*-mannosylation was detected. Instead, we identified two distinct, non-canonical glycosylation (Fig. 4a). First, *O*-fucosylation was detected at $Thr^{242}$, and functional analyses indicated that this modification regulates CCN1 secretion [88]. Second, we observed hydroxylation of $Lys^{203}$ followed by dihexosylation of hydroxylysine ($Hyl^{203}$)—a modification reminiscent of glucosyl-galactosyl-hydroxylation (GGH), typically found in collagen-like proteins. Despite CCN1 lacking a classical collagen-like domain, these findings suggest that CCN1 may exhibit functional properties associated with collagen-like glycosylation. To investigate the role of this collagen-like glycosylation, we generated a glycosylation-deficient mutant (Lys203Arg; K203R) in HT1080 cells. Western blot analysis revealed that the secretion of CCN1 was reduced in K203R-expressing cells compared to wild-type. To further explore the secretory defect, we examined the subcellular localization of wild-type and K203R CCN1 by immunofluorescence microscopy. Both forms were predominantly localized to the Golgi apparatus, suggesting that collagen-like glycosylation may be required for efficient transport from the Golgi to the cell surface, rather than for Golgi retention itself (Fig. 4b).

We next sought to identify the enzyme responsible for this modification. Given the known role of lysyl hydroxylase 3 (LH3) in catalyzing glucosylation of hydroxylysine residues in collagen, we hypothesized that LH3 may also mediate glycosylation at Lys$^{203}$ of CCN1. To test this, we generated an LH3-knockout CCN1-overexpressing HT1080 cell line. LC–MS/MS analysis of purified CCN1 from LH3-KO cells revealed the presence of unmodified, hydroxylated, and mono-glycosylated forms at Lys$^{203}$—but the di-glycosylated form was absent. Western blotting further showed that CCN1 secretion was impaired in LH3-deficient cells. These results indicate that LH3 catalyzes the second glycosylation step—likely glucosylation—at Lys$^{203}$ (Fig. 4a), and that this glucosylation is important for CCN1 secretion (Fig. 4b). The identification of collagen-like glycosylation in CCN1 provides compelling evidence for the presence of GGH-like modifications in non-collagen-like proteins. However, while structurally similar, we were unable to conclusively define this modification as canonical GGH based solely on the CCN1 data. Definitive evidence for GGH in non-collagen-like proteins was obtained through enzyme identification in a subsequent study involving a different protein, as described in the following section.

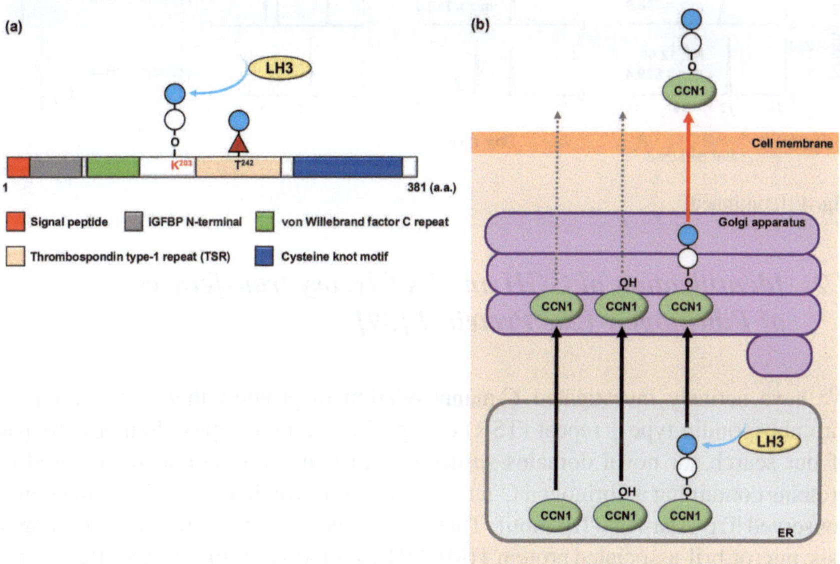


**Fig. 4** **a** Schematic of human CCN1 and its glycosylation sites, **b** collagen-like glycosylation of CCN1 affects the transport from the Golgi apparatus to the cell surface, **c** schematic of human FGL1 and its glucosyl-galactosyl-hydroxylation site, **d** LC–MS analysis of recombinant FGL1 from LH3- and GLT25D1-KO cells showed that glucosylation and galactosylation of GGH in FGL1 are catalyzed by LH3 and GLT25D1, respectively

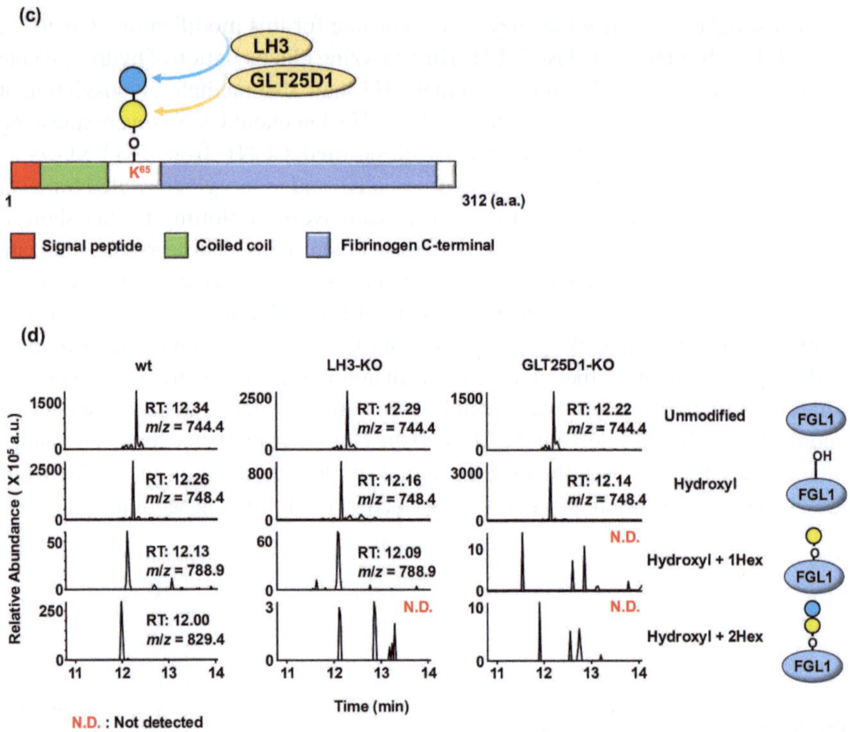


**Fig. 4** (continued)

## 4.2 *Identification of GGH and Its Glycosyltransferases of Fibrinogen-Like Protein 1 [89]*

We have actively investigated *C*-mannosylation in proteins that lack the typical thrombospondin type 1 repeat (TSR) or type I cytokine receptor domains. As part of our search for novel domains susceptible to *C*-mannosylation, we focused on proteins containing a fibrinogen C-terminal domain, which often includes the highly conserved Trp-Xaa-Xaa-Trp motif. Three candidate proteins were selected for analysis: microfibril-associated protein 4 (MFAP4), angiopoietin-like 4 (ANGPTL4), and fibrinogen-like protein 1 (FGL1). HT1080 cell lines overexpressing each protein were established, and the respective recombinant proteins were purified from conditioned media and subjected to mass spectrometry. *C*-mannosylation was detected only at a low level in MFAP4 and was absent in ANGPTL4 and FGL1, suggesting that, despite the presence of the consensus sequence, the fibrinogen C-terminal domain is generally not a *C*-mannosylation target [8]. Unexpectedly, mass spectrometry of FGL1 revealed hydroxylation at $Lys^{65}$, followed by mono- or di-hexosylation of hydroxylysine ($Hyl^{65}$). This modification closely resembled glucosyl-galactosyl-hydroxylation (GGH) observed in collagen-like proteins (Fig. 4c). Importantly, like

CCN1, FGL1 does not possess a collagen-like domain, indicating the presence of GGH in non-collagen-like contexts.

FGL1 is a secreted protein predominantly expressed in the liver and functions in multiple physiological and pathological contexts [90–92]. In normal hepatocytes, it promotes proliferation via EGFR/ERK signaling [91, 93]. Conversely, its expression is often reduced in highly malignant hepatocellular carcinomas (HCCs), where ectopic expression of FGL1 suppresses proliferation and colony formation in FGL1-negative HCC cell lines (e.g., BEL-7402) [94–96]. In addition to hepatic functions, FGL1 acts as an immune checkpoint ligand by binding LAG-3, thereby inhibiting T cell activation [97]. Blocking the FGL1–LAG-3 interaction has been shown to reduce tumor growth in vivo, and FGL1 is upregulated in several cancers, including NSCLC and melanoma, where it is associated with poor prognosis. However, the role of post-translational modifications—such as GGH—in regulating FGL1 function remains poorly understood.

As our previous work on CCN1 could not definitively classify its lysine-linked glycosylation as GGH, we turned to FGL1 with the specific goal of establishing that the modification at $Lys^{65}$ was, in fact, GGH. To identify the glycosyltransferases responsible for this modification, we generated LH3 and GLT25D1 knockout (KO) HT1080 cell lines overexpressing FGL1 using CRISPR-Cas9 technology. Recombinant FGL1 from these lines was purified and analyzed via LC–MS/MS. In control cells, four modification states of $Lys^{65}$ were observed: unmodified, hydroxylated, mono-glycosylated, and di-glycosylated hydroxylysine. In LH3-KO cells, the di-glycosylated form was absent, while in GLT25D1-KO cells, both mono- and di-glycosylated forms were undetectable. These findings indicate that the first hexose is galactose, added by GLT25D1, and the second is glucose, added by LH3—consistent with the definition of GGH (Fig. 4c, d). To explore the functional consequences of GGH in FGL1, we generated a glycosylation-deficient mutant (Lys65Arg; FGL1/KR) and evaluated FGL1 expression in wild-type (FGL1/wt), FGL1/KR, and glycosyltransferase-KO cells. While extracellular FGL1 secretion remained unchanged across all lines, intracellular levels were significantly reduced in both GLT25D1-KO and FGL1/KR cells, but not in LH3-KO cells. As mRNA expression was comparable in all conditions, these results suggest that galactosylation of $Hyl^{65}$, but not glucosylation, is important for maintaining FGL1 protein stability.

This study represents the first definitive identification of GGH in a non-collagen-like protein. Recent structural studies have shown that GGH biosynthesis is facilitated by the formation of a hetero-tetramer composed of an LH3 homodimer and a GLT25D1 homodimer [98], and that GLT25D1 dimerization is essential for enzymatic activity [99]. However, no further discoveries of GGH in non-collagen-like proteins have been reported since our identification of GGL1. Further exploration of this modification in non-collagenous contexts will be critical for understanding its broader functional roles in cell biology and disease.

## 5 Conclusions

Since our initial identification of *C*-mannosylation in HYAL1 over a decade ago, we have pursued a broad range of investigations into the mechanisms, functions, and biological significance of this atypical glycosylation. Through our work and that of others in the field, the understanding of *C*-mannosylation—which was once largely unexplored—has gradually expanded. Recent advances, including the development of antibodies [100] and lectins [101] capable of detecting *C*-mannosylated proteins, have facilitated more precise and large-scale analyses. With this expanding toolkit and increasing interest, *C*-mannosylation is now poised for broader applications, including therapeutic strategies targeting *C*-mannosylated proteins and their corresponding *C*-mannosyltransferases.

During the course of our *C*-mannosylation studies, we unexpectedly identified a novel form of glycosylation—glucosyl-galactosyl-hydroxylation (GGH)—in proteins lacking collagen-like domains. Initially, mass spectrometric analyses of CCN1 and FGL1 were conducted to assess their potential *C*-mannosylation. Although *C*-mannosylation was not detected, we serendipitously discovered GGH-like modifications in both proteins. This unexpected finding prompted further functional studies, particularly in FGL1, culminating in the definitive identification of GGH through characterization of the responsible glycosyltransferases.

To date, CCN1 and FGL1 remain the only known non-collagen-like proteins reported to undergo GGH, and numerous questions remain regarding the prevalence, structural specificity, and physiological relevance of this modification. However, the trajectory of *C*-mannosylation research—from obscurity to functional significance—provides a promising model for the future of GGH research. We anticipate that continued efforts, both from our group and from the broader scientific community, will uncover new substrates, refine mechanistic insights, and ultimately reveal the biological importance of GGH in diverse cellular contexts.

## References

1. Dwek RA, Butters TD, Platt FM, Zitzmann N (2002) Targeting glycosylation as a therapeutic approach. Nat Rev Drug Discov 1:65–75. https://doi.org/10.1038/nrd708
2. Helenius A, Aebi M (2004) Roles of N-linked glycans in the endoplasmic reticulum. Annu Rev Biochem 73:1019–1049. https://doi.org/10.1146/annurev.biochem.73.011303.073752
3. Dricu A, Carlberg M, Wang M, Larsson O (1997) Inhibition of N-linked glycosylation using tunicamycin causes cell death in malignant cells: role of down-regulation of the insulin-like growth factor 1 receptor in induction of apoptosis. Cancer Res 57:543–548
4. Contessa JN, Bhojani MS, Freeze HH, Rehemtulla A, Lawrence TS (2008) Inhibition of N-linked glycosylation disrupts receptor tyrosine kinase signaling in tumor cells. Cancer Res 68:3803–3809. https://doi.org/10.1158/0008-5472.CAN-07-6389
5. Hofsteenge J, Müller DR, de Beer T, Löffler A, Richter WJ, Vliegenthart JFG (1994) New type of linkage between a carbohydrate and a protein: C-glycosylation of a specific tryptophan residue in human RNase us. Biochemistry 33:13524–13530. https://doi.org/10.1021/bi0025 0a003

6. KriegJ HS, Vicentini A, Gläsner W, Hess D, Hofsteenge J (1998) Recognition signal for C-mannosylation of Trp-7 in RNase 2 consists of sequence Trp-x-x-Trp. Mol Biol Cell 9:301–309. https://doi.org/10.1091/mbc.9.2.301
7. Ishizawa Y, Niwa Y, Suzuki T, Kawahara R, Dohmae N, Simizu S (2019) Identification and characterization of collagen-like glycosylation and hydroxylation of CCN1. Glycobiology 29:696–704. https://doi.org/10.1093/glycob/cwz052
8. Osada Y, Suzuki T, Mizuta H, Mori K, Miura K, Dohmae N, Simizu S (2020) The fibrinogen C-terminal domain is seldom *C*-mannosylated but its *C*-mannosylation is important for the secretion of microfibril-associated glycoprotein 4. Biochim Biophys Acta Gen Subj 1864:129637. https://doi.org/10.1016/j.bbagen.2020.129637
9. Wang LW, Leonhard-Melief HCRS, Apte SS (2009) Post-translational modification of thrombospondin type-1 repeats in ADAMTS-like 1/punctin-1 by *C*-mannosylation of tryptophan. J Biol Chem 284:30004–30015. https://doi.org/10.1074/jbc.M109.038059
10. Niwa Y, Suzuki T, Dohmae N, Simizu S (2016) Identification of DPY19L3 as the *C*-mannosyltransferase of R-spondin1 in human cells. Mol Biol Cell 27(5):744–756. https://doi.org/10.1091/mbc.E15-06-0373
11. Morishita S, Suzuki T, Niwa Y, Dohmae N, Simizu S (2017) Dpy-19 like 3-mediated *C*-mannosylation and expression levels of RPE-spondin in human tumor cell lines. Oncol Lett 14:2537–2544. https://doi.org/10.3892/ol.2017.6465
12. Otani K, Niwa Y, Suzuki T, Sato N, Sasazawa Y, Dohmae N, Simizu S (2018) Regulation of granulocyte colony-stimulating factor receptor-mediated granulocytic differentiation by *C*-mannosylation. Biochem Biophys Res Commun 498:466–472. https://doi.org/10.1016/j.bbrc.2018.02.210
13. Hamming OJ, Kang L, Svensson A, Karlsen JL, Rahbek-Nielsen H, Paludan SR, Hjorth SA, Bondensgaard K, Hartmann R (2012) Crystal structure of interleukin-21 receptor (IL-21R) bound to IL-21 reveals that sugar chain interacting with WS*X*WS motif is integral part of IL-21R. J Biol Chem 287:9454–9460. https://doi.org/10.1074/jbc.M111.311084
14. Buettner FFR, Ashikov A, Tiemann B, Lehle L, Bakker H (2013) *C. elegans* DPY-19 Is a *C*-mannosyltransferase glycosylating thrombospondin repeats. Mol Cell 50:295–302. https://doi.org/10.1016/j.molcel.2013.03.003
15. Shcherbakova A, Tiemann B, Buettner FFR, Bakker H (2017) Distinct *C*-mannosylation of netrin receptor thrombospondin type 1 repeats by mammalian DPY19L1 and DPY19L3. Proc Natl Acad Sci USA 114:2574–2579. https://doi.org/10.1073/pnas.1613165114
16. Yamauchi M, Sricholpech M (2012) Lysine post-translational modifications of collagen. Essays Biochem 52:113–133. https://doi.org/10.1042/BSE0520113
17. Spiro RG (1967) The structure of the disaccharide unit of the renal glomerular basement membrane. J Biol Chem 242:4813–4823. https://doi.org/10.1016/s0021-9258(18)99529-3
18. Perdivara I, Yamauchi M, Tomer KB (2013) Molecular characterization of collagen hydroxylysine *O*-glycosylation by mass spectrometry: current status. Aust J Chem 66:760–769. https://doi.org/10.1071/CH13174
19. Richards AA, Stephens T, Charlton HK, Jones A, Macdonald GA, Prins JB, Whitehead JP (2006) Adiponectin multimerization is dependent on conserved lysines in the collagenous domain: evidence for regulation of multimerization by alterations in posttranslational modifications. Mol Endocrinol 20:1673–1687. https://doi.org/10.1210/me.2005-0390
20. Heikkinen J, Risteli M, Wang C, Latvala J, Rossi M, Valtavaara M, Myllylä R (2000) Lysyl hydroxylase 3 is a multifunctional protein possessing collagen glucosyltransferase activity. J Biol Chem 275:36158–36163. https://doi.org/10.1074/jbc.M006203200
21. Shinkai H, Yonemasu K (1979) Hydroxylysine-linked glycosides of human complement subcomponent C1q and of various collagens. Biochem J 177:847–852. https://doi.org/10.1042/bj1770847
22. Sricholpech M, Perdivara I, Nagaoka H, Yokoyama M, Tomer KB, Yamauchi M (2011) Lysyl hydroxylase 3 glucosylates galactosylhydroxylysine residues in type I collagen in osteoblast culture. J Biol Chem 286:8846–8856. https://doi.org/10.1074/jbc.M110.178509

23. Salo AM, Cox H, Farndon P, Moss C, Grindulis H, Risteli M, Robins SP, Myllylä R (2008) A connective tissue disorder caused by mutations of the lysyl hydroxylase 3 gene. Am J Hum Genet 83:495–503. https://doi.org/10.1016/j.ajhg.2008.09.004
24. Goto Y, Niwa Y, Suzuki T, Dohmae N, Umezawa K, Simizu S (2014) C-mannosylation of human hyaluronidase 1: possible roles for secretion and enzymatic activity. Int J Oncol 45:344–350. https://doi.org/10.3892/ijo.2014.2438
25. Overall CM, López-Otín C (2002) Strategies for MMP inhibition in cancer: innovations for the post-trial era. Nat Rev Cancer 2:657–672. https://doi.org/10.1038/nrc884
26. Simizu S, Ishida K, Wierzba MK, Osada H (2004) Secretion of heparanase protein is regulated by glycosylation in human tumor cell lines. J Biol Chem 279:2697–2703. https://doi.org/10.1074/jbc.M300541200
27. Simizu S, Niwa Y (2013) Molecular targets for suppression of metastasis: recent cellular observations. Onco Ther 4:43–51. https://doi.org/10.1615/ForumImmunDisTher.2013008306
28. Goto Y, Niwa Y, Suzuki T, Uematsu S, Dohmae N, Simizu S (2014) *N*-glycosylation is required for secretion and enzymatic activity of human hyaluronidase1. FEBS Open Bio 4:554–559. https://doi.org/10.1016/j.fob.2014.06.001
29. Zhang L, Bharadwaj AG, Casper A, Barkley J, Barycki JJ, Simpson MA (2009) Hyaluronidase activity of human Hyal1 requires active site acidic and tyrosine residues. J Biol Chem 284:9433–9442. https://doi.org/10.1074/jbc.M900210200
30. Kamata T, Katsube K, Michikawa M, Yamada M, Takada S, Mizusawa H (2004) R-spondin, a novel gene with thrombospondin type 1 domain, was expressed in the dorsal neural tube and affected in Wnts mutants. Biochim Biophys Acta 1676:51–62. https://doi.org/10.1016/j.bbaexp.2003.10.009
31. Carmon KS, Gong X, Yi J, Thomas A, Liu Q (2014) RSPO–LGR4 functions via IQGAP1 to potentiate Wnt signaling. Proc Natl Acad Sci USA 111:E1221-1229. https://doi.org/10.1073/pnas.1323106111
32. Gong X, Yi J, Carmon KS, Crumbley CA, Xiong W, Thomas A, Fan X, Guo S, An Z, Chang JT, Liu QJ (2015) Aberrant RSPO3-LGR4 signaling in Keap1-deficient lung adenocarcinomas promotes tumor aggressiveness. Oncogene 34:4692–4701. https://doi.org/10.1038/onc.2014.417
33. Hofsteenge J, Huwiler KG, Macek B, Hess D, Lawler J, Mosher DF, Peter-Katalinic J (2001) *C*-mannosylation and *O*-fucosylation of the thrombospondin type 1 module. J Biol Chem 276:6485–6498. https://doi.org/10.1074/jbc.M008073200
34. Veeman MT, Slusarski DC, Kaykas A, Louie SH, Moon RT (2003) Zebrafish prickle, a modulator of noncanonical Wnt/Fz signaling, regulates gastrulation movements. Curr Biol 13:680–685. https://doi.org/10.1016/S0960-9822(03)00240-9
35. Mizuta H, Kuga K, Suzuki T, Niwa Y, Dohmae N, Simizu S (2019) *C*-mannosylation of R-spondin2 activates Wnt/β-catenin signaling and migration activity in human tumor cells. Int J Oncol 54:2127–2138. https://doi.org/10.3892/ijo.2019.4767
36. Fujiwara M, Kato S, Niwa Y, Suzuki T, Tsuchiya M, Sasazawa Y, Dohmae N, Simizu S (2016) *C*-mannosylation of R-spondin3 regulates its secretion and activity of Wnt/β-catenin signaling in cells. FEBS Lett 590:2639–2649. https://doi.org/10.1002/1873-3468.12274
37. Okamoto S, Murano T, Suzuki T, Uematsu S, Niwa Y, Sasazawa Y, Dohmae N, Bujo H, Simizu S (2017) Regulation of secretion and enzymatic activity of lipoprotein lipase by *C*-mannosylation. Biochem Biophys Res Commun 486:558–563. https://doi.org/10.1016/j.bbrc.2017.03.085
38. Braun JE, Severson DL (1992) Regulation of the synthesis, processing and translocation of lipoprotein lipase. Biochem J 287:337–347. https://doi.org/10.1042/bj2870337
39. Eckel RH (1989) Lipoprotein lipase. N Engl J Med 320:1060–1068. https://doi.org/10.1056/NEJM198904203201607
40. Goldberg IJ, Merkel M (2001) Lipoprotein lipase: physiology, biochemistry, and molecular biology. Front Biosci 6:D388-405. https://doi.org/10.2741/goldberg
41. Ohira M, Endo K, Saiki A, Miyashita Y, Terai K, Murano T, Watanabe F, Tatsuno I, Shirai K (2012) Atorvastatin and pitavastatin enhance lipoprotein lipase production in L6

skeletal muscle cells through activation of adenosine monophosphate-activated protein kinase. Metabolism 61:1452–1460. https://doi.org/10.1016/j.metabol.2012.03.010
42. Mori K, Suzuki T, Waki U, Hayashi S, Kadono S, Kawahara R, Takeuchi M, Mizuta H, Dohmae N, Katayama R, Simizu S (2024) Identification of *C*-mannosylation in a receptor tyrosine kinase AXL Glycobiology 34:cwae096 https://doi.org/10.1093/glycob/cwae096
43. O'Bryan JP, Frye RA, Cogswell PC, Neubauer A, Kitch B, Prokop C, Espinosa R 3rd, Le Beau MM, Earp HS, Liu ET (1991) *Axl*, a transforming gene isolated from primary human myeloid leukemia cells, encodes a novel receptor tyrosine kinase. Mol Cell Biol 11:5016–5031. https://doi.org/10.1128/mcb.11.10.5016-5031.1991
44. Feneyrolles C, Spenlinhauer A, Guilet L, Fauvel B, Daydé-Cazals B, Warnault P, Chevé G, Yasri Z (2014) Axl kinase as a key target for oncology: focus on small molecule inhibitors. Mol Cancer Ther 13:2141–2148. https://doi.org/10.1158/1535-7163.mct-13-1083
45. Myers SH, Brunton VG, Unciti-Broceta A (2016) AXL inhibitors in cancer: a medicinal chemistry perspective. J Med Chem 59:3593–3608. https://doi.org/10.1021/acs.jmedchem.5b01273
46. Hütte HJ, Tiemann B, Shcherbakova A, Grote V, Hoffmann M, Povolo L, Lommel M, Strahl S, Yakhrushev SY, Rapp E, Buettner FFR, Halim A, Imbetry A, Bakker H (2022) A bacterial mannose binding lectin as a tool for the enrichment of C- and O-mannosylated peptides. Anal Chem 94:7329–7338. https://doi.org/10.1021/acs.analchem.2c00742
47. Chang H, An R, Li X, Lang X, Feng J, Lv M (2021) Anti-Axl monoclonal antibodies attenuate the migration of MDA-MB-231 breast cancer cells. Oncol Lett 22:749. https://doi.org/10.3892/ol.2021.13010
48. Taniguchi H, Yamada T, Wang R, Tanimura K, Adachi Y, Nishiyama A, Tanimoto A, Takeuchi S, Araujo LH, Boroni M, Yoshimura A, Shiotsu S, Matsumoto I, Watanabe S, Kikuchi T, Miura S, Tanaka H, Kitazaki T, Yamaguchi H, Mukae H, Uchino J, Uehara H, Katayama K, Yano S (2019) AXL confers intrinsic resistance to osimertinib and advances the emergence of tolerant cells. Nat Commun 10:259. https://doi.org/10.1038/s41467-018-08074-0
49. Folberg R, Hendrix MJ, Maniotis AJ (2000) Vasculogenic mimicry and tumor angiogenesis. Am J Pathol 156:361–381. https://doi.org/10.1016/s0002-9440(10)64739-6
50. Maniotis AJ, Folberg R, Hess A, Seftor EA, Gardner LM, Pe'er J, Trent JM, Meltzer PS, Hendrix MJ (1999) Vascular channel formation by human melanoma cells in vivo and in vitro: vasculogenic mimicry. Am J Pathol 155:739–752. https://doi.org/10.1016/s0002-9440(10)65173-5
51. Lim D, Cho JG, Yun E, Lee A, Ryu H, Lee YJ, Yoon S, Chang W, Lee M, Kwon BS, Kim J (2021) MicroRNA 34a–AXL axis regulates vasculogenic mimicry formation in breast cancer cells. Genes 12:9. https://doi.org/10.3390/genes12010009
52. Yoshimoto S, Suzuki T, Otani N, Takahashi D, Toshima K, Dohmae N, Simizu S (2023) Destabilization of vitelline membrane outer layer protein 1 homolog (VMO1) by *C*-mannosylation. FEBS Open Bio 13:490–499. https://doi.org/10.1002/2211-5463.13561
53. Kawahara R, Usami T, Arakawa S, Kamo H, Suzuki T, Komatsu R, Hara H, Niwa Y, Shimizu E, Dohmae N, Shimizu S, Simizu S (2023) Biogenesis of fibrils requires *C*-mannosylation of PMEL. FEBS J 290:5373–5394. https://doi.org/10.1111/febs.16927
54. Mizuta H, Takakusaki A, Suzuki T, Otake K, Dohmae N, Simizu S (2023) *C*-mannosylation regulates stabilization of RAMP1 protein and RAMP1-mediated cell migration. FEBS J 290:196–208. https://doi.org/10.1111/febs.16592
55. Dickinson ME, Murray SA et al (2016) High-throughput discovery of novel developmental phenotypes. Nature 537:508–514. https://doi.org/10.1038/nature19356
56. Groza T, Gomez FL, Mashhadi HH, Muñoz-Fuentes V, Gunes O, Wilson R, Cacherio P, Frost A, Keskivali-Bond P, Vardal B, McCoy A, Cheng TK, Santos L, Wells S, Smedley D, Mallon A, Parkinson H (2023) The International mouse phenotyping consortium: comprehensive knockout phenotyping underpinning the study of human disease. Nucleic Acids Res 51:D1038–D1045. https://doi.org/10.1093/nar/gkac972
57. Niwa Y, Nakano Y, Suzuki T, Yamagishi M, Otani K, Dohmae N, Simizu S (2018) Topological analysis of DPY19L3, a human *C*-mannosyltransferase. FEBS J 285:1162–1174. https://doi.org/10.1111/febs.14398

58. Shen H, Chou JJ (2008) MemBrain: improving the accuracy of predicting transmembrane helice. PLoS ONE 3:e2399. https://doi.org/10.1371/journal.pone.0002399
59. Bernsel A, Viklund H, Hennerdal A, Elofsson A (2009) TOPCONS: consensus prediction of membrane protein topology. Nucleic Acids Res 37:W465-468. https://doi.org/10.1093/nar/gkp363
60. Käll L, Krogh A, Sonnhammer ELL (2005) An HMM posterior decoder for sequence feature prediction that includes homology information. Bioinformatics 21:i251-257. https://doi.org/10.1093/bioinformatics/bti1014
61. Bernsel A, Viklund H, Falk J, Lindahl E, von Hejine G, Elofsson A (2008) Prediction of membrane-protein topology from first principles. Proc Natl Acad Sci USA 105:7177–7181. https://doi.org/10.1073/pnas.0711151105
62. Viklund H, Elofsson A (2008) OCTOPUS: improving topology prediction by two-track ANN-based preference scores and an extended topological grammar. Bioinformatics 24:1662–1668. https://doi.org/10.1093/bioinformatics/btn221
63. Nugent T, Jones DT (2012) Detecting pore-lining regions in transmembrane protein sequences. BMC Bioinform 13:169. https://doi.org/10.1186/1471-2105-13-169
64. Apweiler R, Bairoch A, Wu CH, Barker WC, Boeckmann B, Ferro S, Gasteiger E, Huang H, Lopez R, Magrane M, Matin MJ, Natale DA, O'Donovan C, Redaschi N, Yeh LL (2004) UniProt: the universal protein knowledgebase. Nucleic Acids Res 32:D115-119. https://doi.org/10.1093/nar/gkh131
65. Käll L, Krogh A, Sonnhammer ELL (2004) A combined transmembrane topology and signal peptide prediction method. J Mol Biol 338:1027–1036. https://doi.org/10.1016/j.jmb.2004.03.016
66. Krogh A, Larsson B, von Heijne G, Sonnhammer ELL (2001) Predicting transmembrane protein topology with a hidden markov model: application to complete genomes. J Mol Biol 305:567–580. https://doi.org/10.1006/jmbi.2000.4315
67. Hirokawa T, Boon-Chieng S, Mitaku S (1998) SOSUI: classification and secondary structure prediction system for membrane proteins. Bioinformatics 14:378–379. https://doi.org/10.1093/bioinformatics/14.4.378
68. Li H, Zheng X, Che M, Hu H (2012) A redox-sensitive luciferase assay for determining the localization and topology of endoplasmic reticulum proteins. PLoS One 7:e35628. https://doi.org/10.1371/journal.pone.0035628
69. Kuroda M, Funasaki S, Saitoh T, Sasazawa Y, Nishiyama S, Umezawa K, Simizu S (2013) Determination of topological structure of ARL6ip1 in cells: identification of the essential binding region of ARL6ip1 for conophylline. FEBS Lett 587:3656–3660. https://doi.org/10.1016/j.febslet.2013.09.017
70. Ota T, Sugano S et al (2004) Complete sequencing and characterization of 21,243 full-length human cDNAs. Nat Genet 36:40–45. https://doi.org/10.1038/ng1285
71. Bloch JS, John A, Mao R, Mukherjee S, Boilevin J, Irobalieva RN, Dabre T, Scott NE, Reymond J, Kossiakoff AA, Goddard-Borger ED, Locher KP (2023) Structure, sequon recognition and mechanism of tryptophan *C*-mannosyltransferase. Nat Chem Biol 19:575–584. https://doi.org/10.1038/s41589-022-01219-9
72. Mori K, Sun H, Miura K, Simizu S (2021) Involvement of DPY19L3 in myogenic differentiation of C2C12 myoblasts. Molecules 26:5685. https://doi.org/10.3390/molecules26185685
73. Montarras D, L'Honoré A, Buckingham M (2013) Lying low but ready for action: the quiescent muscle satellite cell. FEBS J 280:4036–4050. https://doi.org/10.1111/febs.12372
74. Zammit PS, Golding JP, Nagata Y, Hudon V, Partridge TA, Beauchamp JR (2003) Muscle satellite cells adopt divergent fates: a mechanism for self-renewal? J cell Biol 166:347–357. https://doi.org/10.1083/jcb.200312007
75. Kokabu S, Nakatomi C, Matsubara T, Ono Y, Addison WN, Lowery JW, Urata M, Hudnall AM, Hitomi S, Nakatomi M, Sato T, Osawa K, Yoda T, Rosen V, Jimi E (2017) The transcriptional co-repressor TLE3 regulates myogenic differentiation by repressing the activity of the MyoD transcription factor. J Biol Chem 292:12885–12894. https://doi.org/10.1074/jbc.M116.774570

76. Kim HB, Seo HG, Son S, Choi H, Kim BG, Kweon TH, Kim S, Pai J, Shin I, Yang WH, Cho JW (2020) *O*-GlcNAcylation of Mef2c regulates myoblast differentiation. Biochem Biophys Res Commun 529:692–698. https://doi.org/10.1016/j.bbrc.2020.06.031
77. Kitakaze T, Sakamoto T, Kitano T, Inoue N, Sugihara F, Sakamoto T, Kitano T, Inoue N, Sugihara F, Harada N, Yamaji R (2016) The collagen derived dipeptide hydroxyprolyl-glycine promotes C2C12 myoblast differentiation and myotube hypertrophy. Biochem Biophys Res Commun 478:1292–1297. https://doi.org/10.1016/j.bbrc.2016.08.114
78. Lin F, Wang A, Dai W, Chen S, Ding Y, Sun LV (2020) Lmod3 promotes myoblast differentiation and proliferation via the AKT and ERK pathways. Exp Cell Res 396:112297–112297. https://doi.org/10.1016/j.yexcr.2020.112297
79. Ohno Y, Oyama A, Kaneko H, Egawa T, Yokoyama S, Sugiura T, Ohira Y, Yoshioka T, Goto K (2018) Lactate increases myotube diameter via activation of MEK/ERK pathway in C2C12 cells. Acta Physiol 223:e13042–e13042. https://doi.org/10.1111/apha.13042
80. Lacour F, Vezin E, Bentzinger CF, Sincennes M, Giordani L, Ferry A, Mitchell R, Patel K, Rudnicki MA, Chaboissier M, Chassot A, Le Grand F (2017) R-spondin1 controls muscle cell fusion through dual regulation of antagonistic Wnt signaling pathways. Cell Rep 18:2320–2330. https://doi.org/10.1016/j.celrep.2017.02.036
81. Kawahara R, Niwa Y, Simizu S (2018) Integrin β1 is an essential factor in vasculogenic mimicry of human cancer cells. Cancer Sci 109:2490–2496. https://doi.org/10.1111/cas.13693
82. Hayashi S, Osada Y, Miura K, Simizu S (2020) Cell-dependent regulation of vasculogenic mimicry by carcinoembryonic antigen cell adhesion molecule 1 (CEACAM1). Biochem Biophys Rep 21:100734–100734. https://doi.org/10.1016/j.bbrep.2020.100734
83. Nakajima M, Kawahara R, Simizu S (2023) Cofilin promotes vasculogenic mimicry by regulating the actin cytoskeleton in human breast cancer cells. FEBS Lett 597:1114–1124. https://doi.org/10.1002/1873-3468.14594
84. Shimizu S, Kawahara R, Simizu S (2021) Methionine aminopeptidase-2 is a pivotal regulator of vasculogenic mimicry. Oncol Rep 47:31. https://doi.org/10.3892/or.2021.8242
85. Kamo H, Kawahara R, Simizu S (2022) Tyrosinase suppresses vasculogenic mimicry in human melanoma cells. Oncol Lett 23:169. https://doi.org/10.3892/ol.2022.13289
86. Kawahara R, Simizu S (2022) ErbB4-mediated regulation of vasculogenic mimicry capability in breast cancer cells. Cancer Sci 113:950–959. https://doi.org/10.1111/cas.15258
87. Leu S, Lam SC, Lau LF (2002) Pro-angiogenic activities of CYR61 (CCN1) mediated through integrins αvβ3 and α6β1 in human umbilical vein endothelial cells. J Biol Chem 277:46248–46255. https://doi.org/10.1074/jbc.M209288200
88. Lau LF (2011) CCN1/CYR61: the very model of a modern matricellular protein. Cell Mol Life Sci 68:3149–3163. https://doi.org/10.1007/s00018-011-0778-3
89. Niwa Y, Suzuki T, Dohmae N, Simizu S (2015) *O*-fucosylation of CCN1 is required for its secretion. FEBS Lett 589:3287–3293. https://doi.org/10.1016/j.febslet.2015.09.012
90. Mori K, Suzuki T, Miura K, Dohmae N, Simizu S (2021) Involvement of LH3 and GLT25D1 for glucosyl-galactosyl-hydroxylation on non-collagen-like domain of FGL1. Biochem Biophys Res Commun 560:93–98. https://doi.org/10.1016/j.bbrc.2021.04.128
91. Hara H, Uchida S, Yoshimura S, Aoki M, Toyoda Y, Sakai Y, Morimoto S, Fukamachi H, Shiokawa K, Hanada K (2000) Isolation and characterization of a novel liver-specific gene, hepassocin, upregulated during liver regeneration. Biochim Biophys Acta 1492:31–44. https://doi.org/10.1016/S0167-4781(00)00056-7
92. Hara H, Yoshimura H, Uchida S, Toyoda Y, Aoki M, Sakai Y, Morimoto S, Shiokawa K (2001) Molecular cloning and functional expression analysis of a cDNA for human hepassocin, a liver-specific protein with hepatocyte mitogenic activity. Biochim Biophy Acta 1520:45–53. https://doi.org/10.1016/S0167-4781(01)00249-4
93. Zou Z, Cai Y, Chen Y, Chen S, Liu L, Shen Z, Zhang S, Xu L, Chen Y (2015) Bone marrow-derived mesenchymal stem cells attenuate acute liver injury and regulate the expression of fibrinogen-like-protein 1 and signal transducer and activator of transcription 3. Mol Med Rep 12:2089–2097. https://doi.org/10.3892/mmr.2015.3660

94. Gao M, Zhan Y, Yu M, Ge C, Li C, Zhang J, Wang X, Ge Z, Yang X (2014) Hepassocin activates the EGFR/ERK cascade and induces proliferation of L02 cells through the Src-dependent pathway. Cell Signal 26:2161–2166. https://doi.org/10.1016/j.cellsig.2014.04.013
95. Cao M, Xu W, Li C, Cao C, Wang Z, Yao J, Yu M, ZhanY WX, Tang L, Chen H, Li W, Ge C, Yang X (2011) Hepassocin regulates cell proliferation of the human hepatic cells L02 and hepatocarcinoma cells through different mechanisms. J Cell Biochem 112:2882–2890. https://doi.org/10.1002/jcb.23202
96. Nayeb-Hashemi H, Desai A, Demchev V, Bronson RT, Hornick JL, Cohen DE, Ukomadu C (2015) Targeted disruption of fibrinogen like protein-1 accelerates hepatocellular carcinoma development. Biochem Biophys Res Commun 465:167–173. https://doi.org/10.1016/j.bbrc.2015.07.078
97. Yan J, Yu Y, Wang N, Chang Y, Ying H, Liu W, He J, Li S, Jiang W, Li Y, Liu H, Wang H, Xu Y (2004) LFIRE-1/HFREP-1, a liver-specific gene, is frequently downregulated and has growth suppressor activity in hepatocellular carcinoma. Oncogene 23:1939–1949. https://doi.org/10.1038/sj.onc.1207306
98. Wang J, Sanmamed MF, Datar I, Su TT, Ji L, Sun J, Chen L, Chen Y, Zhu G, Yin W, Zheng L, Zhou T, Badri T, Yao S, Zhu S, Boto A, Sznol M, Melero I, Vignali DAA, Schalper K, Chen L (2019) Fibrinogen-like protein 1 Is a major immune inhibitory ligand of LAG-3. Cell 176:334–347. https://doi.org/10.1016/j.cell.2018.11.010
99. Peng J, Li W, Yao D, Xia Y, Wang Q, Cai Y, Li S, Cao M, Shen Y, Ma P, Liao R, Zhao J, Qin A, Cao Y (2025) The structural basis for the human procollagen lysine hydroxylation and dual-glycosylation. Nat Commun 16:2436. https://doi.org/10.1038/s41467-025-57768-9
100. Marco MD, Rai SR, Scietti L, Mattoteia D, Liberi S, Moroni E, Pinnola A, Vertano A, Iacobucci C, Santambrogio C, Colombo G, Forneris F (2025) Molecular structure and enzymatic mechanism of the human collagen hydroxylysine galactosyltransferase GLT25D1/COLGALT1. Nat Commun 16:3624. https://doi.org/10.1038/s41467-025-59017-5
101. John A, Järvå MA, Shah S, Mao R, Chappaz S, Birkinshaw RW, Czabotar PE, Lo AW, Scott NE, Goddard-Borger ED (2021) Yeast- and antibody-based tools for studying tryptophan C-mannosylation. Nat Chem Biol 17:428–437.https://doi.org/10.1038/s41589-020-00727-w

MIX
Papier aus verantwortungsvollen Quellen
Paper from responsible sources
FSC® C105338

If you have any concerns about our products,
you can contact us on
**ProductSafety@springernature.com**

In case Publisher is established outside the EU,
the EU authorized representative is:
**Springer Nature Customer Service Center GmbH**
**Europaplatz 3, 69115 Heidelberg, Germany**

Printed by Libri Plureos GmbH
in Hamburg, Germany